AF711876

Worldwide Wireless Communications

Advances in the Information Industry Series

Presented by the
International Engineering Consortium

Copyright © 1995 by Professional Education International, Inc. All rights of reproduction, including photocopying, electronic storage/transmission, or translation into foreign languages, are reserved. Requests for republication privileges should be addressed to: Publications Department, International Engineering Consortium, 303 East Wacker Drive, Suite 740, Chicago, Illinois, USA 60601-5212.

All opinions expressed in the Worldwide Wireless Communications papers are those of the authors and are not binding on the International Engineering Consortium.

ISSN 0886-229X

ISBN 0-933217-17X

International Engineering Consortium
303 East Wacker Drive, Suite 740
Chicago, Illinois 60601-5212
Phone: +1-312-938-3500
Fax: +1-312-938-8787
publications@iec.org

Worldwide Wireless Communications

Executive Advisory Editor: **Frank S. Barnes**
Professor, University of Colorado

Advisory Editors: **Theodore P. Ehling**
Vice President, EDS

Anthony Ephremides
Professor, University of Maryland

Phillip R. Evans
Director, Perot Systems

Eugene B. Lotochinski
Vice President, Northern Telecom

Jagdish N. Sheth
Charles H. Kellstadt Professor,
Emory University

International Engineering Consortium

The nonprofit International Engineering Consortium is dedicated to advancing the information industry through the development and dissemination of knowledge. The Consortium sponsors the National, Eastern, and Western Communications Forums as well as ComForums and TecForums focusing on industry-specific issues. The Consortium's vital work can only be accomplished through the cooperation of the industry and university communities. The International Engineering Consortium is affiliated with 70 high-tech universities worldwide.

The annual National Communications Forum and the prior National Electronics Conference have made major contributions to the advancement of the information industry. For more than 50 years, tens of thousands of technical papers have been presented, covering virtually every facet of the electronic sciences. In recent years, emphasis has been placed upon the latest developments in telecommunications and computers.

University Program

The Consortium provided grants to more than 500 university professors in 1994, enabling participation in continuing education Forums. Professors use both information obtained at Forums and contacts with professionals from industry to develop and improve courses, laboratories, research, and university infrastructures. Corporations valuing improved university educational programs sponsor the University Program. Through sponsorship, a corporation has a unique opportunity to influence academic programs and interact with professors and students.

Sustaining Sponsors

Alcatel Network Systems — *Ameritech Foundation* — *Andrew Corporation* — *AT&T Foundation* — *Bell Atlantic*

BellSouth Corporation — *Fujitsu* — *GTE* — *Hewlett-Packard Company* — *IBM Corporation*

MCI Foundation — *Motorola* — *NEC* — *Northern Telecom Inc.* — *NYNEX Foundation*

Pacific Bell — *Siemens Stromberg-Carlson* — *Southern New England Telecommunications* — *Southwestern Bell Foundation* — *Sprint*

U S WEST Communications — *International Engineering Consortium*

Associate Sponsors

AirNet Communications Corporation — *ArrayComm* — *Arthur J. Schmitt Foundation* — *TDS Telecom*

Consortium Affiliated Universities

The University of Arizona
Arizona State University
Auburn University
University of California at Berkely
University of California, Davis
University of California, Santa Barbara
Carnegie Mellon University
Case Western Reserve University
Clemson University
University of Colorado at Boulder
Columbia University
Cornell University
Drexel University
University of Florida
Georgia Institute of Technology
Howard University
Illinois Institute of Technology
University of Illinois at Chicago
University of Illinois at Urbana/Champaign
Iowa State University
The University of Kansas
University of Kentucky
Lehigh University
Marquette University
University of Maryland at College Park
Massachusetts Institute of Technology
University of Massachusetts at Amherst
Michigan State University
The University of Michigan
University of Minnesota
University of Missouri-Columbia
University of Missouri-Rolla
North Carolina State University at Raleigh
Northwestern University
University of Notre Dame
The Ohio State University
Oklahoma State University
The University of Oklahoma
Oregon State University
The Pennsylvania State University
University of Pennsylvania
University of Pittsburgh
Polytechnic University
Purdue University
Rensselaer Polytechnic Institute
University of Southern California
Stanford University
Syracuse University
University of Tennessee, Knoxville
Texas A&M University
The University of Texas at Austin
Virginia Polytechnic Institute and State University
University of Virginia
University of Washington
University of Wisconsin-Madison

New Worldwide University Affiliates

Ecole Supérieure d'Electricité

Ecole Nationale Supérieure des Télécommunications de Bretagne

Ecole Nationale Supérieure des Télécommunications de Paris

Imperial College of Science, Technology & Medicine

Institut National Polytechnique de Grenoble

Instituto Tecnologico y de Estudios Superiores de Monterrey

McGill University

Technische Universität München

The Queen's University of Belfast

Universidad Nacional Autónoma de México

Université d'Ottawa

University of Edinburgh

University of Southampton

University of Glasgow

University College London

University of Toronto

Table of Contents

Foreword

Frank Barnes
Executive Advisory Editor

This volume is the first in a series of books designed to cover advances in various key areas of the information industry. Sponsored by the International Engineering Consortium, the series aims to provide coverage at a level that will be useful to industry executives and middle management. This line of books differs from other series in the field in that it is intended to provide coverage of subjects ranging from technical information that is needed for systems development to discussions of developing markets and regulatory environments. The industry has a need for information on a wide range of subjects and at various levels of detail. Some of this information is already available in the technical literature published by organizations such as the *IEEE Transactions* and in textbooks. Other material is available from the trade press. Nevertheless, we believe there is need for authoritative material on a broader range of topics written at an intermediate level and which is more current than most textbooks.

Wireless communications is one of the most rapidly developing segments of the information industry and is already changing the way we communicate. It was only a few years ago that the thought of calling a large fraction of the world from a hand-held portable phone would have been inconceivable. Today it is done on a routine basis. In this volume we review some of the major developments and issues associated with the development of cellular phones, personal communication systems (PCS), data over wireless systems, and low-orbit satellites.

The book begins with an introduction and overview of cellular communications followed by several reviews of current business and marketing considerations associated with the wireless industry. This provides a background for Section II, which reviews some of the technical issues facing the development of wireless technology and its applications. The first review by Richard Chandler examines some of the issues involved in taking voice and data from the base station onto the public switched network, and the essay by Dick Goehring covers applications of cellular systems in the rural environment. Coding and modulation schemes are undergoing rapid evolution as users move from AMPs to digital systems such as GSM and CDMA. Problems of propagation for mobile and portable radio are reviewed by Theodore Rappaport and Dennis Sweeney, and Ernest Smith reviews propagation effects in satellite systems. This is followed by a discussion of smart antennas, which provide an additional means by which to handle more customers from a base station. Technical issues in the interoperability of satellite and terrestrial networks are also reviewed in this section, along with a comparison of proposed near-earth satellite systems. A major problem in the use of hand-held cellular phones is the development of efficient power supplies. The use of an energy management system that minimizes this problem is discussed by Robert Erickson and Dragan Maksimovic. The issues raised by biological effects of modulated radio waves in the cellular band are also examined in this second section, and some of the potential legal issues are discussed. This is a topic that is likely to become increasingly important as a larger fraction of the public becomes regular users of hand-held cellular phones.

Section III reviews the development of wireless communications outside the United States. These essays highlight the rapid development of wireless communication on a worldwide basis. As one goes from country to country, significant differences become apparent in the development and openness of various markets to foreign suppliers of both wireless services and equipment. In this section we have sampled a sufficient number of countries to provide an indication of the variety of circumstances that exist at the current time.

In the closing section, we present some opposing views of the future of cellular and PCS. The essay by Carl Aron projects hard times for wireless, with many players competing for a limited number of customers, while the discussion by Jagdish Sheth sees wireless undergoing a large growth in its customer base as it competes with wireline services for a much larger market. The members of this volume's editorial advisory board and the International Engineering Consortium welcome feedback from readers on the usefulness of this volume as well as suggestions for topics to cover in future books.

Introduction and Background

Cellular Communications: The First Decade

Jagdish N. Sheth
Professor of Marketing
Emory University

Rajendra S. Sisodia
Associate Professor of Business
George Mason University

History

Nearly a century ago, Italy's Guglielmo Marconi transmitted the first radio signals. Fifty years later, AT&T's Bell Laboratories invented cellular technology. However, even ten years ago, the cellular industry in the United States was virtually nonexistent; commercial service on a wide scale did not commence until the FCC's license lottery in 1983. Today, cellular telephony is probably the world's fastest growing major industry.

In a rare instance where a new technology's prospects have been wildly underestimated, AT&T's Bell Labs a decade ago predicted that there would be a mere 900,000 cellular users in the United States by the year 2000. The cellular industry today has an estimated 20 million subscribers in the United States and 40 million worldwide. The U.S. cellular industry's growth rate "dipped" to 32 percent in 1992, but astoundingly rebounded to 46 percent in 1993. Growth in 1994 is conservatively estimated at 30-35 percent; growth rates are even higher in many other countries.

Surprisingly, the first country to implement a cellular system was not the United States; rather, it was Japan in 1979, followed by the Scandinavian countries in 1981, and then the United States and several other countries in 1983. In 1985,

Before 1960	"Push-to-talk" mobile radio – single frequencies
1960	Full duplex VHF mobile – transmit and receive simultaneously
1964	IMTS introduced (improved mobile telephone service)
Late 1970s	Cellular AMPS trials
1979	Japan launches world's first cellular service
1981	Cellular service launched in Scandinavia
1982	AMPS standard adopted
1983	Commercial cellular service started in Chicago and Washington, D.C.
1987	CTIA studies digital cellular
1989	TDMA digital cellular standard adopted
1990	CDMA enters cellular standards bodies
1996+	PCS service rollout commences

Table 1 *Chronology of wireless technology*

cellular service was initiated in ten additional countries, including Hong Kong, Canada, France, and Brazil. From 1989 to 1992, almost all the countries in Latin America implemented cellular. Since 1991, many countries, including several from the Commonwealth of Independent States (the former Soviet Union), have been added to be international cellular arena.

Mobile Radio History

Cellular telephony has been arguably the most successful innovation in the last 25 years; certainly it has been one of the fastest to achieve broad adoption. Cellular telephony has achieved its level of market penetration much faster than the microwave or the camcorder, and as fast as, if not faster than, videotape recorders. In both cases, it only took nine years to reach the benchmark number of 10 million customers. By comparison, it took 22 years for fax machines to find 10 million customers; 25 years for cable TV, and 38 years for conventional phones.

Despite its stunning growth so far, we believe cellular communications has an even more spectacular future. The number of people who have so far been touched by this and other wireless technologies remains a small fraction of the population, even in developed countries. By 1998, the industry is widely expected to have well over 100 million subscribers worldwide – an estimate we think is conservative. Huge growth will continue for years, even decades to come, as the technology for wireless communications gets better, more versatile, and more affordable. As a result of miniaturization and digitalization, cellular technology is evolving rapidly. The economics of the cellular business are now closely tied to the astonishing economic characteristics of semiconductors, suggesting dramatic improvements in price-performance and form factors. Already, the internal volume of cellular phones has been cut drastically; between 1992 and 1994, the weight of a typical cellular phone fell by more than half. Accompanying this reduction in mass has been a precipitous decline in costs. For example, the cost of a handheld cellular phone a decade ago was about $4,000, and pagers were about $450. Today, the prices of low-end portable cellular phones and pagers have dropped to less than $100. Many (if not most) are given away for "free" to induce customers to sign up for service. Similar (though not as drastic) improvements in price-performance for network equipment are fast driving the industry towards becoming a mass market provider rather than a service for the wealthy.

As a result, cellular technology will soon become a viable alternative to the local wired phone company. This transition, surprisingly to some, is already happening in some of the least developed countries of the world. According to Herschel Shosteck Associates, the average per capita GDP of countries launching cellular service was $14,500 in 1984, and will likely drop to $600 by 1998. In other words, cellular is fast becoming an affordable, even economical alternative to a wire-based telecommunications infrastructure, especially for voice communications. If current technology evolution vectors continue, wireless technologies may well allow for interactive broadband (i.e., video) communications as well.

Industry Strategies in the First Decade

As discussed above, cellular has had a brief but spectacular history. So far, the industry has primarily been concerned with creating and managing growth. It has focused on the following:

Creating Capacity

Throughout its short history, the cellular industry has focused on deploying infrastructure equipment as fast as possible: adding cells, increasing density through sectorization, and moving toward digital technology. The first commercial cellular system was built in Chicago in 1983. By 1985, cellular service was available in more than 50 of the largest MSAs (metropolitan statistical areas), encompassing

170 million Americans	use payphones regularly
40 million	use cordless phones
20 million	use pagers
20 million	use cellular phones
16 million	use private radio
12 million	use private dispatch radio systems
10 million	use portable computers or PDAs
1.5 million	use specialized mobile radio
1 million	use mobile data services

Table 2 *A country on the move*

more that half the U.S. population. In June 1992, the industry reached a milestone: all 734 U.S. markets (as defined by the FCC) were being serviced by at least one of the two operators licensed in each market.

The FCC originally allocated 40 MHz of frequency spectrum for cellular service (20 MHz to each licensee); it added another 10 MHz in 1986. Cellular operators, especially in large metropolitan areas, met growth in demand through continuous system construction and the integration of new technology. A commonly used method has been cell-splitting, where a cell is divided by adding more, less powerful base stations. By increasing the number of cells, the system operator is able to increase the level of frequency reuse. Of course, there is a limit to which this can be done without introducing channel interference. As a result, the industry slowly began a partial move to higher-capacity digital technology in 1993.

The cellular industry is now making the transition from simply getting networks up and running to focusing on better managing its operations and becoming more competitive, profitable, and responsive to customers. Cost-efficiency has been improved over time through the introduction of smaller and cheaper base stations and the development of a more intelligent infrastructure. By lowering its costs, the industry is inching toward greater profitability.[1]

Rationalizing Territorial Coverage

At the outset, the cellular industry was highly fragmented, especially on the nonwireline side. Few companies enjoyed economies of scale in network construction, operations, marketing, or customer service. Companies survived and grew year-to-year by targeting those customers with the greatest willingness and ability to pay its extremely high tariffs.

However, the business model for the cellular industry is now fast moving from one based on high margins and low volume to a low-margin, high-volume one. This makes it imperative that cellular companies lower their cost of operations. Many companies are achieving cost savings through the acquisition of regional clusters of operating licenses and the consolidation of high-density markets in particular, allowing for efficiencies in network operations and marketing. This strategy has been followed by most of the large cellular operators in the country.

On the wireline side, the company that has achieved the most in terms of efficiency improvements is Southwestern Bell, which averages $405,000 in revenues per employee (double the industry average) and 600 subscribers per employee (also double the industry average). The company also has the highest penetration rate of the RBOCs and the largest number of subscribers. On the nonwireline side, McCaw moved early on to acquire other cellular companies and eventually became the largest provider in the world.

Even with all the consolidation that has occurred, the industry remains highly fragmented, and it is likely that major consolidation (along the lines of the recently announced deals between Bell Atlantic and NYNEX, and between U S WEST and AirTouch) will continue.

Moving from Local to Regional to National Coverage

Despite the movement within the industry to consolidate regionally, customers have demanded more: the ability to use their cellular handsets outside their regions. Companies have thus come under enormous pressure to create seamless national networks, especially since the basic cellular infrastructure is now virtually complete.

The vision of a seamless system has been widely pushed by McCaw Cellular for a number of years. In this system, cellular subscribers in one region may "roam" in any other region and still use their phone in the same way. The Cellular One Group, led by McCaw, developed the North American Cellular Network (NACN), which operates with nationwide standards and provides a consistent level of quality. Customers are able to dial in many cities as though they were in their own calling area.

In response to the Cellular One NACN, fifteen Canadian and U.S. wireline companies have put together a similar seamless national network under the name of MobiLink. Introduced in the summer of 1993, it now reaches approximately 83 percent of the population. Interestingly, MobiLink competes against its own members in some markets, since some phone companies have bought the nonwireline carriers outside their base operating region. For example, PacTel Cellular, a MobiLink member, has a presence in the San Francisco area, but the MobiLink licensee there is GTE.

Some RBOCs appear to be moving toward creating their own regional, even national wireless networks. Bell Atlantic has indicated its interest in creating a nationwide network; after its deal with NYNEX to combine cellular operations, Bell Atlantic suggested that it was looking for more partners so it can expand its "footprint" to places like California, the Midwest, and the Southeast.

Creating Brand Identities

To achieve more seamless coverage from the customer's viewpoint, the industry has moved towards creating greater brand identity and fewer distinct brands. The cellular operations of several of the RBOCs, for example, now use "mobility" as part of their name. The wireline carriers have joined together to create the MobiLink brand, in response to the Cellular One brand. McCaw plans to use the AT&T brand, while the Sprint name will be used for the PCS offerings being planned by a group of cable companies.

Targeting New Customers

As the pent-up demand for mobile telephony has been satisfied, cellular service providers are starting to aggressively target mass market customers, who are much more price sensitive. The early customer base was mostly heavy users from business, especially owner-managed businesses. Increasingly, new customers are light users, and many are buying for personal reasons. In 1988, 71 percent of cellular phones sold in the United States went to business users and 25 percent went to personal users, according to industry figures. By 1994, personal users accounted for 46.7 percent of cellular customers, while business buyers declined to 45.6 percent (government and law enforcement make up the rest).[2] Despite the rapid growth, cellular service subscribers still account for only 10 percent of the U.S. population. The potential exists to greatly expand the market by making the service as affordable as basic wireline telephone service.

Expanding Distribution Channels

As it grew, the nature of marketing activities within the cellular industry changed rapidly. The primary target market has changed from mobile executives to a much broader array of users, including families and other casual users. The average monthly spending has declined steadily as such customers have become more numerous. It became important to be able to acquire such customers at lower costs than those incurred through a heavy reliance on direct sales.

Reflecting this, the industry has moved from using only specialized distributor/dealers to using a variety of channels, including mass merchandisers, direct marketers, etc. The use of third-party distribution channels is now widespread, and as the industry becomes more competitive, the importance of resellers continues to grow. This has greatly increased market coverage, but has also led to issues of hybrid channel design and management, and some problems with channel conflict.

In 1985, the typical U.S. cellular subscriber generated $197 in revenues per month; that figure is now about $70. The cost of acquiring new customers has been rising recently, from $400 two years ago to about $500 today. The challenge of lowering this acquisition cost has led the industry into new and innovative distribution arrangements.

The retail channel is expected to grow the fastest, as the consumer market grows in importance and the need to lower the cost of customer acquisition becomes more acute. Though growing in importance, the retail channel has thus far been mostly seen as complementary to dealers. The belief is that with new technologies coming in, many customers will feel the need to have access to specialists, while other customers will feel quite comfortable dealing with retail channels.

As third-party distribution has grown in importance, the speed and convenience with which service activation can occur has become increasingly crucial. NYNEX is working toward having every customer—whether retail or dealer—activated within five minutes of the purchase. PacTel offers paperless activation, which takes only a few minutes. BrandsMart customers wait less than 10 minutes for activation at that company's outlets.

Getting retailers to sell cellular service requires education and a mindset change. "Cellular is a completely different animal from (other products sold via) retail," according to Faith Seiders, general manager of indirect distribution for BellSouth, Atlanta. "The normal product allows the retailer to buy a box from a manufacturer, mark the product up to the appropriate market price, and sell it. Cellular requires them to change their thought process and to buy a product from a manufacturer, mark it down, sell it to the customer, and accept commission on the back end."[3]

Bundling Handsets and Service Contracts

Using the "razor and blade" analogy, the industry has achieved rapid growth by subsidizing equipment purchase with service contracts (except in California, where the practice is not allowed). This lowered the barrier to adoption, but also brought in many low-revenue customers at high acquisition cost. Many of these customers are minimally profitable or cause the industry to lose money. Customer selectivity, retention, and efficient customer servicing have thus become key profitability drivers in the industry.

Relative Price Stability

Back in 1982, when the FCC set up rules for the cellular markets, the agency divided the country into 734 territories and permitted no more than two companies to operate in each region. The theory was that competition between two companies in each market would eventually bring prices down.[4]

In reality, the duopoly structure resulted in far less price competition than expected. Within at least two-thirds of the regions, the prices charged by the two cellular carriers have been found to be almost the same or even identical. The General Accounting Office, the investigative arm of Congress, made that finding in July 1993 but said it could not learn why, in part because the industry was not required to provide detailed revenue and cost data. Critics and some industry analysts charge that consumers are paying too much for service and are being forced to underwrite the billions paid by one cellular company to acquire another.[5]

In addition to the escalating prices paid for the acquisition of territories, especially by nonwireline players, another reason for continued high prices has been the reluctance of wireline providers to lower prices due to an expected substitution effect between wireless and wireline service. In other words, if the RBOCs made cellular too affordable, it could cannibalize their bread-and-butter wire-based business.

Few U.S. cellular carriers have made any significant profits in their decade or so of existence. The high prices paid by the nonwireline providers for territories have had the impact of creating a profitable pricing umbrella for the wireline companies. Within the industry, only the latter have made any significant profits from cellular operations.

Prices are expected to come under severe downward pressure from two different sources in coming years. First, with network capacities expanding rapidly, more cellular companies are trying to induce greater usage at the low end, sometimes by bundling in discounted calling time along with the monthly fee. Second, the advent of PCS services will lead to a veritable price war as the aggressive new entrants attempt to buy market share and establish themselves in the market.

Competing on Quality and Reliability

As the industry has matured, customer expectations have soared. Users now want higher call quality, call completion wherever they roam, a high level of reliability, itemized, timely, and accurate billing, responsive customer service, and lower costs. Companies are competing fiercely on customer service, advertising, and distribution channels. They are also competing aggressively on enhanced services, such as voice mail, fax capabilities, etc. The cellular handset, which is a far more sophisticated device than a wireline handset, incorporating a great deal of microprocessor technology, is rapidly shrinking in its form factor while adding advanced capabilities such as voice recognition, handwriting input, etc. CPE functionalities in turn drive the network, which must add capabilities to allow the terminal equipment to operate. Optional service features continue to grow in popularity. Among respondents to a survey by Cellular Marketing, all offer call waiting and call forwarding. Ninety-six percent offer conference calling, 92 percent offer busy-no answer-transfer, 87 percent offer voicemail, 43 percent offer paging and traffic information, 39 percent offer weather information, 30 percent offer sports information, 21 percent offer stock quotes, and 11 percent offer "other" services, including soap opera updates, ski reports, and horoscopes.[6]

Lowering the Cost of Doing Business

As competitive forces and changing customer profiles put downward pressure on prices, the industry has been faced with the need to lower the cost of doing business. In addition to achieving operational and marketing efficiencies through the rationalization of territories, the industry has also worked to lower the cost of acquiring new customers. In part, this has been achieved through the deployment of new, lower cost acquisition channels, such as kiosks in shopping malls (GTE) and grocery stores (U S WEST New Vector). Carriers are also seeking to tie the cost of luring new customers to the revenues that the customer is expected to generate. This is being achieved through the aligning of an agent's commission to the customer's usage level. The formation of national brands is also seen as a way to lower customer acquisition costs.

It appears as if carriers are succeeding in this. According to Cellular Marketing survey respondents, the average activation commission paid is $174.26, down from $ 203.70 in the previous survey and $193.73 from the survey conducted in the beginning of 1992. The lowest activation commission paid is $20 versus a high of $400. In addition to the commission fee, 52 percent of respondents say they are paying residuals (commission tied to a customers usage level and longevity with the provider), versus 53 percent in the previous survey.[7]

Current Industry Competitive Structure

As discussed above, though the rivalry between competing cellular carriers has been quite intense at times, rapid growth and the duopoly structure of the industry have kept competitive forces at a relatively low level in the industry. That is changing rapidly.

The Competitive Analysis Framework

The framework used in this section for assessing the competitive environment is the well-known Five Forces model of industry structural analysis developed by Michael Porter and depicted in *Figure 1*.[1]

This model has become an important strategic planning tool for many industries. The focus of structural analysis is on identifying an industry's underlying structure, apart from the many short term factors that can affect competition and profitability in a more transient way. According to Porter's model, there are five forces that affect competition in an industry:

- rivalry between existing competitors
- bargaining power of buyers
- bargaining power of suppliers
- threat of new entrants
- threat of substitute technologies

The strength of these five competitive forces reflects the underlying competitive structure of an industry. Having evaluated the forces and resulting industry structure, a firm can then identify its own strengths and weaknesses relative to the

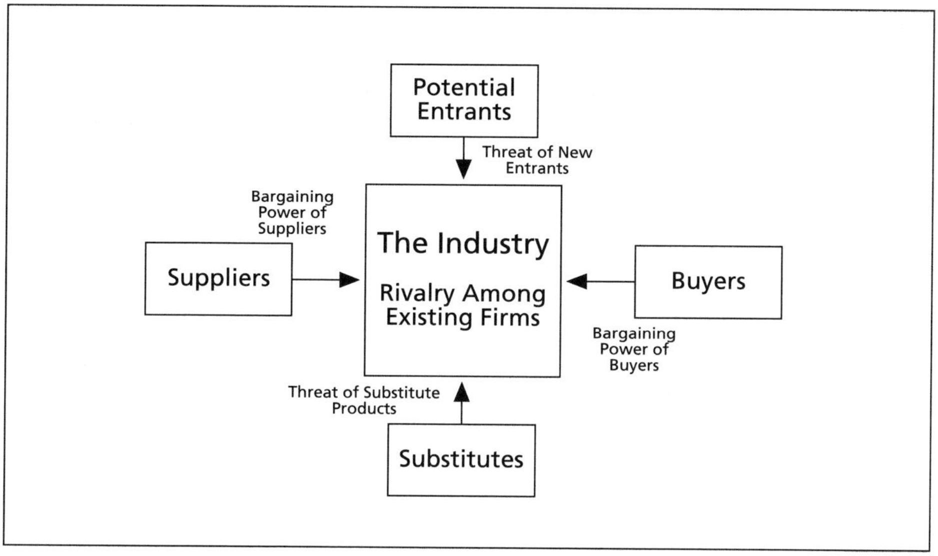

Figure 1 *Forces driving industry competition*

industry. Where does it stand against substitutes? Against the sources of entry barriers? In coping with rivalry from established competitors?

All five forces determine the intensity of industry competition and thus profitability, but the strongest forces are crucial. For example, even a company in a strong market position in an industry where potential entrants are no threat will earn low returns if it faces a superior, low-cost substitute. Even with no substitutes and blocked entry, intense rivalry between existing competitors will limit potential returns.

The effect of heightened competition in an industry is to drive down the rate of return. When the rate of return falls below the yield on long-term government securities (adjusted upwards for the risk of capital loss), investors will not tolerate it for long.

Structural analysis can be used to chart a future course of action. Companies can use an adaptive approach, which takes the structure of the industry as a given and matches the company's strengths and weaknesses against it. A more aggressive option, usually available only to large, even dominant players, is to use an alteration approach, designed to alter the balance of competitive forces. In the following sections, we consider the competitive forces in the cellular industry using this framework.

Industry Rivalry

The degree of rivalry between incumbent players is starting to intensify. Rapid industry consolidation on a national level is leading to the creation of ever larger players. These players have enormous resources that they can bring to bear on customers and competitors. Further industry consolidation is likely, leading to larger rivals facing one another in every major market.

The intensity of the fight for market share is reflected in the increasing problem of churn (currently estimated at approximately three percent per month).[9] Firms are competing especially aggressively for the acquisition of newer customer groups. At the same time, they are struggling to retain existing desirable customers.

Buyer Power

With greater competition will come greater customer choice. The power of buyers is thus rising; this includes both end users and intermediaries.

Channels – As discussed earlier, the trend in the industry is toward the increased use of third-party distribution channels and away from dealers and internal sales. This transition will lead to lower costs for acquiring new customers, but it will also result in a reduction in control over the selling process and the customer base. Large new retailer customers, such as Wal-Mart and Circuit City, have enormous buying power and sophistication. They will use their bargaining power shrewdly.

Corporate Accounts – Until now, most corporate users have really been individuals acting on their own. As procurement specialists get involved in acquiring wireless capabilities, they will be much more sophisticated and demanding customers. As they have done with telecommunications services, they will try to manage and control the communication resource more centrally. As the technology moves towards digital, data processing managers will become more important.

End Users – Users today are more accustomed to higher quality service and better value. They now have higher expectations and are less tolerant of poor service, low technical quality, etc.

Supplier Power

The cellular industry has two kinds of suppliers: those manufacturing terminal equipment (handsets) and those providing infrastructure equipment. More suppliers are starting to specialize; for example, Motorola is increasingly focusing on terminal devices such as pagers and cellular phones.

In the short run, due to a shortage of certain components (such as semiconductors), manufacturers of terminal equipment appear to have some power over cellular operators. In some parts of the world, customers are being put on allotment. There are also several different standards in use worldwide. This situation is not expected to persist. In the long run, terminals will increasingly be built to world standards and will be largely interchangeable. Supplier power will thus be limited.

For infrastructure equipment, we believe the reverse to be true. In other words, infrastructure supplier power is low in the short run, since a large number of competing vendors are attempting to establish greater market presence. However, we expect that there will be a shakeout in the industry in years to come, and only a handful of suppliers will survive as major players. Greater consolidation in the already globalized infrastructure equipment business will lead to a smaller number of major suppliers, potentially increasing their advantage over the still-fragmented cellular industry.

Also, we believe that in the long run, suppliers to the cellular industry have no particular loyalty to cellular service providers. They will look more broadly to the wireless industry and at opportunities such as wireless cable, wireless PBX, digital cordless, etc. Given their ability to serve the needs of cellular rivals, often with superior technology that does not have to factor in an installed base of users of previous-generation technologies, the power of infrastructure suppliers is likely to increase.

Threat of New Entrants

There are a number of markets where already-licensed cellular operators have yet to commence operations. Some of these will initiate service in the next few years. Other than these, there will be no new entrants in the cellular business, as the duopoly structure remains firmly entrenched. However, new players are coming into the industry through acquisitions (e.g., AT&T-McCaw) or mergers/joint ventures (e.g., Bell Atlantic-NYNEX; AirTouch-U S WEST). These players will bring different cultures and much greater resources to the industry.

Threat of Substitute Technologies

As a result of technological change, the industry definition is expanding from the cellular industry (predicated on today's cell-based technology) to the wireless industry (independent of any particular technology or architecture). This change is coming faster than most people realize.

Several technological vectors bear watching:

Digital cellular technology – Digitalization has several important benefits for wireless communications providers. First, it enables the provision of multimedia services on the same platform, since digital bits do not distinguish between different types of information (audio, text, data, etc.). Eventually, this will allow for wireless interactive video capabilities as well. Digitalization also adds features and capabilities not afforded by analog technology, such as number identification. Though high upfront costs are involved in making the transition to digital technology, this is offset by greater reliability and capacity and lower maintenance costs, resulting in lower life cycle costs.

Capacity efficiency technologies, i.e., compression – The ability to compress information is an important by-product of digitalization. Various compression approaches are being developed, which offer up to a ten-fold improvement in the efficiency of spectrum utilization.

Enhanced specialized mobile radio (ESMR) – This new technology, developed by Motorola, enables providers of dispatch-type services to offer cellular-style communications, adding paging and data capabilities as well.

Personal communications services – Cellular was primarily designed as a narrowband, voice channel technology primarily for drive-around applications. PCS will be digital from the outset, and may eventually have broadband capability. The expense of building a PCS system and technological advances in the cellular alternative raise the question of whether PCS represents a true threat.

Declassified military technologies – Many cutting-edge wireless technologies previously directed exclusively for military applications are now becoming available for commercial application. For example, Hughes Electronics is leveraging its defense industry experience in satellite technology to enter a number of communications businesses. Likewise, Steinbrecher's spread spectrum technology offers impressive promise for digital wireless communications.

VSAT (very small aperture terminals) – VSAT systems are already heavily deployed in the specialty retailing business, and their capabilities are being expanded continuously by sophisticated suppliers such as Hughes and others. VSAT systems will carry a lot of data that might otherwise be carried by cellular systems using cellular digital packet data; they may soon add voice functionality as well.

Satellite technology – A number of satellite-based wireless telephony projects have been announced, including Motorola and partners' Iridium, TRW's Odyssey, the Teledesic project proposed by Craig McCaw and Bill Gates, and the Hughes Communications-led American Mobile Satellite Corporation. While these projects all have their skeptics, they do offer the potential of significant competition for traditional wireless players.

Conclusion

The first decade of the cellular industry has been eventful indeed, and we believe that the industry has many years of growth ahead. However, the next decade will see major competition in the wireless communications arena, many new technological advancements, and continued opening of world markets.

Endnotes

1 Martyn Warwick, "Cellular Mobile Radio Changes Colors; Service Providers Adapt To Changing Market," *Communications International,* January, 1994, Vol. 21; No. 1; Pg. 4.

2 Ronald E. Yates, "Popularity Of Cell Phones Far Exceeds Expectations," *Chicago Tribune,* April 19, 1 994 Pg. 7.

3 Dana L. Spoor, "Scoreboard Watching: Find Out Which Side Is Winning As Dealers And Retailers Clash In The Cellular Sales Game," *Cellular Business,* December 1993, Vol. 10; No. 13; Pg. 50.

4 Anthony Ramirez, "Mapping Out The Wireless-Phone Future," *The New York Times,* November 12, 1992, Section A; Page 1; Column 2.

5 Ramirez, op. cit.

6 Maxine Carter-Lome, "Going Direct; 1993 Cellular Carriers Survey," *Cellular Marketing,* March 1993, Vol. 8; No. 3; Pg. 43.

7 Carter-Lome, op. cit.

8 Adapted from Michael Porter, "Industry Structure and Competitive Strategy: Keys to Profitability," *Financial Analysts Journal,* July/August 1981.

9 Carter-Lome, op. cit.

Introduction to Personal Communications Services

Christine Yovnello
Systems Engineer
AT&T Bell Laboratories

Abstract

PCS, for the purpose of this essay, are personal communications services. Beyond the acronym, PCS are not yet explicitly defined. There are different interpretations of the market needs and the capabilities available from the evolving technologies to fill those needs. Various standards bodies and industry groups are currently working on definitions. This essay provides an introduction to PCS by defining its various aspects and by presenting an architecture model for understanding PCS. This essay also provides insight into AT&T's vision of PCS and the technology challenges that lie ahead to make that vision reality.

What Is PCS?

PCS can be defined in many ways, but essentially it entails allowing the end user to communicate from anywhere, to anyone, at any time, and in any form. From the end user's perspective, it provides individual service, with the individual controlling these services anytime and anywhere. As a mobile service, PCS allow end users to obtain their service on the street, in buildings, and in vehicles. The end user's access to services is independent of the terminal equipment and network location. With PCS, the key to this personalization and the anytime/anywhere concept is a personal telecommunications number (PTN) that lets the end user call people instead of places.

The Need for PCS

Two forces working together are creating the need for PCS: a mobile market push and a technology pull. Mobile subscribers today are pushing for more and better service. Also, more and more subscribers want to be mobile. Even within the office, people are more mobile; a segment of today's business world has mobile offices. The bottom line is the need for more access away from home and office, and this mobile market is looking for the same service quality and feature set available from traditional landline service. Pulling on the technology side are microelectronics developments in integrated circuits supporting more portable and powerful terminals, digital radio and antenna technology delivering better service quality and greater availability, and intelligent networking allowing new and expanded features.

Evolution, Not Revolution

PCS are an evolutionary concept, not a revolutionary one. The first mobility services were cordless phones in the residential environment and cellular, which was initially used primarily in business. It is mobile subscribers' increasing demands for capacity, quality, and new capabilities that started the trend toward PCS. Digital standards for cordless and cellular filled some of the need for increased capacity and quality. New markets like wireless private branch exchange (PBX) and Centrex also materialized. Included in these early stages were new services and capabilities made available with the intelligent network (IN), which, in conjunction with the standards for cellular intersystem communications, bring about seamless roaming. Another mobility service available from the IN is personal number service, which provides subscribers a unique, nongeographic number – a personal telecommunications number (PTN) – and allows the network to direct incoming calls to subscribers regardless of their geographic location.

These demands for increased capacity and increased quality are still there, and new markets will continue to appear. The combination of these continuing demands, new markets, digital cordless, digital cellular, wireless PBX and Centrex, and the personal number service is what will evolve into PCS.

Providing PCS

A number of key players will provide all or part of the end-to-end services, or will be creating partnerships to provide PCS. Each brings something to the table:

- Cellular service providers have an embedded base of mobile customers, expertise in wireless voice services, and a substantial wireless infrastructure deployed.

- Local exchange carriers have been providing voice and data services for residential and business customers over broad regional serving areas. They also have an embedded wireline switching and transport network and are now starting to deploy switch-based feature platforms and IN services platforms.

- Cable TV providers have a widespread, broadband distribution network suitable for video and voice services.

- Interexchange carriers offer nationwide switching, transport, signaling, and intelligent networks.

Different Aspects of PCS

To gain different perspectives of PCS, view it from four different perspectives: service, personal, mobility, and network.

Service Aspects of PCS

Services provided by PCS will be tailored to the individual and by the individual. PCS are not offering just a personal number, but a personal service. Some market projections call for a 50 percent penetration; therefore, PCS will have to be affordable to gain that broad market appeal. To satisfy today's end users, the quality of services offered by PCS will have to be comparable to the quality of wired services, and the coverage, again like that of the landline network, will have to be ubiquitous.

Personal Aspects of PCS

A PCS subscriber will not need a home phone, an office phone, and a car phone, but a personal phone. This does not necessarily have to be just a phone; it can be a personal device supporting electronic mail, facsimile, personal computing, etc. Similarly, with PCS the subscriber will not have a home phone number, an office phone number, and a car phone number, but a personal telecommunications number (PTN). This PTN is a unique, nongeographic number that is not associated with a specific terminal or telecommunications service area. Further, PCS will offer the subscriber a menu of services uniquely specified by the subscriber and dynamically reconfigurable by the subscriber. The subscriber has real-time control over his/her services.

Mobility Aspects of PCS

A big part of PCS is mobility, and there are multiple aspects of mobility: terminal mobility, personal mobility, and service mobility.

Terminal mobility is the capability available with cellular today. The network maintains constant communication with the terminal and tracks the terminal's location as it moves throughout the network. Another aspect of terminal mobility is managing network and radio resources so the end user can move about while a call is in progress.

The key to PCS is also to track and locate the subscriber; this is personal mobility. Personal mobility uncouples the subscriber from any given network terminal. With preprogrammed decision criteria such as time of day, day of week, calling party, etc., the subscriber, not the network, decides when and how a call is to be delivered. The network implements the subscriber's decision.

In addition, the subscriber's services move with the subscriber for both incoming and outgoing calls. This service mobility aspect of PCS gives subscribers service management capabilities so they can modify their services from any location, making their services available over any line and to any terminal.

Network Aspects of PCS

PCS allow subscribers to access their services from both wired and wireless terminals, for voice, data, and in the future, imagery. Therefore, PCS require connectivity to multiple networks: wired and wireless, public and private, voice, data, signaling, etc. This connectivity provides subscribers seamless, ubiquitous coverage. The PCS access networks and the connected networks need to use the intelligent network to provide the mobility and the features PCS offer.

Understanding PCS

To understand PCS, look first at cellular and the original key objectives set for cellular:

- efficient use of the limited spectrum available
- affordability, to serve a large number of subscribers
- available throughout the country

	Cellular	PCS
Market Penetration	~ 5-10%	~ 50%
Service Quality	~ 2-5%	<< 1%
Coverage	90%	95-98%

Table 1 *Cellular and PCS Attributes*

- traditional telephone quality of service

These objectives still apply today, and for PCS they are essentially the same, only more and better. Areas for improvements are efficiency of spectrum usage, suitability for handheld terminal service, affordability and availability for broad market appeal, and wireline quality of service. Two additional objectives for PCS over cellular are a higher per-subscriber usage and additional services.

Some key indicators in comparing PCS capabilities with cellular are market penetration, service quality, and coverage *(Table 1)*.

Cellular Architecture Model

To understand the proposed architecture model for PCS, look first at an architecture model for cellular. This discussion of the architecture models draws on the work of two industry standards bodies: the T1P1 committee of the American National Standards Institute (ANSI), which has provided a technical report,[1] and the TR45 committee of the Electronic Industries Association (EIA)/ Telecommunications Industry Association (TIA), which has provided the mobile application part (MAP) standard for cellular intersystem communications.[2] What is presented here is a simplification of the reference architectures of these standards committees. Here, the PCS architecture model is based on the current cellular architecture as a means for a tutorial comparison. Furthermore, these architecture models are intended to represent functionality, not necessarily physical implementations of network elements. As *Figure 1* shows, the key components of the cellular architecture are: wireless terminals, base stations, mobile switching centers (MSC), mobility management for wireless terminals, and operations, administration, maintenance, and provisioning (OAM&P) to support the cellular system. Today's wireless terminals may be either analog or dual mode digital, and can be car-mounted or portable.

The base stations contain the radio transceivers, antennas, etc., and form the cell

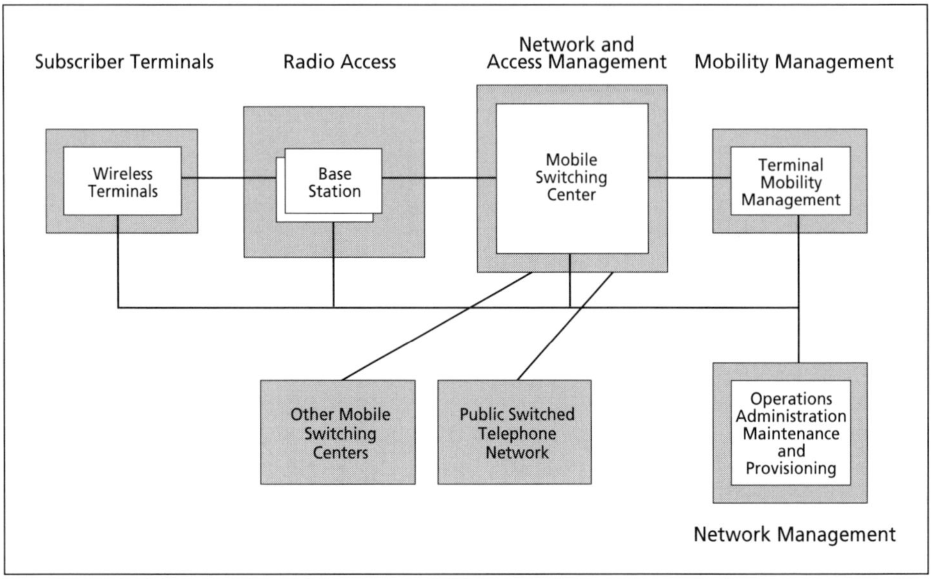

Figure 1 *Cellular architecture model*

sites that perform much of the radio functions. The cell sites use current macrocells, which may have microcells providing extended coverage for the base stations.

The MSC serves as the central control and switch for the entire system. It provides the switching fabric for the traffic connection between the base stations and other network elements, such as other MSCs and network elements in the public switched telephone network (PSTN). In addition, the MSC serves as the wireless access control manager, which manages radio resources and handles interMSC handoffs. It also maintains a terminal or user profile database for validating terminal or subscriber status and supporting terminal mobility management.

Terminal mobility management is the functionality of tracking the cellular terminal as it moves from one MSC service area to another, to know when the terminal is available, and where it is. Whenever a terminal moves to a new MSC area, it registers at the visiting area. This visiting location is reported to the home area. Calls directed to the wireless terminal trigger the home serving area to initiate queries to the visiting serving area near the terminal for call routing instructions. The visiting MSC controls a paging procedure that informs the terminal of the call attempt. In cellular networks today, terminal mobility management is typically part of the MSC's implementation; however, some carriers are deploying separate network elements to support terminal mobility management.

The PSTN is essentially the local exchange carrier (LEC) network for call completion. Depending on the call, this traditional land telephone network may also include an interexchange carrier network. The MSC link to the LEC is required for virtually all call completions.

Connection to other MSCs is needed for handoffs and validations. Handoff is the management of the radio and network resources that allows a wireless terminal with a call in progress to roam from one MSC's service area to another. Validation is required for a subscriber to initiate a call from a visiting service area. InterMSC communication for both handoffs and validations uses the MAP as specified by the TIA. For the most part, OAM&P consists of the tools to facilitate engineering and operation of the cellular system, and provisioning subscribers and terminal profiles.

Proposed PCS Architecture Model

PCS are broader in scope than cellular. Key additions that the PCS architecture model must provide are:

- more efficiency in the wireless access area to accommodate greater subscriber and traffic demands and better quality
- the capability to connect people, not just places
- increased network connectivity to provide ubiquity and additional features

One added functional element over the cellular architecture is wired terminals *(Figure 2)*. Although the market is highly mobile, most of the places a PCS subscriber moves to have wired terminals readily available. Therefore, subscribers should have access to their PCS capabilities through the wired terminals and the traditional wireline access interfaces.

Another difference is that radio access is separated into two (or possibly more) functional elements. The radio port provides transmission and reception of radio signals over the air interface to the subtending radio terminals and interworks the air interface and the network transport interface. The radio port is an air interface-dependent functional element.

Other functional elements related to radio access are intended to be air interface-independent. The radio access and radio port controller functional element

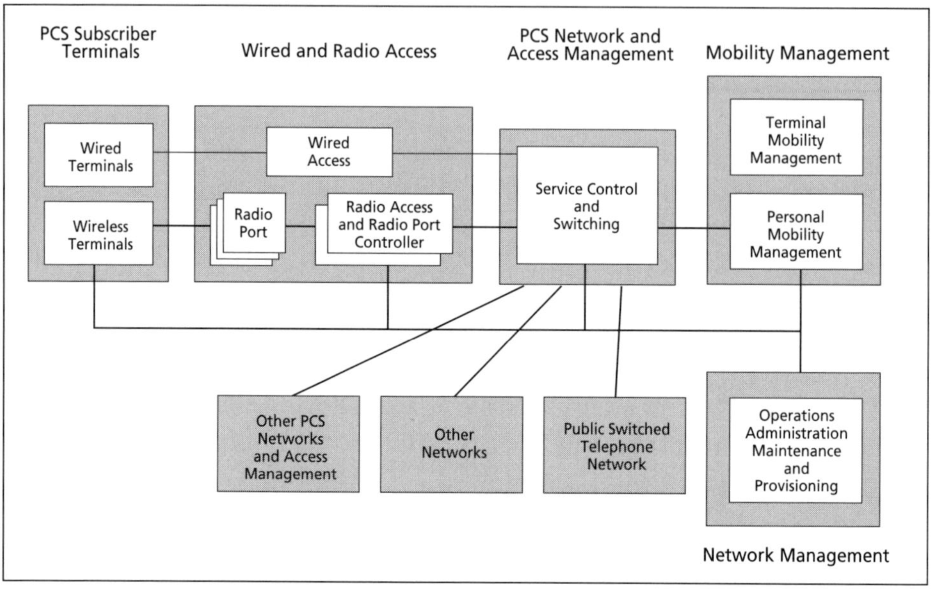

Figure 2 *PCS architecture model*

provides the interface between subtending radio ports and the service control and switching functional element. On the radio port side, it allocates radio facilities, speech coding, etc. On the service control and switching side, it provides wireless call setup, supervision/release and inter- and intra-radio port handoffs. This functional element serves one or more radio ports.

The service control and switching functional element supports access-independent call and service control functions, and is responsible for interconnecting the access systems and the network systems to support end-to-end services. Service control and switching could serve multiple radio access and radio port controller elements. Note that this function may be realized by multiple network elements, such as a switch with an adjunct.

Terminal mobility management provides the control logic and data necessary for terminal authentication and terminal location management, and controls terminal registrations, call delivery, etc. Personal mobility management provides the control logic and data necessary for user authentication, service request validation, user location management, user access to service profile, user registration, and call management, i.e., routing to user-specified destinations. OAM&P represent a set of capabilities typically distributed over both operations systems and network elements that provide end-to-end configuration, fault, performance, security, and accounting management. Other networks represent other voice, digital data, packet data, and broadband data networks, including existing and future networks such as wireline, cellular, interexchange carrier, and those of non-PCS providers.

Numerous physical architectures can be developed from this architecture model. Different industry participants have proposed physical architectures, and different industry players can provide part or perhaps all of these functions for an end-to-end PCS offering.

Design Principles

Designers of architecture elements for PCS should keep in mind some key principles.

Modularity

A modular design is essential for a long-term viable system architecture. For example, with the radio access method isolated, the network infrastructure can evolve independent of changes in the radio access technology. A uniform and spectrum-independent protocol can be provided over the air so that multiple radio access schemes can coexist.

Low Cost, High Utility, Best Quality, and Friendly Human Interfaces

PCS should achieve low cost for mass market deployment, high utility with extensive mobility and services, high quality comparable to traditional wired telephony, and user-friendly interfaces for service activation and reconfiguration.

Common Interfaces and Platforms

Common human interfaces provide common service activation and allow for interaction among services. Common interfaces between network elements, such as the air interfaces, network interfaces, signaling interfaces, etc., could lower development costs and reduce service providers' installation and operating costs.

Network Compatibility

PCS will not be built as an independent network or an individual island of services but as an extension or expansion of existing networks, such as the cellular, wireline, signaling, and intelligent networks. It is economical and efficient to design

PCS architecture elements that are compatible with the existing network interfaces and their future evolution. Further, the embedded network infrastructure can be used to provide enhanced services to the PCS end users.

Intelligence Distribution

To provide flexible implementation, intelligence should be distributed throughout the different elements in the architecture so bottlenecks can be minimized (examples today include terminal-assisted handoff, terminal automatic registration, etc.). This distribution will improve the overall service performance and system responsiveness, because the closer the intelligence is to the element where it is needed the quicker the response. It also improves reliability, because a failure results in only partial loss of intelligence and not complete loss of control.

The Personal Terminal

Looking at the functional elements in the PCS architecture model separately, let us consider the personal terminal. Many different visions of the personal terminal are associated with PCS; however, some key attributes are common among these different visions. To have the mass market appeal predicted for PCS, the personal terminal must be small, lightweight, portable, low-cost, and easy-to-use. Attributes of high performance and feature richness are required to keep the PCS terminal on par with today's wireline terminals.

Some industry participants envision that on the road the personal terminal would act like a cellular phone. While at home or in the office, the personal terminal would handshake with the in-place network to register the subscriber and redirect incoming calls. Thus, a cellular-like personal terminal would become a cordless handset connected with the household telephone when the subscriber is at home and with the office PBX when the subscriber is at work. Also, some envision the personal terminal supporting multiple radio frequencies and/or multiple modulation schemes.

Wireless Access Using Microcells

In the wireless access piece of the PCS architecture model, one of the key activities is the emergence of microcells. Microcells are emerging due to the capacity limitation of macrocells and the market push for ubiquitous coverage and larger capacity. Furthermore, the push for longer talk-time calls for smaller cells with lower transmit power. With lower transmit power, interference is easier to control and battery size is reduced allowing for smaller, lightweight terminals. These handheld portables have a higher statistical call success and quality level. The higher frequency band allocated for PCS also makes smaller cells more attractive. A large number of radio ports will need to be deployed to get the radio port as close to the end user as possible; therefore, the radio ports need to be low-cost.

In reality, an overlay or layer of cell sizes will provide ubiquitous coverage and support the different ranges of mobility. Macrocells are predominantly used today in cellular. They support high-speed vehicular mobility and have a fairly large cell size. As the user density increases in the cell, the radius of the coverage area must decrease. Microcells serve slower-moving users over smaller distances. They are good for urban, high-traffic areas and indoor coverage for small industrial parks, plazas, transportation centers, convention centers, etc. These microcells can be engineered into picocells for in-building PBX and local area network (LAN) coverage to serve more densely populated areas and slower-moving or quasi-stationary users.

For PCS to provide ubiquitous coverage for the highly mobile, local pedestrian and slower business and residential environments, a single layer of cell coverage may not be sufficient to ensure good quality of service. Therefore, a layered cell coverage approach must be adapted where different sizes of cells overlap each other. Effectively, microcells provide primary coverage to serve high-density on-street traffic areas, with macrocells providing extended radio coverage and picocells supporting indoor wireless communications.

Interconnecting Access and Network Systems

The PCS service control and switching (SCS) functional element interconnects the access and network systems to support end-to-end services. This functional element is also responsible for providing access-independent call/ service/connection control functions. It allocates access facilities and network resources and provides the network routing and switching functions. The SCS may also include auxiliary services such as voice mail and paging, and interworking functions.

The PCS SCS is responsible for providing service to subscribers based on their personal identity rather than a line or terminal identity, as in traditional switching systems. Therefore, it requires interaction with the personal mobility management and terminal mobility management functions for subscriber and terminal information.

The PCS SCS must interface to other PCS SCS elements to support intersystem handoff, which is currently provided in cellular networks using MAP. The PCS SCS must also interface to the PSTN to support the general call setup procedures currently provided in many networks using the signaling system 7 (SS7) integrated services digital network user part (ISUP).

Various implementation alternatives can exist for the PCS SCS element, such as adding more wireless access control and mobility management type functions behind a traditional wireline switch. Another implementation may be to add more IN functions on a traditional wireless switch. For either of these implementation alternatives, an increase in the capacity of the switching fabric and the real-time processing capabilities may be needed to support the projected traffic and functions.

Mobility Management

PCS must provide mobility management for both terminals and persons. Terminal mobility management (TMM), as currently provided by cellular networks, is essentially the same in PCS. Wireless terminals must register with the TMM function. This is usually on power-up or with a change of service area. Then the TMM function continues to track the location of the terminal as it moves through the network. When a switch receives an incoming call for that terminal, it queries the TMM function that translates the terminal identification into a network routing address.

Currently, in cellular, TMM is implemented using the TIA mobile application part (MAP) standard for cellular intersystem communications. MAP is an SS7 transaction capabilities application part (TCAP)-based protocol. Therefore, the associated interfaces for functional elements supporting TMM should be based, at least initially, on the cellular MAP.

Personal mobility management (PMM) provides a similar function, but for tracking and locating persons. A person can be located based on a personal telecommunications number (PTN), regardless of their network location or equipment used. PMM translates the PTN into a terminal identification. First the user has to register his/her association with a terminal, either wired or wireless, with the PMM function. If the terminal is wireless, it must be registered with the TMM function. Then the user's PTN is temporarily associated with a network point of attachment for incoming call routing and for personalizing the user-network interface for outgoing calling. The user controls the disposition of calls directed to him/her.

Currently, the standards bodies are considering PMM as an application of the intelligent network (IN). Therefore, SS7 TCAP-based IN application protocol (INAP) may apply.

AT&T's Vision of PCS

AT&T is committed to offering the products and services that provide end users the freedom to communicate from anywhere, to anyone, at any time, and in any form. Some of the key concepts and capabilities of PCS are available today. Terminal mobility is available with cellular, and cellular is responding to the capacity demands with recent deployment of microcells. Some forms of personal mobility are available today with personal telephone numbers. AT&T believes what is needed now to make a PCS offering is for these capabilities to be:

- integrated
- ubiquitous
- provided at the traditional landline quality

Thus, the PCS networks will be fully interconnected and interworking with the existing networks provided by carriers today.

A PCS network can be considered as consisting of the PCS switching element and its associated cluster of micro/macro/pico cells. Because there will be multiple service providers offering PCS, for ubiquity these different PCS networks need to appear uniform to the end user. Furthermore, for both inbound and outbound call completion, these PCS networks need to interconnect and interwork with the in-place backbone switching and transport networks provided by the local and interexchange carriers. This network connectivity is illustrated in *Figure 3*.

Because mobility, both terminal and personal, requires intelligence, Figure 3 illustrates both the TMM and PMM functions implemented on service control points (SCPs) in the intelligent network (IN). The PCS networks require links to the IN, with the PCS switching element actually requiring some service switching point (SSP) functionality to query these SCPs. The IN is responsible for locating PCS subscribers for inbound calls and providing the capability for subscribers to interact with the mobility management functions to dynamically change their service parameters for both inbound and outbound calling.

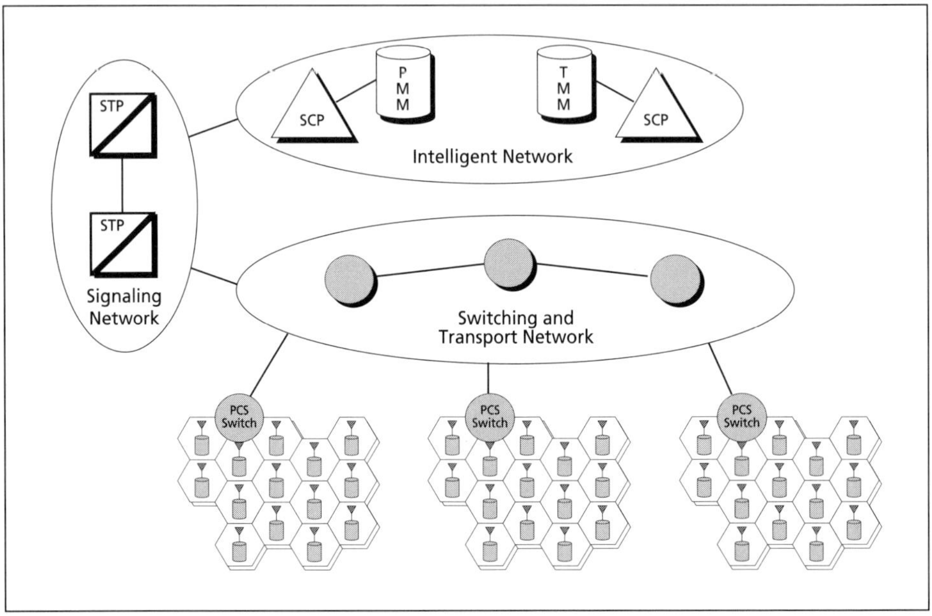

Figure 3 *Network connectivity for PCS*

The IN also allows for the proper distribution of intelligence and a hierarchical system of data management for the mobility functions. By centralizing the PMM and TMM functions into network databases, the concept of a home service area is not needed; therefore, the intelligence needed for these functionalities, although stored centrally in the network, can be distributed locally on a temporary needs basis.

The underlying transport network for the message flows associated with mobility management, connection setup and release, etc., among the PCS network, the switching and transport network, and the intelligent network will be the SS7 network with its fully interconnect pairs of signal transfer points (STPs). The signaling network is effectively serving as the glue for this network connectivity.

Service Mobility

Having full-network connectivity and properly integrating both terminal and personal mobility provides the end user with service mobility. The objective of service mobility is that the subscriber can be associated with any terminal on a dynamic basis and has specific features available from any point in the network. Service profile features include originating calling features, such as preferred carrier and speed dialing lists; terminating call routing instructions, such as voice mail, home/office/portable phone, no answer treatment; and other features such as call waiting, call screening, distinctive ringing, etc. Once this user-to-terminal association is made, the user should then be able to make calls and access other services, all of which will be charged to the user rather than the phone.

Ways of providing this personalized service to the user may include real-time downloading the switch and/or the terminal with specific features and capabilities resident in the service profile data in the SCP. This is all within the limitations of the switch and terminal. Another way may be for the user to personalize the terminal with a smart card.

Smart Cards

Smart cards provide convenient storage of a large amount of user and system profile data, such as system access codes, security codes, classes of service, billing information, originating and terminating calling features, etc., which can simplify and customize the user-network interface. The smart card not only adds features to the terminal but also provides security with user authentication, controls access to the subscriber's services and charging capabilities, provides terminal independence, and allows the caller to use any smart card-compatible terminal. Through the intelligent network, the smart card can link with network databases, thereby combining the user and network profiles, further realizing the concept of service mobility.

Technology Challenges

PCS still face some technology challenges to make all this happen. One challenge entails ensuring the TMM and PMM functions are integrated to provide true service mobility. This requires interworking between the network elements that are in place today and the new network elements that will be developed and deployed to support these functions. This may require interworking the INAP capabilities planned for PMM with the MAP procedures used for TMM today.

Second, having the different networks supporting PCS end-to-end services linked together requires a unified approach to network management. End-to-end OAM&P will improve the quality of service to end users, reduce the network operating costs, and increase the utilization of the networks for both the service providers and the end users. To achieve these goals, both network elements and operations systems should be migrating toward the industry standard object-oriented interface referred to as the telecommunications management network (TMN).

Next, achieving the projected market penetration and providing the features and customization users are expecting from PCS requires low-cost terminals, possibly using advanced displays and smart card technology. Enabling technologies for this next-generation personal communicator include advances in signal processing systems, high-density batteries, high-resolution displays, and voice recognition capabilities.

The digital radio access technology for PCS demands that service providers, vendors, and other PCS players agree in the industry standards forums. This radio access technology must be a spectrally- efficient access scheme with advanced coding and modulation and provide speech compression algorithms for high-quality communications.

Finally, the high-traffic capacities and quality of service demanded require low-cost micro/pico cellular base station technology. In addition, to get radio ports as close to the end users as possible, cost is a major factor because a large number of radio ports will need to be deployed. Also, traffic flow control between microcells and macrocells could be somewhat complex, especially if vehicular traffic is included along with nonvehicular traffic.

Endnotes

1 *A Technical Report on Network Capabilities, Architectures, and Interfaces for Personal Communications, Committee T1 - Telecommunications, T1P1/93-062R2,* Draft Technical Report, September 1993.

2 *Cellular Radio - Telecommunications Intersystem Operations: Functional Overview,* EIA/TIA Interim Standard (IS) - 41-B, December 1991.

Wireless: A New World of Possibilities

Jack M. Scanlon
Executive Vice President, General Manager-Cellular Infrastructure Group, Motorola

We are in a communications revolution where wireless technology has become a cost-effective and, in many instances, a better alternative to wired technology. We are at a point in the technology cycle where wireless infrastructure is coming down rapidly in its cost of deployment, while wireline costs continue to increase due to the high amount of labor content involved in laying cable and wire.

In 1992, there were approximately 56 billion minutes of worldwide use in wireless compared to wire line minutes of use that were four trillion. As seen in *Figure 1*, it is predicted that by 2003 the total minutes of telephone use will approach eight trillion and wireless will account for two trillion (25 percent) of worldwide minutes. That will drive an incredible increase in wireless systems around the world. However, it is more than a simple multiplication story. Rather, it is one that is "wired" with incredible possibilities for both developed and developing nations to establish and expand wireless telecommunications as we know it today.

Cellular Industry Status

Let us look at wireless telecommunications today to propose an expected scenario for the future of the industry. Today, we are witnessing five major trends: (1) accelerated worldwide subscriber growth, (2) increased emphasis on fixed wireless systems, (3) adoption of digital technology, (4) aggressive pricing, and (5) worldwide consolidation of operators. Each will have a significant influence on the evolution of the industry.

Reprinted with permission from New Telecom Quarterly, *1Q95. Copyright © 1995, Technology Futures, Inc.*

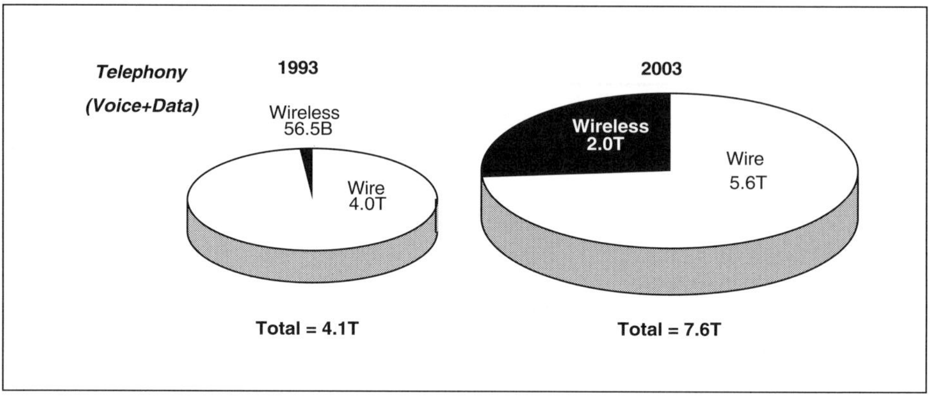

Figure 1 *Worldwide minutes of use: wired and wireless*

Accelerated Worldwide Subscriber Growth

As we look at the current figures of worldwide subscribers, it is clear that we are in an accelerated growth phase. Just a few years ago, the number of cellular telephone subscribers was growing at a 40 percent annual rate. Most recently, the number of subscribers has escalated to more than 50 million, growing at a 50 percent rate. That number of subscribers is expected to increase to more than 450 million subscribers by the year 2004.

Markets that present significant potential for increased penetration rates over the next 10 years include the United States, Japan, China, Germany, France, India, Italy, Great Britain, Russia, Brazil, Mexico, Indonesia, Canada, Spain, and South Korea. Countries like the United States, Japan, Germany, Italy, Great Britain, Canada, and South Korea – which today account for a substantial percentage of worldwide subscribers – are predicted to add significantly to their subscriber bases. Countries like China, France, Brazil, Mexico, Indonesia, and Spain have just welcomed cellular telephony into their markets and also represent areas of great growth potential. Currently, there is less than 2 percent cellular penetration worldwide and a strong demand for phone service.

Increased Emphasis on Fixed Wireless

For developing countries with little or no telephone service, a new focus is on providing basic residential service via cellular technology. In countries such as India, where only 1 percent of the 900 million people have telephone service today, wireless offers tremendous communication growth opportunities.

One prominent example is fixed wireless, in which emerging economies such as India, Indonesia, Malaysia, Sri Lanka, Russia, and China can finally get "in the loop" of emerging telecommunications service. This is accomplished by attaching wireless tails to cable infrastructure to provide wireless telephony from that infrastructure. Used in both rural and urban areas, fixed wireless has been instrumental in providing telephone service in emergency situations and national disasters, as well as more general commercial use. One example of how fixed wireless has been successful in providing needed telephone service is WiLL, wireless in the local loop, developed by Motorola and used in more than eight nations.

Adoption of Digital Technology

In addition to subscriber growth and a worldwide focus on wireless, another key trend is the adoption of digital technology. The commercial introduction of time division multiple access (TDMA) and the current trials of code division multiple access (CDMA) technology have given operators more than just new cellular standards. Digital technology will provide a number of advantages for both the operator and the subscriber, including improved voice quality, fewer dropped calls, better coverage, enhanced privacy, increased network capacity, new messaging features, and wireless data applications.

Only in its gestation, digital technology has already made an incredible impact on the cellular industry. We have seen an explosive growth in the adoption of the European digital standard, global system for mobile communications (GSM), not only in Europe, but also in other parts of the world. For example, in Germany, there are more than twice as many subscribers to digital mobile services as analog subscribers. Overall, GSM is in commercial service supporting approximately four million subscribers in 52 countries worldwide.

And just what has the introduction of digital technology meant to the analog market? Interestingly enough, it has increased penetration of analog-based cellular systems around the world. Every place where digital technology has been introduced, there has been one absolutely predictable result-more subscribers flock to both the digital and analog technology.

Currently, 90 percent of cellular subscribers worldwide are served by analog equipment. Most recently, in markets where analog systems perform well, there were surprising increased growth rates, especially in Pan America, Spain, the United Kingdom, and Japan. After GSM digital networks were introduced in the United Kingdom, not only was there increased competition and lower priced service packages, but also a substantial increase in analog growth.

Why the positive impact of digital on analog growth? First, analog works, and works well. Second, with the introduction of digital, analog providers become very competitive with their pricing, and rates for minutes of cellular use go down. This causes the subscriber population to accelerate at a more rapid pace. Clearly, the more competition, the better.

Aggressive Pricing and Open Architecture

What are the forces behind this competitive pricing? In the cellular industry, one significant influence has been the advent of open architecture standards. Until recently, analog systems were based on proprietary architectures and interfaces between network elements. As a result, operators had little opportunity to integrate the best system elements available, leaving them captive to a single vendor for fulfilling their end-to-end product requirements.

When these proprietary interfaces open up, it is then possible to mix-and-match radio and switch elements from various vendors. The result is incredible competitive pressure on the suppliers leading to aggressive pricing practices for cellular infrastructure.

Driven by the adoption of GSM, open architecture standards are in place in Europe and more than 50 countries around the world. *Figure 2* highlights cellular penetration in Japan and Germany. As illustrated, when you introduce a new competitor, you create more demand among the population. Germany, which grew slowly for a long time because of their outdated system, had reached their subscriber capacity of one million and was impacted by high prices from one cellular equipment provider. This scenario quickly changed when two GSM providers were introduced in the country. The result is an established competitive environment and heightened cellular growth.

Operator Consolidation

If you look around in the telecommunications industry today, in addition to a competitive stance, you will see operator consolidation. This progression toward centralization of operations is driven, in part, by the demand for seamless wireless service in which a cellular subscriber can make a telephone call from anywhere in the world and by the development of the economies of scale required to compete.

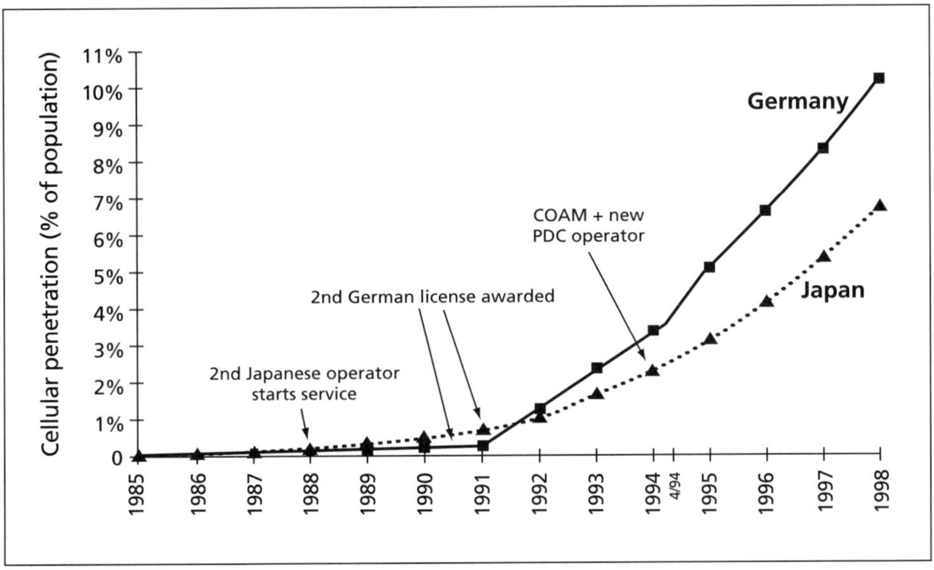

Figure 2 *More competition = more demand*

There are a number of examples of this operator consolidation that span the globe. One such scenario is two cellular operators teaming up to offer service in a more expansive area. Or perhaps you have seen the consolidation of a local exchange carrier with an interexchange carrier (IXC) and even a cable company with a local exchange carrier. The merging of such companies contributes to upgraded service for subscribers, more expansive territory coverage, and accelerated cellular growth for the entire industry.

Global Factors Driving the Communications Revolution

When we talk about the status of the cellular industry today and where it is moving in the future, there are four key driving forces behind this revolution:

- Countries emerging to market-driven economies.
- Rapid and effective communications as a necessity to compete in today's world.
- The allocation of significant quantities of spectrum for wireless systems and the evolution of technology for use of the spectrum.
- Telecommunications monopolies in developed countries are giving way to a true competitive market.

Countries Emerging to Market-Driven Economies

Across the world today, 80 percent of the population has recently emerged from controlled economies to market driven economies. Such an environment drives the need for communications significantly higher. This is in contrast to the few remaining countries that operate in a controlled economy in which a central committee plans the production of items and therefore the need for interactive communication is rather minimal.

Now, we have countries, like China with a 1.3 billion population, emerging from being a centrally-controlled economy to a market-driven economy. Today, China has about 15 million phones for 1.3 billion people – slightly more than one phone per 100 people. They have a five-year plan to add approximately 85 million new telephone lines in the country or 5.6 phones per 100 people. In such an environment, wire and wireless telecommunications are absolutely essential to the growth of the country.

Rapid and Effective Communications as a Necessity to Compete in Today's World

More and more people are using cellular telephones. For instance, in Buenos Aires, 95 percent of subscribers' telephones are wireless. A significant percentage of people using mobile communications systems today do so to give themselves an edge in conducting their business. There is an underlying presence that says wireless communications will allow you to operate and be an effective and competitive force in today's world. Telephone service is more than just an appeal by many to be kept "in the loop," but rather a staple for any country that wants to upgrade its competitive standing in today's information-driven world.

Allocation of Spectrum for Wireless Systems

Another key driving force behind the future of wireless telecommunications is personal communications services (PCS) which combines wireless voice and data service. In the United States alone, the FCC has set aside 140 MHz of spectrum for PCS 1900 MHz and recently (1994) began auctioning licenses for PCS.

This allocation of new spectrum for wireless service is more than a story about MHz or a particular frequency. It is about bandwidth available for new and additional services. PCS allows for new competitors to enter the market, and that inevitably drives down prices and creates demand for new products and services.

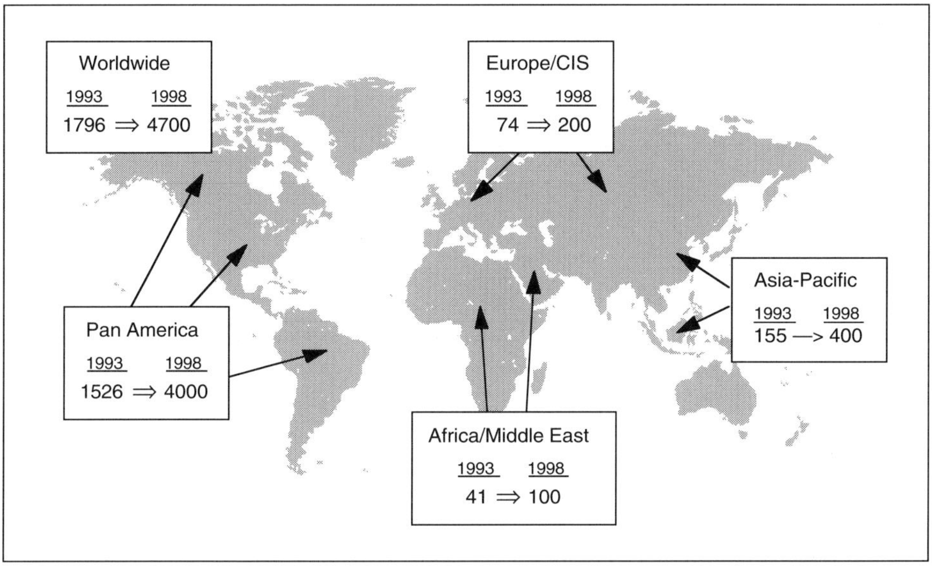

Figure 3 *Cellular and PCS licenses: a three-fold increase between 1993 and 1998*

It is estimated that, by 1998, there will be 25 million PCS subscribers worldwide. This significant growth is based upon the expected three-fold increase in cellular and PCS licenses from the 1993 figures of approximately 1,800 licenses to a projected 4,700 in 1998 (see *Figure 3*).

Monopolies Giving Way to True Competitive Market

In this movement toward a wireless world, monopoly is no longer the popular board game. Instead, we are part of a worldwide drive toward a competitive environment that spurs aggressive pricing and product development in the cellular industry. We see new alliances being formed almost daily, especially in the United States where there is considerable competitive activity in telecommunications.

In other countries, such as the United Kingdom, Germany, France, and China, we are seeing the telecommunications monopolies break up through the division of adjunct services from the core business. In addition, the core businesses are starting to become decentralized and fragmented. As monopolies continue to break down, the result is more and more competition in every dimension of the telecommunications business. That means equipment providers will work to deliver new and advanced services for operators and subscribers.

Cellular: Reaching Beyond with Open Network Architecture

Competition is clearly a key driving force for the cellular industry. As previously discussed, one of the premises for competition is an open network architecture. With an open architecture, wireless providers will be able to choose the switching equipment and base stations that best meet the needs of their systems – and ultimately their customers. This freedom of choice also forces suppliers of both cell site and switching equipment to design and develop products that clearly compete on the basis of price and performance.

There is more than just talk about open architecture in the cellular industry! Elements of open architecture are in place, including standards such as GSM, intersystem operations (IS-41), and A Interface (IS-634), all of which support the beginning of a wireless intelligent network.

GSM digital technology has been introduced in Europe, the Middle East, Asia, and Africa under an open network architecture. As the GSM experience to date has demonstrated, these defined benefits of open architecture can be realized in the industry. In Europe, where the standardized GSM A Interface is in use, the result has been consistently lower infrastructure prices.

Progress in open infrastructure standards in non-GSM markets has, to date, been limited. One of the most notable advances is the introduction of the IS-41 specification that permits subscribers on one vendor's infrastructure system to roam to another. The IS-41 platform is now being offered more often by equipment suppliers deploying cellular networks that are based on analog technology. The defined roaming capability of IS-41 has opened up the lines of communication not only for the subscriber, but also the operator and supplier.

IS-41 is an important first step in moving toward open architecture for the advanced mobile phone service (AMPS) analog standard. Yet, there is another concept that is getting considerable attention in today's industry. Standardization of the A Interface yields additional benefits for the quest to create a worldwide open network architecture for cellular systems. The proposed A Interface (IS-634) standard is based on the European GSM A Interface and modified to support wireless air interfaces in the Americas. IS-634 defines the essential messaging between the mobile telephone switching office (MTSO) and the RF subsystem for both cellular and PCS applications.

In the quest for an open, competitive network, there is another important factor emerging – the wireless intelligent network. Supported by the described open architecture standards, wireless intelligent network is an evolution in the cellular industry that promises faster and more flexible service delivery. It extends "open" cellular concepts beyond the cell site to the switch interface. By moving call processing functions out of the mobile switching center and onto linked intelligent network computer platforms, the growing needs of the wireless user for advanced services can be more quickly addressed.

Looking into the future, the greatest success in the wireless intelligent network will come to those who can swiftly create, integrate, and deploy new voice, data, video, and multimedia features and services which customers demand. In the next few years, the evolution of intelligent network personal services will deliver news memos, stock memos, fax displays, fax notification, and other services over the cellular network.

The Future of Wireless Telecommunications

Overall, the proprietary infrastructure period for cellular is progressively coming to a close. GSM has pioneered an open infrastructure specification that has already provided operators with a new level of vendor flexibility and intense competition at the voice channel level. The day will come when cellular equipment can be selected with a high degree of freedom. The resultant competition could put voice channel prices on about the same annual learning curve as subscriber unit prices and stimulate the growth of the entire industry.

With open network architecture and advancements in wireless technology leading the way, the cellular industry is transmitting revolutionary messages. As discussed, the status of the wireless telecommunications industry today reflects a period of accelerated worldwide subscriber growth, increased emphasis on fixed wireless systems, adoption of digital technology, aggressive pricing, and worldwide consolidation of operators. It is an exciting time of worldwide cellular penetration, advanced technology, and competition. It is an ever-changing environment that has opened the door to great prospects for the future of cellular. What is the scenario for the future of the wireless world? The following is a sampling of what we may expect.

(1) *Interest in Generic C will initiate a cellular application platform market.* This is the next step in worldwide cellular technology standard. Generic C is an

emerging standard that takes all radio intelligence out of the switch and places it into the radio equipment or intelligent adjuncts. This standard will allow operators to attach radios and radio networks directly to landline switches, thereby forming a closer integration of wireline and wireless. Generic C is nearing completion as a standard for PCS. Although it may take several years before Generic C is implemented worldwide, it promises to be the next generation of open interface as it develops into a universal standard.

(2) *U.S. cellular operators will mandate "open" interfaces for next generation wireless.* In the next several years, we will witness the evolution of companies actually forming consortia or partnerships around standard application interfaces. Comparable to what Microsoft has done for the computer software industry, an open network interface will dictate the future of the cellular industry. Application interfaces will be delivered into the cellular network that allow application vendors to provide new services beyond the plain voice network and into an information era.

(3) *Open applications programming interfaces (APIs) will play an important role in open architecture.* As part of the succession of the wireless intelligent network and open architecture in general, a key to developing faster service delivery is the standardization of open APIs for wireless service development. By opening APIs, it will accelerate the rate of innovation by allowing independent software vendors (ISVs) to develop new customer-driven, open architecture solutions. This provides ISVs with improved access for their products to an expanding wireless market.

(4) *The cellular industry will increasingly provide 911 and other emergency services.* Currently, the cellular network can locate the cellular user within a 2,000-foot radius, depending on the size of the cell. As we begin installing microcells with shrinking cell sizes and antenna advancement, it will be possible to locate anyone and anything within a small radius. This definitive improvement in location granularity translates into a myriad of commercial service and public safety opportunities, such as emergency 911 (now delivered over the landline), safety locator services for children and health care patients, and large business applications .

(5) *Defense contractors will enter the wireless industry.* Today, defense contractors are realizing that the diminishing defense budget calls for expanding into other areas. As a result, we will see more defense contractors taking their technologies (antenna, radio, and so forth) to commercial outlets.

(6) *Flexibility will remain a key element of wireless integration worldwide.* Even with the emphasis on developing standard interfaces, cultural differences in the way people use the telephone will dictate a need for flexibility by operators and equipment manufacturers. In addition to different cultural demands, the design and implementation of cellular networks in other countries will still continue to be affected by government policies.

Conclusion

Cellular technology has made a remarkable evolution since first-generation analog cellular service was launched in the Nordic countries in 1981 as the Nordic mobile telephone (NMT-450) standard, followed by the first U.S. AMPS analog service in Chicago and Baltimore/Washington, D.C. in 1983. More than one decade later, we are able to witness the next generation of wireless telecommunications. The "opening" of the network architecture will facilitate healthy competition among suppliers and operators. This increased competition will lead to advanced technology, additional voice and data transmission features, and an exciting introduction of new products and services. This movement to open network architecture transmits a definitive message of increased subscriber growth for the future of wireless telecommunications.

In addition to open architecture, the allocation of new spectrum for wireless service will spur tremendous growth in the industry. As companies compete for PCS licenses, we will see an environment of increased user demand and constructive competition among suppliers to deliver new services and advanced features for wireless telephony.

In order to compete in the growing cellular market, equipment suppliers will need to direct their efforts toward technology. Wireless technology has taken giant leaps in the last decade, yet there is more that lies ahead in the quest for ubiquitous wireless communications. Digital technology is poised to improve the quality of cellular service that operators can provide subscribers, while also increasing the network capacity and spurring subscriber growth worldwide. With TDMA system deployment on the rise and CDMA trials taking place on an international level, the challenge lies with suppliers to demonstrate the advantages of digital over analog.

In addition to a focus on digital technology, fixed wireless systems have already demonstrated tremendous progress in communications for residents of many developing countries who had little or no phone service. Fixed wireless will remain an integral part in driving the overall growth of the wireless telecommunications industry.

Without a doubt, our need to communicate will continue to play an ever encompassing role in an information-driven world. It is this worldwide need that will place wireless communications in the forefront of the expansive telecommunications industry and advanced technology as the enabler for the next leap. As the percentage of worldwide telephone minutes used shifts from wire to wireless, the overall impact will be an escalated penetration of phone service around the world.

Today, according to Northern Business Information, only 15 percent to 20 percent of the world's population has adequate phone service. Assuming that India and China meet their aggressive telecommunications growth plans, it is predicted that 50 percent to 60 percent of people around the globe will know first-hand the conveniences and advantages of phone service by the year 2004!

A Layman's Guide to Cellular

Martin Cooper
Chairman and Chief Executive Officer
ArrayComm, Inc.

Basics of Cellular

This essay describes the principles of cellular telephony and outlines some of the associated problems and opportunities. The essay does not promise to turn anyone into an engineer; nonetheless, it may provide an insight to the technology for "the educated layman."

Just as you don't need an understanding of electronics and video technology to watch television, you don't need to know cellular radio technology to use a cellular phone. You may, however, find some understanding of cellular radio principles useful to get the most out of your cellular mobile or portable when you want to travel from city to city, when a third or fourth carrier starts service in your area, or when you want to know how to analyze your cellular bill.

The object of cellular telephony is the freedom to make and receive calls while you are on the move as easily as you do at home or at the office. Cellular radiotelephone was born as a result of two conflicting circumstances: first, a demonstrated market need for telephone communications by people who, for one reason or another, don't have access to the wired telephone network, and, second, the shortage of radio frequency spectrum that made it impossible to fulfill this need using traditional mobile radio techniques.

Unlike ordinary two-way radio communications that involve relatively short messages (averaging 10 to 30 seconds for a complete conversation) with press-to-talk operation, telephone conversations take much longer (an average of 2 to 3 minutes) and require more interference protection, better signals, and simultaneous talk-and-listen. The result is that the best two-way radio systems can share a radio channel among hundreds of talkers in a city while the best traditional mobile telephone systems can offer good service to an average of only 20 to 25 subscribers per channel.

Cellular radio technology offers a trade-off between higher complexity and cost, and better use of the radio spectrum. Cellular radio telephone can use radio channels more effectively than traditional mobile telephone because of a concept, already widely used in other radio services, called "geographic sharing."Just as one FM broadcast channel may be used simultaneously in many cities across the country, a cellular channel may be used in different neighborhoods in the same city. Thus, several (or even 50) separate, independent conversations can be held simultaneously on a single cellular radio channel in different parts of the city.

Cellular radiotelephone is beautiful in concept. It is, in fact, simplicity itself. There are only two criteria that define a cellular telephone system in contrast with a pre-cellular automatic radiotelephone system. There must be:

1) geographic reuse of the radio frequency channels, and

2) automatic hand-off of a subscriber from area to area as the subscriber moves about a city.

If this sounds too simple, you're right. There's a little more to it, although not much more. Let's start by briefly describing some technical principles; then we'll be ready to actually step through a cellular call.

Frequency Modulation and Frequency Reuse

Voice communications in cellular systems are not unlike common two-way radio that is used today by police, taxis, and literally millions of other people in the world. Transmission is in the ultra-high frequency (UHF) radio band that is directly adjacent to the UHF TV band. Cellular radio uses frequency modulation (FM) such as that on your home or car radio, except that about ten cellular stations can transmit in the same frequency space as one broadcast FM channel.

Why FM? Because it has a characteristic called FM capture that is the single most important technical basis of cellular radio. When two different stations are received at the same time by an FM receiver, and one is stronger than the other, the weaker signal is suppressed and the stronger enhanced. This is in contrast with

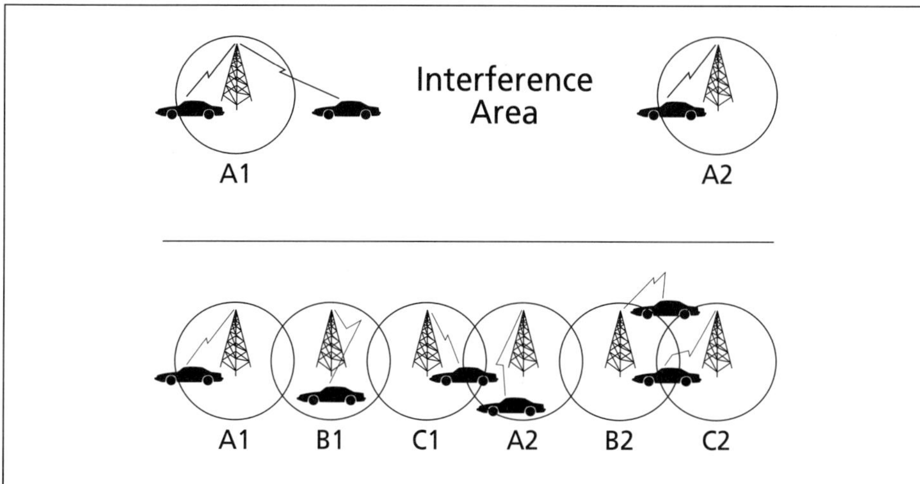

Figure 1 *(top) The strength of the FM signal in area A1 blocks out the weaker signal on the same frequency in area A2. (bottom) A simple cellular system using three frequencies, A, B, and C.*

amplitude modulation (AM) where, in some cases, both stations are heard, with the stronger one louder than the other. Thus, we can use a specific FM channel in one part of a city and re-use the same channel in another part of the city so long as the cellular subscribers are always closer to the station to which they want to listen.

In the top half of *Figure 1*, the subscriber in area A1 can only hear transmitter A even though transmitter A2 is sending information on the same channel to a different subscriber. In the area between Al and A2, the strength of the two signals can be comparable and the chances of interference are high. Interference may be the sound of garbled voices, or perhaps the wrong signal will capture the receiver.

Now that we have discovered that we can transmit effectively on the same frequency in two different areas of the same city, it's only logical that we build some new transmitters on different frequencies to fill in the spaces between Al and A2. In the bottom half of *Figure 1*, we have a simple but workable cellular system with three sets of frequencies: A, B, and C. Simultaneous conversations can be held in areas Al and A2, Bl and B2, and Cl and C2.

In this example, we get double duty out of each frequency with very little interference. Cities are rarely stretched out in a line (although there will be many heavily traveled corridors between cities where the system will look just like *Figure 1*).

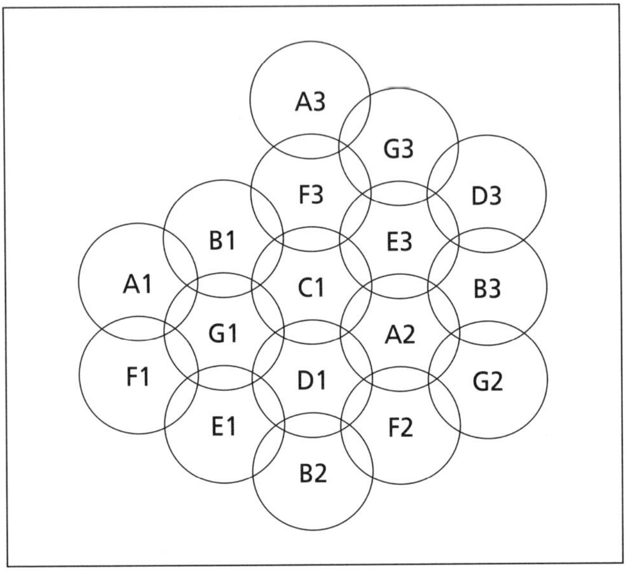

Figure 2 *A 17-cell system using 7 sets of frequencies but with the ability to use and reuse capacity.*

It is simple to blanket any shape of city by just adding new transmitters and new areas, as shown in *Figure 2*, and still retain the ability to use channels over and over again.

In practice, these areas are irregular in shape and overlap. Because hexagons fit together so neatly, it has become traditional to draw the cellular system as a honeycomb and to call each area a cell. The result is the familiar pattern seen in newspapers, magazines, and countless cellular trademarks.

Hand-off

It is clear by now that when a subscriber in the midst of a call drives out of the interference-free range of a transmitter, he or she will have to start talking through a different transmitter and on a different radio channel. When this happens, the mobile unit must switch to a new channel, and the system must also cause a change to a new land-based transmitter.

Hand-off is the process of all this happening automatically. For all practical purposes, the cellular system keeps track of the location of every subscriber who is talking on any channel. When the system determines that a subscriber has moved out of one cell into another, it executes a hand-off. A control center called the mobile telephone switching office (MTSO) notifies the mobile telephone inaudibly to change the frequency to a channel that is available in the new cell. The MTSO also transfers the conversation to that channel. All this happens in a fraction of a second, and the subscriber barely notices a muted click.

What Happens During a Cellular Call?

When a subscriber turns a cellular radiotelephone on, the microcomputer in the unit directs the radio to tune to the strongest signal on one of a number of data-only channels known as *paging channels*. These channels are never heard; they just broadcast data throughout the city that can only be understood by the computers in the cellular radiotelephones.

To make a call, the subscriber enters a number or recalls it from the unit's memory and depresses the SEND button. In less than a second, the radiotelephone automatically transmits the number being called, the telephone number of the subscriber, and the electronic serial number of the subscriber's unit. This information is digested by a much larger computer in the MTSO.

The MTSO is connected to each cell in the system and directs all voice and data movement in the system. The MTSO orders the radiotelephone to change to a voice channel, connects that voice channel to the landline telephone network, and dials the number into the network. The rest of the call is conducted as though it were an ordinary landline phone call until the subscriber hangs up, at which time an electronic disconnect is transmitted and the MTSO disconnects from the landline network.

A call from a landline subscriber to a cellular subscriber in the same city is conducted in much the same way. Once again, the cellular unit is tuned to a signaling channel. The MTSO transmits the cellular telephone number on all signaling channels, the called unit responds with a data signal, and the MTSO sends a signal to the cellular unit to cause the phone to ring. When the phone is answered, it is directed to a talking channel and the conversation can begin.

As a subscriber moves about the city during a phone call, the MTSO measures the quality of the signal. If there is a deterioration because of movement from one cell to another, a hand-off to a new channel occurs with no interruption to the call. The data signal that orders this hand-off sent to the cellular unit is on the voice channel but at a frequency that cannot be heard. It is not unusual for several hand-offs to occur in a minute if the subscriber is traveling along the boundary between cells.

There are over 400 radio channels assigned to each of the two carriers in a city. In any given cell it is possible to talk on one of a group of channels that may be as few as two and as many as hundreds. The mobile or portable phone can tune to any one of these in the process of finding one that is not in use.

Cellular radiotelephone is different from other forms of land-mobile communications because it includes a combination of geographic sharing, cell-to-cell hand-off, and channel trunking. These techniques are not new; in fact, the creation of the cellular industry was not stimulated by a technological breakthrough. Rather, it was the need expressed by the FCC for a system that would fill a large market need for mobile telephone but would avoid uncontrolled use of the scarce radio frequency spectrum.

Telephone differs from traditional radio communications in that the talking parties literally own their own private communications channel until they choose to give it up. There is rarely interference, social or technical, to encourage the talkers to conserve their channel time – in other words, to talk less. Thus, if traditional methods were used, 5,000 simultaneous talkers in a city would use 5,000 radio channels. That is equivalent to the radio spectrum needed for 50 television stations and several times the maximum number of channels allocated for all land mobile services. Further, if this system were to expand beyond 5,000 simultaneous talkers, a new channel would be needed for each new conversation.

Even if the 5,000 channels were shared, as they most certainly would be, reasonable telephone service could be made available to a maximum of about 100,000 subscribers in one city. Clearly, the FCC could not condone an approach that would, if it became popular and economically successful, consume unprecedented segments of radio spectrum – a spectrum that already is widely sought after.

Enter cellular technology. Geographic reuse, as discussed earlier, permits use of a radio channel simultaneously by two or more people in one city at any given time. In *Figure 3*, we show an example of a multiple cell system using a seven-cell repeating pattern. Seventeen cells gives us an average repetition of each frequency 2.4 times (i.e., 17÷7) . Since there are about 416 channels available to each of the two operators in a city, the system would have 2.4 times 416, or 832 channels in which simultaneous conversations could be held. When a large number of channels is accessible, between 22 and 25 subscribers can be served by each channel.

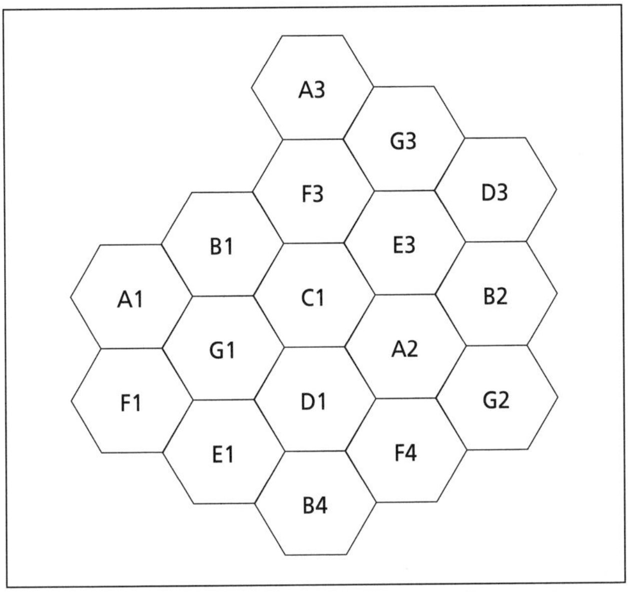

Figure 3 *A multiple-cell system using a 7-cell repeating pattern.*

The ultimate system capacity depicted is then 22 × 832, or 18,304 subscribers.

In practice, the operator will run out of capacity well before this ultimate capacity is achieved because traffic will not be distributed evenly throughout the city. Radiotelephone conversations happen where people congregate—along expressways in cars or in downtown areas with a combination of portable and car telephones. The operator will discover that the system has run out of capacity, not everywhere, but in a specific group of cells at certain times. No matter how the 400+ channels are juggled, service starts deteriorating. The only solution now is splitting the cells to achieve greater reuse.

Cell Splitting

The simplest way to look at the splitting of cells is by uniformly splitting all cells in the system. Start with the three-cell example of *Figure 4*. Each cell in this example has a transmitter site placed at its center. If an additional transmitter site is placed at every corner of our three cells, the results are as shown in *Figure 5*. Notice that we now have 17 small cells instead of three and that certain frequencies, those in small cell A for example, can now be used three times instead of just once. If this kind of

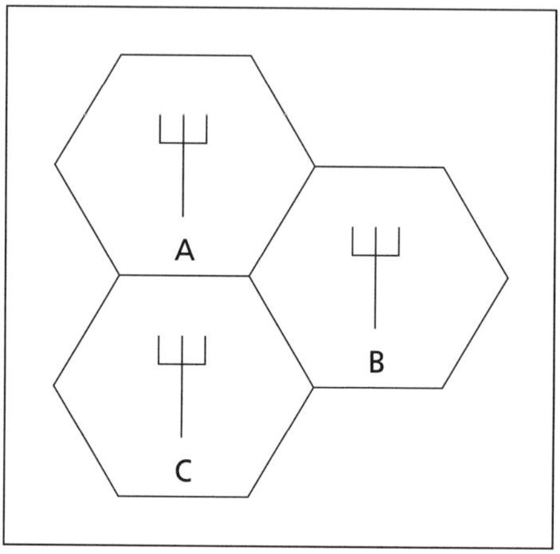

Figure 4 *3-cell pattern*

splitting were applied to the cell pattern in *Figure 3*, the theoretical capacity of that system would be increased seven times to over 100,000 subscribers.

It doesn't make sense to do that. We want to split cells where there is little traffic. That can be done but only with great care and lots of engineering. Once a single cell is split, the frequency patterns of all the cells surrounding it are affected.

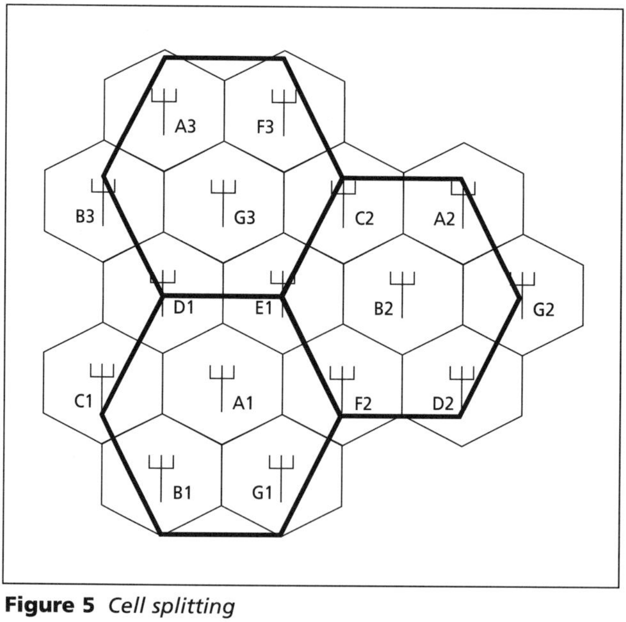

Figure 5 *Cell splitting*

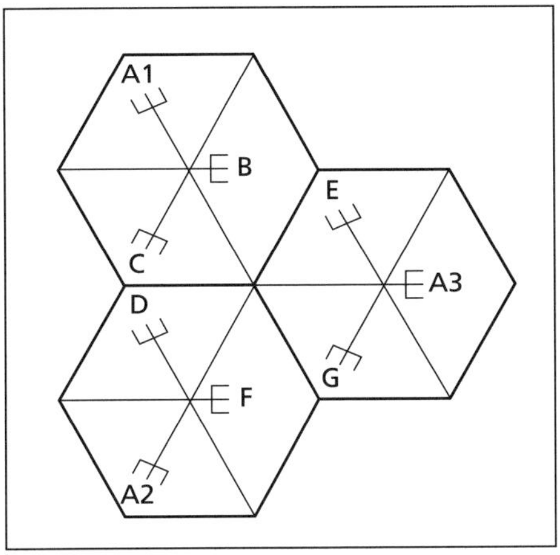

Figure 6 *Corner antennae cell splitting*

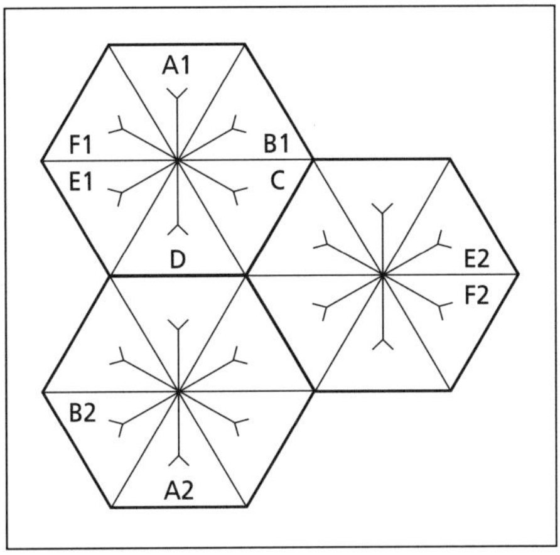

Figure 7 *Sector cell splitting*

Careful planning and analysis can match the growth in capacity of a system to the habits and growth of the industry.

There are some ways of splitting cells that are less disruptive and less costly than just adding transmitter sites by brute force. Cells don't have to be round or hexagonal. Furthermore, antennas don't have to radiate or receive equally in all directions; they can cover some segments of the compass and not others. Antennas that do this are called directional or sector antennas.

Figure 6 shows how the three cells of *Figure 4* can be split into cells by making each antenna become three 120-degree corner antennas. (In practice, a single tower mounts these separate antennas.) Note that we don't have to add any cell sites, that the new cells are parallelograms, and that we have multiplied use of the channels by three.

Figure 7 shows the use of sector antennas to create triangular cells, again with no new sites, to allow channel reuse six times. (Actual reuse patterns are a bit more complex than shown.) Note that it is not necessary to reduce power output of the transmitter or the height of the antenna when cells are split.

Serving Portables

Portable cellular phones work identically to mobile units but have some additional capabilities and limitations. Naturally, portable units can be carried anywhere a person can go, including into buildings, elevators, and airplanes. The limitations include:

- The portable will work only where signals can penetrate. In a building, for example, it will usually work near the windows and may not work in the center of the building, on lower floors, or in elevators.

- The portable power output and antenna efficiency are lower than those of a mobile unit.

- The portable battery must be recharged or replaced at times, depending on usage.

- The portable cannot legally be used on airplanes.

A portable cellular telephone will work reasonably well in any cellular system within these limitations. On the streets, it will work about as well as a mobile unit. In a car, it will work not quite as well, unless the portable is coupled with an antenna on the car. Coverage in buildings depends on how far away the nearest cell site is located and how much metal shields the portable from the radio signals. The smaller the cells, the more likely the signals will penetrate buildings. It does not matter how the cells were split; small cells foster better portable performance.

An expedient used by some manufacturers to enhance portable performance is the use of sector-receive antennas at the base station. In this case, the transmitter sends its signal in all directions, but the receiving antennas reach out on 60-degree sectors to literally magnify the power of the portable (by six times) so that it is now comparable with mobile power. (Of course, mobile power is also magnified, but there is rarely a problem with too much power.)

Sector-receive antennas are generally an interim expedient; the system evolves into full-sector antennas or into one of the configurations discussed earlier. When the cells have dimensions of one or two miles, portables will work superbly in most places.

Roaming

The original concept of cellular was that it would be a nationwide, compatible mobile and portable telephone service. Cellular compatibility has been achieved in the United States. Every cellular radiotelephone in the United States can work in every city, and every city can accommodate each subscriber (new competitive wireless communications systems may not be compatible with existing systems). But this is not enough. Because there are so many different carriers offering service in the United States, and because there will be millions of subscribers, successful subscriber roaming requires mechanisms for:

- Authorization – when a subscriber enters a city away from home, the foreign carrier must receive some assurance that his cellular telephone bill will be paid.

- Billing transfers – once the foreign carrier allows a roaming subscriber to make calls, some means must exist for the costs of those calls to appear on the bill from the subscriber's home carrier.

- Clearing and settlement – a process required to settle accounts between carriers

- Charge card accommodations

These functions have been achieved to some extent in most places, but the advent of fraudulent ways of using cellular has impeded true seamless roaming. Nevertheless, in many areas of the Unites States, it is possible to travel from city to city and to call freely in each city. In some areas, a subscriber can be called on his or her home number and be reached in another city automatically and quickly.

Even as the traditional cellular service as described in this essay evolves, new technology promises to make a wide variety of wireless services available to more people at affordable prices with the introduction of new services and the improvement of the traditional services. Cellular systems are evolving into digital technology, which promises the ability to serve more subscribers in a given amount of radio channels (albeit at the expense of voice quality). A number of data services allow computers to talk to each other, and computers are in themselves becoming personal communications devices. Most importantly, in the United States and elsewhere new carriers are entering the business of providing these services. Competition can only accelerate the availability of options to the communications consumer.

In the long term, it is now clear that wireless systems will ultimately be lower in cost and greater in utility than the wired network for voice communications. This will have enormous impact in underdeveloped countries where most people have no telephone service at all. Nevertheless, it will affect all of us in important ways.

The cellular industry has spawned a new era in communications – the wireless age. Combined with the convergence of computers and communications, cellular technology created a new definition of mobility and of personal productivity. While the technical principles we have brushed upon in these pages provided the foundation of this new era, the real beginnings lie in the faith and commitments of dozens of companies and hundreds of engineers, marketers, executives, and entrepreneurs who believed in and devoted themselves to the cellular dream.

Section I

Business and Marketing Considerations

Wireless Information Networks

Kaveh Pahlavan
Associate Professor, Department of Electrical Engineering
Worcester Polytechnic Institute

Allen H. Levesque
Staff Scientist
GTE Laboratories

Introduction

Wireless information networks and services represent a rapidly growing segment of the communications industry. This essay provides an overview of the four major segments of this industry: cellular telephone, future cordless telephone, mobile data, and wireless LANs. Each of these industry segments can be related directly to a particular category of users' requirements for wireless communications.

The mid-1980s saw major initiatives in all sectors of the wireless information industry. To increase the capacity of cellular telephone systems, which have reached their technical limits in some large market areas, the transition to digital cellular technology was begun. The Pan-European GSM standard was followed by the EIA-TIA North American digital cellular standards initiatives and the Japanese Digital Cellular standard.[1] The extraordinary success of the cordless telephone market spurred new standardization efforts for digital cordless and CT-2 TelePoint in the UK,[2] wireless PBX, DECT in Sweden,[3,4] the advanced cordless phone in Japan ,[5] and the concept of a universal digital portable communicator in the United States.[6-9] The success of the paging industry led to development of private wireless packet data networks for commercial applications requiring longer messages.[10-12] Motivated by the desire to provide portability and to avoid the high

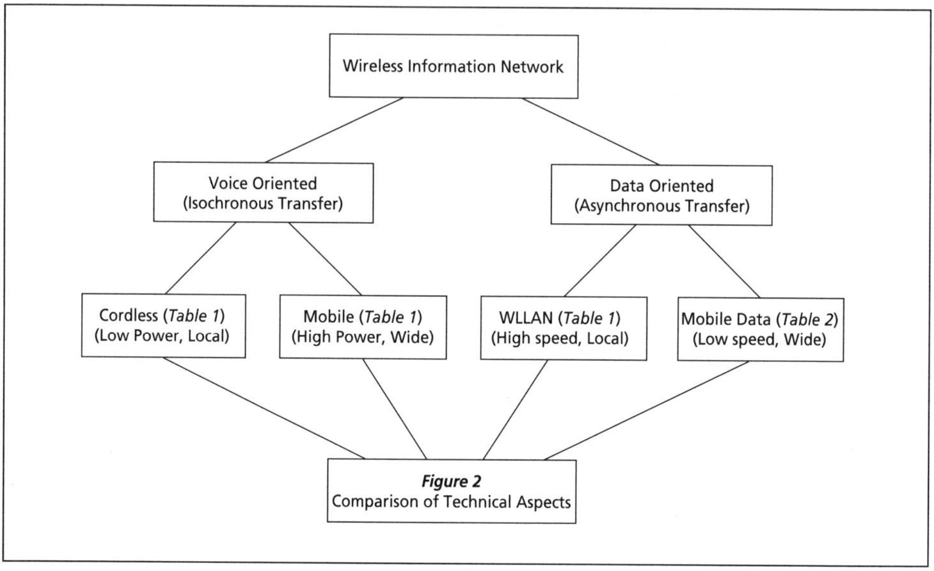

Figure 1 *Wireless information networks*

costs of installation and relocation of wired office information networks, wireless office information networks were suggested as an alternative.[13,14] Another major event in this period was the May 1985 FCC announcement on unlicensed ISM bands.[15-17] This announcement cleared the way for development of a wide array of commercial devices from wireless PBXs and LANs to wireless fire safety devices using spread-spectrum technology.[18-20]

Figure 1 distinguishes the various categories of wireless networks that we discuss in this essay. We first define two broad categories of networks as voice-oriented or isochronous networks and data-oriented or asynchronous networks. Under each main category of networks we distinguish further between low-power, local-area networks and high-power, wide-area networks. As will be seen subsequently in this paper, each of the resulting four sub-categories of networks has a set of characteristics that leads to certain design choices specific to the sub-category.

Figure 2, which is structured according to the categories and sub-categories defined in *Figure 1,* depicts various dimensions of today's voice and data communications industries.[21] The figure compares local cordless personal communication with wide-area cellular and wireless LANs with wide-area, low-speed data services. *Figure 2(a)* compares various dimensions related to the wireless voice

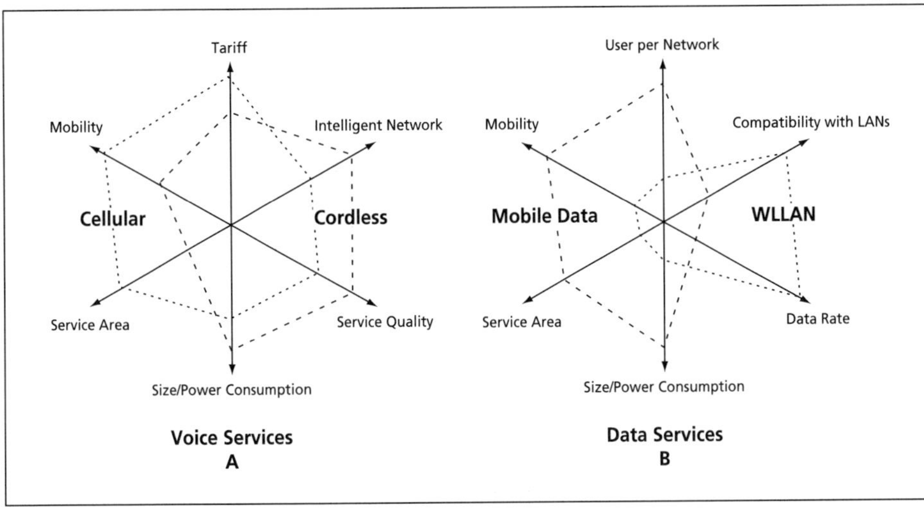

Figure 2 *Comparison of technical aspects*

industry, and *Figure 2(b)* provides an analogous comparison between wireless data systems. Although from the user's standpoint the service characteristics and the appearance of the handset will be similar for digital cellular and PCS systems, there will be a major difference in the operation of the networks supporting the two types of systems.

The digital cellular system is designed to support mobile users roaming over wide geographic areas; thus, coverage is provided by an arrangement of cells with cell size typically more than 2 miles in diameter. The radio cell sites for this system are large and expensive, particularly in urban areas where the cost of property is high. The handset requires an average power of around 1W, which is reflected in limited battery life and the need for frequent recharging. The number of users per cell is large. To provide as many user channels as possible in the allocated bandwidth, complex speech coding techniques are used. The speech coding techniques minimize the digitized speech transmission rate but consume a significant amount of electronic power, which places a high demand on battery power.

The PCS systems will be designed for small, low-power devices to be carried and used in and around office buildings, industrial complexes, and city streets. The size of each cell will be less than a quarter-mile in diameter, and the relatively small base stations will be installed on utility poles or attached to city and suburban business buildings. The average radiated power will be 10-20 mW, leading to a relatively long battery life. The PCS services are to replace cordless phones, and the quality of voice service is intended to be comparable to wireline phone service. As a result, simple but high-quality speech coding algorithms are to be used. At this writing, the leading candidate for adoption is 32 kbps adaptive differential pulse code modulation (ADPCM).[22] While a high-quality voice coding algorithm of this type does not provide the spectral efficiency of the lower-rate (4- to 8-kbps) vocoders used in the digital cellular standards, it is far less demanding of digital signal processing complexity. It permits the use of low prime power in the portable units. It is expected that PCS service will distinguish itself from digital cellular service by higher voice quality and higher-rate data services (up to 64 kbps).[22] *Table 1* provides a summary of the wireless technologies employed in several worldwide standards for the wireless voice industry.[23] For more details on the digital cellular and PCS initiatives, the reader can refer to works cited at the end of this essay.[24-29]

Mobile data networks operate at relatively low data rates over well-understood urban radio channels using familiar multiple-access methods. The technical challenge is the development of a system that makes efficient use of the available bandwidth to serve large numbers of users distributed over wide geographical areas. The transmission technology used in mobile data networks is generally rather simple and similar from one network to another. We discuss the major existing and planned mobile data networks and services in Section 4.

Wireless local area networks (WLANs) and mobile data networks serve somewhat different categories of user applications and give rise to different system design and performance considerations. A WLAN typically supports a limited number of users in a well-defined indoor area, and system aspects such as overall bandwidth efficiency and product standardization are not crucial. The achievable data rate is generally an important consideration in the selection of a WLAN; therefore, the transmission channel characteristics and the application of signal processing techniques are important topics.[30]

Access methods and network topologies used in WLANs are much the same from one system to another, but the transmission technologies can be quite different. Efficient design of these systems requires evaluation of various transmission techniques, an understanding of the complexities of indoor radio propagation, and the analysis of the effects of interference. WLAN manufacturers currently offer a number of nonstandardized products based on conventional radio modem technology, spread-spectrum technology in the ISM bands, and infrared technology. We discuss the various WLAN technologies in Section 5.

	Digital Cellular				Low-Power Systems				
System	IS-54	GSM	JDC	IS-95	DECT	PHP	CT-2	CT-3	Bellcore UDPC
Multiple access:	TDMA/ FDMA	TDMA/ FDMA	TDMA/ FDMA	FDMA CDMA-SS	TDMA FDMA	TDMA/ FDMA	FDMA	TDMA/ FDMA	TDM/TDMA/ FDMA
Frequency band Base TX (MHz) Mobile TX (MHz)	869-894 824-849 (U.S.)	935-960 890-915	810-826 940-956 1447-1489 1429-1441 1501-1513 1453-1465	869-894 824-849 (U.S.)	1800-1900 (Europe)	1895-1907 (Japan)	864-868 (Europe & Asia)	862-866 (Sweden	Emerging Technology (U.S.)
Duplexing:	FDD	FDD	FDD	FDD	TDD	TDD	TDD	TDD	FDD
Ch. spacing (kHz):	30	200	25	1250	1728	300	100	1000	350
Modulation:	$\pi/4$- QDPSK	GMSK	$\pi/4$- QDPSK	BPSK/ QPSK	GFSK	$\pi/4$- QDPSK	GFSK	GFSK	$\pi/4$- QDPSK
Portable transmit power, Max/Avg:	600 mW/ 200 mW	1W/ 125 mW	–	200 mW	250 mW/ 10 mW	80 mW/ 10 mW	10 mW/ 5 mW	80 mW/ 5 mW	200 mW/ 20 mW
Frequency assignment	Fixed	Dynamic	Fixed	–	Dynamic	Dynamic	Dynamic	Dynamic	Autonomous
Power control Handset: Base:	 Yes Yes	 Yes Yes	 Yes Yes	 Yes Yes	 No No	 – –	 No No	 No No	 Yes No
Speech coding and rate (kbps)	VSLEP 8	RPE-LTP 8	VSLEP 3	QCELP 1-8 (var.)	ADPCM 32	ADPCM 32	ADPCM 32	ADPC 32	ADPCM 32
Speech channels per RF channel	3	8	3	–	12	4	1	8	10
Channel rate (kbps)	48.6	270.833	42	1,228.8 (chip rate)	1152	96	72	640	500
Channel coding:	Rate-1/2 Conv.	Rate-1/2 Conv.	–	R-1/2 fwd. R-1/3 rev. CRC	CRC	CRC	None	CRC	CRC
Frame duration	40	4.625	20	20	10	5	2	16	2

Table 1 *Summary of wireless technologies employed in several standards for the wireless voice industry*

The future of the wireless data communication industry is toward multi-media, multi-rate, and multi-power applications. Multi-media developments will support the ever growing demand for mixed data, voice, and imagery applications and will be used to connect the pen, pad, and lap-top devices to backbone information resources and computational facilities. Multi-rate modems will support high-speed, wide-area as well as low-speed, local applications with a single device. Multi-power terminals will allow the selection of power resources such as AC power lines, car battery, or small batteries based on availability. These terminals will adapt the quality of the service to the consumption of the power resource.

Wireless Data: Market and User Perspectives

From the data user's perspective, the minimum satisfactory service requirement is low-speed access in wide areas and high-speed access in local areas. The low-speed, wide-area access will serve a variety of short-message applications such as notice of electronic or voice mail, while the local-area access will support high-speed local applications such as long-file transfers or printing tasks. In the current literature, low-speed, wide-area wireless data communication is referred to as mobile data, while local high-speed data communication systems are called wireless LANs. As we discussed in the previous section, the relationship between WLANs and mobile data services is analogous to the relationship between PCS and digital cellular services. While PCS is intended to provide high-quality local voice communication, the digital cellular services are aimed at wider-area coverage with less emphasis on the quality of the service.

Low-Speed, Wide-Area Systems (Mobile Data)

Mobile radio data systems have grown out of the success of the paging- service industry and the increasing customer demand for more advanced services. Today, several hundreds of thousands of customers are using mobile data services, and the industry expects 13 million users by the year 2000. This could be equivalent to 10 percent to 30 percent of the revenue of the cellular radio industry. Today, mobile data services provide length-limited wireless connections with in-building penetration to portable user terminals in metropolitan areas. The future direction is toward wider coverage, higher data rates, and capability for transmitting longer data files.

The transmission rates of existing mobile data systems are comparable to voice-band modem rates (up to 19.2 kps). However, the service typically has a limitation on the size of the file that can be transmitted in each communication session.

The coverage of the service is similar to standard land-mobile radio services. The difference is that the mobile data service must provide in-building penetration. Land mobile radio users typically use a telephone unit inside a vehicle and usually while driving. Mobile data users typically use the portable unit inside a building and in a stationary location. Therefore in-building penetration is an essential feature of mobile data services.

Mobile data services are used for transaction processing and interactive, broadcast, and multicast services. Transaction processing has applications such as credit card verification, taxi calls, vehicle theft reporting, paging, and notice of voice or electronic mail. Interactive services include enterprise applications such as database access and remote LAN access. Broadcast services include general information services, weather and traffic advisory services, and advertising. Multicast services are similar to subscribed information services, law enforcement communications, and private bulletin boards.

There are other low-speed data products using voiceband modems over radio systems originally designed for voice communications. Some of these products are used in the land-mobile radio bands around 100-200 MHz for low-speed local data communications in or around buildings. Other products such as portable facsimile devices and voiceband modems are used over the analog cellular telephone network to provide wide-area data communications for mobile users without any restrictions on the connection time. The term "mobile office" is sometimes used to describe these applications.

High-Speed, Local-Area Systems (Wireless LANs)

Today, most large offices are already equipped with wiring for conventional LANs, and the inclusion of LAN wiring in the planning of a new office building is done as a standard procedure along with planning for telephone and electric-power wiring. The WLAN market will likely develop on the basis of the appropriateness of the wireless solution to specific applications. The target markets for the wireless LAN industry include applications in manufacturing facilities, in offices with wiring difficulties, and in branch and temporary offices. In manufacturing facilities, ceilings are typically not designed to provide a space for the distribution of wiring. Manufacturing floors are also not usually configured with walls through which wiring might otherwise be run from the ceiling to outlets. Underground wiring is a solution that suffers from expensive installation, relocation, and maintenance. As a result, the natural solution for networking on most manufacturing floors is wireless communications.

Other wide indoor areas without partitioning, such as libraries or open-architecture offices, are also suitable for the application of WLANs. In addition, buildings of historical value, concrete buildings, and buildings with marble interiors all pose serious problems for wiring installation, leaving WLANs as the logical solution. WLANs are well suited to unwired small offices such as real estate agencies, where only a few terminals are needed and where there may be frequent relocations of equipment to accommodate reconfiguration or redecoration of the office space. Temporary offices such as political campaign offices, consultants' offices, and conference registration centers define another set of logical applications for WLANs. The WLAN industry expects to capture 5 percent to 15 percent of the LAN market in the near future.

Although the market for personal computers is not growing as it has in previous years, the market for portable devices such as lap-top and pen-pad computers and personal digital assistants (PDAs) is growing rapidly. The market for networked portables is growing much faster than the overall market for portable computing. Wireless is obviously the communication method of choice for portable terminals. Mobile data communication services discussed earlier provide a low-speed solution for wide-area coverage. For high-speed and local communications, a portable terminal with wireless access can bring the processing and data base capabilities of a large computer directly to specific locations for short periods of time, thus opening a wide horizon for new applications. For example, one can take portable terminals into classrooms for instructional purposes, or to hospital beds or accident sites for medical diagnosis.

The existing technologies for wireless LANs are licensed cellular systems operating at 18 GHz to 19 GHz,[42] unlicensed spread spectrum systems operating in ISM bands,[20] and diffused and directed-beam infrared (IR) systems.[43] *Table 2* summarizes the features of the systems currently available in the market. Diffused IR LANs provide moderate data rates and coverage and are suitable for moderate-size offices, short-distance battery-oriented applications (portable to printer), and environments with significant amounts of radio interference. Directed-beam IR offers higher- data rates with reasonable coverage for applications employing fixed terminals. Directed-beam IR is suitable for large file transfers between main frames and servers as well as large open offices with many fixed terminals.

WLANs at tens of GHz are suitable for high-speed communications in large partitioned areas such as larger offices or similar open areas such as some libraries. The spread-spectrum systems provide the largest coverage and are suitable for applications where penetration through building floors is desired. Spread-spec-

Technique	DF/IR	DB/IR	RF	DSSS	FHSS
Data rate	1-4 Mbps	10 Mbps	5-10 Mbps	2-20 Mbps	1-3 Mbps
Mobility	Stationary/ mobile	Stationary with LOS	Stationary/ mobile	Stationary/ mobile	Mobile
Range	50-200 ft.	80 ft.	40-130 ft.	100-800 ft.	100-300 ft.
Frequency/wavelength (in nm)	l = 800-900	l = 800-900	f = 18 GHz or ISM	ISM bands	ISM bands
Modulation technique	OOK	OOK	FSK/QPSK	QPSK	GFSK
Access method	CSMA	Token ring	Reservation	CSMA	CSMA

Table 2 *Wireless LAN technologies*

trum technology is suitable for small-business applications where a few terminals are distributed over several floors of a building. An important factor in selection is the ease of installation and flexibility of the software to accommodate various protocols for interconnecting with backbone-wired LANs.

The evolving next generation of WLANs is designed to be incorporated into the lap-top, notebook, and pen-pad computers, where significant reductions in size and power consumption are needed. It will be ideal if these devices also provide low-speed wireless access with wide-area coverage.

Frequency Administration Issues

The major long-standing problem facing the radio communications industry is the fundamental limitation on availability of frequency spectrum. The history of the industry has been characterized by a steady migration toward higher-frequency bands as new systems and services have come into the market. When cellular telephone service was launched in the late 1970s, the FCC allocated 40 MHz of bandwidth in the 800 MHz band by moving prior occupants (educational TV channels) out of the bands. An additional allocation of 10 MHz was made in 1986 as the cellular industry grew. The current state of the cellular industry in many major market areas is that the analog cellular systems have reached capacity. This is the primary motivation for the digital cellular initiatives, which will increase capacity by factors of about 3 to 10 over the analog systems.

In the land mobile radio (LMR) bands (150 MHz, 350 MHz, and 850 MHz), capacity limits are also being reached, though the market growth there is not as strong as in the cellular market. In the LMR industry, plans are being made to migrate from 25 kHz channels to 12.5 kHz, with plans for further migration in the future. All of this means that in these bands efficient spectrum utilization is of the utmost importance, and these systems must be designed to use the available bandwidth to serve the greatest number of users over wide service areas. Thus the emphasis is on bandwidth-efficient modulation, efficient frequency management schemes, and, in the case of data services, efficient multiuser access protocols.

The principal technological problems for implementation of WLANs are: data rate limitations caused by the multipath characteristics of radio propagation, the difficulties associated with signal coverage within buildings, and the need for low-power electronic implementations suitable for portable terminals. These technical difficulties can be resolved, and effective solutions are being developed for all of them. The greatest obstacle to achieving wireless multimegabit data communication rates is the lack of a suitable frequency band for reliable high-speed communication. The existing ISM bands assigned for multiple-user applications are suitable for WLANs, but they are restricted to spread-spectrum technology and can suffer from unnecessary interference caused by careless users.[15-17] More widespread use of high-speed wireless data communication technology will depend upon cooperation from frequency administration organizations in providing wider bandwidth allocations without restricting the adopted technology. Such organizations must also administer rules and etiquette for cooperative use of these bands.

At frequencies around several GHz, the technology is available for wireless implementations having reasonable size, power consumption, and cost. Moving to higher frequencies is the solution for the future. As frequency increases, the prospect for obtaining a wider bandwidth from spectrum regulatory agencies will improve. However, with today's technology, implementation at a few tens of GHz with reasonable product size and power consumption is challenging, particularly when wideband portable communication is considered. At higher frequencies, signal transmission through walls is more difficult. This feature is advantageous in WLAN applications where confinement of the signal within a room or building is a desirable privacy feature. Also, at higher frequencies the relationship between cell boundaries and the physical layout of the building is more easily determined by facilitating the planning of WLAN cell assignments within the building. The technology at higher frequencies is highly specialized and not commonly available within the computer industry. This has encouraged joint ventures among semiconductor, radio, and computer companies to develop new WLAN products at these frequencies.

A mobile data service is shared by a number of users distributed over a large geographical area. Development of a mobile data network requires a large investment affordable only by a major industrial organization. When the need for the network is justified, the band allocation can then be obtained from the appropriate administrative agency. The transmission techniques and access methods used in these systems are very similar from one network to another, and the decisions on these elements are overshadowed by the system-level decisions that provide a compromise between the cost of implementation and the coverage provided by the network. For example, the concept of overlaying a packet data service onto the existing analog cellular networks is the single most important technical decision for the development of that system. Motivation for this decision is to provide a system that uses the infrastructure of the existing AMPS systems in the same band. It can therefore be deployed at low cost and provide a wide area of coverage without any demand for additional band from the frequency allocation agencies. The second most important issue is the development of a standard. Development of the elements of the network are rather routine engineering tasks not involving major decisions.

Wireless LANs are designed for a small number of users usually operating in indoor areas. The range of coverage is small, which leaves many options open for the transmission technology. As a result, the major decision here is on the choice of the transmission method. The development of a wireless LAN does not involve a large infrastructure that requires extensive investments. As a result, small groups in large companies or small start-up companies usually start development of these products. Convincing the frequency administration agencies to provide a band for such applications has been more difficult. That has encouraged the companies to align the transmission technology with the existing bands rather than ask for new bands. An overview of wireless data networks, with emphasis on wireless LANs, is given by Pahlavan and Levesque.[31,32]

Mobile Data Networks

Packet data services currently available for mobile applications include ARDIS, formed by IBM and Motorola, and the RAM Mobile Data Network, which uses Ericsson MOBITEX data technology.Recently introduced into the market is cellular digital packet data (CDPD), which is designed to transport data as a supplementary service overlaid onto existing North American analog cellular telephone networks. The new digital cellular standards, IS-54 and IS-95, will eventually include an array of data services, including both circuit-mode and packet-mode services, for the next generation North American digital cellular systems. In Europe, the European Technical Standards Institute (ETSI) has begun developing a public standard for trunked radio and mobile data systems, designated as TETRA.

ARDIS

ARDIS is a two-way radio service developed as a joint venture between IBM and Motorola and was first implemented in 1983. The ARDIS network consists of four network control centers with 32 network controllers distributed through 1,250 base stations in 400 cities in the United States. The service is suitable for two-way transfer of data files less than 10 kbytes in size, and much of its use is in support of computer-aided dispatching as used by field service personnel often while they are on customers' premises. Remote users access the system from lap-top radio terminals, which communicate with the base stations. Each of the ARDIS base stations is tied to one of the 32 radio network controllers, as shown in *Figure 3*. The backbone of the network is implemented with leased telephone lines. The four ARDIS hosts, located in Chicago, New York, Los Angeles, and Lexington, KY, serve as access points for a customer's mainframe computer, which can be linked to an ARDIS host using async, bisync, SNA, or X.25 dedicated circuits.

The operating frequency band is 800 MHz, and the RF links use separate transmit and receive frequencies 45 MHz apart. The system was initially implemented with an RF channel data rate of 4,800 bps per 25-kHz channel, but this has been upgraded to 9.6 kbps, with a user data rate of about 8,000 bps. The system architecture is cellular, with cell coverage areas overlapped to increase the probability that the signal transmission from a portable transmitter will reach at least one base station. The base station power is 40 W, which provides line-of-sight (LOS) coverage up to a radius of 10 miles to 15 miles. The portable units operate with 4W of radiated power. The overlapping coverage, combined with designed power levels and error-correction coding in the transmission format, insure that the ARDIS can support portable communications from inside buildings as well as on the street. This capability for in-building coverage is an important characteristic of the ARDIS service. The modulation technique used is frequency-shift keying (FSK), the access method is frequency division multiple access (FDMA), and the transmission packet length is 256 bytes.

While the use of overlapping coverage, almost always on the same frequency, provides reliable radio connectivity, it poses the problem of interference when

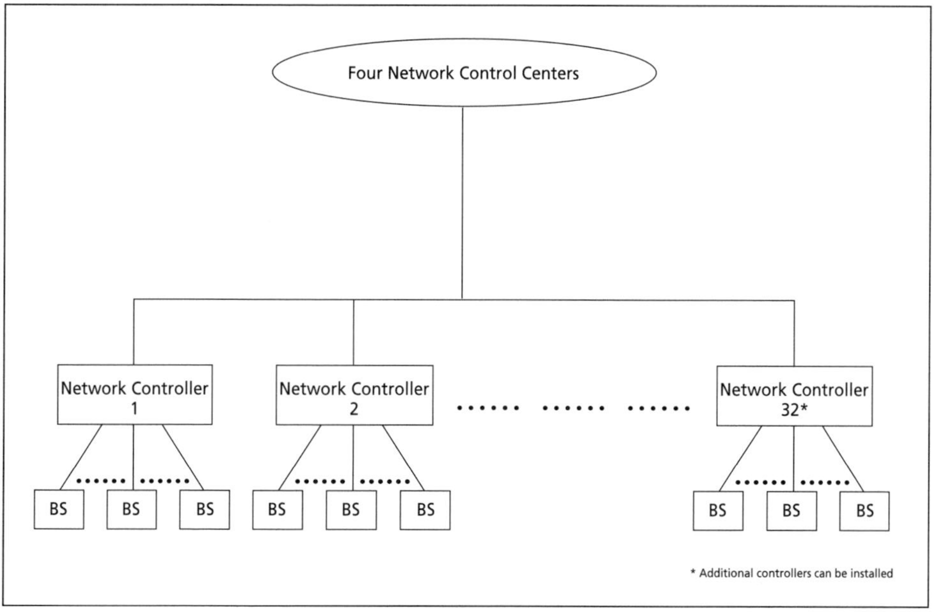

Figure 3 *ARDIS network architecture*

signals are transmitted simultaneously from two adjacent base stations. The ARDIS network deals with this by turning off neighboring transmitters, for 0.5- Is, when a base station is transmitting. This scheme has the effect of constraining overall network capacity.

The lap-top portable terminals access the network using a random access method called data sense multiple access (DSMA). A remote terminal listens to the base station transmitter to determine if a busy bit is on or off. When the busy bit is off, the remote terminal is allowed to transmit. However, if two remote terminals begin to transmit at the same time, the signal packets may collide, and retransmission will be attempted, as in other contention-based multiple access protocols. The busy bit lets a remote user know when other terminals are transmitting, which reduces the probability of packet collision.

Mobitex

The Mobitex system is a nationwide, interconnected trunked radio network developed by Ericsson and Swedish Telecom. The first Mobitex network went into operation in Sweden in 1986, and other networks have been implemented in Norway, Finland, Canada, the United Kingdom, and the United States. In the

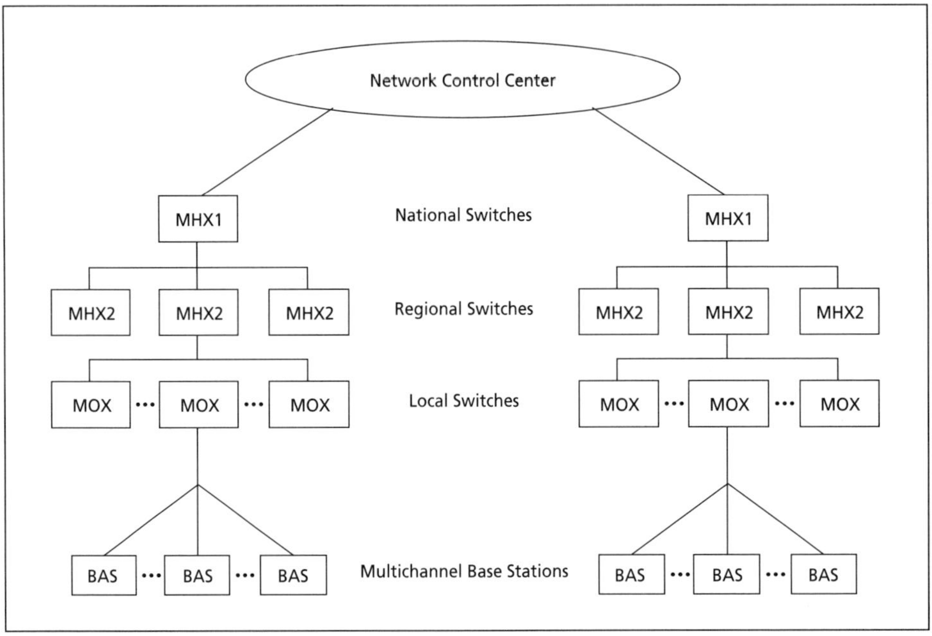

Figure 4 *Mobitex network architecture*

United States, Mobitex service was introduced by RAM Mobile Data in 1991 and now covers about 100 major metropolitan areas. While the Mobitex system was designed to carry both voice and data services, the U.S. and Canadian networks are used to provide data service only. Mobitex is an intelligent network with an open architecture which allows establishing virtual networks. This feature facilitates the mobility and expandability of the network. The system supports a number of standardized network interfaces, including TCP/IP and X.25, and interfaces are being added in response to market demands.

The Mobitex network architecture is hierarchical, as shown in *Figure 4.* At the top of the hierarchy is the network control center (NCC) from which the entire network is managed. The top level of switching is a national switch (MHX1) that routes traffic among service regions. The next level comprises regional switches (MHX2s), and below that are local switches (MOXs), each of which handles traffic within a given service area. At the lowest level in the network, multichannel trunked-radio base stations communicate with the mobile and portable data sets. Mobitex uses packet-switching techniques, as does ARDIS, to allow multiple users to access the same channel at the same time. Message packets are switched at the lowest possible network level. If two mobile users in the same service area need to communicate with each other, their messages are relayed through the local base station, and only billing information is sent up to the NCC.

The base stations are laid out in a grid pattern using the same system engineering rules as are used for cellular telephone systems. In fact, the Mobitex system operates in much the same way as a cellular telephone system, except that hand-offs are not managed by the network. That is, when a radio connection is to be changed from one base station to another, the decision is made by the mobile terminal, not by the network computer as in cellular telephone systems.

To access the network, a mobile terminal finds the base station with the strongest signal and then registers with that base station. When the mobile terminal enters an adjacent service area, it automatically re-registers with a new base station, and the user's whereabouts are relayed to the higher-level network nodes. This provides automatic routing of messages bound for the mobile user, a capability known as roaming. The Mobitex network also has a store-and-forward capability.

The mobile units transmit at 896 MHz to 901 MHz and the base stations at 935 MHz to 940 MHz. About 10 to 30 frequency pairs are used in each service area. The system uses dynamic power setting in the range of 100 mW to 10 W for mobile units and 100 mW to 4 W for portable units. The Gaussian minimum shift keying (GMSK) modulation technique is used with noncoherent demodulation. The transmission rate is 8,000 bps half duplex in 12.5-kHz channels, and the service is suitable for file transfers up to 20 kbytes. The packet size is 512 bytes with 1 s to 3 s delay. Forward error-correction, as well as retransmissions, is used to ensure the bit error-rate quality of delivered data packets. The system uses the dynamic slotted-ALOHA random access method.

RAM Mobile Data has installed approximately 800 base stations in 100 metropolitan areas throughput the United States, providing coverage to about 90 percent of the urban business population. By locating its base stations close to major business centers, the RAM mobile system provides a significant degree of in-building signal coverage.

CDPD

The cellular digital packet data (CDPD) system is designed to provide packet data services in an overlay to the existing analog cellular telephone network. CDPD has been developed by IBM in collaboration with nine cellular carriers: McCaw, GTE, Contel Cellular, Ameritech, Bell Atlantic, NYNEX, Pacific Telesis, Southwestern

Bell, and U S WEST. These carriers will cover 95 percent of the United States, including all major urban areas. A basic goal of the CDPD system is to provide data services on a non-interfering basis using the same 30-kHz channels used by the existing cellular telephone services. To do this, CDPD is designed to make use of cellular channels that are not being used for voice traffic and to move to another channel when the current channel is allocated to voice service. As an alternative, some carriers may choose to simply dedicate some number of channels in each cell to CDPD service.

The compatibility of CDPD with the existing cellular telephone system allows it to be installed in any analog cellular system in North America, thus providing data services that are not dependent upon support of a digital cellular standard in the service area. A preliminary field demonstration system was operated during the second half of 1992, and subsequently the participating companies issued a specification for the CDPD system.[33] The specification was updated in late 1994. Intended applications for CDPD service include: electronic mail, package delivery tracking, inventory control, credit card verification, security reporting, vehicle theft recovery, traffic and weather advisory services, and a potentially wide range of information retrieval services. At this writing (mid-1995), CDPD service is implemented in more than a dozen major market areas, and deployment in an additional 55 markets is planned to take place by the end of 1995.

Although CDPD cannot increase the number of channels that can be used in a cell, it can provide an overall increase in capacity if data users employ CDPD instead of voice channels. This capacity increase would result from the inherently greater efficiency of a connectionless packet data service relative to a connection-oriented service, given bursty data traffic. That is, a packet data service does not require the overhead associated with setup of a voice traffic channel in order to send one or a few data packets. In the following paragraphs, we briefly describe the CDPD network architecture and the principles of operation of the system. Our discussion follows Quick and Balachandran closely.[34]

The basic structure of a CDPD network (*Figure 5*) is similar to that of the cellular network with which it shares transmission channels. Each mobile end system (M-ES) communicates with a mobile database station (MDBS), using the protocols defined by the air-interface specification, to be described below. The MDBSs are expected to be collocated with the cell equipment providing cellular telephone service to facilitate the channel-sharing procedures. All the MDBSs in a service area will be linked to a mobile data intermediate system (MD-IS) by microwave or wireline links. The MD-IS provides a function analogous to that of the mobile switching center

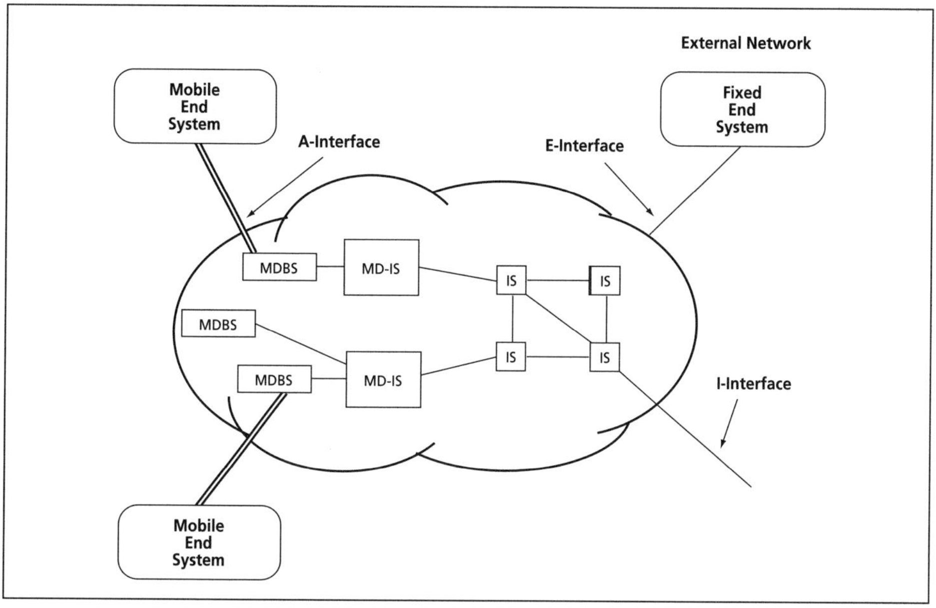

Figure 5 *CDPD data service network*

(MSC) in a cellular telephone system. The MD-IS may be linked to other MD-ISs and to various services provided by end-systems outside the CDPD network. The MD-IS also provides a connection to a network management system and supports protocols for network management access to the MDBSs and M-ESs in the network.

Service endpoints can be local to the MD-IS or remote, connected through external networks. An MD-IS can be connected to any external network supporting standard routing and data exchange protocols. An MD-IS can also provide connections to standard modems in the PSTN by way of appropriate modem interworking functions. Connections among MD-ISs allow routing of data to and from M-ESs that are roaming, that is, operating in areas outside their home service areas. These connections also allow MD-ISs to exchange information required for mobile terminal authentication, service authorization, and billing.

CDPD employs the same 30-kHz channelization as is used in existing analog cellular systems throughout North America. Each 30-kHz CDPD channel will support channel transmission rates up to 19.2 kbps. However, degraded radio channel conditions will limit the actual information payload throughput rate to lower levels and will introduce additional time delay due to the error-detection and retransmission protocols.

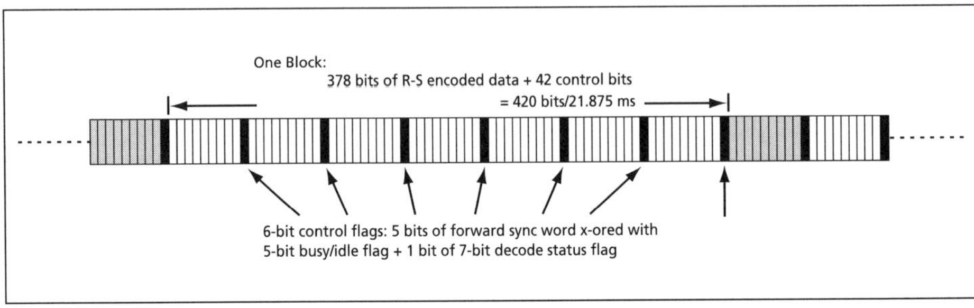

Figure 6 *CDPD data service – forward-link block structure*

The CDPD radio link physical layer uses GMSK modulation at the standard cellular carrier frequencies on both forward (base-to-mobile) and reverse (mobile-to-reverse) links. The Gaussian pulse-shaping filter is specified to have bandwidth-time product $B_bT = 0.5$. The specified B_bT product assures a transmitted waveform with bandwidth narrow enough to meet adjacent-channel interference requirements while keeping the intersymbol interference small enough to allow simple demodulation techniques. The choice of 19.2 kbps as the channel bit rate yields an average power spectrum that satisfies the emission requirements for analog cellular systems and for dual-mode digital cellular systems.

The forward channel carries data packets transmitted by the MDBS while the reverse channel carries packets transmitted by the M-ESs. In the forward channel, the MDBS forms data frames by adding standard HDLC terminating flags and inserted zero bits and then segments each frame into blocks of 274 bits. These 274 bits, together with an 8-bit color code for MDBS and MD-IS identification, are encoded into a 378-bit coded block using a (63,47) Reed-Solomon code over a 64-ary alphabet (6-bit symbols). A 6-bit synchronization and flag word is inserted after every 9 code symbols. The flag words are used for reverse link access control. The forward link block structure is shown in *Figure 6*.

In the reverse channel, when an M-ES has data frames to send, it formats the data with flags and inserted zeros in the same manner as in the forward link. That is, the reverse link frames are segmented and encoded into 378-bit blocks using the same Reed Solomon code as in the forward channel. The M-ES may form up to 64 encoded blocks for transmission in a single reverse channel transmission burst. During the transmission, a 7-bit transmit continuity indicator is interleaved into each coded block and is set to all ones to indicate that more blocks follow, or all zeros to indicate that this is the last block of the burst. The reverse channel block structure is shown in *Figure 7*.

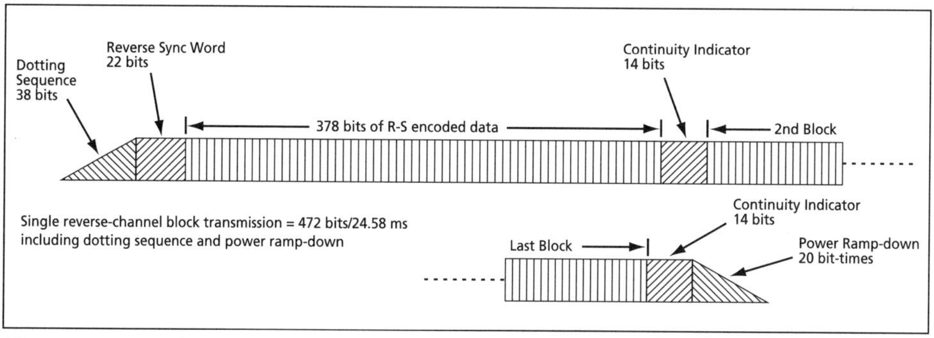

Figure 7 *CDPD data service – reverse channel block structure*

The media access control (MAC) layer in the forward channel is relatively simple. The receiving M-ES removes the inserted zeros and HDLC flags and reassembles data frames that were segmented into multiple blocks. Frames are discarded if any of their constituent blocks are received with uncorrectable errors.

On the reverse channel, access control is more complex since several M-ESs must share the channel. CDPD uses a multiple access technique called digital sense multiple access (DSMA), which is closely related to carrier sense multiple access with collision detection (CMA/CD).

The network layer and higher layers of the CDPD protocol stack are based on standard ISO and Internet protocols. The earliest CDPD products, now coming into the market, use the Internet protocols.

Digital Cellular Data Services

In response to the rapid growth in demand for cellular telephone service throughout the United States and Canada, the Cellular Telecommunications Industry Association (CTIA) and The Telecommunications Industry Association (TIA) have been developing standards for new digital cellular systems to replace the existing North American analog cellular system (called the advanced mobile phone service, or AMPS). Two air-interface standards have now been published. The IS-54 standard specifies a 3-slot TDMA system, and the I-95 standard specifies a CDMA spread-spectrum system. A variety of data services are being planned for both systems.

The general approach taken in the definition of IS-95 data services has been to take the previously specified physical layer of the IS-95 protocol stack as the physical layer of the data services, with an appropriate link-layer radio link protocol (RLP) to be overlaid.[35] The current standardization effort is directed to defining three primary services: asynchronous data, Group-3 facsimile, and short-message service. Later, attention will be given to other services, including contention-access packet data, synchronous data, and other primary services.

IS-95 asynchronous data will be structured as a circuit-switched service. For circuit-switched connections, a dedicated path is established among the data devices for the duration of the call. It is used for connectivity through the PSTN, requiring point-to-point communications to the common PC or fax user. There are several applications that fall into this category. For file transfer involving PC-to-PC communications, the asynchronous data service is the desired cellular service mode. The service will employ a radio link protocol (RLP) to protect data from transmission errors caused by radio channel degradations at the air interface. The RLP employs automatic repeat request (ARQ), forward error correction (FEC), and flow control. Flow control and retransmission of data blocks with errors are used to provide an improved error performance in the mobile segment of the data connection at the expense of variations in throughput and delay. Typical raw channel rates for digital cellular transmission are measured at approximately a 10^{-2} bit-error rate. However, acceptable data transmission usually requires a bit-error rate of approximately 10^{-6}, and achieving this will require the design of efficient ARQ and error-correction codes to deal with error characteristics in the mobile environment.

In parallel with the CDMA data services effort, another TIA task group has been defining standards for digital data services for the TDMA digital cellular standard IS-54.[36,37] As with the IS-95 data services effort, initial priority is being given to standardizing circuit-mode asynchronous data and group-3 facsimile services.[38]

TETRA

As is the case in the United States and Canada, there is interest in Europe in establishing fixed wide-area standards for mobile data communications. While the Pan-European standard for digital cellular- termed global systems for mobile (GSM) will provide an array of data services, data will be handled as a circuit-switched service, consistent with the primary purpose of GSM as a voice service system.[39] Therefore, the European Telecommunications Standards Institute (ETSI) has begun developing a public standard for trunked radio and mobile data systems. The standards, which are known generically as Trans-European trunked radio (TETRA), are the responsibility of the ETSI RES 6 sub-technical committee (*Figure 8*).[40]

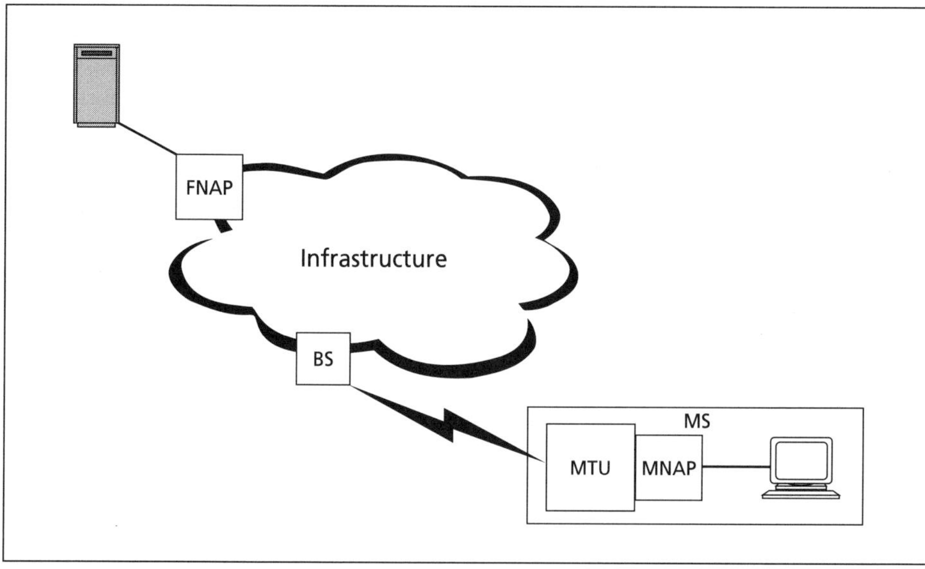

Figure 8 *TETRA network interfaces*

TETRA is being developed as a family of standards. One branch of the family is a set of radio and network interface standards for trunked voice (and data) services. The other branch is an air-interface standard optimized for wide-area packet data services for both fixed and mobile subscribers and supporting standard network access protocols.[41] Both versions of the standard will use a common physical layer, based on $\pi/4$-shift differential quaternary phase-shift keying ($\pi/4$- DQPSK) modulation, operating at a channel rate of 36 kbps in each 25-kHz channel.

Table 3 compares the chief characteristics and parameters of the five wireless data services described above.

Wireless LANs

For wireless LANs, the distances are short and, therefore, there is no need for a universal standard to develop a product. As a result, a variety of technologies have emerged in the marketplace. Technologies used in the existing wireless LAN products are divided into five categories: diffused IR (DFIR), directed beam IR (DBIR)],[43] standard radio (RF),[42] direct sequence spread-spectrum (DSSS),[20] and frequency hopping spread- spectrum (FHSS). *Table 3* shows a comparison among

System	ARDIS	Mobitex	CDPD	IS-95[b]	TETRA[b]
Frequency Band Base to mobile (MHz): Mobile to base (MHz):	(800 Band. 45 kHz sep.)	935-940[a] 896-901	869-894 824-849	869-894 824-849	(400 and 900 Bands)
RF channel spacing	25 kHz (U.S.)	12.5 kHz	30 kHz	1.25 kHz	25 kHz
Channel access/ Multiuser access:	FDMA/ DSMA	FDMA/ Dynamic- S-ALOHA	FDMA/ DSMA	FDMA/ CDMA-SS	FDMA/ DSMA & SAPR
Modulation method:	FSK, 4-FSK	GMSK	GMSK	4-PSK/DSSS	π/4-QDPSK
Channel bit rate (kbps):	19.2	8.0	19.2	9.6	36
Packet length:	Up to 256 bytes (HDLC)	Up to 512 bytes	24 to 928 bits	(Packet service-TBD)	192 bits (short) 384 bits (long)
Open architecture:	No	Yes	Yes	Yes	Yes
Private or public carrier:	Private	Private	Public	Public	Public
Service coverage:	Major metro areas in U.S.	Major metro areas in U.S.	All AMPS areas	All CDMA cellular areas	European trunked radio
Type of coverage:	In-building & mobile	In-building & mobile	Mobile	Mobile	Mobile

Table 3 *Characteristics and parameters of five mobile data service*
Notes: a = frequency allocations in the U.S. (in the U.K., 380-450 MHz band is used)
b = IS-95 and TETRA data services standardization in progress
DSMA = Data-Sense Multiple Access, DSSS = Direct Sequency Spread Spectrum
S-ALOHA = Slotted ALOHA, SAPR = Slotted-ALOHA Packet Reservation

various features of these technologies collected from the products available in the market. These technologies have evolved around the availability of the channel and the suitability of the transmission technique to provide high data rates in the wireless media. The IR products are designed to operate at optical frequencies, which are not regulated by the FCC. The spread-spectrum wireless LANs are designed to operate in the ISM bands. The RF products are implemented either in the 18 GHz to 19 GHz licensed bands or in the ISM bands using very low power. The ISM bands allow non-spread-spectrum transmission devices if the transmitted power is low.

Data rate is an essential ingredient of local communication networks and the most important aspect for marketing and sale of the product. The higher the data rate, the greater the impact of the product in the market. Another important feature affecting the market for wireless LANs is the mobility of the terminal, which is a function of the power consumption and the size of the product, except for DBIR, in which the terminal must remain stationary to keep the radiation pattern of the device effective. If the technology can be implemented with small batteries and lightweight packaging, it will be suitable for mobile applications and can be used for personal portable computers. Power consumption is a function of the electronic implementation of the device, and some technologies can be implemented either with or without battery operation for mobile and stationary applications, respectively. The access method for most wireless LANs is CSMA/CA, though some use reservation-slotted ALOHA or TDMA with packet reservation.

Although none of the standards for wireless LANs are complete at this writing, there are numerous wireless LAN products in the market. Wireless LANs are stand-alone products that can be manufactured without conformance to a widely accepted standard. As the penetration of wireless LANs grows and the standards are completed, this situation will change.

Currently all standard activities for wireless LANs use unlicensed bands. There are two approaches to regulating use of an unlicensed band. One approach is to develop a standard that will allow different vendors to communicate with one another using a set of interoperability rules. This approach is taken by IEEE 802.11 and ETSI's RES10, HIPERLAN. The second approach is to provide a minimum set of rules or spectrum etiquettes that allow terminals designed by different vendors to use a fair share of the available channel-frequency time resources and coexist in the same band.[44] The second approach does not preclude the first approach, and it is being pursued by the WINForum. In a coexisting environment, a vendor can interoperate with another vendor by using the same protocol and transmission scheme. The trend of this industry is toward interoperability; however, a discovery period, in which the technology and the market settle themselves, is being planned.

An analogous situation existed in the development of the voiceband modems. In the early days of modem development, there was no standard, and as the industry evolved, the CCITT adopted standards based on successful modem designs.

The three major activities related to standards for wireless LANs are IEEE 802.11, HIPERLAN, and WINForum. IEEE 802.11 is developing a standard for DSSS, FHSS, and DFIR technologies for use in the ISM bands. The HIPERLAN standard is concerned with the 5.2- and 17.1-GHz bands recently released in the EC, and it focuses on standard transmission techniques using adaptive equalization,[45-47] sectored antenna,[48] or multicarrier modulation.[49,50] The WINFORUM is an alliance among major computer and communication companies to obtain parts of the PCS band for unlicensed data and voice applications. The technical contribution of

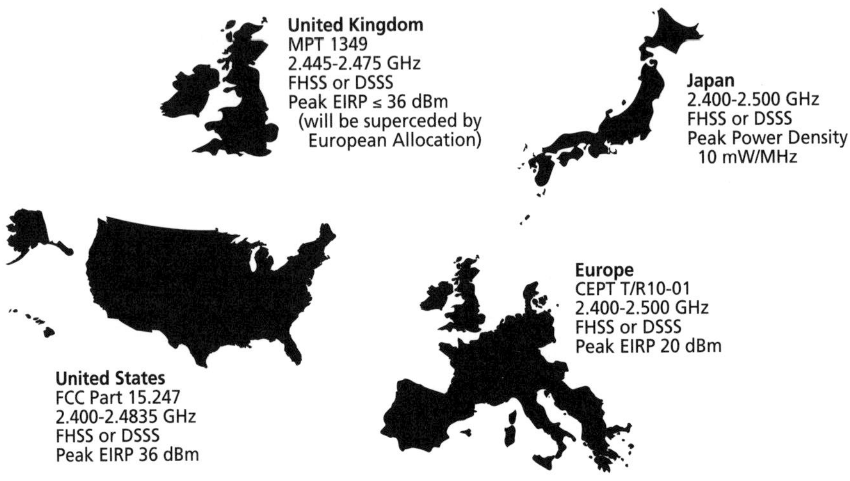

Figure 9 *Worldwide wireless LAN spectrum allocation in the 2400-2500 MHz band*

WINFORUM is the development of spectrum etiquettes for multivendor coexistence in unlicensed bands.

IEEE 802.11 and ISM Bands

The IEEE 802.11 subcommittee focuses on the physical and media access control (MAC) layers for peer-to-peer and centralized topologies, accommodating DFIR as well as DSSS and FHSS in the 2.4-GHz ISM bands. The 2.4-GHz ISM band is selected over 0.9- and 5.7-GHz ISM bands because it is widely available in most industrialized countries. *Figure 9* shows four major countries with their position toward the bands at 2.4 GHz - 2.5 GHz. In this band, about 80 MHz of bandwidth is available. It is attractive for high-speed data communications because the implementation at this band is cost-effective.

The standards support DSSS with BPSK and QPSK modulation at data rates of 1 Mbps and 2 Mbps, respectively; the FHSS with GFSK modulation and two hopping patterns at data rates of 1 Mbps and 2 Mbps; and DFIR with OOK modulation at a data rate of 1 Mbps. For DSSS operation, the band is divided into five overlapping 26-MHz sub-bands centered at 2.412, 2.442, 2.47, 2.427, and 2.457 GHz, the last two overlapping the first three. This setup provides five orders of frequency-selectivity for the user, which is very effective in improving the transmission reliability in the presence of interference or severe frequency- selective multipath fading.

For FHSS operation, the channel is divided into 79 sub-bands, each with 1-MHz bandwidth, with three patterns of 22 hops left as options for the users. A minimum hop rate of 2.5 hops/second is assigned to provide an opportunity for slow frequency hopping in which each packet can be sent in one hop. If it is destroyed, the following packet can be sent another hop at which the channel condition would be different. This approach will provide an effective time-frequency diversity that takes advantage of a retransmission scheme to provide robust transmission. The standard supports CSMA and is intended to provide interoperability among all users.

WINForum and PCS Bands

An important issue that was not addressed in the ISM bands was the enforcement of a time limit for the transmission time of a terminal. Suppose that a low-power wireless LAN and a device operate in the ISM bands near one another, and the device continually radiates at the maximum allowable power. All information bursts generated from the wireless LAN will suffer from maximum power interference while the wireless LAN produces only bursts of low-power interference to the device. This is not fair to the wireless LAN because it is not given a fair share of the available resources. The resource is not the bandwidth alone; rather, it is the bandwidth over time, which is measured in Hz-sec. Furthermore, the wireless LAN creates less interference to the other users. A fair method of allowing several devices to operate in the same band is to restrict their Hz per second share of the channel according to their transmission power. The technical contribution of WINForum is the development of a spectrum etiquette that allows several users in a multivendor environment to each receive a fair share of the channel-frequency time resources in accordance with their individual radiated powers.

WINForum was initiated by Apple Computer as an industry alliance, with the goal of obtaining frequency bands for so-called data PCS. Today, WINForum comprises about 40 members from leading information technology companies, and its objective is to obtain and effectively employ radio spectrum for unlicensed user-provided voice and data personal communication services (referred to as User-PCS). Currently, WINForum intends to foster technical advances in applications such as wireless LANs and wireless PBX services. Originally, WINforum was looking for a 40-MHz unlicensed band for data PCS, and the FCC has shown some inclination to provide them with a 20-MHz band divided into two 10-MHz sub-bands for voice and data applications.

The technical innovation initiated by the WINForum is the development of so-called spectrum etiquette which is a means of providing fair access to an unlicensed band to widely different applications and devices.[44] The etiquette is not intended to preclude any common air-interface standards or access technologies. Spectrum etiquette demands listen-before-talk (LBT), which means a device may not transmit if the spectrum it will occupy is already in use within its range. The power is limited in order to keep the range short. This allows operation in densely populated office areas. The power and connection time are related to the occupied bandwidth in order to equalize the interference and provide a fair access to frequency time resources. In May 1993, WINForum filed its spectrum etiquette with the FCC.

The view of WINForum distinguishes two classes of information type, generated from asynchronous and isochronous transmissions. The asynchronous transmission, typified by wireless LAN-like applications, is bursty, begins transmission within milliseconds, uses short bursts that contain large amounts of data, and releases the link quickly. The isochronous transmission, typified by voice services such as wireless PBX, exhibits long holding times, periodic transmissions, and flexible link-access times that may be extended up to a second. The asynchronous sub-bands may range from 50 KHz to 10 MHz, while the isochronous sub-bands may be divided into 1.25-MHz segments. The two types are technically contrasting and cannot share the same spectrum. Since connection-oriented access always leaves unoccupied bands available, it is possible to use the unused parts of the isochronous band for transmission of bursty data. Dividing an isochronous band into one unique sub-band should be done carefully because it may deny other experimental systems a fair access to the market.

A comparative study of various approaches for implementing a wireless system is extremely difficult. In long-spirited debates, such as that addressing CDMA versus TDMA or optical versus radio LANs, the WINForum has reached the conclusion that an environment must be created that allows for a variety of ideas to appear in the market. The market will decide from among these ideas in the long term. The WINForum approach appears fair, particularly in regard to the asynchronous devices that are just beginning to find their place in the market.

HIPERLAN and 5.2 and 17.1 GHz Bands

Europeans are approaching the wireless LAN developments in a different way.

They intend first to establish a standard and then develop products based on the standard. ETSI has directed the subtechnical committee RES10 to develop a standard for high-performance radio local area networks (HIPERLAN). The committee has secured two bands at 5.12 GHz - 5.30 GHz and 17.1 GHz - 17.3 GHz for the development of HIPERLAN, to operate at a minimum useful bit rate of 20 Mbps for point-to-point data communications with a range of 50 m. They expect that at this rate and range they can provide 500 Mbps - 1000 Mbps for a standard building floor of around 100 meters square, which is comparable to FDDI. RES 10 is chartered to define a radio transmission technique that includes type of modulation, coding, and channel access as well as the specific protocols. The first meeting of the HIPERLAN took place in December 1991. They secured the bands in 1992-1993, and their target date for submitting the standard for the approval of the ETC was December 1994.

Conclusions

Current trends in the evolution of wireless data services are either in the direction of high-speed wireless LANs or low-speed, wide-coverage mobile data services. The direction for the future is toward flexible multirate services for multimedia applications, adjusting the specifications of the service to the user requirement and the environment. We have provided an overview of the rapidly expanding field of wireless data services and systems. After a brief overview of the wireless data market and the user perspective of wireless data services, we briefly described the major mobile data networks and wireless LAN activities.

References

1 Thomas Haug, "Overview of GSM: Philosophy and results", *Int. J. Wireless Info. Networks*, vol. 1, Plenum Publishing Co., Jan., 1994.

2 A.J. Motley, "Advanced cordless telecommunication services,", *IEEE J. Selected Areas Commun.*, vol. SAC-5, June 1987, pp. 774-782

3 C. Buckingham, G.K. Walterink, and D. Akerberg, "A business cordless PABX telephone system on 800 MHz Based on DECT technology," *IEEE Commun. Mag.*, vol. 29, no. 1, Jan. 1991, pp. 105-110.

4 H. Ochesner, "DECT-Digital European cordless telecommunications", in Proc. 39th IEEE Veh. Tech. Conf: San Francisco, CA, May 1989, pp. 718-721.

5 T. Hattori, A. Sasaki, and K. Momma, "Emerging telephony and service enhancement for cordless telephone systems," *IEEE Commun. Mag.*, vol. 26, Jan. 1988. pp. 53-58.

6 D.C. Cox, "Universal portable radio communications," *IEEE Trans. Veh. Technol.*, vol. 34, pp. 117-121, Aug. 1985.

7 D.C. Cox, "Universal digital portable radio communications," *Proc. IEEE*, vol. 75, pp. 436-477, Apr. 1987.

8 D.C. Cox, H.W. Arnold, and P.T. Porter, "Universal digital portable communications", *IEEE J. Selected Areas Commun.*, vol. SAC-5, pp. 764-773, June 1987.

9 D.C. Cox, "Portable radio communications," *IEEE Commun. Mag.*, vol. 28, no. 11, Nov. 1990, p 30-40.

10 I. Brodsky, "Wireless MANs/WANs Offer Data to Go," *Business Communications Review*, Feb. 1991, pp. 45-51.

11 J. A. Kilpatrick, "Update of RAM Mobile's packet data radio service," Proc. 42nd IEEE Veh. Technol. Conf:, Denver, CO, 1992, pp. 898-901.

12 K. Parsa, "The MOBITEX packet-switched radio data system," Proc. Third IEEE Int. Symp. on Personal, Indoor and Mobile Radio Comms.(PlMRC'92), Boston, MA, pp. 534-538, 1992.

13 K. Pahlavan, "Wireless office information networks," *IEEE Commun. Mag.*, vol. 23, no. 6, June 1985, pp. 19-27

14 K. Pahlavan, "Wireless intra-office networks," *ACM Trans. Office Info. Sys.*, July 1988.

15 M. J. Marcus, "Recent U.S. regulatory decisions on civil uses of spread spectrum," Proc. IEEE GLOBECOM'85, pp. 16.6.1-3, 1985.

16 M. J. Marcus, "Regulatory policy considerations for radio local area networks," *IEEE Commun. Mag.*, vol. 25, no. 7, pp. 95-99, July 1987.

17 M. Marcus, "Regulatory policy considerations for radio local area networks," Proc. IEEE Workshop on Wireless LANs, Worcester, MA, May 1991, pp. 42-48.

18 M. Kavehard and P.J. McLane, "Performance of low-complexity channel coding and diversity for spread spectrum in indoor, wireless communications," *AT&T Tech J.*, vol. 64, no. 8, Oct. 1985, pp. 1927-1965.

19 Proc. IEEE Workshop on Wireless LANs, Worcester, MA, May 1991.

20 B. Tuch "An ISM band spread spectrum local area network: WaveLAN," Proc. IEEE Workshop on Wireless LANs, Worcester, MA, May 1991, pp. 103-111.

21 K. Ogawa, K. Kohiyama, and T. Koboyashi, "Toward the personal communication era," *Int. J. Wireless Info. Networks*, vol. 1, no. 1, Jan. 1994.

22 Joint Technical Committee for PCS, Service Requirements Description, 1993.

23 D.C. Cox, "Wireless access: An overview, presented at PIMRC 92, Boston, MA, Oct. 1992.

24 D. J. Goodman, "Trends in cellular and cordless communications," *IEEE Commun. Mag.*, vol. 29, no. 6, June 1991, pp. 31-40.

25 R. Steele, "Deploying personal communication networks," *IEEE Commun. Mag.*, vol. 28, no. 9, pp. 12-15, Sep. 1990.

26 T.S. Rappaport, "The wireless revolution," *IEEE Commun. Mag.*, vol. 29, no. 11, Nov.

1991, pp. 52-71.

27 P.A. Ramsdale, "Personal communications in UK — Implementation of PCN using DCS-1800," *Int. J. Wireless Info. Networks*, vol. 1, Jan. 1994.

28 Crump, Jr. S., "How to make the wireless PBX a reality," *Personnel Commun.*, Oct. 1985, pp. 8-16.

29 A.J. Viterbi, "A vision of the second century of wireless communication," *Int. J. Wireless Info. Networks*, vol. 1, Plenum Publishing Company, Jan. 1994.

30 K. Pahlavan, "Signal Processing in telecommunications," Chapter 22, *Handbook of Signal Processing*, C.H. Chen, ed. (Marcel Dekker Inc., 1988).

31 K. Pahlavan and A. Levesque, "Wireless Data Communications," *Proc. IEEE,* vol. 82, Sept., 1994, pp. 1398-1430.

32 K. Pahlavan, "Wireless LANs," Chapter 11, *Personal Communication*, Gardiner and West, eds., (Artech House, 1994).

33 "Cellular Digital Packet Data Specification, Release 1.0," CDPD Industry Coordinator, P. O. Box 97060, Kirkland, WA, USA.

34 R. R. Quick, Jr., and K. Balachandran, "Overview of the Cellular Packet Data (CDPD) system," Proc. PIMR 93, Yokohama, Japan, 1993, pp. 338-343.

35 E. Tiedemann, "Data services for the IS-95 CDMA standard," presented at PIMRC 93, Yokohama, Japan, Sept. 1993.

36 A. D. Sacuta, "Data standards for cellular telecommunications - A service-based approach," Proc. 42nd Veh. Technol. Conf., Denver, CO, May 1992, pp. 262-266.

37 D. Weissman, A. H. Levesque, and R. A. Dean, "Interoperable wireless data," *IEEE Commun. Mag.*, vol. 31, no. 2, Feb., 1993.

38 TIA TR45.3 Subcommittee, Project No. PN-3123, Async Data and Fax, and Project No. PN-3306, Radio Link Protocol 1, both issued Nov. 14, 1994.

39 ETSI, "Bearer Services Supported by GSM PLMN," Rec. GSM 02.02, version 3.1.1, Jan., 1990.

40 J. L. Haine, P. M. Martin and R. L. A. Goodings, "A European standard for packet-mode mobile data," Proc. Third IEEE International Symposium on Personal, Indoor and Mobile Radio Communications (PIMRC'92), Boston, MA, 1992, pp. 513-519.

41 International Standards Organization, "Protocol for providing the connectionless mode network service," Publication IS 8473, 1987.

42 T. Freeburg, "A new technology for high speed wireless local area networks," Proc. IEEE Workshop on Wireless LANs, Worcester, MA, May 1991, pp. 127-139.

43 R. E. Kahn, et al., "Advances in packet radio technology," *Proc. IEEE*, vol. 66, no. 11, Nov. 1978, pp. 1468-1496.

44 D.G. Steer, "Coexistence and access etiquette in the United States unlicensed PCS band," *EEE Personal Commun. Mag*, vol. 1, no. 4, 4Q94.

45 K. Pahlavan, S. Howard, and T. Sexton, "Adaptive equalization of indoor radio channel," *IEEE Trans. Commun.*, vol. 41, Jan. 1993, pp. 164-170.

46 S. W. Wales, "Modulation and equalization techniques for HIPERLAN," Proc. of IEEE PIMRC 94, Sept. 1994, pp. 959-963.

47 A. Nix, M. Li, J. Marvill, T. Wilkinson, I. Johnson, and S. Barton, "Modulation and equalization considerations for high performance radio LANs (HIPERLAN)," Proc. of IEEE PIMRC 94, Sept. 1994, pp. 964-968.

48 G. Yang and K. Pahlavan, "Sector antenna and DFE modems for high speed indoor radio communications,"IEEE Trans. Vehi. Technol., vol. 43, Nov., 1994, pp 925-933.

49 G. Yang and K. Pahlavan, "Performance analysis of multicarrier systems in office environment using 3D ray tracing," Proc. IEEE GLOBECOM, Dec., 1994.

50 M. Aldinger, "Multicarrier COFDM scheme in high bit rate radio local area networks," Proc. IEEE PIMRC 94, Sept. 1994, pp. 969-973.

The Origins of Cellular Communications

Evan B. Richards
Vice President, Network Planning
Ameritech Cellular Services

In this essay, the author discusses the development of the cellular telephone, including the regulatory actions, long before any equipment was designed, built, or operated. The first developmental system that proved the theory is also discussed along with the construction of the nation's first commercial system in Chicago.

Origins of the Cellular Telephone

In this era of rapid changes in technology, it is hard to believe that the cellular telephone's roots began over 25 years ago. The lesson to be learned from this is that technology alone does not set the pace of progress.

Cellular technology got started in 1967. At that time, mobile radio was a sleepy little backwater at Bell Laboratories. Mobile telephone service was provided in two different frequency bands: 11 channels at 150 MHz and 12 channels at 450 MHz. Mobile sets could only access either band, and the service was not profitable.

In 1967, a typical mobile system operated with high-power transmitters located at the top of the tallest building in the center of a metro area and covered an approximate radius of 25 miles, depending on the terrain. Power from the mobile unit was limited, so transmission was not equal in both directions. Although the coverage area was defined by usable signal, there was still interference beyond that point for a distance of approximately 75 miles. Nearby cities, therefore, interfered with one another, and all channels could not be used in any one city. For example, New York interfered with Newark and Asbury Park in New Jersey, White Plains and Stamford, Connecticut, to the north, and Hempstead, Long Island, to the east. All these locations needed to use different channels. The result was that only three channels could be assigned to New York, and customers could only have three calls in progress at any one time while the fourth caller would be blocked.

With such low capacities, revenues could not cover costs. So the newer 450-MHz system was implemented in only a few cities. A large block of spectrum was required for sufficient capacity to have an economically viable service, but spectra were scarce. All major blocks of spectra were already allocated to other uses.

The history of additional spectrum requests goes back to 1947. The Bell System petitioned for the 450+ MHz band to provide a high-capacity system. The FCC reluctantly denied this position and instead allocated spectra to UHF TV as the more important use. At this time they believed there was a great need for additional TV channels. Subsequently, these additional channels were not fully activated and were partially released for cellular use at a later date.

In 1969, 22 years later, the FCC surprised the Bell System by issuing a Notice of Inquiry. They reminded AT&T of their previous filing and asked for details for a high-capacity mobile telephone service. They were considering re-allocating channels 70-84 of UHF TV. AT&T was caught off-guard. It was not clear if there ever was a specific plan back in 1947, and after 22 years there was no one left at the company who worked on the original team.

At this time, three young members of the technical staff at Bell Labs were engaged in doing a paper design of a cellular system. Dick Frenkiel, Phil Porter, and Joel Engel worked closely as a team in developing the concept of cellular communications. In December 1994, Frenkiel and Engel were honored with the United States Department of Commerce National Medal of Technology for their contributions to the development of cellular communications.

The cellular concept was not revolutionary. It had been reinvented many times. As described earlier, the interference area of a radio transmitter is much greater than the coverage area, so nearby cities had to use different channels. Once far enough away, a different cellular system can then use the same channel again. This is a familiar concept because the assignment of TV channels is a large-scale example of cellular communication. New York City has channels 2,4,5,7,9,11 & 13. Philadelphia must use channels 3,6,8,10, & 12. Baltimore can use New York channels, Washington, D.C., uses Philadelphia channels, and so on.

The cellular concept shrinks the whole geometry. Instead of assigning channels to entire cities as the television system does, cellular assigns them to small cells within the city with much lower-power transmitters. The channels are then reused many times within the city in cells far enough apart not to interfere with one another. Adequate space between cells is measured not in absolute miles but in the ratio of spacing between cells to radius of cells.

While Frenkiel, Porter, and Engel were analyzing radio fading and FM capture to determine how far apart cells had to be in order to use the same channels, they recognized that vehicular mobile telephone traffic would not be uniform over a metro area. Cells would have to be smaller in areas of peak traffic, so cells of different size would have to be accommodated. They were also studying how to start with large cells and split them as the system grew over time. Different methods of locating mobiles were analyzed so the system would know what channel to assign. Methods of handing off calls as mobiles crossed cell boundaries had to be developed. Between 1967 and 1969, there were many memos and papers published in the *IEEE Transactions* on these system design problems.

The Bell System's reply to the FCC Notice of Inquiry was very clever. AT&T acknowledged that the FCC had asked an interesting question and that it would take about 18 months to develop a detailed response based on exploratory development and field experiments. They also indicated that the Bell System would not make the investment in financial resources unless the FCC gave reasonable assurance that it was serious about allocating spectra. The Bell Labs estimated it would take three years to formulate a response, but AT&T did not believe the FCC would accept that time frame so they reduced it to 18 months. Fortunately, Bell Labs started work immediately, before the FCC responded, which took 1-1/2 years! It took three years to develop solutions and an overall plan. A comprehensive report, which AT&T filed with the FCC, contained a complete system design that was essentially the analog system design in use today.

Nevertheless, the FCC did not actually allocate spectrum until more than ten years later because there was opposition to the AT&T plan. As expected, TV broadcasters opposed the idea because they did not want to lose spectra. In addition, Motorola was a major and extremely effective opponent. This is hard to imagine now when we all know how successful Motorola has been as a participant in the cellular business, but at the time, they evidently felt threatened by this new concept. The major use of mobile radio was for fleet dispatch, taxi fleets, delivery trucks, and so forth. These were private systems owned by users, and Motorola was in the business of manufacturing, engineering, constructing, and even operating and maintaining systems for customers. They were evidently concerned that a common carrier system, with base stations owned and operated by carriers, would leave them with the mobile set business. Further, if the FCC type accepted and approved mobile sets, they would become a commodity. Motorola's reputation would not have the same value against competitors, so they opposed AT&T's plan vigorously.

Significant opposition also occurred within Bell Labs and AT&T. A number of engineers at Bell Labs thought the system would never work. Some said that radio waves "didn't travel in hexagons" and, therefore, the concept would fail. In addition, AT&T hired a market research firm to study customer willingness to pay for mobile services. Amazingly, they reported "no market at any price" and recommended that AT&T withdraw the request to the FCC for spectra to deploy cellular technology. This report indicated a final customer penetration of 1 percent in the year 2000 (actual penetration in the United States is 8 percent as of the end of 1994).

Developmental System

Finally, in March 1977, the FCC authorized Illinois Bell Telephone to construct and operate a developmental AMPS (advanced mobile phone system), using AT&T equipment in the Chicago area and a second developmental system using Motorola equipment in Washington, D.C. The Chicago developmental system had 10 large cell sites containing 136 voice channels. This was all controlled by a mobile telephone switching office (MTSO) that was located in Oak Park, Illinois. The system covered approximately 2,100 square miles in the Chicago area and included an equipment test of approximately 100 mobile users in mid-1978 followed by a customer service test that involved approximately 2,000 customer mobile units and ran for five years. Mobile phone transceivers at the time were so large and heavy (greater than 100 lbs.) that many vehicles participating in the trial had to have their rear suspensions beefed up.

Throughout this trial, basic call processing using early versions of software designs and all the system shakedown and debugging activities were necessary. Also required were designs of Bell System supplied components, confirmation of some non-Bell System mobile units and cell site radio equipment, verification of the basic engineering procedures to layout the system, verification of the radio serving signal quality including acceptable location and hand-off procedures, and the verification of system recovery procedures, call-processing sequences, and the overall signaling plan. Most important, a demonstration of co-channel operation was necessary, using several of the cells spaced at appropriate frequency-reuse distances for start-up cellular systems.

During the customer service test, data were collected to support the market study activities, including verifying the sales prediction procedures that were currently in use to determine the future market for cellular telephones. Ten years after the basic concept had been developed, the market studies still indicated that a total penetration of only 1 percent of the population could be expected.

All of the cellular towers in use at the time cellular communications became a reality were a minimum of 150 feet high with most averaging 250-325 feet and one escalating to 550 feet in height. Compared to today's cellular system, extremely high omni-directional antennas were used and were appropriate at the time because of the minimal impact of interference. The major engineering problem in the initial design was not interference but to assure that mobile users were covered in as wide an area as possible. Original cells used a cable trough (designed by Bell Labs) that went underground and connected the tower to the hut. The idea here was to protect the coax by keeping it underground and inside the monopole tower, but the trough would fill up with water that would become stagnant, freeze, and cause damage. Current design has cabling suspended above ground. A van the size of a mobile home was used to gather data on system performance, but it gathered more data than could be analyzed and used. There were not enough people available to analyze the data effectively, and sophisticated computerized tools were not yet available. Much of the data were unused.

Commercial Systems

While the technology was being fine tuned in Chicago and Washington, the FCC worked on plans to allocate spectrum and decided to create two bands, A and B, in each of the 306 MSAs (metropolitan service areas) and 428 RSAs (rural service areas). The B band was allocated to the local telephone company, and AT&T orchestrated a nationwide negotiation that resulted in the MSAs being divided up between the Bell System companies and the independent telephone companies. The A band was open to all comers. For the largest MSAs, the FCC received detailed applications and made an allocation based on the merits of the applications. These were all contested in various courts, so in the smaller MSAs and RSAs, the FCC conducted a lottery with thousands of companies and individuals participating.

While this work was going on, a tremendous upheaval occurred in the business world of telecommunications. The breakup of the Bell System that culminated in the Modified Final Judgment (MFJ) was completed in 1983. The first offer by AT&T to Judge Harold Green called for AT&T to keep Bell Laboratories, the long distance company, Western Electric manufacturing company, National AMPS (which was, of course, the cellular companies), and the Yellow Pages operations of the telephone companies. Many would say that AT&T saw the tremendous potential of mobile communications and wanted on their side of the MFJ balance sheet. They also wanted the Yellow Pages operations because of its tremendous profitability. Judge Green decided that the cellular and Yellow Pages companies should be broken up into seven different companies corresponding to the RBOCs. He also decided that it would be necessary to have structural separation between the cellular and Yellow Pages companies and that the local telephone company operations and transactions had to be at arm's length. This resulted in seven strong mobile telecommunications companies being formed during the 1983-1984 time frame.

Coincidental with this business upheaval, the first commercial service was deployed in Chicago on October 13, 1983. There was a race among Washington/Baltimore, New York, and Los Angeles to see which market would come in second to Chicago. Equipment shortages existed, and all companies fought against poor infrastructure and subscriber mobile equipment. The initial construction in Chicago was difficult because of the tremendous amounts of rain and mud. The cellular properties were located far off the paved roads; therefore, it was difficult to pour concrete foundations for both the building and, more importantly, the tower. Approximately 60 tons of concrete are used in the bases of some cellular towers. Concrete was pumped approximately 400 feet to the back of a property when heavy concrete trucks could not get closer so that construction schedules could be maintained.

The key element in successfully implementing this system was a comprehensive, detailed, and carefully laid out plan. Contingency plans were extremely important as well. It is essential to have a flexible plan of action, especially when dealing with factors such as regulatory requirements, weather conditions, and the schedules of others necessary to complete the work.

This was a detailed planning project that required the cellular carrier to anticipate problems and delays and consider many different issues. Which cell sites should be installed first? In which areas will zoning approval be most difficult to obtain? Where should the work crews be sent and when? How long should the crews be at a particular site? When should the mast – which weighs thousands of pounds – be delivered to the site? As unexpected delays occur, where else can the crew be sent to work in the meantime?

From an operational point of view, everyone involved in the project knew that these systems had to operate 24 hours a day, 7 days a week. In Hickory Hills, Illinois, an ACC (AMPS control center) was established and provided the diagnostic capability for monitoring not only the Great Lakes AMPS (later Ameritech) systems but also the nationwide systems that were in the process of being built. The original concept, since AT&T had planned to operate all of the cellular systems nationally, was to have one center in Hickory Hills that would monitor the entire nation. In addition, an extensive training area was developed so that cellular operations personnel from various parts of the country could be trained. Within a few years, that concept quickly fell apart, and each of the major cellular operators built their own monitoring and maintenance centers in order to have control over this vital piece of the operation.

These early days of cellular communications were exciting not only because all employees had the benefit of participating in mobile communications history but also because they had the challenge of developing a totally new technology. There was a true sense of ownership and pride. The technology, the company, and many of the employees were new, and the whole venture seemed a risk. Would this technology and service be successful? Only if we made it so, and that made it a risk worth taking. Success completely depended on the employees, especially those who were involved in the process from the beginning. It was a true sense of ownership that compelled everyone to work extremely long hours and continue working until answers to problems were found.

Challenges were evident in everything that was touched. The operation of the system, the way in which the system was managed, the measurements, the maintenance messages, and the alarms by the system were all new and suspect. The early employees were part of the design of the technology. They worked side by side with the developers from Bell Labs, Motorola, and other companies. When operators found a problem, Bell Labs asked how the carriers wanted it to work, and laboratory people went to work to develop software to solve the problem.

Also, in early 1983, the seven RBOC cellular companies created an organization called the Cellular Centralized Service Organization (CCSO). In order to develop common systems, a shared data center was funded by all seven cellular companies, and all new systems and enhancements to existing systems were to be decided upon through a consensus process. Auxco Computer Services located in Florida provided these data services. To meet the billing requirements, Auxco cloned an application running at a Caribbean telephone company. They modified the message processing to accommodate air-time rating and called the product ACTION 1.1. The vision was to use this product for the initial turn up of Chicago while the system of the future was being developed.

The CCSO disbanded in 1983 because of some serious allocation issues that were never resolved and because it became apparent that the seven carriers might someday become potential competitors. This left seven regional information technology organizations to determine individually how to deliver these applications and identify a data center to do the processing. The greatest concern and impact caused by this decision was felt by Ameritech Mobile Communications (AMCI) because it was planning to commercially deploy the Chicago MSA in October of that year. There was a pent-up demand of at least 5,000 customers from the Chicago trial, and there was no way to bill them! AMCI had no choice but to proceed with ACTION 1.1 and find a way to make it work.

Since we were never too sure whether changes in the cellular system caused problems, there were many road rallies to test new software and hardware. Conceptually the equipment worked, but there were hundreds or possibly thousands of changes to be made to the system before it could even process a call, and hand-offs were even more difficult. There were many changes to the hand-off algorithms and subsequent mini road rallies to analyze hand-off effectiveness. In fact, when this author joined the National AMPS AT&T cellular company in February 1983, the Chicago system was just being designed. As building started and I moved to Chicago, our engineers fixed me up with a suitcase phone that I carried around in my car and used constantly to test the system for faults. My wife and I would spend all weekend driving throughout Chicago for several purposes: to get acquainted with the city and to make calls throughout our drive time to determine where trouble spots were located and what possible fixes might be made. (After I left cellular in 1987 and returned in 1992, the first question my wife asked me was, "Does this mean we have to spend every weekend driving throughout Chicago testing the system again?")

Zoning was always a problem. The first 17 cells required negotiating with 17 different local governments in 17 different communities with 17 different sets of zoning regulations in order to obtain approval for plans to build 17 cell site buildings and 17 antenna mast structures.

By the time Chicago started commercial service in October 1983 (the first in the United States), the AMPS control center was monitoring 12 other cities around the United States that were in various stages of construction. Chicago was the control point for Dallas, Los Angeles, New York, Philadelphia, Miami, Buffalo, New Orleans, Detroit, Cincinnati, Milwaukee, and a few others.

One of the early requirements was that anyone working on cellular radio equipment needed to have their FCC radio operator's license. The ruling was eventually changed to make the company responsible and not the individual. When AMPS was started in June 1983, however, the responsibility for obtaining an FCC license was the employee's. This was a condition of employment, so employees were told that it was imperative that they study for and pass the FCC exam. This was stressful because they probably could not return to their previous job if they failed. Many of the employees studied on their own for weeks, but as the test date approached, they realized that they had to find a peaceful atmosphere in which to concentrate. Several employees rented a hotel room not far from the office and locked themselves inside for a solid week to study for the exam. They lived on coffee, donuts, and White Castle hamburgers. Their families thought they were crazy, but that is the dedication and commitment to the job that was contagious right from the start. Long hours, late nights, and weekend work characterized the early days of this business. Wait a minute – that is exactly our current situation! As they say, "The more things change, the more they remain the same!"

Early Days of Cellular Service

The early days of engineering in a rapidly growing system like Chicago's was a challenge. The information that was available was found in the early AT&T and Bell Laboratory documents and in articles on cellular design theory. All of these documents were originally used for a trial system lacking a growing customer base. In many cases, the engineers relied on trial and error to feel their way through problems, such as determining and correcting interference. Another issue was determining what parameters needed to be adjusted when a new cell was integrated into the system to balance hand-off, power levels, and the like. This had to be learned on a pay-as-you-go basis, and much of it was trial and error.

In the early days of 1983, RF engineers did not have the luxury of having PC-based propagation models or traffic engineering tools. They had a terminal that was used to input design parameters for the development of coverage of future cells. Several days after the data were entered, a package would arrive with the long-anticipated plots. Heaven help the engineer if they had input an erroneous character for latitude or antenna height because all they had then was useless Mylar, and the process had to start over again. At the time, this type of service was costly, and mistakes could not be afforded. Sometimes, when engineers could not wait for the slow response, they used calculators, paper, pencils, and maps to determine the theoretical coverage of a proposed cell site.

From day one, the engineers had to stay on top of the capacity curve. Customer growth was fantastic from the start, and engineers were always strained to keep up with adding enough radios for capacity. They were added on a daily basis in the early days, and the first cell-splitting for capacity went on-line six months after the commercial service was started in Chicago. The next two split cells came up within weeks of the first. This led to co-channel interference problems. Directionalization needed to be performed. The first directional cell site was no picnic to deal with. Engineers struggled with design and hardware for some time before the final configuration was determined. This required many hours of drive testing and parameter changing before engineers felt comfortable with this new animal called a directional cell. This started a cascade effect throughout the system to directionalize more and more existing omni-cells. Not too long afterward, new cells were turned up in a directional configuration.

Along with the benefits of directionalization, which helped to control co-channel interference, a new beast arose: adjacent channel interference. Because of the inherent design of the reuse pattern employed in many systems, N =7, each cell has at least one or many times two neighbors that have adjacent cell assignments. When in an omni-configuration, there are sufficient channels available in each channel set to achieve segmentation easily even with adjacent channels. When directionalizing cells, more radios are required because of the lack of trunking efficiency on each new face of the cell. This requirement did not allow for successful segmentation of the radio channels to avoid adjacent channel interference, let alone co-channel interference. Hand-off parameters were adjusted and power levels tweaked to help maintain the optimal balance between the adjacent channels.

A new tool was then discovered – sequential trunk hunting – which allowed for voice channels to be assigned in a specific order. Engineers were then able to assign adjacent as well as co-channel faces in reverse order, thus avoiding interference until the cell became busy.

Information from the switch regarding usage and system failures was sketchy at best during this early period, and methods of analyzing and reporting system faults had to be developed. Usage information had to be analyzed without the use of computer programs. Even with this rudimentary way of determining system capacity, it was felt that the system in Chicago would exhaust capacity by 1986 or 1987 based on a customer base of approximately 70,000 customers – many times called "the wall." This determination led to Ameritech's lead in the petition to the FCC for more spectra. It took three years to get the approval and allocation of new spectra, but with the advent of the additional 5 MHz (making a total of 25 MHz) came the need for new equipment, both from the switch and cell site vendors and mobile manufacturers.

New spectra allowed cellular systems to expand further, but that was just the beginning of the problem. Vendors equipment had to be changed in every cell, both hardware and software. The major problem was mobile manufacturers who had difficulty keeping up with the huge demand by new customers for cellular service, and it took manufacturers over a year to develop mobiles that could use the expanded spectra provided by the FCC. As a result, the new spectra sat dormant for at least a year before customers had equipment at their disposal, both from a system viewpoint and a subscriber equipment viewpoint that allowed them to use this added capacity.

During that time, cellular operators could not wait for the equipment to catch up and aid in their efforts to provide capacity for customer usage, so they forged beyond what they thought were the limits of the system. They further expanded by splitting cells at the tertiary level. In Chicago, the first split cell at that level was completed in March 1985 and was another first in the industry. This led to the relatively famous build program called a "Tertiary Farm," which reduced the whole core of the Chicago system to the same level of split. It caused many problems with customers and much anguish in the engineering teams. As with all undertakings, if there is not enough preparation, the plan is destined to fail. We learned that the mass integration of cells is simply not feasible because of the impact that each cell has on the overall system. This step in evolution led to the development of the microcell.

It seemed like Chicago and other major systems were always in a catch-up mode to meet capacity needs, and construction of new cells seemed to take forever. The original cells that were built in Chicago and other markets were designed as brick and block buildings with approximately 480 square feet of interior room. This space was planned to allow for up to 6 radio frames holding 96 radios. Once the system started to expand, they found there was no feasible way to add that much equipment to a single cell location. So why build such a huge space? Why does it have to be so permanent? With those types of questions in engineers' minds, they reviewed the whole approach to cell site construction. They modified their plans and were able to reduce the normal six month's construction interval to less than three months. This was accomplished by using prefab buildings of about 300 square feet of space and placing them on the site with the equipment already installed. This helped system operators to be more responsive to the capacity and coverage needs of their customers.

Cellular equipment was still in an evolutionary period during the first few years of commercial service. Software changes were made once a month or so, and the first customers that were on the system helped the company troubleshoot problems by reporting trouble. This change in the midst of a rapidly expanding system led to frustrating moments at times.

A common joke around Ameritech Cellular was "we only have this slight hump to get over and next year it will all be easier." That phrase has been used for 11 years now, and we currently think there is only one more hump to get over before the digital conversion and everything will be done! Ha! We will likely go on to the next problem as we have for the past 11 years.

Summary

The availability of mobile telephone capability has had a dramatic impact on society, not only on the approximately 20 million people who have cellular telephone service but also on all the others who are able to communicate with users of cellular phones. In the public sector, it has resulted in major improvements in security and safety. In the city of Chicago alone, there are 10,000 calls per month to the cellular equivalent of 911. In the commercial sector, it has enabled new methods of conducting business that have resulted in substantial improvements in productivity and competitiveness. The prestige of the United States is enhanced by its recognition as the originator of this technology and by its continued leadership as a result of this early initiative. Yet growth has not stopped. This has resulted in an industry in the United States that currently generates nearly $10 billion of annual revenue and is growing at a 40 percent annual rate.

References

"Advanced Mobile Phone Service." *Bell System Technical Journal 58*, no. 1 (January 1979).

Engel, Joel. "Origins of the Cellular Telephone." Presentation given at the Institute of Electrical and Electronic Engineers COMSOC Conference, November 14, 1994.

Nelson, Mark B. "The Chicago Story." *Telephony* (February 13, 1984).

Contains numerous conversations among personnel involved in building the first commercial cellular system in the United States: Nick Minarich, Dave Jesson, Dave Fox, Doris Harner, Carol Widerberg, and others.

Major Challenges to Wireless Providers

Morris W. Westerhold
President
TPC Consulting, Inc.

The challenges for any business start with understanding the customer's needs and continuing through service implementation to meet those needs. The ultimate test of success is that the customer's needs are satisfied and that the business is profitable and remains so into the future. This paper addresses the challenges facing wireless providers in the United States, both existing cellular carriers and the new PCS providers.

Background on the U.S. Wireless Industry

For the past decade, commercial wireless service has been synonymous with cellular service. Cellular service was introduced into the United States in 1984. This was the same year as the breakup of the old Bell System, one of the defining milestones in the transition to a more competitive telecommunications environment in this country. After years of industry debate in the late 1970s and early 1980s, the Federal Communications Commission (FCC) decided that cellular should be competitive and issued two licenses in each service area. One license was initially assigned to the local exchange carrier (LEC) or to a coalition of LECs. The other license in each area was assigned by lottery among the qualified non-LEC applicants.

One of the defining characteristics of the U.S. cellular industry is its rapid growth, typically growing the customer base at the rate of 40 percent to 60 percent per year for the past decade. The early customers were predominantly business people who spent a substantial portion of the day in the car and needed to stay in touch with customers and clients. These early business customers saw the mobile phone as a means of significantly improving their productivity and were very willing to pay several thousand dollars for the car phone and hundreds of dollars per month in service fees. As manufacturing volumes increased, the prices of car phones dropped to the point where now they are available for about $100, or "free" with a service contract of one year. As the price of car phones dropped, the market was further stimulated, first by new segments of the business market and later by the consumer.

The consumer had different needs and saw the phone as a means of improving safety and security as well as a way to stay in touch. Today there are over 20 million cellular customers in the United States. Industry experts project rapid growth to continue for the next ten years with an ultimate penetration of perhaps 80 million to 100 million customers for cellular and PCS.

The initial product was the phone installed in a car with an antenna mounted outside on the roof or glass. These phones could transmit with a maximum power level of 3 watts, powered by the car battery. With an outside antenna, these phones could communicate with a cell site 10 to 20 miles away, depending on terrain and other factors.

A second product was the bag phone that required no installation and usually was powered from the car via the cigarette lighter. The bag phone also transmitted with a maximum of 3 watts and had an antenna attached to the electronics in the bag. Placing the antenna inside the car was not optimum, and the phone could effectively communicate with a cell site 8 to 12 miles away.

The third and most popular product was the portable phone, many models of which can easily fit in a pocket, especially when equipped with a small battery. These portables transmit at 0.6 watts in the United States and have a self-contained antenna. With the lower power level and often non-optimum antenna placement, these phones can reliably communicate with a cell site perhaps 4 to 6 miles away.

Cellular service providers have been continuously expanding their coverage over the last ten years. Cellular started in major metropolitan areas with a small number of cell sites. A service provider in a flat city such as Chicago could introduce service to much of the population in the area with less than 20 cell sites. From the beginning, customers wanted better coverage. In the early years the challenge was to provide good coverage to car phones in metro areas and along major highways between cities. By the late 1980s this objective was essentially achieved in the more populated areas of the country, especially in the eastern half of the United States. In the more lightly populated areas, the expansion of coverage continues today.

Since the late 1980s the challenge in the major metro areas has shifted to providing good coverage for portable phones. As a result of expanding both the coverage for car phones and portables, metro area service providers typically have 10 to 20 times the number of cell sites they had when service was introduced in the mid-1980s. In the mid-1990s, in-building coverage for portables remains a major challenge for most cellular carriers.

The characteristics of the typical user changed as the industry grew to 20 million customers. Many of the original users were business customers with hundreds of minutes of usage per month. In recent years, the consumer often makes up a substantial portion of the new customer mix. Many of these consumers are infrequent callers, often with usage levels of less than 50 minutes per month.

Cellular Cost Structure

To a great extent, declining phone prices are driven by manufacturing economies of scale. However, another factor is that service providers have historically subsidized the price of phones to attract more customers. The service provider typically would sell the phone for several hundred dollars less than they paid the manufacturer if the customer agreed to a multi-year service contract. It was good business to subsidize the phone and grow the market when many of the customers were profitable heavy users. It was assumed that other customers would increase their usage as they became accustomed to the extra convenience of a cellular phone. Some service providers were also motivated to increase market share with subsidized phones because the larger customer base would increase the value of their cellular business in any subsequent sale.

In addition to phone subsidies, there are several additional costs of acquiring customers. Advertising is a significant expense for most cellular carriers. The sales and provisioning process is complex, labor intensive, and very costly. Combining these components, it is not unusual for a cellular provider to incur expenses of $300-$400 per customer added. The assumption is that these customers will remain loyal and generate profit to the carrier for many years. However, the cellular industry has experienced customer churn levels of perhaps 30 percent to 40 percent per year, which exacerbates the issue of customer acquisition cost.

Another key component of cellular cost structure are the network costs to provide coverage and usage. In most cellular systems and at current usage levels, the majority of the network cost is the capital investment needed to provide the coverage that customers want and expect, especially with portable phones. The cellular industry originally thought it would experience a rapidly declining investment per customer as new customers were added. So far, this decline in investment per customer has not been nearly as rapid as expected, primarily because of the large increase in cell sites needed to provide good coverage to portable phones. As the customer base continues to expand, the potential is still there for a rapidly declining investment per customer, unless there are substantial new investments required in the future. A much smaller portion of the network cost is related to usage.

An industry executive was recently quoted as saying it costs about $30 per unit of population (POP) to establish a cellular network. Costs will vary from carrier to carrier based on density of population in service areas and quality of coverage provided. When measured per customer, the range seems to be $600 to $1,200 per customer, with an average of approximately $800 at a current level of 20 million customers.

Cellular Revenue Structure

For metro area carriers, the major sources of cellular revenue are the monthly service fee and the usage, or air time, charge. Cellular carriers typically offer a variety of monthly packages, many with some usage level included. Usage above the level stipulated in the package is billed at a per-minute rate. For heavy users, the majority of revenue comes from the usage component of the bill. For light users, the majority of revenue typically comes from the monthly service fee.

Roaming is another source of revenue to cellular carriers. Carriers in major markets typically derive a small portion of their revenue from roaming. However, some rural service area carriers, serving heavily traveled highways, derive a large portion of their revenue from roaming charges.

According to a recent Cellular Telecommunications Industry Association (CTIA) report, the average cellular bill for the first six months of 1994 was $58.65 per month, down from $67.31 for the first six months of 1993. The number of customers grew by 48 percent during this 12-month period. On average, roaming revenue comprised 12 percent of the monthly bill. Averages can be deceiving, especially in the cellular industry where there are few average customers. Although the carriers are periodically adjusting rates, the primary reason for this decline in the average monthly bill is that most new customers are much lighter users of air time than the embedded base. The National Cellular Reseller Association recently reported that rates for light users actually had risen 33 percent in the past six years.

Customer Perceptions of Cellular

Several customer perceptions and expectations of cellular are important in understanding the challenges facing the wireless (cellular and PCS) industry. The first one is coverage. Customers want their cellular phones to work wherever they are when they want to make a call. This is the case regardless of whether the customer is using a 3-watt car phone or a 0.6-watt portable phone inside a car or building. As might be expected, most customers do not understand the difference in power-transmit levels or the importance of an unobstructed antenna for maximum range. Customers pay more for a portable phone than for a car phone and believe the "premium" portable should provide good coverage everywhere.

Business employees with a need to communicate while moving about generally perceive cellular service as an essential business tool. The cost of cellular is easily justified based on the productivity improvements obtained. Most employees recover the cost of cellular service from their company and are not price sensitive. In contrast to the business user, many consumers perceive cellular service to be a premium service and believe they are paying too much, especially for air time. This is undoubtedly one of the reasons why many consumers make few and short calls. The perception of high air time charges is also a factor in customers churning from one carrier to another when their service contract expires.

Customers are not particularly loyal to their service provider for a variety of reasons. They may not be completely satisfied with current service and price levels. There also is very little to differentiate one cellular carrier from another. These same consumers apparently perceive the monthly charge to be a reasonable value in helping to meet their needs for safety and security.

Because of phone subsidies, customers do not see the true costs. It may be cheaper for the customer to get a new phone and account number than to repair a broken phone, further adding to churn. A related issue is that the customer cannot understand why a car adapter kit for a portable, to boost power and connect to an external antenna, should cost more than either the portable itself or a separate car-mounted phone. Faced with these choices, the customer rarely buys a car adapter kit for a portable and may buy a separate car phone (with a second number) even though this might not have been the best solution to meet the customer's needs.

In recent years, customers have begun to expect better voice quality. Business customers are especially intolerant of dropped calls, scratchy calls, or interference.

Cellular Challenges

Even though the cellular industry has been growing at a rapid rate, it faces a number of major challenges in its second decade. The cellular industry is marginally profitable but not nearly as profitable as might be expected from its growth rate or the prices on recent sales of cellular properties. To illustrate the point, many of the new customers are consumers and light users of cellular service. In this scenario, the customer acquisition cost may be $300-$400, and a monthly bill of $30 might include $20-$25 of gross margin. In such a scenario, it could take 20 months to break even, not counting the time value of money.

One of the FCC's objectives with PCS is to increase the level of competition in the cellular industry. In theory added competition will expand the service choices available to the public and put downward pressure on the prices of existing cellular services. For cellular carriers to succeed in the next decade, they will need to overcome several challenges. Cellular has to do a better job of meeting customer needs. The carriers also need to pay close attention to their cost structure, especially the cost of acquiring new customers. Another challenge is that digital cellular must be repositioned to better meet the needs of the carriers.

Meeting Customer Needs

Today, the cellular industry offers some degree of product variation through different models of phones, different service packages, and features such as voice mail, call forwarding, and so forth. However, when compared to many other industries, cellular has a relatively small amount of product variation and differentiation. Current levels of customer churn and customer perceptions suggest that many carriers can do a better job of meeting their clients needs, both for existing users and potential customers they wish to attract. The new competition from PCS will make this a necessity.

The challenge of meeting customer needs should start with a classical segmentation of the market and comprehensive market research of each of the segments. It is clear that the needs of the business user are very different from the needs of the consumer. What is unknown is how many distinct segments are present and how their needs may differ from each other and from the current product mix. Results of this market segmentation and research can then be used to develop modified or new products more targeted to the needs of specific market segments. The following are areas to address in developing these new products:

- Align revenue sources more closely with customer perceptions of value.
- Avoid subsidies that distort the customer's perception of value.
- Seek to differentiate products more in terms of coverage, quality, privacy, and data capabilities.
- Differentiate on price as a last resort.
- Sell applications as solutions to customer problems.
- Excellent customer service will be increasingly important.
- Package cellular with other communications services such as long distance and landline.
- Improve brand identity, in parallel with other improvements, to increase customer loyalty.
- Industry-wide alliances will be essential in order to offer seamless service availability.
- Wireless data is expected to be important to certain market segments.

Cost Control

Cost control will be another major challenge to cellular providers. Increased competition will squeeze prices and margins; carriers with the lowest cost levels will have a distinct advantage. The most important area for cost control is the cost of acquiring new customers. With better knowledge of the base of existing and potential customers, carriers can improve their target marketing to the most attractive prospects. Carriers should also revise their packages to minimize the chances of light users receiving highly subsidized phones. Smart cards and over-the-air registration are two alternative techniques for reducing the sales and provisioning expenses. These techniques will also allow the customer to purchase cellular phones off-the-shelf, similar to other consumer electronic products. Hopefully, the carriers can then find more effective and more economical distribution channels.

Most cellular carriers have already spent the money to provide good coverage for portable phones. Looking to the future, these carriers must decide how to position digital cellular and the type of digital coverage that will be required (more on this in the next section). Any substantial increase in coverage costs for digital should be justified by the associated benefits.

Fraud is currently a major problem in the cellular industry. Fraud costs the industry hundreds of millions of dollars per year, and this figure is growing, according to published reports. Solutions must be found to reverse this costly expense.

The following cost areas are not as critical but still need attention as margins are squeezed:

- Customer service and billing
- Roaming expenses
- Network interconnection
- Coverage of new venues such as in-building
- Advertising effectiveness
- Economies of scale by attracting more customers, especially profitable ones.

Repositioning Digital Technology

The cellular industry has been planning to replace analog radios with new digital radio technology for most of the past decade. However, this goal is still illusive. The industry initially believed that analog use would reach a plateau in major metro areas by the early 1990s; until recently, this has been the primary motivation for the new digital technology. In retrospect, this focus on capacity has mispositioned the cellular carriers for the competitive future. Complexity was encouraged as digital technology was pushed to maximize digital capacity. For example, the industry pushed for extremely low-bit-rate voice coders and was oblivious to the negative effect these coders would have on voice quality. The digital phones themselves needed to be dual mode to also work with analog since areas without a capacity problem would not convert to digital for a long time. The focus on capacity was also a major factor in the development of two competing digital technologies (TDMA and CDMA) and the division of the industry in a religious war over which technology was better.

Today, capacity remains a concern for cellular providers in a few metropolitan areas. However, it is not nearly as serious a problem as was forecast in the 1980s. Use per customer turned out to be a lot less than initially forecast as new, low-use customers signed up. Also, analog capacity exploded in metro areas as many more cells were added to support portable phones. Further cell splitting and sectorized antennas can also be used selectively in the few remaining hot spots to increase analog capacity almost without limit. Some carriers have adapted a modified analog technology (N-AMPS) to further increase ultimate analog capacity.

The overemphasis on digital capacity has only recently been recognized. In most markets, capacity is not approaching exhaust levels, and no other trigger has yet been identified for a digital conversion. Concerns have been raised about digital voice quality, and manufacturers now have accelerated programs to introduce higher bit-rate voice coders. The requirement for dual mode phones, and their associated extra cost, is an increasing concern as cellular carriers must compete against the new PCS providers. Customers also have very little reason to convert from analog phones to dual mode digital phones.

The cellular industry must reposition digital technology to better meet its needs in the coming decade. Digital technology can facilitate new differentiated products and services to help the industry better meet customer needs. For example, better voice quality, superior privacy, and wireless data improvements are all possible with digital technology and should be important to the heavy users. Such new services could be a trigger for carriers to convert to digital over a shorter period of time. New services might also be an incentive for certain customers to upgrade to the digital phone. Digital technology would also help carriers simplify the labor-intensive service provisioning process by using smart cards or over-the-air activation. Finally, digital can help the industry overcome the current fraud problem.

The bottom line is that digital technology has the potential to better meet the needs of customers and also help the cellular providers maximize their profitability in the more competitive environment ahead. The key word is "potential," and much work is required to make sure this potential is achieved. The right services/features need to be developed based on solid market research. Seamless availability of the new (digital) services may require nationwide deployment of the new technology. Digital roaming agreements will be a challenge as long as the industry is divided on the best digital technology and on the triggers for implementing it. With increased competition coming from PCS, the cellular carriers need to reposition digital technology as quickly as possible.

PCS Challenges

The FCC is auctioning six new PCS licenses, for a total of 120 MHz, to expand the availability of wireless services to the public and to provide more competition to both the cellular and LEC industries. The FCC rules for PCS are sufficiently broad to allow a "cellular like" service definition but will also accommodate different service definitions. If the ultimate market is as large as 80 million to 100 million wireless customers,that will be four to five times the size of the current cellular market. There should be plenty of opportunity for many variations of both cellular and PCS.

Some of the PCS challenges are very similar to those facing cellular; other challenges are distinctly different for PCS, primarily because of different starting points. For example, cellular already provides analog coverage over a large portion of the country and has an embedded base of customers. PCS has no customers but is not encumbered by an embedded analog technology, existing processes, procedures, or organizational structures. The new PCS spectrum is currently assigned to various point-to-point microwave users; the PCS provider must arrange to relocate these radio systems in the PCS serving area.

The major challenges facing the new PCS providers are as follows:

- meeting customer needs
- providing coverage at a reasonable cost
- paying a reasonable amount for the license
- keeping operating costs low
- selecting the right technology

Meeting Customer Needs

The challenge of meeting customer needs is similar in many respects for PCS as for cellular. One significant difference is that cellular has a customer base of 20 million customers while PCS currently does not have any customers. If the PCS provider targets the existing cellular customers, the challenge will be to convince them that PCS is better or more economical than cellular. Even for dissatisfied cellular customers, this may be difficult because the customer will need to obtain a new PCS phone.

The PCS rules allow mobility services to be combined or integrated with fixed services. Any type of PCS provider may choose to offer both mobility services and wireless local loops. The LEC does not need a separate subsidiary to offer PCS, as is required for cellular. Thus the LEC can offer PCS together with traditional landline service and/or can use PCS spectrum for a wireless local loop, again when ancillary to a mobile service. The non-LEC PCS provider can also offer mobility and/or fixed services and may choose to offer PCS services that compete with both cellular and the LEC. The rules also allow any provider to create new PCS services that are a hybrid between traditional landline service and traditional mobility services. There are undoubtedly a number of possible hybrid services. For some of these hybrids, the technical challenges could be substantial. To maximize the chances of success, any such hybrid should be based on solid market analysis.

Because of the high fixed cost of wireless businesses, it will be very important for the PCS provider to rapidly grow an adequate customer and revenue base. Cellular will be going after many of the same potential customers, and time to market will be important. Market segmentation and market research will be at least as important to PCS as to cellular. The following are areas to address in developing successful products to rapidly grow the business:

- Align revenue sources with customer perceptions of value.
- Look for non-traditional ways (coverage, quality, privacy) to meet needs and increase value with differentiated products.
- Target segments not being served well with cellular.
- Differentiate on price as a last resort.
- Consider packaging PCS with other communications services.
- Establish a strong brand identity.
- Offer seamless availability.
- Provide excellent customer service.

Cost Control

Cost control will be very important to PCS providers, and none of the cost elements are currently sunk costs. Prudent planning can help the PCS provider avoid some of the cost obstacles encountered in cellular.

Cost of Coverage

PCS at 1900 MHz is more than twice the frequency of cellular. Also, the PCS rules only allow a maximum of 0.2 watts for portables as opposed to 0.6 watts for cellular portables. Both of these factors lead to more cell sites for PCS than for cellular. PCS providers with no infrastructure in place have more flexibility than cellular in deciding how to deploy new technology. For example, it will be easier for PCS to take maximum advantage of digital radio technology and new high-gain antennas. The use of such advanced technology will somewhat mitigate the number of additional cell sites needed for PCS. Also, since PCS cells are smaller than cellular, these smaller increments should allow PCS to save a few cells by not covering as much area with no customers. Some PCS providers may have access to existing infrastructure (e.g., towers, site locations, rights-of-way) and may be able to reduce somewhat their cost of PCS deployment. The bottom line of all the coverage variables is that, for an identical service and without access to existing infrastructure, the cost of coverage is somewhat higher for PCS than for cellular.

In theory, many market segments do not require ubiquitous coverage and may be satisfied with lower coverage and lower prices than cellular. PCS providers might choose to target such segments and limit coverage to the needs of those segments. Some PCS providers may succeed with such a strategy, but the risks are higher that customers will not really be satisfied with limited coverage service.

The relocation of point-to-point microwave is another consideration in providing coverage. Before PCS can gain exclusive use of the spectrum, any microwave systems in the license frequency band and in or near the service area must be relocated to other spectrum. The PCS provider must negotiate such an agreement with each microwave licensee in the band and should expect to pay all or part of the cost of such relocation, depending on the desire of each party for a speedy or slow transition. Extremely low-power PCS applications may be able to share spectra with microwave under certain conditions. The cost of microwave relocation depends on the specifics of each serving area but generally should be a small portion of the overall coverage costs.

Paying a Reasonable Price for the License

The initial cellular licenses were obtained for the cost of a $35 application fee to the FCC. The market soon recognized that the license value was significantly higher; many such cellular licenses were sold in the after market and before system construction for millions of dollars. For PCS, Congress decided that the government and not speculators should get the money for the full value of spectrum licenses; thus, the FCC decided to auction the PCS licenses to the highest bidder starting in December 1994. When the first auction ended in March 1995, the average price of the 30-MHz MTA licenses was $13.90 per POP. The highest prices were for the two licenses of the Chicago MTA at $31.90 and $30.88 per POP, respectively. Approximately half of the licenses sold for less than $10.00 per POP. At the low end, the cost should be less than the cost of installing infrastructure for coverage. At the high end, the cost of the license is comparable to the cost of coverage. PCS providers need a good knowledge of the economics of their planned services to avoid paying too much for a license.

Low Operating Costs

New digital technology can help PCS with operating costs. The service provisioning process can be simplified and made less costly with smart cards or over-the-air activation. Target marketing may also open the door to better distribution channels and allow for more efficient advertising. Hopefully, with these types of improvements, PCS will find a way to reduce significantly the cost of customer acquisition. Fraud control will be easier with new digital technology. The cost of adding more radio capacity should also be less with digital than with analog cellular.

Because of the high fixed costs, economies of scale will also be as important for PCS as for cellular. Again, the challenge will be to increase rapidly the customer base with profitable customers and at a low cost of customer acquisition.

Selecting the Right Technology

With analog cellular, there was only one technology. For digital cellular, the service provider must choose between two technologies, and at the present time there is no clear winner. For PCS, there are currently seven competing technologies being standardized. Again, there is no clear winner although some offer advantages in one area or another. The three technologies that seem to be gaining momentum are IS-95-based CDMA, IS-54-based TDMA, and GSM/DCS-based TDMA. Picking the ultimate winner can yield advantages in lower cost from volume production and make it easier to enter into roaming agreements. Picking the wrong technology can leave the provider and customer base stranded. Because this decision is critical, it is assumed that PCS providers will enter into nationwide alliances after the auctions, and the alliance will select a technology. Even so, different nationwide alliances may pick different technologies. Many PCS providers will be looking for their nationwide alliance to provide rapid build-out so they can avoid dual-mode phones.

Conclusion

Both cellular and PCS providers will face many very substantial challenges in the second decade of commercial wireless service. Each has advantages in some areas and disadvantages in other areas. The winners in 2004 will be the wireless carriers doing the best job of understanding the market, targeting the most attractive customers, and then succeeding in meeting those customers needs. In the very competitive environment expected in the next decade, this will be a marketing challenge much more than a technical challenge. Cost control and other technical challenges will be important, but the winner will be the company doing the best job of marketing and meeting customer needs.

Implementing Smart Cards in the USA: A Wireless Perspective

Lorraine M. Pirro
Executive Director, TOMCOM L.P.

Leo E. Legaspi
Director, NYNEX Mobile Communications

Abstract

Smart cards are fast becoming an indispensable element of the business landscape in many parts of the world. The smart card is now used extensively in public telephone systems in Europe and has also made significant inroads into mobile telephony through the success of GSM (global system for mobile communications). However, the U.S. has yet to see any significant, large-scale applications of smart cards. While there are some potential benefits to U.S. mobile telephone carriers in implementing smart card technology, especially in fraud reduction, the benefits for customers are not as clear cut. The probability of success of smart cards in mobile telephony will be greatly enhanced if telephony applications are packaged with other complementary services such as banking. This will require much closer cooperation within the mobile telephone industry to develop common platforms and standards, and ultimately, cooperation with other industries to develop multi-application smart cards that provide real value to the customer.

Introduction

1994 marks the 10th anniversary of the launch of the first commercial smart card by France Telecom in 1983. The aggregate production capacity of smart card manufacturers worldwide is expected to reach 500 million cards per year by

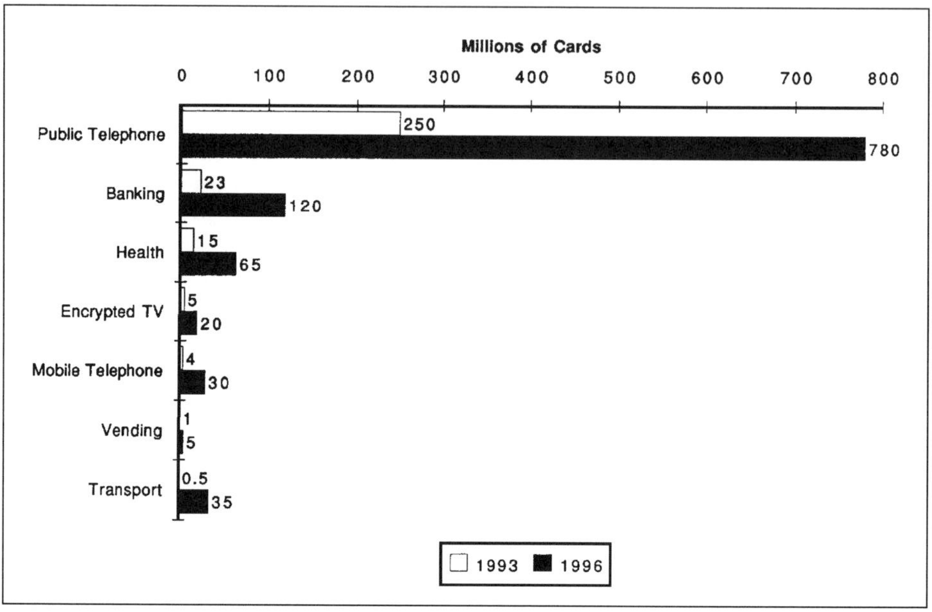

Figure 1 *Smart Cards worldwide market*

mid-decade and 1 billion cards per year by the end of the decade. Smart cards are now in use in numerous applications in every continent and in virtually every developed country. By most standard indicators, the smart card industry is now a stable part of the business landscape.

The most visible, and certainly the most well-known application of smart cards has been in telephony. The prepaid telephone card accounts for over 80 percent of smart cards issued in the world (see *Figure 1*). Indeed, it can be rightfully said that telephony applications have been most responsible for the widespread consumer acceptance of smart cards, particularly in Europe. The prepaid telephone smart card is now an integral part of life in many European countries, as indispensable as the charge card and the ATM card.

The growing acceptance of GSM, the pan-European digital mobile telephone standard, has provided yet another vital platform for growth of the smart card industry. The GSM SIM (subscriber identity module) card is an integral and inherent element of the GSM standard; in the U.S., every GSM phone today is equipped with a smart card reader. Consequently, there are now 4 million GSM subscribers who are all accustomed to smart card usage and predisposed to using these cards in other applications outside telephony.

U S WEST Public Telephone Market Trial - Based on the European technology of prepayment cards, U S WEST intends to apply this technology initially in specific markets to reduce the cost of current pay phone operations as well as raise revenue. Further, U S WEST is a partner with Time Warner in an interactive two-way home services program which may use a smart card in the near future

Money Access Service (MAC) will launch a program next year initially in the Northeast, but then grow to its over 30 million cardholder base an electronic purse "cash card" for making small value purchases.

Chemical Bank, in cooperation with AT&T, will offer a cash card based on the AT&T smart card to consumers in the New York City area. The first steps will be an internal trial followed by commercial application in the New York City area.

The State of Ohio will advance its smart card pilot test to a statewide rollout for electronic welfare benefit delivery. This application may also include other citizen services and benefit programs above food stamps.

The Western Governors Association will be issuing a tender for a health passport which will allow benefit recipients access to numerous health care services in 5 western states along with social welfare benefits including WIC, food stamps and immunization programs.

Source : "Smart Cards: Where is the Industry Going?" Henry Dreifus, Dreifus Associates Ltd., From "Building Foundations for Innovation" - Conference book of Card Tech/SecureTech '94 Conference, April 10-13, 1994, Washington DC

Table 1 *Smart Card trials*

The widespread availability of smart cards in Europe causes many an American visitor to wonder why this technology has not been as widely adopted in the United States. Indeed, there are today no large scale, commercial applications of smart cards in the U.S. There are certainly many major market and technical trials in the U.S. (see *Table 1*). However, we have yet to see a full-scale commercial launch of a smart card application by the private-sector.

Given the success of the SIM card in the GSM system, it would appear that a smart card with equivalent functionality would be an ideal application for the U.S. mobile telephony market. So why has this not happened?

There are, of course, some obvious, albeit incomplete answers. For one, the U.S. cellular communications network infrastructure is predominantly analog, based on the original AMPS (advanced mobile phone system) standard. The smart card is basically a digital device and it could therefore be argued that until a digital radio access platform is widely adopted, smart cards cannot be well integrated into mobile telephony. However, there are in fact smart card implementations in analog mobile telephone networks. For example, C-Netz, the German mobile telephone network, is analog, but uses smart cards that can be used not only within the C-Netz mobile network, but in the German public telephone system as well.

It should also be borne in mind that the first applications of smart cards were in public telephony and there is today a rich history of experience in European public telephone systems to draw from. Why then have smart cards not penetrated the U.S. public telephony market? Some will argue that the superior telecommunications networks in the U.S. have, at least in the past, obviated the need to have the off-line or decentralized intelligence that smart cards provided for the less robust European telecommunications networks. However, smart cards are now proving their value to European telephone companies beyond network operations and are becoming quite useful marketing tools. Still, smart cards have not penetrated the marketplace in public telephony in the U.S.

Perhaps the more important question to ask is not why smart cards have failed thus far to penetrate the market but rather when and how they can be integrated into the business fabric. Smart cards gained acceptance in Europe through telephony applications. It appears that in the United States, the driving applications, at least in the initial stages, will be in financial services. Several key financial services giants, notably Mastercard, Visa, and Citibank, have already committed to using smart cards. Citibank NA believes the cornerstone to the consumer banking markets lies in your wallet. Citibank wants consumers to use a single card that provides debit, credit, and automated teller machine (ATM) services. The card would be used at ATMs, telephones, interactive TVs, and point of sales terminals in retail outlets. The goal, says a Citibank consumer banking division spokesman, is to give consumers the ability to deal with the bank anytime they want, any way they want, anywhere they want.[1]

Interestingly enough, the words used by the Citibank executive – "anytime they want, anyway they want, anywhere they want" – are rather similar to the "anytime, anywhere" slogan now being used by the mobile communications industry. There is no doubt that the major financial institutions are committed to using smart cards. "Having endorsed microchip technology as the acceptance platform of the future, Mastercard International estimates that virtually all of Mastercard-branded payment products and acceptance locations will include chip technology in some form by the year 2000."[2] There is also no doubt that the rest of the financial services industry will eventually follow the lead of Citibank and Mastercard as they pave the way for smart card-based financial services.

To answer the question, how and when will smart cards be integrated into the mobile communications industry will require the examination of the trends in other complementary industries that have been actively investigating smart card applications, including financial services, retail trade, transportation and health care. These industries, particularly financial services, will probably play the pioneering role that the public telephone industry played during the early years of smart cards in Europe. But before proceeding to paint the likely scenarios for implementing smart cards in mobile telephony, it is important to first understand the trends in the core smart card technology.

Definitions

First, what is a smart card? On the surface, this would seem to be a trite question and not worthy of any discussion. However, it is important to understand that smart card technology is not limited to use in credit card-sized devices such as the GSM SIM card. The International Standards Organization (ISO) has developed a suite of standards, commonly referred to as ISO 7816, that define a card having the dimensions of a regular credit card. However, ISO has also defined a smaller size for the smart card that can be more easily accommodated by smaller-sized pocket cellular phones. In addition, there are other devices that do not have a card shape, such as smart keys, smart tokens, and smart badges.

Jerome Svigals, a prominent authority on smart cards, gives the following definition of a smart card:

> A smart card consists of an integrated circuit chip or chips packaged in a convenient form to be carried on one's person. The package may be a bank card with integrated circuit chips or it may be a replacement for a soldier's dog tag. It may be almost any shape, size and thickness that is convenient for the user to carry, or insert into a device and retrieve.[3]

The use of a card form factor is quite understandable. Cards, after all, are perhaps the most widely used method of verifying the identity of a person all over the world. Nevertheless, the credit card form factor is not the only way to house the intelligence of a microchip. For example, the chip could very well be embedded in a smart car key. It is quite conceivable to devise such a smart car key that would not only authenticate the driver in order to gain access to the vehicle but would also, at the same time, initiate the authentication process for the car phone to gain access to the mobile telecom network. Combined with voice recognition technology, this could be a very useful application of smart card technology in mobile telephony.

For that matter, it is important not to limit the scope of discussion on smart cards to microprocessor cards. As the definition of Svigals states, the smart card consists of "integrated circuit chip or chips," and is therefore not restricted to integrated circuits that are not true microprocessors. In fact, the vast majority of smart cards are "memory-only"; they do not have a true microprocessor, and their computing and logic functions are quite limited. In contrast, microprocessor-based cards such as the GSM SIM card are capable, on their own, of complex algorithmic calculations and are closer in capabilities to their highly integrated counterparts such as the 386 chip.

Technology Trends

Increased Processing Power and Memory

The capabilities of today's smart card are indeed quite remarkable. The GSM SIM card, for example, has most of the basic functional elements of a personal computer. It has a CPU, (central processing unit), RAM (random access memory), ROM (read only memory) and the capability of executing complex algorithms such as encryption and decryption. The most advanced smart cards, such as the super smart card, actually have keypads for input and LCD (liquid crystal displays) for output, making them true "workstations on a card."

It is also clear that the capabilities of these smart cards are expanding rapidly, both in processing power as well as memory capacity. While most smart cards today use microcontrollers with an 8-bit bus, work is already underway to develop a 32-bit RISC chip for smart cards. Memory capacities are also expanding by orders of magnitude as chip densities increase geometrically. Thus, while today's smart cards still work with fairly limited memory of up to 16 kilobytes, it is easy to envision smart cards with megabyte memories in the future.

Open Systems

The implications of these tremendous improvements are profound. The first smart cards were limited by computing and memory constraints to single applications only, i.e., public telephony. With this expanded power, smart cards will eventually have the capability of easily handling multiple applications, all in the same card.

For mobile telephone service providers, this means the possibility of combining or packaging other services, aside from mobile telephony, into the same card.

Indeed, we are already seeing some early examples of these multifunction cards. Mercury One 2 One, for example, has already teamed up with Barclay's Bank in Great Britain to develop a smart card that works like a SIM card for accessing the former's PCN service, and also provides some rudimentary banking functionality. It is easy to imagine a smart card application such as the "wireless mobile ATM" which would be aimed at the banking needs of people on the move. This application would allow a user to conduct banking chores which would normally be performed on an ATM – such as account balance verification, transferring of funds from money market to savings account, etc. – by using a mobile phone and a smart card. Indeed, it is conceivable that downloading funds to a smart card via mobile telephone will eventually become as commonplace as withdrawing cash from an ATM.

Notwithstanding the exciting potential of multiple applications on a card, the vast majority of smart cards today work for one application, for one company's service in a one-to-one correspondence. In these "closed" systems, the smart card issuer is also the service provider. The simplest example is the prepaid telephone card. The expansion of smart card capabilities is paving the way for "open systems" and interoperability of services. In an open system, a smart card can be used to make telephone calls on a mobile or public telephone system, to pay for gasoline purchases, or for public parking or for tolls. This of course assumes that gas pumps and parking meters, as well as mobile phones and payphones, are equipped with the necessary smart card readers, which itself will require substantial investments and more importantly, close cooperation between the various service providers.

Benefits of Smart Cards

Customer Benefits

Multiple applications on a card provide tangible benefits to the end user. But do single-application cards provide equal benefits? For example, what are the benefits to the user of a GSM SIM card? GSM operators point to the following as benefits to their customers in using their SIM cards:

- Enhanced calling features – users can store frequently called numbers on a directory in the card
- Alphanumeric message storage – via the short message service (SMS) feature in

GSM, users can receive information such as alerts, news or advertisements which they can read on the mobile phone display while continuing to talk

- Calls always billed to smart card user's account – customers can operate other people's telephones with their smart card and have their calls billed to their account

- Single telephone number convenience – via the international roaming agreements in the GSM community, users can be reached anywhere there is an affiliated GSM network, using the same telephone number issued by their home carrier

- Multiple cards – a household can have several smart cards for use in one phone, each customized to the unique calling needs of each household member

Upon closer investigation, however, one might conclude that most of these benefits are not really compelling reasons for customers to use a smart card. In fact, it can be argued that most of these benefits – stored message service, directory, single number – could be implemented in a central switch using intelligent network capabilities. In fact, one particular benefit, namely a directory of frequently used numbers, is being implemented on current analog phones without the advantage of any advanced network intelligence. The main reason smart cards are used in GSM systems is the fact they are part and parcel of the GSM standard to begin with. GSM customers really have no choice but to use these cards.

Carrier Benefits

The real beneficiaries of mobile telephone smart cards – at least current GSM-type smart cards – are the mobile telephone service carriers. Consider the benefits:

- Enhanced security/fraud prevention – The authentication procedure in GSM provides one of the most "hack-proof" systems in mobile telephony today; the potential savings in fraud reduction can be quite substantial

- Advertising/branding – Advertising on the card face can be used to keep the carrier's brand in front of the customer as often as possible

- Enhanced marketing – The smart card can be an effective tool in direct marketing, and can be used for highly targeted marketing campaigns; for example, the short message service can be used to deliver highly customized messages to individual customers

- Incidental revenues – Space on the card face as well as on the chip's memory can be rented out to other companies, providing an additional source of revenues

- Cash flow improvement – The float on prepaid cards can be a tremendous boon to cash management

- Minimizes carrier involvement in hardware business – Since the card is the key access point for service, carriers do not have to sell the phone to the customer and can therefore minimize investment in hardware inventory

- Deeper channel – Smart cards have the potential of expanding the number of "doors" where mobile service can be obtained. Because telephones need not be sold by the establishments that sell smart cards, the number and types of channels will expand. A deeper channel means less power for existing channel members and potentially lower customer acquisition costs

The potential reduction in fraud alone is a significant financial benefit. It is well known that fraud in the GSM system is miniscule in comparison to that experienced in the U.S. cellular industry. The current authentication and verification procedures in GSM using smart cards make it extremely difficult and definitely unprofitable for hackers to try to crack the code.

Customer Issues

Deterrents to Customer Acceptance

Notwithstanding all the apparent benefits, there are some very basic deterrents to acceptance of smart cards in mobile telephony in the United States:

- Existing analog phone base – There are close to 20 million analog phones in the U.S. today. Customers would have to change these analog mobile phones to smart card-equipped phones. Unfortunately for both the mobile telephone and smart card manufacturers, and fortunately for the customer, these analog phones are quite sturdy and reliable. In fact, replacement cycles for these phones are actually lengthening, in part because of their high degree of reliability. To the average customer, there is little incentive to replace these perfectly working analog phones.

- Price – The addition of a smart card reader will raise the price of a mobile phone by between $30 to $75, according to smart card manufacturers. The difference between the older generation analog phones and the new smart card phones will be a deterrent. For the customer, it is not unlike having to decide between retaining an older generation 386-based personal computer or upgrading to a Pentium-based machine.

- Ease of use – A smart card phone call requires at least two steps. First, the smart card is inserted, then the call digits are dialed. With current analog phones, the call can be placed immediately. The imposition of the extra step of inserting card could become an inconvenience to users accustomed to current phone procedures. Indeed, many users today already feel inconvenienced by some carriers' requirement of keying in a PIN code to initiate a call on analog networks.

- Too many cards – Most customers already have too many charge cards, ATM cards, department store cards, gasoline cards, etc. In fact, many customers have already taken the step of consolidating their cards.

Keys to Customer Acceptance

It is clear that there are legitimate, practical concerns that may deter customer acceptance. The following factors are deemed necessary to overcome these deterrents and to smooth the way for customer acceptance:

- Make it a benefit to carry the card – not just a requirement. If mobile telephone smart cards are limited in their functionality to merely providing access to the network, then their perceived value to the customer will be diminished. Electronic ID badges, for example, provide physical access to office buildings and rooms; yet they are often left behind by their owners. These badges are a requirement for access, but they provide no real benefit to the user.

- Give the customer real value in carrying and using the card. The customer must derive real value, particularly financial value, in using the smart card. Airline frequent flyer programs use tie-ups with credit card companies that generate points for customers every time the card is used. These points represent stored value that can be redeemed for tangible financial benefits, i.e., free travel.

- Make carrying the card a habit. Link it to everyday habits and routines. The use of the smart card must be linked with day-to-day routines such as buying gas, paying for groceries, or buying subway tokens. Using a smart card on a mobile telephone can be looked at as purchasing time on the network, not unlike buying a theatre ticket or paying for a hotel room. Why not link the telephone calling process as closely as possible with these everyday routines?

Some Implementation Scenarios

Given the foregoing discussion, it would not be unreasonable to say that if mobile telephony cards are packaged with other applications, their chances of success will be greatly enhanced. The ideal scenario would be to launch multi-application smart cards at the same time as digital phones are launched. However, this would mean having a fully developed smart card spec embedded in the various digital access standards and also a shared set of standards with other industries. It is a fact that most U.S. cellular companies today are only in the very early stages of investigating the potential applications of smart cards. This is obvious in the fledgling state of development of the basic smart card specs in both the TDMA (T1-P1) and CDMA digital standards. In both cases, a stable set of standards for smart cards has yet to emerge. The stark reality is that CDMA or TDMA phones with smart card readers will not be manufactured in any significant quantities until these specs are developed. In fact, some carriers are reportedly investigating the possibility of testing smart cards in the current analog AMPS environment in order to gain some marketing intelligence on their likely acceptance even before the smart card specs are agreed to and ratified by the various standards bodies.

GSM SIM Card

There is, of course, the distinct possibility of carriers launching a variant of GSM, and consequently smart cards, using the new 2 GHz band. Some carriers have in fact publicly expressed interest in GSM as a digital platform for PCS, citing its availability, its lead in terms of experience over TDMA and CDMA, its inherent interoperability, and perhaps most importantly, its rich service features. Some cable TV consortiums vying for broadband PCS licenses have also expressed interest in GSM.

Thus, the GSM SIM card may well be the first implementation of smart cards in the U.S. mobile telephony industry. Indeed, both Ameritech and the North American Cellular Network (NACN) have announced separate initiatives that will allow customers to use SIM cards when they travel outside the United States, thereby extending their ability to roam into any country in the world that is covered by a GSM network.

Smart Card for High-End Customers

It is difficult to forecast how and when carriers that adopt CDMA or TDMA will integrate smart cards in their operations. However, it is probably safe to say that smart cards will likely be offered as an option, much like advanced auxiliary hardware. First of all, the additional cost of the smart card reader and the smart cards will add to the costs of a digital phone, whether CDMA or TDMA based. These phones will therefore not be priced within the reach of average users, at least not initially.

The likely initial targets will be high-end users. Every carrier has a key segment – the proverbial 20 percent of customers that accounts for 80 percent of revenues – that must be retained at all costs. This segment currently represents the largest source of value for the carriers and could be the prime targets for smart card phones.

Some carriers are already implementing joint affinity card programs with financial institutions, using the older generation magnetic stripe cards. For example, NYNEX Mobile Communications Company's MobilePerks card is basically a Visa card and is used like many airline frequent flyer cards. The logical extension of this card is to add the smart chip. This would open up a world of services that would not be feasible with current magnetic stripe technology. The card could in fact still use the magnetic stripe in the back for credit card and ATM use. The card could also be extremely personalized, or "mass customized," through the use of digital color photos and by adding personal data to the card such as frequently called numbers, emergency numbers, and key medical data.

Carriers can also opt to adopt multi-application cards such as that of Mercury One-2-One and Barclays Bank. Currently, this card uses the smart chip to display a series of menu options on the mobile telephone's display. This allows a Barclays customer to be directly routed to an appropriate customer service agent for balance inquiries, authorization of credit limit increases, and confirmation of receipt of payment or review of transactions. The user can also charge the usage on Mercury One-2-One to their Barclays accounts.

Smart Card for Multiple Wireless Services

Research by smart card manufacturers has established that closely allied and complementary applications are best combined in a smart card. From trials, it was learned that the best multi-application programs are based on close applications access – inside the same general application : telecommunication. Today, SIM cards may support GSM and DCS1800 access, but also international phone access. Tomorrow, SIM cards will be able to support multi-subscription private and business usage – DECT, DCS1900 and Iridium.[4]

The ever-widening variety of wireless service choices in different regions of the spectrum will lead to a proliferation of different, incompatible access devices ranging from pagers to handsets to personal digital assistants. These devices can be integrated within the framework of a smart card. Thus, a customer can use a smart card with a GSM phone and insert the same smart card in a DCS 1800 phone.

Prepaid Smart Card for Low-Usage Customers

Another market segment that can be targeted would be light or occasional users. Carriers could issue prepaid smart cards with preset calling limits, say $50 or $100. After the customer uses the stored value in the card, he or she can "refill" the card by making an additional prepayment. By doing this, the customer establishes a spending threshold and the carrier eliminates the risk of nonpayment, which today is a considerable expense. The phones themselves could probably be leased to the customer to minimize the dollar outlay and also to limit the perceived risks of obsolescence. This type of smart card could be used for the 2-GHz PCS market which will likely have many of these light users.

Other Advanced Applications

The serious financial impacts of cellular fraud will probably focus more attention on smart cards. Time and again, supposedly hack-proof systems have been hacked and cracked. Smart card manufacturers and service providers will therefore have to stay one step ahead of hackers and fraudsters in terms of technology. One of the intriguing technological enhancement possibilities is the use of neural network technology to provide biometric authentication of mobile telephone users. Southampton-based Neural Computer Sciences is developing a version of its neural network artificial intelligence system that can be embedded in a smart card chip. The Advanced RISC Machines Ltd. ARM 32-bit RISC processor is being incorporated into the chip to enable the portable intelligence device to perform algorithms to enable improved security and data compression. The new technology will enable the verification of the user's identity through unique characteristics such as fingerprints, signature recognition, voice, or facial characteristics. For example, a purchaser could speak a random word over the phone and in real time the terminal would run a signal processing algorithm on the speech to generate a profile, amounting to a sequence of a couple of hundred bytes of data, which is sent back in encrypted form to the user. The integrated network within the smart card then analyzes this profile to determine whether the caller is the rightful user of the card and enables the transaction be accepted or rejected.[5]

Other Issues and Conclusion

Inter-Industry Cooperation

As already mentioned, the probability of success of smart cards in mobile telephony will be greatly enhanced by integrating multiple applications on a card. This will ultimately require considerable cooperation between the different industries that could benefit from such multi-application cards. To this end, organizations such as the Smart Card Forum, an inter-industry association that seeks to advance the smart card agenda, will be extremely valuable. The mobile telephone industry must learn from the interminable TDMA vs. CDMA debates and proceed to accelerate the development of a smart card platform through closer cooperation.

Standards

As always, it is vital that a stable set of standards be developed for the digital mobile telephony platforms. Perhaps lessons can be learned from the rich experience of GSM. At present, smart cards are clearly not a high priority in the standards development bodies and advisory groups. It may be in the best interest of these bodies to look beyond the basic network access functions of smart cards and examine the potential marketing and financial benefits that can accrue to the industry as they flesh out the standards.

Reterminalization

The issue of reterminalization – or swapping out the existing analog infrastructure – is absolutely critical. In public telephony, only AT&T has phone booths that are equipped with smart card readers. Virtually all the ATMs and cash machines of the financial services companies are magnetic-strip-based. Unless there is a concomitant push by the financial services companies, in particular, to convert to a smart card infrastructure, the chances for other smart card applications will be somewhat diminished. The investment requirements to replace the existing plant and equipment will be quite substantial.

Awareness

Ultimately, the smart card community, which hopefully will shortly include the mobile telephone industry, will have to intensify its efforts to raise the awareness levels of both consumers and businesses regarding smart cards. The vast majority of people in the U.S. have never heard of, let alone used, a smart card. In fact, it is probably fair to say that the awareness levels regarding smart cards within the cellular community itself are probably not very much higher than the average consumer.

A concerted effort among the potential smart card industries has to be undertaken to generate awareness and excitement over the many potential uses of this technology. This effort should be aimed at both consumers and companies. Once this initial awareness hurdle is overcome, it can be expected that the smart card will finally be on its way becoming an integral element in the business fabric of the United States.

References

1 Alice LaPlante, "Cashless Society", *Information Week,* September 12, 1994.
2 *Business Wire,* July 19, 1994.
3 Jerome Svigals, "Smart Cards: The New Bank Cards", Macmillan Publishing Company, 1987.
4 Jerome Traisnel, "Maximizing the Marketing Potential of Smart Cards to Help Meet Your Customers' Needs", Paper for Schlumberger Transactions Systems.
5 "Smart Cards With Integral Neural Chips Being Developed by Neural Computer Sciences," *Computergram International,* May 18, 1994.

Domestic Spectrum Regulation in the United States

Charles M. Oliver, Esq.
Cohn and Marks

The focus of this essay is the regulation of radio spectra as it is done, or can be done, by national governments using the existing U.S. system of regulation as a point of departure. The U.S. Federal Communications Commission (FCC) regulates domestic use of radio spectra through a two-step process. First, it allocates spectra, i.e., it designates bands of spectra for specified categories of use. Second, it assigns radio licenses to specific users: companies, private individuals, state and local governments, and so forth. The National Telecommunications and Information Administration (NTIA), which manages the federal government's use of radio spectra, operates in much the same manner. Since many bands are shared among federal and non-federal users, the two agencies must coordinate with one another.

Both the FCC and NTIA must operate within the constraints of international agreements on spectrum regulation. For most services, this means that the United States must have close and continuing conversations with Canada and Mexico for radio operations in border regions. Having completed such discussions, the United States has broad discretion to deviate from allocation tables laid down by the International Telecommunications Union (ITU), provided that it does not create interference with any other countries that follow the ITU tables.

Assigning Licenses

Traditionally, the FCC had one means of resolving mutually exclusive applications for radio licenses: comparative hearings. In 1981, Congress authorized the Commission to use lotteries. In 1993, Congress authorized the use of auctions.

Copyright 1994 by Cohn and Marks. This article is printed with permission of the copyright holder.

Comparative Hearings

In principle, comparative hearings were supposed to select the best-qualified and best-intentioned applicants. Sometimes they did. On other occasions, they assumed the rhetorical tone of parole board hearings. Even when participants avoided unctuous and unprovable claims of good character and beneficent purpose, outcomes were often decided on the basis of distinctions that bordered on irrelevance. One cellular telephone applicant bested a competitor by promising to deploy an additional minicell in a place with no roads and no inhabitants.

Trees were often the biggest losers in comparative hearings. When the FCC announced its intention to convene comparative hearings for the first round of cellular telephone applications, the agency received so much paper that it had to call in a structural engineer to evaluate the ability of the commission's headquarters building to support the legal verbiage and consultants' reports that accompanied the applications. The engineer advised the agency to distribute the applications away from the corner of the building where they had initially been shelved, lest the structure fall down.

Given the opportunities for corruption afforded by comparative hearings, it is remarkable that they have been accompanied by few public scandals. During the Eisenhower administration, an FCC commissioner was accused of accepting a bribe for his vote on a VHF television license but escaped punishment by arguing that he was too stupid to know that he was being bribed. The public – and future potential clients – believed him. He was later found dead in a cheap Miami flophouse with all the money to his name stacked in neat piles of coins on a bedside table. This embarrassment probably served as a warning to his successors. (For those who may be incredulous at the level of physical detail in this account, it is fully documented in an official report by the U.S. Senate Commerce Committee.)[1]

Lotteries

Congress eventually concluded that comparative hearings are not always the best means to choose among competing applicants and authorized the use of lotteries for non-broadcast license proceedings. The random selection process was faster than comparative hearings and reduced appearances of favoritism, but it produced anomalous results. Lucky lottery winners were rarely qualified to build and operate complex businesses like cellular telephone franchises. After holding their licenses for whatever minimum period was required by the Commission, they would usually flip the licenses to someone else, typically a big phone company with enough money to outbid any other interested parties. A group of partners who won a cellular license for Cape Cod later benefitted from a bidding war between the cellular licensees for Boston and Providence and walked away with $40 million after building nothing.

The Cape Cod episode was an unusual case, but the total value of spectra given away by lotteries amounted to real money. In 1992, NTIA estimated that the market value of licenses given away for cellular telephone licenses alone, most of them in lotteries, amounted to between $40 billion and $80 billion, depending upon what appraisal method one used. Either figure, it noted, was more than the U.S.-Government spent to defeat Iraq in Operation Desert Storm.

The initial response among Congressmen with oversight responsibility was to urge the FCC to strengthen its so-called anti-trafficking rules to prohibit new licensees from selling out too soon. It quickly became apparent that this remedy would create an even bigger problem: unqualified and underfinanced lottery winners sitting on their licenses and engaging in the minimum amount of construction required until the waiting period expired.

Auctions

Despite the proliferating absurdities, Congress resisted auction proposals for 12 years after authorizing lotteries, on the ground that auctions would grant licenses to deep pockets. Auctions were finally authorized in 1993 as part of a Budget Reconciliation Act, when an incoming administration needed money and, for the first time in recent memory, was of the same party as the Congressional leadership. Since then, the FCC has raised more than $7 billion through spectrum auctions.

Besides raising money for the public treasury, auctions are faster than lotteries because they can put radio licenses immediately into the hands of the companies that are most likely to develop them and bring service to consumers. The FCC has optimized the process within the constraints of existing law by arranging simultaneous auctions for multiple, potentially interdependent licenses.

In other respects, the FCC's auction procedures have fallen far short of their potential for improving consumer welfare. The essence of the problem is that while the Commission's auctions have facilitated license assignments, they have done little or nothing to improve the efficiency of spectrum allocations.

Allocating Spectrum Bands

Many of the people involved in spectrum regulation have colorful charts mounted on their office walls showing the hundreds of bands allocated to various narrowly defined service categories. This system would be relatively harmless if the amount of spectra allocated to each service bore any rational relationship to the amount of spectra needed. The problem is that it does not and cannot bear such a relationship.

The Existing Allocation Process

To appreciate the level of detail practiced by the FCC in its regulation of spectrum allocations, imagine that a federal agency had been put in charge of real estate zoning throughout the United States and that it had referred to headings in Yellow Pages telephone directories for guidance on the establishment of appropriate zoning categories. Certain parcels of land would be earmarked for dry cleaning establishments, for example. If space in a commercial building were vacated by a florist shop and you wanted to use it for a book store, you would have to ask the agency to change its allocation table or issue a waiver.

This kind of procedure applied to spectrum regulation has engendered an understandable degree of frustration among policy analysts to the extent that some of them have called for the complete elimination of spectrum allocations and total reliance on market processes. Others have called for the abolition of the FCC. Either of those remedies would be overkill, but the situation does warrant consideration of significant changes.

As senior policy adviser to the NTIA Administrator, I once asked the agency's spectrum engineers how many allocation categories the FCC would need if its only policy objective were to prevent electrical interference. The answer was "about six." From an engineering standpoint, the FCC could establish six broadly defined spectrum categories. Within those allocations, it could conduct auctions and tell winning licensees how much electrical energy they could discharge in specified frequency bands and geographic areas. Subject to those constraints, licensees could do whatever they wanted within broadly defined service definitions. Incumbent licensees could be grandfathered but would be liberated either to sell out or to provide different services to themselves or others. The purposes for which spectra were used would depend primarily upon market demand, changing technology, and licensees individual needs and circumstances.

The FCC implicitly recognized the advantages of such an approach when it established ground rules for the so-called personal communications service (PCS). Frequencies with the PCS moniker can be used for almost anything the licensee wants, other than free broadcasting.

The FCC rarely provided such service flexibility in earlier allocation decisions. Part of the explanation is technological. Before computers became inexpensive and easy to use, it was simpler for spectrum engineers to coordinate like users than to coordinate disparate kinds of uses within a given frequency band. Pre-1990s radios had limited means to screen out unwanted signals, unlike the smart transceivers being developed for the PCS market, some of which incorporate technologies developed by defense contractors to combat purposeful enemy jamming. The FCC also had political reasons to avoid service flexibility: many interest groups believed, and still believe, that they could obtain more spectra by investing in lobbyists than by bidding for spectra in a competitive market.

Some of the people who oppose simplification of the allocation table are motivated by genuine public interest concerns. If television stations were forced to bid against mobile radio operators for spectra, most of them would go off the air. Many people believe that free over-the-air broadcasting produces public benefits that exceed the money that broadcasters can collect from advertisers. This view is especially strong among broadcasters themselves. Amateur radio hobbyists are similarly convinced that they provide significant public benefits, even though they collect no money at all for their services. Ambulance drivers and policemen also raise arguments for special treatment. Some of these arguments make sense; in other cases, the proponents' needs would be better met with subsidies in cash instead of implicit spectrum subsidies.

In analyzing the political dynamics, it is important to recognize that the people who lobby spectrum allocation matters do not necessarily approach the Commission on bended knee; in many cases, they have persuaded the agency to delegate coordination within designated bands to industry advisory groups. The Commission rubber-stamps many of the decisions made by such groups and is grateful to have off-loaded a significant amount of work. The designated representatives of the advisory committees in turn take satisfaction in their contribution to the public welfare and have occasionally amassed a significant amount of power. Many of these individuals would oppose a restructuring of the allocation process that took them out of the loop.

Proposals to Improve the Allocation Process

Given the Commission's sensitivity to spectrum coordinators and other sources of political pressure, the agency is unlikely to adopt sweeping proposals for reform in the short run. For the time being, the best we can realistically hope for is a gradual reduction in the number of allocation categories and a concomitant broadening of the degrees of freedom permitted within those allocations. Fortunately, progress on spectrum coordination issues is not an all-or-nothing proposition. Significant improvements can occur through incremental changes, even while the Commission and its critics continue to debate arguments for fundamental reform.

Although spectrum regulators operate in a high-technology context, they can use many of the same political techniques that successful reformers have used for centuries. They should avoid the temptation to preach to the choir: the people who need to be won over are the unconverted. Instead of trumpeting the extent of change, they should emphasize continuity with past precedents. Concrete assurances should be provided to stakeholders who fear that incremental changes will lead to more dangerous departures down the road. Regulators should buy off politically influential opponents with targeted concessions but give away as little as possible. They should engage in well-timed saber rattling to intimidate opponents but should avoid stabbing anybody unnecessarily.

Congress, the FCC, and NTIA can apply those techniques as follows:

Stealth Deregulation
Before it was merged into the newly created Wireless Bureau, the FCC's Private Radio Bureau pursued an unstated policy of retaining an alphabet soup of narrowly defined allocation labels but allowed sharing among licensees in different categories. The Bureau cultivated a retrograde demeanor while pursuing a progressive deregulatory policy and gradually transformed the Specialized Mobile Radio (SMR) Service into the functional equivalent of cellular telephone service. The cellular telephone industry was enraged when it discovered what was happening and persuaded Congress to impose cellular-style regulation on SMR operators – but Congress did not roll back the *de facto* reallocation that had already been accomplished.

Property Rights
Today, many incumbent radio licensees get nervous when somebody proposes to offer them more freedom because they fear that freedom will be a preamble to expropriation. The licensees would face the prospect of change more confidently

if Congress allowed the Commission to grant property rights in spectra so that radio licensees would receive the same constitutional protections against uncompensated expropriation available to other forms of property. In that context, broader allocation categories would provide incumbent licensees with an opportunity to profit from change by attuning their services more closely to changing demands of technology and markets or selling out to others seeking to do so.

Bribery

Today, whenever the FCC conducts a spectrum auction, the Commission must return the proceeds to the U.S. Treasury. Congress should authorize the agency to keep some of the proceeds and use them to buy incumbents out, clearing bands congested with obsolete uses and making them available for refarming.

Warehousing

Today, whenever markets or technology change, innovators seeking to provide new kinds of radio service must buy out incumbent licensees or persuade the Commission to do so. There is rarely an adequate supply of unused spectra available in inventory, as there is for almost any other commodity. One of the reasons is that Congress has required the FCC to be vigilantly watchful against spectrum warehousing and take licenses away from anybody who is not making immediate use of them. Congress should allow private entrepreneurs to buy radio spectra and use their own judgment to decide when to put it to use or hold it in reserve for possible future needs.

Technological Fixes

One legitimate reason why the FCC relies on spectrum allocations is that it is easier to prevent interference among comparable uses than among disparate, commingled uses of spectrum, even though doing so inevitably results in under-utilization of a scarce asset. Either the Commission or private spectrum brokers should be encouraged to apply more computer hardware, software, and engineering analysis to spectrum coordination, using proceeds from the auction process.

Accounting for Government Use of Spectra

Some of the biggest spectrum hogs are federal agencies. All such agencies should be required to put their licenses into a spectrum bank and lease them back – for cash – at prevailing market prices. Those payments should be treated like any other expenditure, appearing on the agencies budgets and subject to regular Congressional review in the authorization and appropriations process. Congress could fund the entire process by selling spectra to private brokers – in effect, privatizing a resource that was nationalized in the 1920s.

More Bribery
Some of the most politically sensitive reallocations involve public safety users. Congress should authorize the FCC to pay hush money to some of these groups. Public safety organizations need to communicate with police cars and ambulances, but they might find ways to do so more efficiently if they were offered cash instead of spectra. They could use some of the cash to acquire spectra at market-clearing prices, with the rest going for other things they might need.

Spectrum Parks
Increased reliance on market processes need not preclude some continuing set-asides of spectra for public purposes, such as public broadcasting. There is a viable market in real estate even though government has set aside some lands for public parks and applies zoning regulations to property in private hands.

Talk About Abolishing the FCC, but Don't Do It
Eliminating the FCC, as some have proposed, would not really solve existing problems with the spectrum regulation process, but talking about eliminating it might facilitate the decision-making process. As Samuel Johnson said, "The prospect of being hanged in a fortnight has a wonderful ability to concentrate one's powers of attention."

The underlying premise of all these proposals is that markets are usually more efficient than government agencies in deciding how to use scarce resources but that market-based transactions between private actors do not necessarily take into account all the side effects that private actions can have on society as a whole. Some government intervention in the market can therefore be justified on the basis of economic theory.

These suggestions also reflect a philosophy of governance: that the most efficient solutions can sometimes be approached, but rarely achieved, and that good-hearted guile often achieves more than hard-headed sanctimony. This advice comes not from Machiavelli but from an earlier and even wiser commentator who said, "Lo, we are as sheep among wolves. Be ye therefore wise as serpents, and gentle as doves."

Reference

1 Senate Comm. on Commerce, 94th Cong., 2 D Sess., Appointments to the Regulatory Agencies 69 (Comm. Print 1976) Prepared by J.M. Graham and Victor H. Kramer)

Cellular/PCS: Complement or Substitute for Wireline Telephony?

Ray L. Hodges
Senior Consultant
Technology Futures, Inc.

Wireless telephone service continues to grow at astounding rates worldwide. In 1993, cellular subscribership increased 245 percent in South America, 71 percent in Australia and Asia, 63 percent in Western Europe, and 47 percent in North America.[1] With a base of over 20 million, growth is projected to continue at about 30 percent in the United States for the next several years. Presently, two out of three new phone numbers go to wireless customers.[2] Contrast these growth rates with 3.5 percent for wireline access, and it is easy to understand why so many companies are anxious to enter the market and compete for subscribership. This essay addresses the prospects for wireless access displacing wireline access. The scope of this essay is limited to a discussion of the likelihood of access line substitution even though the impact on wireline usage-based access revenues due to traffic diverted entirely to the cellular network is very significant.

The displacement of wireline access lines was forecast by Lenz and Vanston in 1993 and by Vanston and Rogers in a 1995 update.[3,4] *Table 1* shows the more recent forecasts of cellular/PCN subscribers and a baseline projection of access lines as well as the percent of cellular/PCN customers expected to forego wireline service. The last column shows the number of access lines lost to cellular/PCN that is implied by the forecast.

The projections of percent of wireline access lines lost were based on two limiting factors:

Reprinted with permission from New Telecom Quarterly, *1Q95. Copyright © 1995, Technology Futures, Inc.*

Year End	Wireline Access Lines (millions)	Cellular/PCN Subscribers (millions)	% of WL Access Lines Lost (Cellular/PCN Subscribers Only)	"Lost" Wireless Access Lines (millions)
1994	154	22	0%	0
1995	159	30	0%	0
1996	163	39	1%	.4
1997	168	49	2%	1.0
1998	173	60	4%	2.4
1999	179	73	8%	5.8
2000	184	86	13%	11.2
2001	189	101	19%	19.1
2002	195	115	26%	29.9
2003	201	130	33%	42.7
2004	207	141	41%	59.0
2005	213	158	49%	77.6
2006	220	172	57%	98.2
2007	226	186	64%	118.8
2008	233	199	70%	139.2
2009	240	211	76%	160.7
2010	247	223	81%	181.0

Table 1 *Cellular/PCN subscriber forecasts*

- Wireless is not a perfect substitute for wireline telephones.
- Cellular prices compared with wireline service remain relatively high.

Also, the forecasts did not account for potential price changes by wireline carriers in response to competition or new wireline services such as ISDN that would tend to counteract the abandonment of wireline service.

Based on the forecasts, 1 percent in 1996 is approximately 240,000 access lines lost, 19 percent in 2001 is approximately 19 million, and 81 percent translates into 181 million access lines lost by 2010. To obtain these numbers, however, significant improvements in both service and pricing must be made since there is little evidence of direct substitution to date. We will review the outlook for wireline quality service and competitive prices.

Complement or Substitute Service?

Cellular service is primarily valued for its portability, and customers have demonstrated that they are willing to pay a premium for the portability. The ability to originate and receive calls while in transit is an adjunct to the wireline network since the wireline network cannot perform this function. Cellular enables wireline customers to hold timely conversations with other cellular customers, which would not be possible over the wireline network. The services are therefore complementary. Cellular increases the value of wireline access and vice versa.

While wireline access cannot provide mobility/portability, cellular can substitute for fixed wireline service. In this case, it is a competitive offering rather than a complementary one. Yet, to date, there has been very little direct substitution. This reflects both higher prices and the fact that cellular has been less than a perfect substitution for wireline in both service quality and versatility (full-time service independent of battery charge and the ability to accommodate extensions, answering machines, and other CPE). Also, with the present cellular duopoly enjoying rapid growth, there has been little incentive for cellular companies to lure wireline customers with attractive pricing structures. This may all change dramatically in the near future.

First, digital technology, coupled with better coverage of the areas where people live and work instead of just where they travel, will greatly diminish the service quality disadvantage of cellular. At least one of the largest cellular companies is adding additional cell sites specifically to improve coverage in residential areas to position its network for highquality wireless residential service. Also, cellular products are now available that connect to a standard telephone jack (RJ11). The wireline service can be discontinued at the point of demarcation between the customer and the telephone company. Conventional customer premise equipment (CPE) – standard telephones, answering machines, fax machines, and computers – interfaces seamlessly with the wireless service just as it did with the wireline service.

Alternative Wireless Access Products

Two products from Tellular Corporation, PhoneCell SX and Celdock, create the same dialtone interface as wireline service. PhoneCell SX is a direct replacement for the telephone company loop (the connection between the customer location and the telephone company central office), but Celdock does this and much more. It is an intelligent cellular interface built into a portable cellular phone charger base. It connects to a standard electrical outlet (110 V/AC) and telephone jack (RJ11) and provides complete emulation of a land line over the cellular network. It also offers the portability of a single number anywhere the customer travels. It is important to note that these products are different from others in that they require no special equipment or modification to the cellular network. The decision to forego wireline service is strictly up to the customer and not the cellular provider that may also have an interest in the wireline network. The customer simply buys the CPE, plugs it into a phone jack, and discontinues their wireline service.

Other manufacturers offering a variety of wireless loop (radio links directly from the customer location to the cellular network or the telephone company central office) products are Ericsson, Motorola, Qualcomm, InterDigital, and Hughes Network Systems.

The Ericsson RAS 1000 wireless loop also provides interfaces to other communications devices such as answering machines, fax machines, and modems. Motorola has developed a system specifically designed as a wireless alternative. The system uses conventional cellular frequencies and channel assignments, but requires only part of the regular cellular band because of antenna re-use patterns and narrowband analog technology (NAMPS). Only two bands of 5.5 MHz of spectrum are required for 50,000 subscribers when a six-sector antenna pattern and NAMPS technology are utilized.[6] The significance of this method is that it employs more efficient channel utilization and antenna technology to provide the additional capacity required for fixed location subscribers rather than reducing cellular capacity. The subscriber unit provides what appears to be conventional service, including dialtone, which interconnects with all standard CPE.

Hughes Network Systems GMH 2000 is being marketed for both conventional cellular and fixed local loop applications. Qualcomm is marketing a wireless local loop system using its code division multiple access (CDMA) digital technology.[7] This system transmits across a 1.25 MHz segment spectrum and provides 10 times the capacity of the competitive AMPS systems presently in service.[8,9] The Qualcomm System also supports a variety of standard wireline type interfaces. It also has the added capability of an integrated services digital network (ISDN) primary rate interface that can be easily integrated with existing digital end offices or switches.[10] Motorola personal communications systems has already introduced a new product for quick, cost-effective wireless loop services at PCS frequencies for the United States and international markets – even before PCS is operational. The new product, called Teledensity, is designed to be quickly installed and provide low-cost, high-capacity wireless voice and data service in urban and suburban environments.

Most of these wireless alternative products are enjoying greater success in the markets of developing nations, where 80 percent of the world's population has

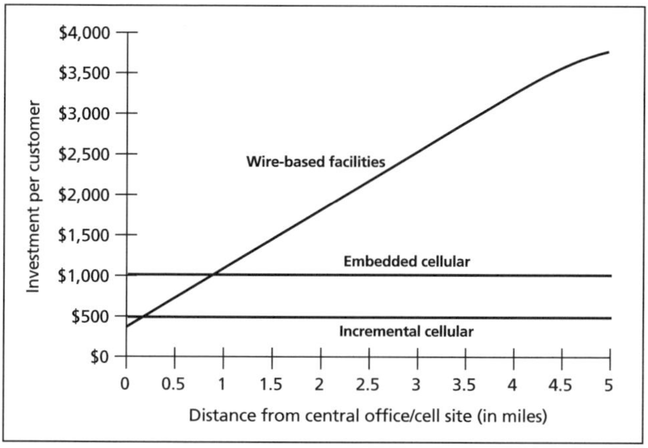

Figure 1 *Comparison of capital expenditure per customer (wireline versus wireless)*
Source: *Telephony,* June 20, 1994

less than 20 percent of the phones,[11] than in the United States where essentially everyone has wireline service available and at very low (often below costs) prices. Wireless is a quick and cost effective way to implement telephone service especially in less dense areas where longer loops are required. *Figure 1* shows that even at embedded investment levels, cellular is less costly at distances greater than one mile from the central office.

Cellular Cost Versus Price

Today, cellular customers are being added to existing systems for about $500 per customer, and the cost is expected to continue to decline while the cost of wireline facilities (especially the copper loop) are expected to increase. In addition, the cost of maintaining the copper-based network has been estimated at two times that of wireless systems.[12] On a pure costs basis, the time may be near when the total costs of wireless fall below the variable costs of wireline loops. Admittedly, regular wireline subscribers use their telephones about three times as much (in terms of busy hour traffic on the network) as do cellular customers. Usage is the primary driver of costs on the cellular network, since it requires no physical connections from the network to the customer. Thus, the costs of using cellular as a substitute for wireline service is about three times the current $500 per incremental cellular customer cost with today's AMPS technology. Even so, the capital costs would be less for cellular than wireline loops over 9,000 feet. Also, digital technology, both TDMA and CDMA, and advanced antenna technology will continue to reduce the costs for the additional traffic carrying capacity needed for fixed location subscribers.[13] I expect the costs of adding fixed wireline customers utilizing digital technology to actually be less than $500 per customer in the near future even with higher usage for fixed service. New technology coupled with the additional PCS spectrum soon to be operational will provide ample capacity at continually decreasing costs. The real question is whether there will be a glut of capacity resulting in airline style price wars. In this scenario, the copper loop will be as economically obsolete as a fleet of DC-3s.

Wireless Trials

GTE Mobilnet

The current situation is that technology is available and on the market that should make wireless telephone service an economical alternative, but will the customer accept it? GTE Mobilnet tested an offering in the Tampa Bay area in which customers were asked to give up their wireline service. In turn, they were provided a specially designed OKI handset and base equipment that functioned very much like the Tellular CelDock. Their monthly charges included flat rate usage from their home cell. When the handset was used outside the home cell, regular cellular usage charges applied. The base unit connected to the telephone jack and provided an interface for the normal household or office CPE. The results of the particular part of the trial that quantified customer acceptance levels have not yet been publicly released. However, based on discussions with a small number of participants, I believe that it was quite well accepted at the trial prices.

GTE has since introduced a somewhat similar offering, Tele-Go with one critical difference – it encourages continued use of wireline services. Tele-Go utilizes a specially designed OKI handset that also operates like a high-quality 900 MHz cordless phone. When the handset is used in conjunction with the enhanced cordless base station (ECB), calls are routed over the wireline network with no usage charges. When outside the range of the ECB, the regular cellular network is utilized and usage charges are applicable according to a special schedule with lower rates than cellular for a base area and higher rates for an extended premium area. This offering is designed to compete with future new PCS offerings rather than to lure wireline customers. However, it puts wireless handsets in more homes and lets customers become accustomed to using wireless in conjunction with primary telephone service. All that remains to be done to have customers cut the cord is for wireless providers to offer attractive prices.

Plan	Cost	Additional Usage
Digital Plan I—per cellular number, including 400 minutes of usage per month (peak or off-peak)	$18.00 per month	$.09 per minute over 400 minutes
Digital Plan II—per cellular number,including 600 minutes of usage per month (peak or off-peak)	$22.00 per month (limited to one Block)	$6.00 for additional block of 100 min. usage $.09 per minute over 600
Digital Plan III—per cellular number,including 800 minutes of usage per month (peak or off-peak)	$35.00 per month (limited to one Block)	$6.00 for additional block of 100 min. usage $.08 per minute over 800.
Digital Plan IV—per cellular number,including 1,200 minutes of usage per month (peak or off-peak)	$44.00 per month (limited to three Blocks)	$6.00 for each additional block of 100 min. usage $.07 per minute over 1,200

Table 2 *CyberTel Monthly Charges for Digital Cellular Service*
Source: Cybertel Cellular Application to Hawaii PUC, Effective June 24, 1994

CyberTel Cellular

CyberTel Cellular in Kauai, Hawaii, has recently received approval from the Hawaii Public Utility Commission to implement digital cellular rates of $18.00 per month that includes 400 minutes of use. Additional usage beyond 400 minutes is $.09 per minute. Tellular alternative access products are being promoted to directly challenge wireline service, much of which was wiped out by Hurricane Iniki. They also have a total of four plans for increasingly heavy users according to the schedule in *Table 2*.

These prices are being implemented as GTE Hawaii pursues substantial increases for wireline service in an effort to move local wireline service closer to actual costs. Presently, GTE's wireline rates are value-of-service priced where exchanges (islands) with the greatest calling scope in terms of customers accessible within the local calling area have the highest rates. This concept of pricing is known as value-of-service pricing and was almost universally used by telephone companies for residential service during the monopoly era. As an example, the residential rate on the heavily populated island of Oahu is $13.50 per month, while the rate on Lanai is only $9.25. This relationship is the exact opposite of the cost relationship on a per customer basis and is only sustainable in a monopoly environment. GTE is proposing to move to uniform rates throughout the islands. This will help reduce the inverted price/cost relationship, but still will not entirely eliminate it. This move will lessen the threat from fiber-based competitive access providers (CAPs) in the densely populated areas where wireline service may be priced considerably above costs. However, it will make the less densely populated

GTE Residence Individual Line Rates			
	Current	**Proposed**	**% Change**
Oahu	$13.50	$21.95	62.6
Hawaii	12.25	21.95	79.2
Maui	11.70	21.95	87.6
Kauai	11.70	21.95	87.6
Molokai	10.15	21.95	116.3
Lanai	9.25	21.95	137.3
GTE Business Individual Line Rates			
	Current	**Proposed**	**% Change**
Oahu	$33.35	$43.90	31.6
Hawaii	24.70	43.90	77.7
Maui	22.75	43.90	93.0
Kauai	22.75	43.90	93.0
Molokai	18.45	43.90	137.9
Lanai	16.25	43.90	170.2

Table 3 *GTE line rate comparison*

islands much more vulnerable to wireless access providers such as CyberTel with their new digital offerings. *Table 3* show GTE's present and proposed rates as filed with the Hawaii Public Utility Commission in Docket #7579.

This head-to-head competition should answer the question of whether people are truly ready to cut the cord. Cellular is being offered at rates that are comparable to wireline and with equipment that will emulate the service quality and features of wireline service. The Kauai case exemplifies what I believe will become the rule rather than the exception. As local exchange carriers de-average prices and attempt to move to cost-based pricing of local rates, overall prices must increase, with the less densely populated areas increasing the most. While this strategy will reduce the threat of some forms of competition, such as fiber-based CAPs, it will increase the vulnerability to wireless bypass.

If GTE is successful in obtaining its proposed rates, we will have a real test of customer willingness to abandon wireline service for wireless at comparable prices. This case will be followed closely and will be reflected in future forecasts since it should be a precursor of things to come. LECs will move prices upward to match actual costs to reduce competitive threats in other business segments. Wireless will reduce prices in response to increased wireless competition and technology driven costs decreases.

Conclusions

Customers' desire for mobility/portability and their willingness to pay for it has generally exceeded expectations. Their willingness to forego wireline service is just now being tested with new products that better emulate the service and features of wireline service. The substitution rate will now depend on marketing strategies and pricing policy of the wireless providers. Costs will favor wireless and, if current trends continue as expected, wireless will become much less costly than the embedded copper-based wireline service. As new PSC providers bring additional wireless capacity on-line, it is likely that prices will be driven to costs that will at some point undercut wireline service. Situations similar to the one on the island of Kauai will become more common over the next several years if wireline facilities are not upgraded to do more than provide voice and voicegrade data. In either case, the economic viability of the copper-based wireline distribution is limited. Wireline is generally priced below costs today but will likely rise. Wireless alternatives' costs are falling; therefore, prices will decrease. This leaves little opportunity to increase wireline prices.

The forecast of the percent of customers foregoing wireline for wireless service is quite low in the near term of 1995 through 1998. However, competition among wireless providers will heat up as PCS providers bring more capacity on-line, and wireless prices begin to approach costs. The projection of 81 percent lost by 2010 may appear high to many but reflects the logical conclusion of these trends. Of course, a clearer picture will emerge based on actual results of preliminary trial offerings, such as Tele-Go and CyberTel Kauai.

Notes

1 Telecommunications, *IFFF Spectrum* (January 1995).

2 CTIA Survey Marks 30% in Cellular Subscribers, *TR Wireless News,* vol 4, no. 19 (September 22,1994).

3 R. C. Lenz and L. K. Vanston, *Personal Communications: Perspectives, Forecasts, and Impacts* (Austin, TX: Technology Futures, Inc., 1994).

4 L. K. Vanston and C. Rogers, *Alternative Voice Communications: Competitiue Impacts on the Wireline Network* (Austin, TX: Technology Futures, Inc., 1995).

5 Narrowband-AMPS or N-AMPS is an interim FM analog solution bridging the gap between today's analog and tomorrow's digital systems. This technology splits the 30 kHz RF channel into three discrete 10 kHz channels, resulting in a potential three-fold increase in system capacity. Source: *IEEE Communications* (April

1993):78.

6 A Niche Market in the U.S.," *Telephony* June 20, 1994).

7 CDMA is a spread spectrum technology which digitally modulates signals from all channels in a broad spectrum. This may be accomplished by a method known as frequency hopping spread spectrum (FH/SS), or by adding a high-speed digital bit stream to the digital voice channel. This is known as direct sequence spread spectrum (DS/SS). As with TDMA, spread spectrum has several implementations including narrowband and broadband versions based on both FH/SS and DS/SS. Qualcomm is a front-runner with a narrowband (1.25 MHz) DS/SS system which utilizes code division multiplexing as its access methodology. *Source: The Mobile Revolution:* U.S. Cellular: From Analog to Digital.

8 Advanced mobile phone systems (AMPS) is the original analog 30 kHz RF channel cellular phone system. Source: H. Mine, Latin American Cellular Markets, *New Telecom Quarterly,* vol. 2, no. 4 (November 1994).

9 "A Niche Market in the U.S."

10 Ibid.

11 Ibid.

12 G. Calhoun, *Wireless Access and the Local Telephone Network* (Norwood, MA: Artech House, 1992).

13 TDMA is a technology where several voice channels are digitized and multiplexed using time division. The channels are then separated and sent to individual subscribers using a multiple access method in which subchannels contain packetized addressing data. There are several iterations of TDMA, including North America's interim standard (IS) 54, Europe's global system for Mobile (GSM) communication, and a version developed by InterDigital Corporation. These differ in circuits per channel, timing, and channel width, but they all use a similar access methodology. The U.S. TDMA system requires multiplexing digital voice circuits within the 30 kHz channel. Initially, IS-54 will assign two slots (13 Kb/s) to each user, deriving three circuits per band from a voice coder standard of 8 Kb/s per circuit. TDMA's long-term objective is to halve this coding rate enabling six 6.5 Kb/s slots per channel to be multiplexed (in 40 millisecond frames). Extended-TDMA (E-TDMA) increases the number of mobile users who can share a given number of circuits by coding voice signals. Multiple access is achieved by adding a packet control subchannel to the received multiplexed information. This provides addressing within the voice packets. The request and assignment process is very rapid and causes no perceptible impact on quality. *Source: The Mobile Revolution:* U.S. Cellular: From Analog to Digital.

Section II

Wireless Technology and Applications

Network Issues for Personal Communications Services

Richard Chandler
Senior Vice President
Hatfield Associates

Introduction

Much has been written about the radio-related aspects of personal communication services – modulation, channel sharing mechanisms, voice coding techniques, propagation issues, signal processing, and so forth. These considerations are obviously vital to the proper design and functioning of commercial PCS systems. Beyond these critical air interface concerns, however, there are many equally vital network-level issues that must be addressed for the commercial deployment of personal communications services. These systems will be network-based and will rely on complex network infrastructures to deliver services to end users. In many cases, PCS providers will develop their own network infrastructures to support their service offerings, while others may rely on established carriers, such as local exchange and interexchange companies (LECs and IXCs), and on new entrants, such as competitive access providers (CAPs), to provide network services. Some PCS operators will be extensions of some of these entities: IXCs, LECs, and CAPs have figured prominently in the strategic alliances that are forming to bid for broadband PCS licenses.

This essay discusses the network issues that apply to the development and deployment of PCS: interconnection with other networks, including LEC and IXC facilities, numbering plan concerns as they relate to the particular case of wireless terminals and itinerant subscribers, signaling network operation and functional needs, such as subscriber database management and associated communications, and other support services, such as operator services, directory assistance, and network management.

Numbering Plan and Corresponding Network Architecture Concerns

Early descriptions of the PCS concept invariably included what is sometimes known as portable numbering. This term, when used in this context,[1] applies to the ability of the switched network to deliver calls efficiently to PCS subscribers whether they are roaming or active on their home systems.

This feature obviously is independent of the radio technology in use and thus will apply to any wireless network, whether PCS, cellular, enhanced specialized mobile radio (ESMR), or any other radio service that supports interconnection with the public switched telephone network. The necessary network modifications must be made to the switched network and do not directly affect the fundamental architecture of a wireless network, PCS or otherwise.

PCS "telephone" numbers in the United States will comply with the North American numbering plan (NANP), just as cellular numbers (called mobile identification numbers, or MINs) follow the existing NANP format. Today's telephone numbers follow the format NPA NXX-XXXX, where the NPA (numbering plan area, colloquially the "area code") field identifies the geographic area in which the number is assigned, the NXX[2] code identifies the end office switching machine where the number logically resides, and the XXXX field identifies the line number within the NXX code.[3]

The NPA-NXX combination thus serves as a geographically fixed basis for routing calls to a subscriber. Although this works well for fixed subscriber terminals such as conventional telephone sets, it does not allow efficient routing of calls to wireless terminals, which are not geographically fixed.

Under the current numbering plan, numbers (MINs, or mobile identification numbers, in the case of cellular terminals) assigned to wireless terminals are indistinguishable from those assigned to conventional fixed terminals. A call placed to a wireless terminal thus cannot be recognized as such by the originating switch and hence will be routed to the geographic location specified by the NPA/NXX fields in the dialed number.[4] If the wireless terminal happens to be registered on its home system, the MSC completes the call to the terminal directly. If, however, the wireless terminal is elsewhere (and registered as such), the routing becomes considerably more complicated. Note that the call will always be routed first to the home MSC regardless of the current location of the wireless terminal, because the MIN is the basis for routing.

Current practice for delivering calls to roaming wireless subscribers in the absence of nongeographic numbers for wireless terminals is specified in Telecommunications Industry Association (TIA) standards.[5] When a wireless terminal registers in a system other than its home system, the serving MSC reports the registration to its visitor location register (VLR). If the VLR has no previous knowledge of the MIN, it notifies the terminal's home location register (HLR) of the registration (it routes the registration notification according to the NPA/NXX of the MIN). The HLR then records the current serving system and cancels any previous foreign registration. When a call subsequently arrives at the home MSC for the wireless terminal, the MSC requests the current location of the terminal from its HLR, which in turn routes information from the serving VLR, supplied in the form of a temporary local directory number (TLDN), a telephone number from the numbering space of the serving MSC. The home MSC then forwards the call over the switched network to the serving MSC.

This routing is inefficient because all calls must first be directed to the home MSC, which ultimately determines the routing to the desired terminal. If the originating office could recognize the dialed number as a nongeographic entity, it could immediately request routing from some specified database and then send the call directly to the MSC serving the wireless number without first routing to the home MSC.[6]

The specifics of this discussion apply to current cellular mobile services and may well apply to commercial PCS operations as well. There may be other implementations,[7] but the general procedure must involve early identification of a wireless terminal number and a database number translation to obtain routing to the terminal's current location.

Database and related signaling transport and routing services may be provided by a PCS provider's own signaling network and incorporated databases, by another carrier, such as an IXC or LEC, or by a third-party signaling services provider. The latter might be an existing entity or possibly one formed by a consortium or other cooperative arrangement among wireless services providers. There are numerous and often subtle and interrelated regulatory and technical issues underlying the

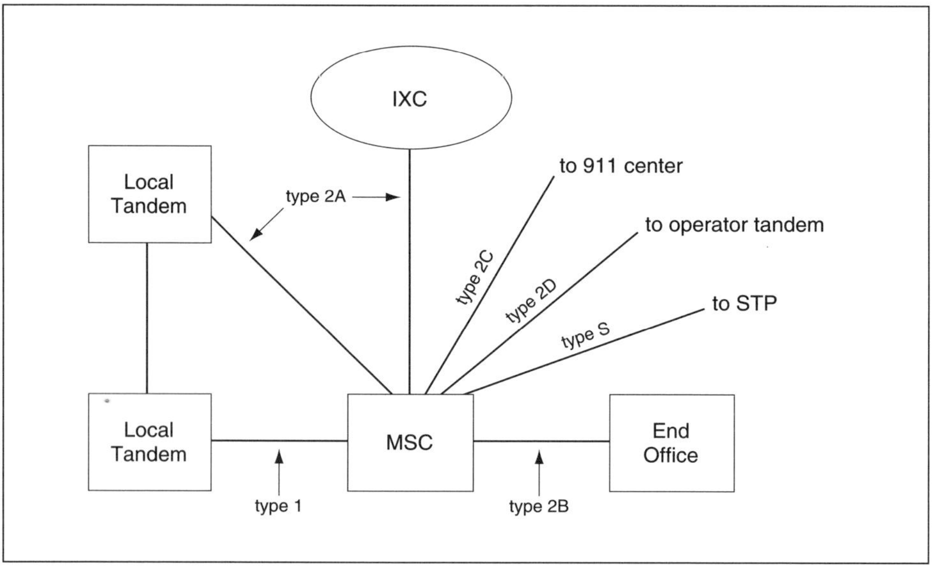

Figure 1 *Bellcore-specified interconnection types*

provision of signaling services to wireless and other local service providers by local exchange carriers.

Interconnection With Other Networks

Existing cellular system interconnection policies provide a useful model for the discussion of PCS interconnection. Cellular service is likewise network-based; cellular systems exist in parallel with local exchange networks to provide wireless mobile and portable telephone service. There are generally five types of interconnecting trunks specified by Bellcore, as shown in *Figure 1*.

The functions of each type are as follows:

Type 1 Sometimes called a "trunk with line treatment," this type is physically a trunk and uses E&M trunk signaling, but is treated like a line by the end office; it allows access to operator services, directory assistance, 911, and may be presubscribed to an interexchange carrier. Central office codes may be dedicated or shared with the serving end office.

Type 2A This is the most common interconnection type and connects the mobile switching center (MSC) directly with the local tandem or to an interexchange carrier; it does not allow operator services access, 911 access, or 411 directory assistance (although DA may be reached by 1-555-1212). It also requires that NXX codes be dedicated and reside in the MSC.

Type 2B A relatively rarely used type, a type 2B trunk supports calls only between the MSC and the end office supporting the trunk.

Type 2C This type supports calls to a 911 center; specific routing depends on local LEC policy.

Type 2D This supports connections with the LEC's operator tandem for access to operator services, including directory assistance.

Bellcore specifies the Signaling System 7 interconnection physical interface for wireless carriers.[8] There are few, if any, such arrangements today between cellular carriers and LEC networks, although they will undoubtedly prevail in the future. Bellcore identifies this as a type S interface and publishes a document describing services that may be supported by a type S interface.[9] Examples of these services follow:

- basic call setup and release using SS7 in conjunction with type 2A and 2B trunks
- passage of information of SS7 parameters for call forwarding service
- passage of messages to support transport of IS-41 transactions from the wireless service provider to another interconnecting SS7 network
- passage of messages to support CLASS features[10]
- passage of messages to support 800 service and message waiting indicators

Interexchange carrier interconnections typically are described as type 2A. If these are supported through an access tandem, LEC-specified signaling will apply. If there are direct interLATA connecting trunks between the resale switch and one or more IXCs, TW will need to determine the preferred signaling method from the IXCs . Generally, large IXCs interconnect with local networks using SS7, with the signaling interface specified once again by Bellcore.[11]

The latter types of interconnecting trunk – Types 2C and D, as well as Type S - have been recently defined by Bellcore. Without the Type 2C and D connections, a cellular carrier must engineer a number of Type 1 connections to allow its subscribers access to operator services (including 411 directory assistance) and 911 emergency services. The availability of these new types of interconnecting trunks, along with the definition of SS7 connections, will give the MSC a status in the local switched network hierarchy (note that local telephone networks, unlike interexhcange networks, remain hierarchical) more nearly resembling that of LEC end offices than was previously possible.

The Type S interface is a physical specification, and such a connection may be used in conjunction with any of the other interconnecting trunk types. Also, Type 2A trunks may support 911 and operator services connections with suitable specifications.[12]

PCS networks will have similar needs. There has been much discussion of the level of services that a PCS service provider might require from local exchange networks. Bellcore has released several publications dealing with the various ranges of services that PCS networks might be provided from LEC networks.

These discussions generally consider the interconnection and related requirements of full-fledged PCS networks separately from those of small PCS service providers, a class of PCS licensees postulated by some as "mom and pop" and other small entrepreneurial entities that presumably would own local PCS base stations, much as operators of coin telephone services operate. Larger entities arguably may need to obtain network services from another carrier (or else over-extended backhaul facilities connecting remote coverage areas to the entity's network components located in some urban or even regional center) in certain small markets where local switching is not economically justified.

Network Services

As noted in the beginning of this essay, personal communications services will be network-based. Unlike simple cordless telephones, PCS terminals will be the endpoints of fully developed wireless network structures. There is far more to the specification of a wireless network than the definition of the common air interface (CAI) between base station and subscriber radio equipment. Beyond the CAI, there are many other interfaces that must be specified and corresponding functions provided to construct a complete operating network. This section discusses these interfaces and required network services.

There have been several wireless network models proposed by various bodies to define all of the functional interfaces in a wireless network and between such networks and other networks, such as local exchange and interexchange carriers'

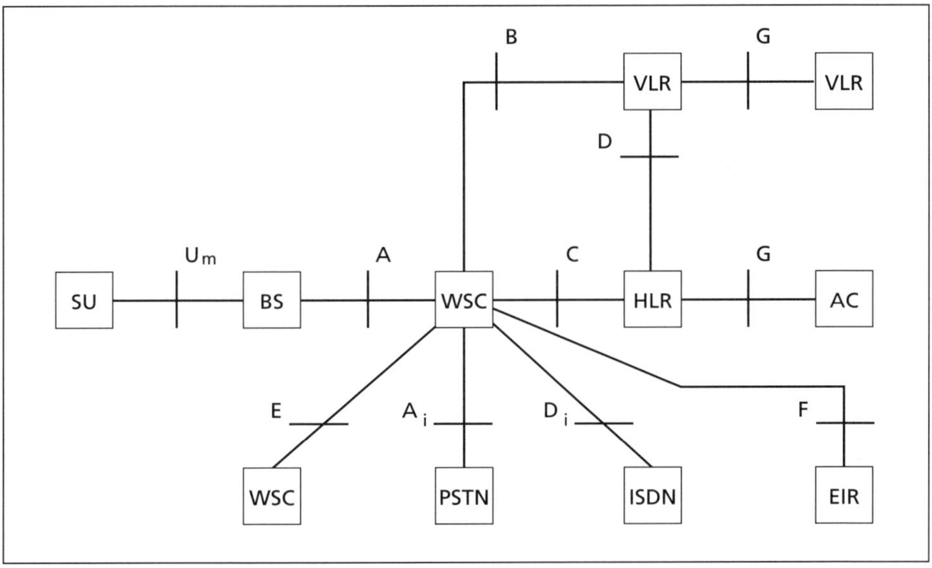

Figure 2 *Generic wireless network model (adapted from EIA/TIA IS-41.1B network reference model)*

SU=subscriber unit
BS=base station
WSC=wireless switching center
HLR=home location register
VLR=visitor location register
AC=authentication center
PSTN=public switched telephone network
ISDN=integrated services digital network
EIR=equipment identity register

facilities. These models are similar in many respects and a useful representative is the Network Reference Model developed by the Telecommunications Industry Association. *Figure 2* shows the components of this structure.

Where the model identifies interfaces and functional divisions of a network, standards must be developed to specify the information content and format that must flow across each of the interfaces. The standards are necessary to allow manufacturers to develop compatible equipment.

The lack of standards can constrain carriers in their choice of equipment. Current AMPS (advanced mobile phone service, the U.S. analog cellular standard) systems do not, for example, have specified a standard interface between the MSC and the cell site controllers (this corresponds to the A interface in *Figure 2*). The fact that these interfaces are nonstandard means that a cellular carrier must obtain infrastructure equipment from a single supplier. An AT&T MSC, for example, will not work with a Motorola radio controller. This limits the carrier's options in dealing with various manufacturers for network components because equipment cannot be "mixed and matched."

There are systems for which more complete sets of interface standards have been accepted. The GSM (global system for mobile communications) system, the European-developed digital cellular architecture, specifies most of the interfaces shown in *Figure 2*, as well as some that are not shown (including one – the Abis interface – between the base station controller and the base station radios) in the figure. Such structures can be readily adapted to different applications. GSM, for example, has been adopted by Motorola for its MIRS (Motorola integrated radio system), the network infrastructure used by Nextel in its cellular-like digital mobile radio system it is deploying in several U.S. cities, including Los Angeles, Denver, Oakland, and Seattle. This flexible architecture allows a carrier to use some or all of the standard interfaces as the system requirements dictate. In the MIRS technology, the common air interface is not the GSM standard but is instead a proprietary interface developed by Motorola for the specific needs of Nextel in accordance with the FCC rules governing the spectrum for which Nextel is licensed. The GSM structure has also been adopted for some proposed PCS network standards.

Some of the interfaces in the model may not be standards, at least as the term applies in strict usage. The A_i interface, for example, in *Figure 2* lies between a wireless switch and the public switched network. Note that there may be several types of interface under this classification, including the types listed previously in the interconnection section of this paper. In the U.S., these interfaces are not standardized by formal standards-making bodies such as the TIA, but they are specified by Bellcore requirements documents. These requirements are adhered to by telephone switching and transmission equipment manufacturers and by carriers.

Beyond the interface specifications and standards required to assemble network infrastructures, PCS service providers will require a range of support functions in order for them to offer a complete range of subscriber services. These include the following:

- operator services
- directory assistance
- network monitoring and maintenance
- network administration and provisioning
- signaling services (including call control functions as well as non-call related services such as those supported by database machines for number translation to support transparent roaming and portable numbering)

These services may be provided by the PCS service provider, by existing carriers such as local exchange and interexchange carriers, possibly competitive access providers (if they have suitable network support elements available), or by third-party sources, possibly including consortia of service providers, much as a consortium of independent telephone companies operates the independent telecommunications network, a vendor of signaling services to independent telephone companies as well as other carriers, including many cellular carriers.

Interexchange carriers have a particularly powerful motivation to become involved in PCS: they pay significant percentages – as much as 45 percent – of their gross revenues to local exchange carriers in the form of access charges that reimburse the LECs for carrying local traffic to and from the interexchange carriers' local points of interconnection. Cable television operators have developed extensive networks that make broadband cable service available to more than 90 percent of U.S. homes (about 62 percent of U.S. households subscribe to cable service). Although significant additions must be made to the existing cable infrastructures to allow operators to offer PCS, the inherently wide bandwidths of the coaxial cable plant makes it attractive for offering services beyond the transmission of one-way video entertainment. Many large cable operators are supplementing their systems with optical fiber to improve system reliability (by reducing the number of active components in the network) as well as to improve signal quality delivered to their subscribers. An obvious side benefit of the fiber enhancement is generally an increase in available system bandwidth, possibly allowing the carriage of two-way digital voice signals to and from PCS base stations located along the cable network.

The large facilities-based interexchange carriers operate highly developed switched networks that include advanced signaling systems; these functions might readily be modified to provide local PCS switching and signaling services in the early stages of PCS commercial deployment. As PCS penetration and hence subscriber traffic increases, local switching will probably need to be added to offload PCS traffic from the interexchange tandem switching machines. The interexchange carriers might well offer these services as third-party service providers, or they might use them for their own local wireless operations.

Cable television operators are often associated with CAPs; the two entities might cooperate in certain markets to use the cable network for PCS base station backhaul connections to a local switch operated by the CAP (if it is so equipped).[13] The CAP may also have access to a common channel signaling network that could support call-related as well as non-call related signaling transactions required for the local PCS function.

Conclusion

Where cellular mobile telephone systems, in their early years of operation, were essentially "islands" that existed more or less in isolation from other systems, PCS systems will, from the inception of commercial service, necessarily offer complete network-based functions. Some of these functions – automatic roaming and intersystem handoff – were added over time to cellular networks as the need for advanced roaming services evolved for cellular subscribers. PCS operators, however, will not be able to evolve such services and begin operations as isolated local services providers. The service expectations of an increasingly sophisticated subscriber population will be such that transparent roaming, including efficient call delivery, plus a full complement of network services that wireline subscribers have enjoyed for decades must be in place when the service is initially offered.

Notes

1 There is another level of portable numbering that applies to the numbering flexibility that allows a subscriber to change carriers and retain a telephone number. In this sense, 800 numbers are now portable, and there are efforts to make portable numbers available at the local level as well to help encourage local competition.

2 In the NANP nomenclature, N refers to any decimal digit 2 through 9, and X means any digit 0 through 9. The NXXX code, or central office code, as it is often called, can thus assume any value over the range 200-999, a total of 800 numbers. Some of these, such as 555, 976, and others, are reserved for specific services and internal telephone company functions. The NPA format has been NO/1X (meaning that the second digit can be either a 0 or a 1) but will change to NXX beginning January 1, 1995.

3 There will inevitably be different types of numbers defined within the evolving NANP, including those that are not geographically fixed, as are existing numbers.

4 The *Local Exchange Routing Guide*, published by Bellcore's Traffic Routing Administration, contains the V and H (geographical) coordinates of all assigned NPA/NXX combinations in the United States, along with other information, such as addresses and capabilities of the switches in which the NXX codes reside. These switches include MSCs and CAP switches as well as LEC offices.

5 EIA/TIA/IS-41.3-B, *Cellular Radiotelecommunications Intersystem Operations: Automatic Roaming*, Washington, D.C.: Electronic Industries Association, December 1991. The automatic roaming procedures outlined in this discussion apply only if the wireless user has subscribed to the corresponding roaming service, typically optional.

6 Note that the geographically based TLDN is still required to serve as the call routing basis.

7 See, for example, *Personal Communications Services (PCS) Network Access Services to PCS Providers*, Bell Communications Research, Special Report SR-TSV-002459, Issue 1, October, 1992.

8 *Common Channel Signaling Network Interface Specification (CCSNIS) Supporting Network Interconnection, Message Transfer Part (MTP), and Integrated Services Digital Network User Part (ISDNUP)*, Bell Communications Research, TR-TSV-000905, Issue 2, October, 1993. See section 6.

9 *Common Channel Signaling (CCS)Network Interface Specification Supporting Wireless Services Providers (WSPs)*, Bell Communications Research, TA-NWT-001434, Issue 1, September, 1993.

10 Cellular carriers generally plan to offer CLASS features (those based on the availability of the calling number at the terminating office) to cellular subscribers, and certain cellular standards specify the mobile telephone display requirements to support such features as calling number identification. Note that SS7 is required to enable the CLASS functions.

11 *Switching System Requirements for Interexchange Carrier Interconnection Using the Integrated Services Digital Network User Part (ISDNUP)*, Bellcore Technical Reference TR-NWT-000394, Issue 3, August, 1991. The ISDN is misleading: this specification is for non-ISDN call routing. The SS7 protocol structure places all call control functions, ISDN and non-ISDN alike, under the ISDN User Part. There is a separate publication for ISDN intraLATA call control (TR-444).

12 The SS7 interface specifications may be found in *CCS Network Interface Specification (CCSNIS) Supporting Wireless Services Providers*, Bell Communications Research, Generic Requirements GR-1434-CORE, Issue 1, April, 1994.

13 Note that CAP-operated switches are relatively rare because local switched services competition is essentially non-existent. This may change over time, with the result that there may be a number of CAPs equipped for local switching (and empowered to do so by state regulators) in the future.

Network Management Systems

Gisela Hertel
VEBA AG

Introduction

In the last few years, competition in mobile communications has lead to a rapid market growth in Europe. Western Europe today has 11 million cellular subscribers, compared to seven million a year ago.

In a country like Germany, all networks operate with national licenses and therefore cover a population of 80 million people. The German DCS 1800-network, operated by E-Plus Mobilfunk GmbH, went into commercial service in May 1994. It is designed for a capacity of several million subscribers and will have approximately 6000 cell sites when fully built.

The network management of such a large number of cellsites, in addition to several dozen switches, thousands of leased lines and microwave links, voice mail systems, and other special network nodes for several million subscribers, taxes the limits of existing network management systems. An additional challenge is the rapid growth of the system. The German GSM systems, for example, had 100,000 subscribers after the first six months and now after two years of commercial operation have 700,000 subscribers each.

Every month, hundreds of new network elements are taken into service, leading to significant changes in the frequency plans and configuration parameters. In addition, the GSM/PCS technology is still rapidly evolving, leading to software upgrades and introduction of new features every few months. Being responsible for network management in such a rapidly growing network can sometimes be compared to working in an intensive care unit.

Some standard network management features that are perfectly adequate in a normal size, technically stable cellular system, can cause severe operational problems as well as significant additional cost once the system exceeds a certain size. Some of the lessons learned in the operation of large-scale cellular systems in Europe may prove useful during the rollout of PCS systems in the United States.

The following sections briefly describe the network management functions:

- alarm handling
- fault management
- security management
- performance management
- configuration management
- subscriber administration
- charging administration

Their importance to the commercial success of a cellular network is discussed as well. Interworking issues between equipment of different manufacturers are addressed in the last section of this essay.

Functions of a Network Management System and Their Importance to the Commercial Success of a Telecommunications Network

Alarm Handling

If a network element – a base station, for example – develops a problem or goes out of service, an alarm is sent to a screen in the control room. Depending on the system, it is displayed graphically or printed as a line in an alarm list. The alarm has to be acknowledged in some way by the operator, and then disappears from the display.

Usually, alarms are grouped into classes according to severity. Critical or service-affecting alarms indicate a failure that will be noticeable to the customer. Failures of a redundant unit or minor aberrations are sorted into lower alarm classes. Although all commercially available network management systems have alarm handling functionality, the exact implementation can make a vast difference to the operation of the network.

Alarm lists may be sufficient for a small network, but beyond a certain number of network elements a graphical display becomes more efficient. The graphical functionality should include zoom capability to move from a total network overview to a particular area. It is also useful to have maps overlaid on the display. The zoom facility requires the use of several layers of maps, because the blowup of a national highway map will not be of any use to a technician trying to find his way to a faulty base station.

The functionality contained within the alarm list is often neglected during the vendor selection process. Unfortunately, some systems require the operator to click through hundreds or even thousands of lines to find a particular alarm. Search capability for certain types of faults, certain regions, certain network elements or types of equipment become extremely useful in larger networks. In some cases, alarm acknowledgment can take a long time to execute, slowing down network operations, especially in emergency situations.

Another important requirement for a network operator is flexibility in the alarm classification. A minor alarm in a particularly critical base station, at an airport for example, may have to be treated differently from the same defect in another base station. In some cases, special treatment is only necessary in a particular time window, such as during a trade fair. If a network management system does not offer flexible alarm classification, the selective observation of particularly critical network elements becomes labor intensive, or even impossible in a large network because relevant alarms have to be filtered from a large list of otherwise irrelevant information.

The above consideration shows how the implementation of such a simple network management function as alarm handling can affect the number of operational personnel necessary to run a network, as well as the quality of the service. To staff one operator terminal 24 hours a day, seven days a week, takes five employees. A user-friendly, intelligent alarm handling functionality determines how fast a fault can be located and service to the customer restored. It also influences the commercial viability of a network by determining the number of operators and terminals necessary in the network management center.

Fault Management

Another important functionality of a network management system is the fault management. After identifying and acknowledging an alarm, the operator accesses the relevant network element remotely and tries to diagnose and hopefully fix the problem. If a remote repair is not possible, a service technician has to be dispatched to the site.

Again, all network management systems offer fault handling capability, but the available functionality varies widely. Useful is a detailed configuration of all equipment installed at a particular location that comes up on the screen automatically when the operator tries to locate the fault. Again, a layered information structure, where the operator can successively go down in detail, is usually required. Fault detection becomes more effective by good help screens or by automatically suggested repair procedures.

Once the defective equipment is located, the fault management, in some systems, can automatically generate a work order for the technician that contains all relevant information (including items like who has the key to the property, which seem trivial at first sight, but save a lot of time during everyday operation).

In some cases, the network management system is connected to an inventory management program. The location of a required spare part can be determined, the part can be shipped to the technician, and a new part can automatically be reordered.

Security Management

The security management of a cellular system covers two different aspects. On the one hand, the network-specific security features, such as authentication, encryption and ciphering keys, and algorithms need to be administrated. More important for the day-to-day operation, however, are the security aspects inherent in any communications network.

As it is important to remotely access a large number of network elements, either from the network management center or from portable PCs carried by the technicians, access control mechanisms and logging and reporting capabilities of security-relevant events and transactions become critical for controlling intrusions into the network. Again, nonoptimal security features can cause severe commercial damage.

Performance Management

Each network element is capable of generating a large number of statistical data. Collecting and compiling this data in an efficient and meaningful way is one of the most important functions of a network management system. Due to the huge investment in infrastructure, network elements need to be dimensioned in such a way that congestion is avoided but excess capacity is minimized. Network planning relies on accurate and meaningful performance data to accomplish this goal.

Performance reports fall into two categories: For regular operation, routine reports of basic network data, like blocked or dropped calls or average availability of network components, need to be generated on a daily or weekly basis. These standard reports should take no effort on the part of the operator and ideally be created at the touch of a single button or even fully automatically.

In large networks, the data transmission rate can become a problem. Even if generating a report does not require operator effort, the collection of data from network elements can congest the data network or interfere with the transmission of alarms or fault management information.

The second category of performance report covers special situations or design studies. In these cases, much more detailed statistical information on certain network elements is required. Flexibility of the performance data collection and presentation, and the ease of designing customized reports then becomes an issue. In many cases, it is necessary to export data to an external information system. The use of a standard database product and the availability of a flexible external interface are an important requirement for any network management system.

Configuration Management

Changes in the network configuration, hardware upgrades, software updates or additions, and reconfigurations of network elements must be performed without violating network integrity. The configuration parameters needed by each network element have to be downloaded from the network management system. Configuration management also covers changes as a reaction to failures or congestion.

Again, in large systems, capacity of the configuration management module becomes an issue because the downloading of new frequency plans or network software can take a long time. Intelligent solutions for change, backup, and recording of parameters, and built-in plausibility and consistency checks reduce the risk of failures due to incorrect configuration data. Of particular importance is the capability to do efficient software downloads and to store and manage different software versions in the network elements in case a fall-back solution is necessary.

Subscriber Administration

If a new subscriber is activated, or an existing subscriber changes his service profile, the relevant subscriber data needs to be downloaded to the relevant registers. In the case of GSM/PCN, these are the HLR, AuC, and EIR. Although subscriber administration is considered a network management function, it is usually executed by a billing system.

In special cases, however, during an HLR split or a major switch failure, for example, large amounts of existing customer data need to be reloaded or transferred between network elements. These transactions are normally performed by the network management system and require good functionality and high throughput of the subscriber management module of a network management system.

Charging Administration

After a call is made, call record information is generated at the switch and transferred to the billing system either by tape or by electronic means. The data generation and transfer are managed either by the network management center or the billing system. In any case, the data network has high security and availability requirements because the loss of billing data for even short time periods has tremendous financial consequences.

Network Management Systems in a Multivendor Environment

The already complex function of a network management system becomes even more complicated if the network elements are supplied by two or more different system vendors. Unless a network is divided into clearly separated single-vendor regions, some degree of integration between the two network management systems is unavoidable. It is difficult for an operator to locate a fault if alarms appear on two different screens, each displaying only the subnetwork supplied by a single vendor.

It is equally difficult to download configuration data into two different network management systems from the same network planning tool. In this case, plausibility or consistency checks performed by the network management system become impossible. In addition, the presentation of performance data becomes difficult, or even impossible, if the base data delivered by both systems is not equal.

Confronted with a situation like this, many network operators contemplate a vendor-independent integrated network management system. Although this may seem the obvious solution, it has several serious disadvantages, apart from the considerable cost and development time necessary to create such a system.

Availability of Interfaces

To perform an integration, the entity developing the vendor-independent network management system must have detailed documentation about the network element interfaces. Often, these interfaces are not documented and supported by the vendor in a way that can be used by outsiders. In most cases, vendors are reluctant, if not unwilling, to make documentation on these interfaces available, as it is practically impossible to keep them confidential considering the large number of external personnel working with it during the integration. For the system integrator, it is absolutely essential to have stable interfaces well before the point where the integrated system is delivered. For most vendors it is practically impossible to guarantee that level of interface stability on a network element that is still changing rapidly.

Time Delay

In a single-vendor network, a new software release for a network element (or the network management system) can be deployed whenever the vendor has it available. In a multivendor environment, the features of the new software can only be used if the integrated network management system has been adapted and tested too. Due to the usually late availability of detailed design information, this leads to a time delay of the new features compared to a single vendor network, depending on the complexity of the changes of up to nine months. In a situation of several competitive networks, this issue alone can rule out the feasibility of an integrated system.

Loss of Features

Integrated network management software can only support features and functionalities that are supported by all network element vendors. Consequently, in a multivendor network, the network management system only supports the smallest common denominator of features offered by all vendors. Even if a feature is available in one vendor's network element, it cannot be administrated from the network management center and in most cases can therefore not be used. The same argument applies if both vendors offer the same features but have a different deployment schedule.

Complexity

If a new software is introduced into a single-vendor network, it is usually tested in one region and then successively rolled out through the entire network. If errors occur due to the new software release, they can be located quite easily. Using an integrated network management system, a possible error can either be in the new network element software, in the network management software, or even in another vendor's network element, if the functionality of its interface to the network management center has now changed. Testing and rollout of new software therefore becomes much more complex. The network management software has to be adapted every time one network element vendor schedules a new release, basically eliminating any stability in day-to-day operation.

Considering the above difficulties, many large cellular networks in Europe are now supplied by a single vendor. As a consequence, operators are torn between the two options of being held hostage by one vendor or risking significant operational problems by introducing a second vendor. The easiest approach, if implemented from the start, is a regional separation. The network is divided into separate regions, each with a homogeneous equipment environment. The regions are supervised by vendor-specific network management systems employing operational staff with vendor-specific training.

The regions should be designed in such a way that traffic and handover across the border are minimized to avoid interoperation problems in network planning and configuration management. If a national network management center is envisioned, it has to operate both systems in parallel, or two separate network management centers should be planned, each operating all regions equipped with one type of system hardware. The costs for purchasing, operating, and updating a second network management system have to be balanced against possible cost reductions resulting from negotiations with a second manufacturer.

In some cases, a regional separation is not possible because one vendor has already installed equipment in parts of all regions. In that case, all regional OMCs have to run two parallel, vendor-specific network management systems. To avoid the complexities of operating mixed networks, a scenario where a region consists of two geographically mixed, but logically separated subnetworks should be explored. If, for example, one manufacturer has deployed equipment in major cities at the time a second vendor is selected, the second vendor can install umbrella sites in rural areas and possibly in capacity sites in the city areas.

If the available spectrum allows it, both subnetworks can have their own frequencies, eliminating the need for integrated network planning. Handover could be allowed between sites with identical equipment only, resulting in totally separate configuration management of both subnetworks. As the subnetworks are sufficiently self-contained, performance reports on the single-vendor subnetworks are meaningful, so an integration of this functionality can be avoided. The same applies for alarm handling and fault management features.

The network then looks, in effect, like two networks built by two operators in the same geographical area, one with nationwide coverage, and the second one with coverage in the most densely populated areas only. Depending on the subscription, the customer may be restricted to one network only or can be allowed to roam between the two subnetworks. For the roaming subscriber, an additional redundancy is provided in the high-traffic areas. If one of the subnetworks fails or is congested, he can still use the second layer. As the release schedules of both manufacturers hopefully are different, service deterioration during major software upgrades can be compensated by the second subnetwork.

The creation of logically separate, self-contained subnetworks for each manufacturer could be a way to introduce two vendors into a network but avoid a large part of the complications caused by developing integrated network management software. As in the case of a regional separation, the additional systems, as well as operations and staffing requirements, need to be balanced against the commercial and strategic advantage of a two-vendor strategy.

Conclusion

As the previous sections have shown, the network management system can determine network quality and cost to a significant extent. The network management systems in place can make it impossible to bring a second or third vendor into the network, even if his network elements themselves have superior functionality.

Although the temptation to concentrate in the early phases of network planning and deployment on cellsites is great, the functionality, capacity, and compatibility of network management systems should be analyzed carefully. A network management system with an intelligent, user-friendly surface can dramatically improve quality and reduce necessary resources. The choice to either implement a single-vendor network or go to a multi-vendor architecture should be considered carefully, and should be decided in the earliest stages of network planning.

PCS Customer Information System Architectures

Donald M. Kiceina
Group Manager–Wireless
Cincinnati Bell Information Systems Inc.

Abstract

The choice of a customer information system architecture is one of the most critical decisions an information systems executive has to make. An architecture that is well-planned, robust, and flexible will affect the company's ability to compete by allowing it to both recognize changes in the market and act to take advantage of these changes. This paper will review the major elements of an architecture for a customer information system supporting personal communications services (PCS). In the context of this paper, PCS refers to services from both existing cellular operators and the future PCS operators.

Introduction

Before discussing architectures, it is necessary to understand what is meant by a customer information system and what business drivers affect the nature of the customer information system architecture. When reviewing customer information systems, there are two perspectives to consider. The first and most important perspective focuses on the customer. The second focuses on supporting internal processes.

Customer-directed perspectives include acquiring new customers (either directly or through dealers or agents), responding to ongoing questions or inquiries from the customer (generally referred to as "customer service"), billing the customer, collecting customer payments, and finally, identifying new customer programs. To facilitate these customer-directed activities, the customer information system must support the following internal processes: activating the customer's service, collecting and disseminating network event and usage data (usually referred to as "call records"), rating and pricing the calls, issuing customer bills, receiving and accounting for customer payments, settling accounts with other companies (e.g., for roaming services), and finally, assessing business performance to identify necessary company actions.

The customer information system architecture defines the framework or structure for the various applications that comprise the customer information system. The architecture should define the platform upon which the applications reside, how data will be maintained in the system (i.e., the data model), how internal and external systems interface, and how end users access the system.

The customer information system architecture is important because if it is well-architected, it will provide flexibility resulting in competitive advantages for the operator while reducing the risk. The increased competitive advantages result from being able to respond to customer requests quickly, efficiently, and accurately, improving access to data that enables better business decisions, and offering new competitive services faster; however, a robust, well-planned architecture really pays off by reducing company risks associated with changing or adding new applications, incorporating new hardware and software technologies, and protecting current information system investments.

A robust customer information system architecture allows a PCS operator to respond to change from two directions. First, from changes in the business. For example, a competitive market requires offering new or enhanced services (e.g., personal numbers) or price plans, integrating or unbundling services, serving new market segments (e.g., the mass consumer market), or reducing costs.

Second, from changes in technology. For example, new wireless technologies, such as cellular digital packet data (CDPD), data message handlers (DMH), home location registers (HLR), and subscriber identification modules (SIM), will impact the customer information systems' interaction with the network. In addition, new information technologies (IT), such as distributed and client/server processing, will allow customer information systems to be more efficient and responsive.

PCS Customer Information System Architecture Elements

This essay will review four elements of a PCS customer information system architecture:

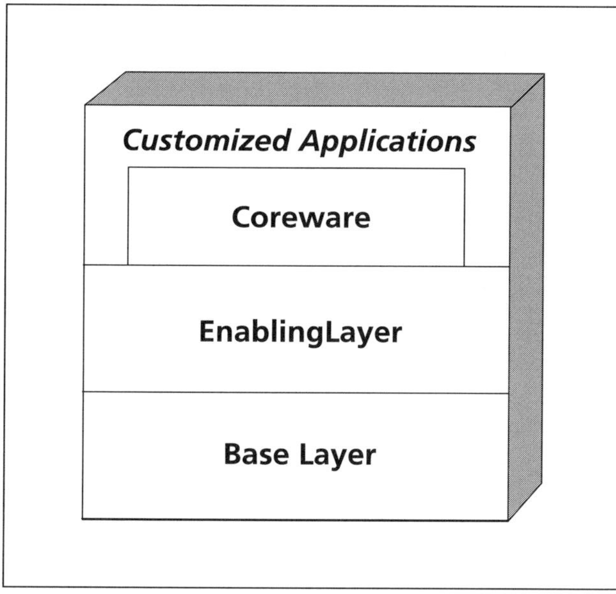

Figure 1 *Application platform architecture*

- The application platform architecture
- The functional architecture
- The systems architecture
- The application architecture for two important applications, data exchange and decision support

Application Platform Architecture

The application platform architecture represents the foundation of the overall customer information system architecture. As shown in *Figure 1,* an application platform architecture will generally have four layers:

The first layer may be referred to as a base layer. This layer defines the underlying information system technologies supported: the operating system, the database management system, the communications protocols, and the programming languages.

The second layer may be thought of as an enabling layer. This layer isolates the business applications from the information system technologies and defines the supported transaction services, the image and graphic services, and system management services. This layer will also sometimes be referred to as middleware or groupware since it separates the software associated with the physical platform from the software providing business or end user applications.

The third layer may be called coreware. This layer defines the core functions or applications that would be common to most communications industry customer information systems. Core functions such as customer service, provisioning, service management, transaction management, and marketing support would be included in this layer. The final layer of the application platform architecture includes the customer-specific applications required by the PCS operator and would typically represent customized modification to the coreware applications.

If the application platform architecture is robust and flexible, this aspect of the customer information system architecture will allow the introduction of new information technologies without requiring extensive changes or replacement of the end user applications, thus protecting the operator's current business application investments.

PCS Functional Architecture

From a functional perspective, as shown in *Figure 2,* the customer information system consists of two major parts: One is the data exchange system, which collects and distributes the event and usage data from the network. The second is the business management system, which performs the major business functions of producing bills, responding to customers, and providing information to business and market decision makers.

A PCS business management system will generally consist of subsystems that perform the following major functions:

- Rating – which is currently performed on a "batch" basis but in the future will be a real-time process
- Pricing – based on a number of different plans available
- Billing – which initiates the formatting of the bill or invoice and is also currently performed on a monthly basis, but in the future will be performed as required, ranging from "on-demand" to periodically

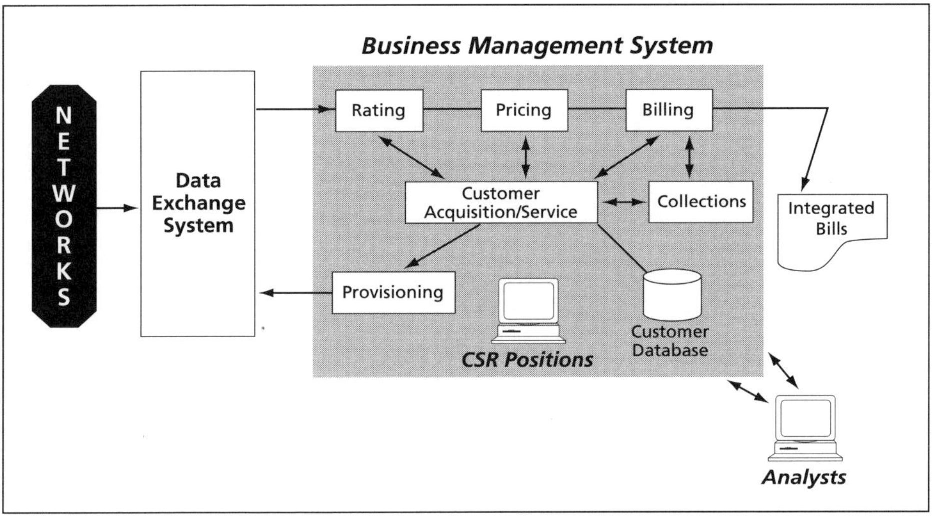

Figure 2 *PCS customer information system functional architecture*

- Customer acquisition and service – which procures and manages new and existing customer accounts

- Provisioning – which activates service in the network and is currently initiated in the business management system, but in the future will enable the network to establish new accounts and update customer account features

- Collections – which manages overdue accounts receivables

PCS Systems Architecture

The PCS environment is a highly distributed environment. Services are provided in geographically-defined markets. In many cases, specific PCS operators will provide service in a number of markets widely dispersed across the U.S. As a result, a distributed, client/server-based systems architecture, as shown in *Figure 3,* is particularly suitable for PCS customer information systems.

A distributed, client/server systems architecture provides several advantages. A distributed architecture allows the strategic placement of applications and their associated data where they are best suited. This placement could be locally, regionally, or centrally, and can be changed as business needs change. In addition, as the PCS market grows, the volume of data to be processed will increase.

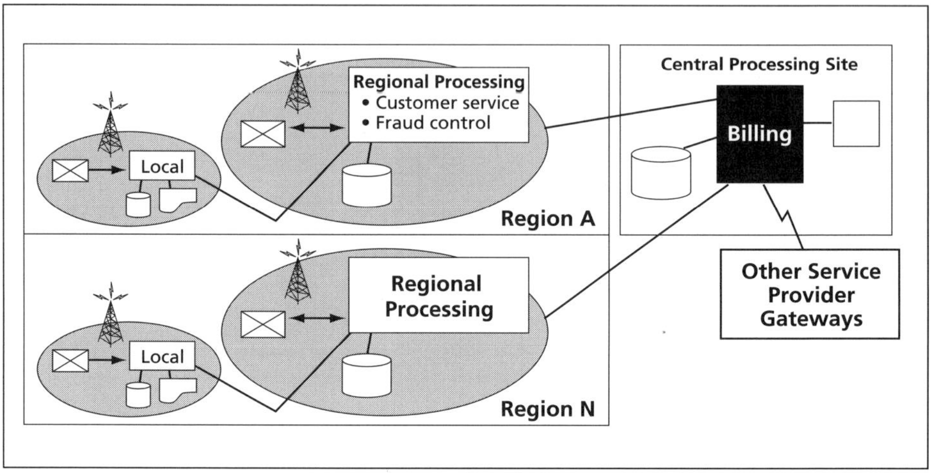

Figure 3 *Systems architecture: strategic application placement*

This will require processing the data early in the system flow and within the context of an overall architecture. Finally, since multiple service providers will participate in PCS, the customer information system will need to integrate their information. The inputs from the various providers need to be received in the system where most appropriate, whether that be locally, regionally, or centrally.

Data Exchange System Architecture

The first specific customer information system application architecture to be reviewed in this essay is for the data exchange system (DES). The DES collects the PCS network event and usage data and distributes it to the appropriate business management applications.

Often new services require new network equipment that will not output event or usage data through existing network (i.e., switch) output ports. For example, CDPD provides data through a new network element called a mobile data intermediate system (MD-IS) and CDPD usage data is not in call records but in "traffic matrices." Even if event/usage data is provided through existing network ports, the format of the call record is usually changed. The commands to activate and deactivate new service (i.e., service provisioning) will likely change also.

To overcome these challenges, event/usage data must be converted to a standard format defined in the overall customer information system architecture for the

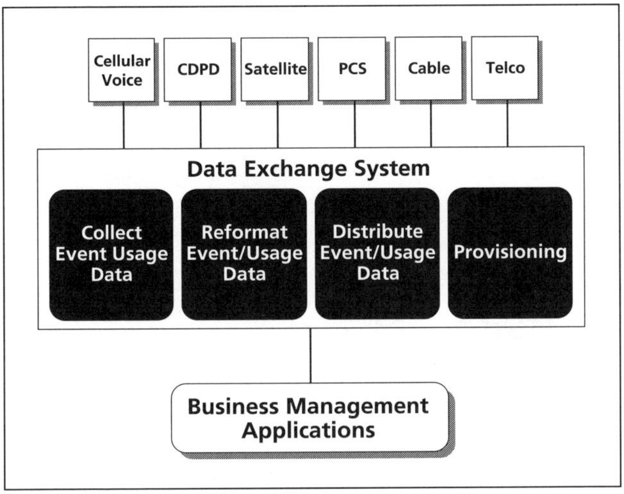

Figure 4 *Data exchange system architecture*

downstream applications. Record formats should be maintained in tables to allow changes without requiring extensive reprogramming. In addition, the customer information system architecture that specifies these formats should incorporate industry standards (e.g., cellular intercarrier billing exchange roamer record [CIBER] formats, IS-41, and DMH for intersystem operations; CDPD for data, and even international standards like GSM). As shown in *Figure 4,* the DES would interface both to the various networks (e.g., cellular, CDPD, PCS, LECs, cable, and satellite) and to various business management applications.

The DES would collect the event/usage data, reformat the data to the format accepted by the business management applications, distribute the data to the business applications, and interface to the networks for service activation/deactivation.

Decision Support Architecture

The second customer information system application architecture to be reviewed in this essay is a decision support architecture (DSA). An important function of the customer information system is to provide a means for assessing business performance. In the highly-competitive environment of PCS, it is essential to have access to information at multiple levels in the enterprise. Information is required at the local level for local marketing teams, rental businesses, and dealers. At the regional level information is needed for customer service and fraud centers, and at headquarters it is required for corporate planning and centralized operations.

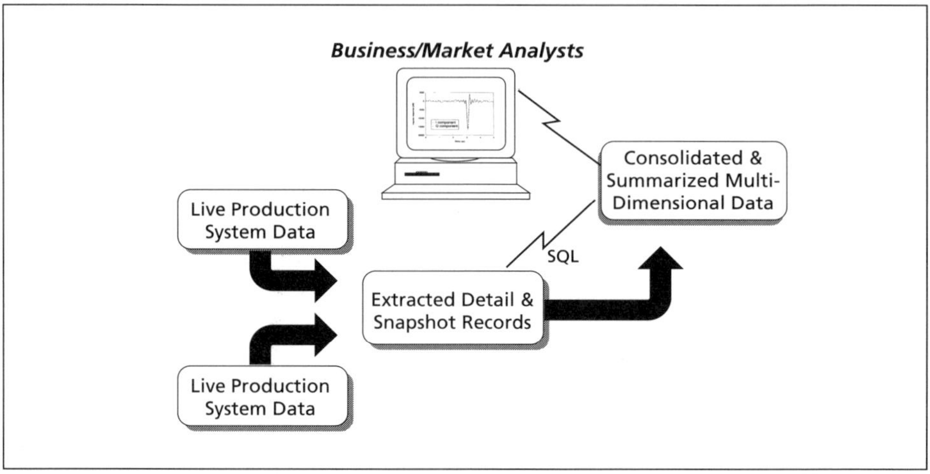

Figure 5 *Decision support architecture*

A gap exists today between end users' requests for information and the system's ability to satisfy those requests; however, with new decision support technologies (e.g., powerful workstations and multidimensional analysis techniques), information systems can satisfy those end-user requests by providing easier access not only to data but to information. The new decision support techniques allow multiple views of information (both in detail and different customer segments), intelligent exception reporting, graphical representations of information rather than numeric, and prior period comparisons and time series analyses.

Figure 5 summarizes a DSA for accessing information to analyze PCS business performance. Operational data is extracted from "live" production systems from multiple sources (whether mainframe or server-based) and consolidated for access at any level in the enterprise. This architecture isolates production systems from analysis queries, which provides fast and consistent end-user response times.

Summary

The decision regarding the PCS operator's customer information system architecture is important. This essay has reviewed several different aspects of a customer information system architecture and two specific application architectures. Implementing the right architecture will ensure that the PCS operator's numerous business applications are properly integrated and will allow the operator to act on a timely basis to changes in the exciting and highly dynamic PCS market.

Rural Cellular

Richard W. Goehring
Vice President, Cellular Engineering
United States Cellular Corporation

The issues that confront a rural cellular operator are comparable to those faced by cellular operators in more urban areas. There are, however, some unique characteristics of the rural markets that can alter the approach taken for the design of the cellular system. The focus of this essay will be the differences and nuances that arise in a rural market, and consideration will be given to the contrasting solutions that are employed in such markets.

The principal difference between urban and rural markets is obvious. Fewer people live in rural markets, or more precisely, there are fewer people per square mile in rural markets. For example, the urban market may have a population density of 500-700 people per square mile, whereas the rural market may have only 10-15 people per square mile. This dictates somewhat different design criteria from the outset. For example, if one assumes a long-term penetration rate of 15 percent to 20 percent, the limiting factor in the urban markets quickly becomes frequency availability. Restricting our consideration temporarily to analog cellular systems, the initial design in an urban area must take into consideration aggressive frequency reuse, either through sectorization, cell splitting, or both. Additionally, microcells may be considered to "deload" the macro-infrastructure in areas where heavy cellular traffic exists.

Conversely, in rural areas the need for frequency reuse is less intense and may not occur for many years beyond the initial deployment of the system. The key issue in this instance is pervasive geographic coverage at an economical cost. This dictates fewer, but much larger, cell sites in a rural cellular system. Whereas in an urban area a cell site may have an effective serving radius of 2-5 miles, the rural cell site may have a 15 or 20-mile radius. As such, it could serve a geographic area 16 times greater than the urban cell site (i.e., a 20-mile radius vs. a 5-mile radius). Since the population density in the urban area may be 30 to 50 times greater than that found in the rural market, the small urban cell is still capable of providing service to more potential customers.

The cellular business requires a relatively high initial investment compared to other industries. The high fixed costs of the cellular business are principally due to the cost of establishing the network infrastructure required to provide service. For the business to be successful, sufficient market penetration must be realized to recover these high fixed costs and the other costs of operation necessary to achieve the economies of scale available. This is more difficult in a rural market where fewer potential customers are covered by each cell site, unless rural cell sites cost less than urban cell sites. Indeed, real estate costs are significantly lower for most rural sites (i.e., $500-$4,000 per rural acre vs. $100,000-$200,000 per urban acre), but real estate costs are not one of the major costs required to establish a cellular infrastructure.

Conversely, the cost of a cellular tower in the rural markets is most likely greater due to its size and lighting requirements. For example, most urban towers are less than 150 feet high whereas many towers in rural markets are over 400 feet. Larger towers are often necessary to obtain the larger cell radius that is common in the rural markets. High power PAs (power amplifiers) are an alternative means of providing a large footprint in some cases, but greater tower height is required to overcome hills at the periphery of the cell by providing a more direct path for the mobile. In short, a higher tower height will usually outperform one with greater power output even though they both may have the same cell contour (32dBu).

In addition, the relatively lower cost of land in the rural markets allows a more prevalent use of guyed towers rather than self-support or monopole towers. At heights over 300 feet, self-support towers become expensive, and the use of monopole towers is virtually nonexistent. As such, the rural cellular operator will secure much more land (five to ten acres per site) to accommodate the tall guyed towers dictated by their cellular system design whereas the urban cellular operator will have relatively small sites (usually well under an acre) or locate on an existing structure or building.

The lower cost and greater availability of land in these rural markets also results in more sites being purchased outright rather than leased from the existing site owner. The lower cost and availability of acceptable sites, however, is not the only reason why this practice is more prevalent in rural areas. As noted earlier, the limiting factor in the more dense urban areas is frequency availability. As such, frequency planning is a more critical criterion, and cell splitting is required more frequently. In many instances, the new system design necessitates that the original site be moved in order for the overall system and the new frequency plan to remain viable. In such cases, a leased site is more applicable. Furthermore, the density of cell sites in the urban area requires a more precise placement of the cell site. This restricts the number of viable site locations and thereby increases the real estate cost of the site. Therefore, the more frequent need for cell splitting caused by the urban operator's more dynamic frequency plan and the limited number of alternatives often results in more sites being leased rather than purchased.

Another aspect of rural operation is the need for suitable buildings to house the equipment. Viable existing structures in such areas are rare. Therefore, the vast majority of rural sites will require a building specifically designed and constructed to meet the needs of the cellular operator. Conversely, the urban operator can frequently find space in an existing building and modify it for their use. Providing a building at each site is not necessarily more costly than the costs of leasehold improvements for urban sites. The need to provide a secure, environmentally controlled area for the equipment can often cost more than a standardized, prefabricated building that is used in the rural markets.

There are different economic criteria for the urban versus the rural cellular operator that result from factors such as the more prevalent need for sectorization or digital technology. Most of these are of less significance. With all factors considered at approximately the same penetration level, two to three times more home subscribers are covered by the urban cell site than those covered by the rural cell site. This dictates that rural operators must aggressively pursue other revenue opportunities to realize the same economic return on their investment. Historically, this other revenue stream has come from roamers, or home subscribers from nearby urban markets. Since this is a significant revenue stream that rural operators depend on to support their operations, attention must be given to the design requirements of these roamer customers in the initial planning of the network.

More attention must be paid to the major highways and recreational areas used by roaming customers to ensure that they receive quality service. While adequate coverage overlap must be ensured to guarantee that reliable hand-off exists along major highways, cell sites must be engineered to minimize the total capital investment. This design criterion must be balanced with the need to provide good, and preferably portable, coverage in the major pockets of population within the rural area. These design criteria sometimes conflict, and the rural cellular operator may have to develop a plan that compromises one or both of these objectives while still maximizing the long-term return on capital investment for the cellular system.

Another difference between urban and rural cellular operations is in the systems interconnect design, both from the cell site to the mobile telephone switching office (MTSO) and with the interface to the public switched telephone network (PSTN). The facilities linking the cell sites to the MTSO in rural areas are often accommodated via microwave radio.

The use of 2 GHz microwave is prevalent because of its propagation characteristics, which accommodate the greater cell site spacing of the rural markets. It is also used because of the high cost and unavailability of digital circuits from the local exchange carrier (LEC). In many instances, the site is positioned in an area where the special construction costs are prohibitive, and the delay in availability due to this construction adds several additional months to the construction interval. In urban markets, digital facilities are often available, and most of the time they can be provided by the LEC within 15 to 45 days.

When rural operators are assessing the various interconnect options available, they should consider the long-term impact (four to five years) of their decision. Although it may be more cost-efficient initially to lease facilities, the advantages of microwave over the full life cycle often merit its deployment through the growth of the network (both traffic and future cell sites). Care must be taken to consider this design criterion initially to ensure that towers are adequately designed to accommodate the loading requirements of the microwave antennas and that there is a clear path between sites. Consideration of this option after the tower is built will impose high costs on the operator for tower modifications.

The interconnect with the LEC is problematic at times in the rural areas. In many cases, a Type 2 or tandem interconnection is not practical because of the location of the LEC's switch. This often imposes multiple Type 1, or PBX-like, trunk groups on the rural operator. Given the more fractured nature of the extended area service (EAS) provided in rural areas, the rural cellular operator may have to consider an interconnect at almost every cell site to avoid the high cost of toll from the multiple LECs serving the rural area. This splits the terminating cellular traffic among many smaller, less efficient trunk groups and increases the cost of the cellular operation.

The services offered by the rural operator closely parallel those offered in urban areas; however, there is a time lag as a result of the relative infancy of the rural market and the two to four-year difference in the licensing date for the market. Voice services, or POTS (plain old telephone service), dominate over data services, but some unique data services are developing in rural markets.

Fixed cellular service in the rural markets offers a unique opportunity. This is because of the higher cost of providing landline service in these areas. In many cases, this cost is in excess of $2,000 per landline subscriber. Cellular can provide this service, plus the benefit of portability, for approximately half of this amount. Some landline telephone companies have recognized this advantage and have pursued business relationships with the cellular operator to resell the cellular service. This can significantly reduce the telephone company's held orders and enhance the customer's satisfaction with their service. By using fixed directional antennas at the customer's residence, the telephone company can dictate the home cell of the rural subscriber. This allows for a fixed monthly charge to be used on the home cell site while allowing for an incremental airtime charge for the mobile's use on other cell sites in the cellular system.

Data services are also being offered in rural markets to increase productivity and enhance the quality of life in these communities. These are often targeted at the agricultural industry, where remote monitoring of irrigation, soil conditions, and other environmental factors that influence the farmer's productivity can most easily be ascertained by using a cellular data link directly to their fields. More pervasive use of this technology is anticipated in the future as the highly competitive agricultural industry continues to seek technological solutions to increase productivity and lower costs of operation.

In summary, the rural cellular operator faces many of the same issues that confront the urban cellular operator, but there are several areas where the nature of the markets they serve is different. These differences will cause the rural operator to alter his design criteria and the nature of the network deployed in order to accommodate the different economic criteria presented in operating a cellular system in a rural market. Accommodating roamer customers, determining a cost-efficient interconnect design, and differing infrastructure costs will all influence the decisions made by the rural cellular operator.

Coding and Modulation: A Tutorial

Thomas A. Sexton, Ph.D
Advanced Receiver Technology Group
Motorola

Introduction

This discussion of coding and modulation is intended for engineers, mathematicians, or physicists working in other areas who wish to become acquainted with the basic terms and ideas. Elaborations are kept to a minimum.

Let us begin with two definitions. *Modulation* may be defined as follows: "The variation of a property of an electromagnetic wave or signal, such as its amplitude, frequency, or phase in a manner determined by another wave or signal, especially for the purpose of transferring information from an audible signal, such as the human voice, to a carrier wave suitable for radio or telephonic transmission (*The American Heritage Dictionary of the English Language* 1976). *Error Control Coding* is protecting digital data against the errors that occur during transmission through a communication channel (see Blahut 1984). This essay treats error control coding and modulation in two parts, qualitative (section II) followed by quantitative (section III). The additional field of source encoding (for example, the mapping of an analog voice signal to 64k PCM) is not covered here (See Sklar 1988). Readers not familiar with the field may wish to read the qualitative section and study the figures in the quantitative section. Someone familiar with this subject who is looking for a quick overview may find the quantitative section useful. Some details of the required demodulating and decoding functions are given.

An illustration of coding and modulation that is well known to most is the black box which connects a computer terminal to a telephone or radio channel. This is called a *modem* (modulator demodulator). The computer terminal is a source of information represented in bits, whereas the telephone line network twisted pair, coax, or free space (for a radio modem) is the channel to which the modem connects the source. The process of modulation is what the modem does to the bits before launching them in some transformed manner onto the twisted pair, coax, or "airwaves." At the other end of the channel there will usually be another modem prepared to reverse the transformation so as to recover the information bits (that is, demodulate them). Usually, this listening end also acts as a source of information bits, so the demodulating box also contains a modulator to get the bits onto the channel. This makes the box, as a whole, a modem. We say usually because there are important examples of one-way and even low-success-rate one-way communication. In a paging system, only a tiny fraction of the total pages transmitted by paging base stations reach the desired pocket-pager. If there is a concern about a noisy channel, a coder/decoder pair is also added. In general, there is an information source and sink pair at each end of a channel. A modem (and, where needed, a coder/decoder) connects the source and sink pair to the channel at either end.

Qualititative Coding and Modulation

The communication problem and solution is illustrated in a serial block diagram in *Figure 1.* Following the information source on the radio link is a coder, modulator, channel, demodulator, decoder, and information sink. The following sections describe these blocks.

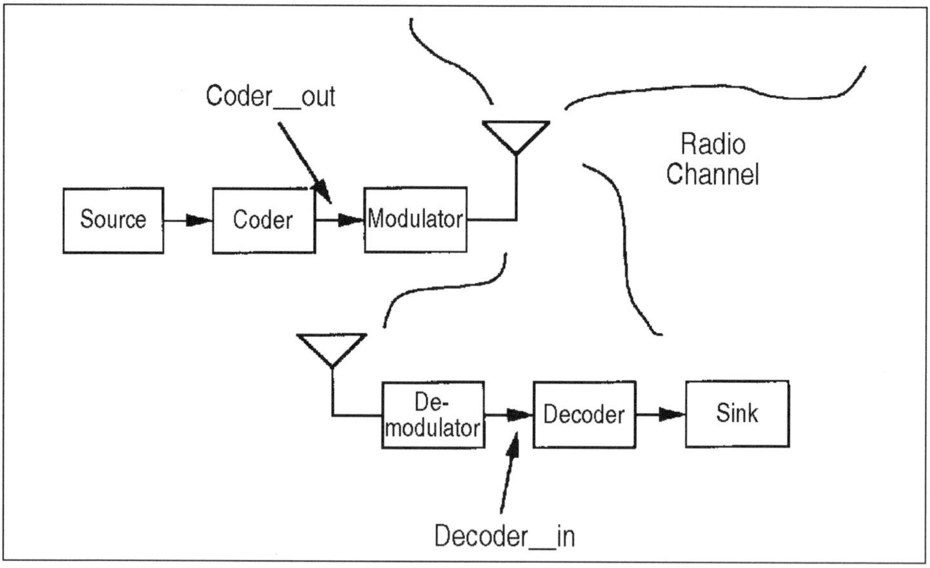

Figure 1 *Radio link from source to sink*

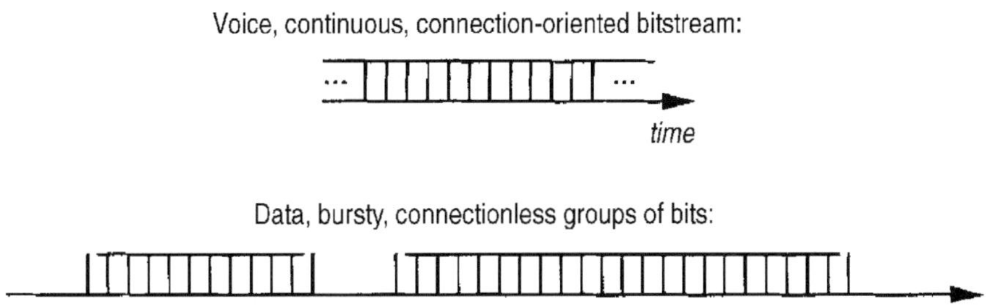

Figure 2 *Voice and data sources*

Source

Computer, telephone, network, or radio systems can be divided into two useful groups: connection-oriented and connectionless. This terminology was created by the telephone industry to distinguish between voice and data communications (see *Figure 2*). Note that although data may be sent over either a connectionless or a connection-oriented link, voice sources cannot use a connectionless link.

Voice communications are unique in that we are basically hearing a face-to-face conversation without seeing face-to-face. Here are some qualities in common between a phone call and a visit to a friend's office: we attempt to reach the person geographically where we expect them to be; if he is there, we almost certainly expect to be able to speak to him unless he is busy; the conversation will be ended when either party sees fit and not against our will; we expect to hear clearly what he says and recognize his voice, and we expect him to hear and recognize our voice without delay (in fact, we expect to hear anything he says even if he speaks at the same time we do). To a customer, these points may seem trite. However, it is important not to lose sight of them in the fascination of the technology developed, which nominally fulfills the customer's wishes while satisfying other motivations. Some motivations that determine the physical characteristics (design, size, material, cost) of the modem and coder are:

- what the paying customer wants it to do for him
- what channels the local government allows the modem to connect to, with what interference levels to other channels
- the technical interests of the manufacturing company
- assorted constraints relating to patent rights, local, and distant government

initiatives, compatibility with older related products, and the performance of older "wired" means that provide similar functionality.

Data communications are akin to sending a letter in many ways except for accuracy. In both forms of communication, information is created, enclosed in an envelope, and the envelope is addressed. The envelope is put in a place that a carrier periodically checks. From there it is taken to a sorting center where it is bundled with other messages heading in the same direction. Ultimately, the far local carrier leaves the envelope in a place periodically checked by the information consumer; the envelope is opened and the information absorbed. The accuracy of communication by letter is much lower than that of an arbitrary decoded data message, however. Consider a letter inviting someone to a lecture on an upcoming referendum; quite often the date or exact meeting room will be left off unintentionally. Now consider a data transaction involving moving funds from one bank to another. The likelihood of funds being transferred to an uncertain destination accidentally is very, very low.

The differing characteristics of voice and data sources are so stark that designs which are commercially viable for both are few and far between. Furthermore, the market sizes are quite unequal. Almost everyone speaks and is a candidate for a radio phone, but only a small fraction create data which they also want to transmit. In a communication system, the source produces a message, M, which is a sequence of k bits. If the source is speech, then M needs to be delivered with small delay, and the person listening will need to recognize both the words and the speaker's voice in most cases. Also, unexpected interruptions to the overall conversation should not occur during the phone call. If the source is data, then M may be delivered with greater delay but with high accuracy. If the transfer of data occurs regularly and at high rates, interruptions to the communication channel may be considered very costly.

The Coder and Delay

The purpose of the coder and decoder pair is to reduce the likelihood of the information sink receiving some bits in error, that is, the wrong information. The coder does this by sending the same message x times in some sense (x not an integer, in general) so that the decoder can operate by a kind of process of elimination to determine what the message most likely was. The simplest codes do this in a straightforward way, so-called repetition codes. Repetition codes send the message n times (where n is an integer). For the simplest receiver, each received copy of a given bit is demodulated, and the decoder tallies the "vote" among the

n received copies to determine the decoded value for that bit. The repetition code selected determines the message delay.

The blocks following the source in *Figure 1* are determined by the nature of the source. An important aspect of a communication link is the throughput delay between the original transmission of the information and the time at which it is finally delivered to the sink. A connectionless link tolerates a high throughput delay relative to a connection-oriented link.

For the connectionless link, the tolerable delay is usually large enough to allow for several attempts at transferring the message. In this case, the task of the coder is to add a field to the message composed of parity bits. These bits, $P = f(M)$, constitute a very high rate code (not adding many bits to the overall transmission). At the receiver's decoder, the parity bits are recomputed on the received message to see if it has been received without error. If the computed parity and received parity field, P, are identical, the message is declared correct at the receiver and passed along to the sink. If the parity fields disagree, a message is sent to the transmitter requesting retransmission. There are many protocols suited for different applications of this automatic repeat request, or ARQ.

For the connection-oriented link, a high cost is placed on throughput delay. There are three prominent reasons why speech cannot be delayed too much. The first is the situation in which the speaker hears his own voice. Telephone sets, both wired and wireless, feedback directly into the earpiece of the local handset a highly attenuated version of what the microphone picks up. This is called *sidetone.* Sidetone gives the speaker the feeling that the communication link is working, without being consciously perceptible. If the delay in creating sidetone is too large, sidetone becomes perceptible and irritating to the speaker.

The second reason that large delay in a telephone link is intolerable is that the spoken message usually ends up in a traditional analog telephone network. As the voice energy travels to the final central office, which connects the call to the land termination, it is carried over four wires, two for each direction. For the sake of economy, two wires carry the audio energy for both directions simultaneously over the final mile from the central office to the terminating phone set. The connection from four wires to two wires is made by an imperfect transformer device called a *hybrid.* When discussing echoes in a telephone link, the speaker is considered to be the near end and the listener is the far end (see *Figure 3*). The result of imperfection in this case is that most of the energy goes in the desired direction while some is coupled into the return pair of the four wires and

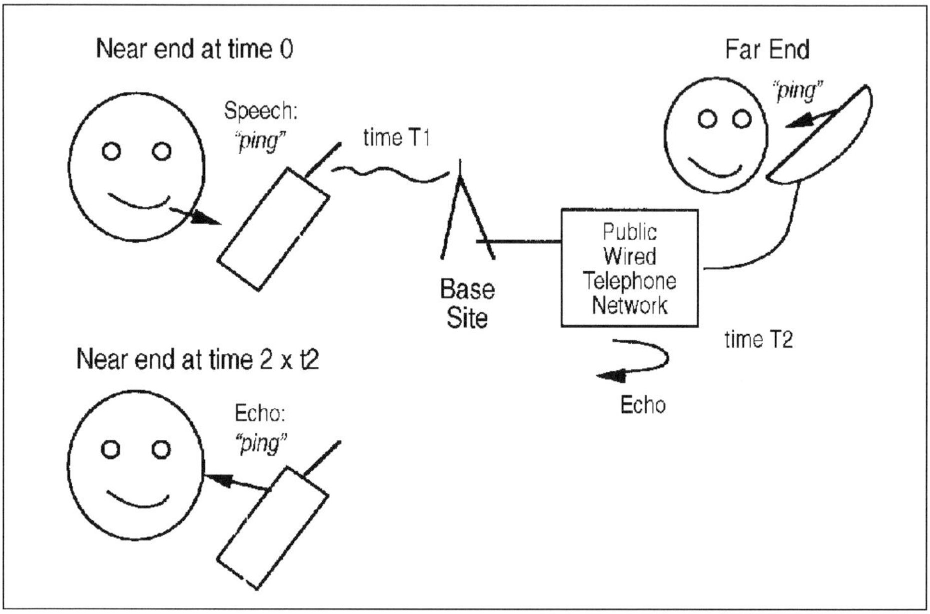

Figure 3 *Delay in voice link*

appears to the phone system as far-end speech. Since the phone system cannot distinguish this echo from far-end speech, the echo energy will be faithfully transmitted back to the near end. At the near end, depending on the strength and delay of the echo, the near-end speaker will hear his own speech. If the delay is large enough, the near-end speaker will find it quite irritating that the phone call is not like a face-to-face conversation but rather like mechanical speech.

The final reason why delay cannot be too large concerns the timing of utterances in a conversation. Suppose that echoes are suppressed somehow and that sidetone is just right. People are used to listening, considering while they're listening, and sometimes speaking when the other person's words seem to stop. The speaker occasionally pauses, giving the other party a chance to speak. If the speaker hears nothing, he may resume within one or two seconds (remember, he has not the benefit of the listener's body language to help in the timing of this interchange). The addition of a one- or two-second delay into every utterance breaks up the conversation flow remarkably. A good example is a call to Europe. The conversation will be bounced off a satellite and will be delayed by about two seconds. The speakers must deliberately wait for long periods to guess, after each idea is communicated, if the other party has started to speak. All of these voice-echo-

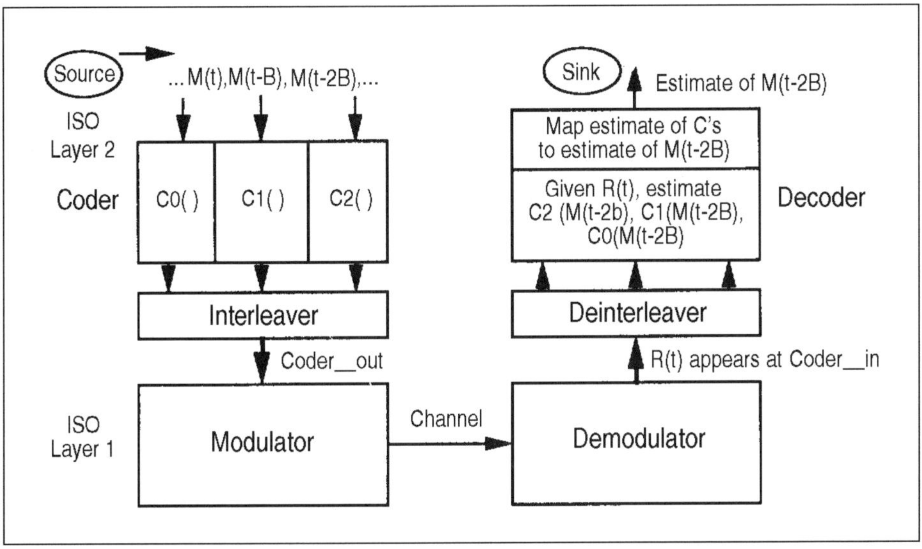

Figure 4 *Coder and Decoder flow at time* t

delay problems go unnoticed if the delay is less than about 5 ms. Delays of 10 ms can be mitigated (in the case of near-end "echo") with a circuit at the near end called an echo canceler.

What has delay got to do with coding? Quite a bit. Coding protects against sudden noise events interspersed with relatively long periods of calm on the communication channel. Consider a time period *B*. At time *t*, we may consider the message to be composed up to this time of M(t), M(t-B), M(t-2B) (See *Figure 4*). The coder may compute three functions of the data and transmit them at time *t*: c0(M(t)), c1(M(t-B)), and c2(M(t-2B)), where the whole group is denoted (C(t)). Consider the receiver at time t =2B and its attempt to decipher M(0). At time 0, it received a noisy version of c0(M(0)). At time 1B, it received a noisy version of c1(M(0)). Now at time t=2B, it receives a noisy version of c2(M(0)). Using these three noisy functions of M(0), the decoder will estimate M(0) at time t = 2B. If the decoding takes place instantaneously upon receipt of the noisy version of c2(M(0)), the throughput delay due to the coding process will be 2B. For continuous transmission, however, the modem accumulates 2B seconds worth of speech, encodes it, and then takes 2B seconds to transmit it, for a total minimum delay of 4B. The more functions c0(), c1(), c2(), c3(), and so on that the coder employs to protect against noise events, the larger the delay will be. Also, using more functions causes a greater number of math operations to be performed on the received data before the message estimate is passed to the sink.

Suppose that M(0) consists of *k* bits and that the total bits sent by the transmitter to represent these *k* bits is *n*. The coder is said to be a rate k/n coder, and the code is said to be an (n,k) code. Counterintuitive terminology thus makes a low-rate coder the one with the better error rate performance, in general.

We can now understand some guidelines for selecting a coder for a voice communications link. The coder delay should be less than 10 ms unless the product price can support an echo canceler. The coder delay should not be much more than 100 ms if the link is to imitate the conversation flow which is possible on a plain old telephone system (POTS) call. The coder must produce a C(t) which is not too large to fit in the bandwidth allocated for the link (we will discuss bandwidth later). The code must be strong enough (k/n must be low enough) not to irritate the listener with too many errors. Errors take on many forms in terms of perception. The code must be simple enough to be decoded within the computational resources of the receiving end.

The guidelines for a data communications link are similar yet with a different emphasis. The tolerable delay for most applications may be significant. The greatest number of users of wireless data links are groups who keep track of inanimate objects such as overnight packages, freight haulers, retail inventory, warehouse inventory, and railroad and mass-transit cars. We do not consider sending binary encoded voice over microwave links furnished by telephone service providers to be a data application. For the data applications mentioned, ARQ is the preferred method for eliminating errors from the final message passed to the sink.

Modulation

After the coder produces the block *C(t)*, the modulator operates on it to reach a form which may be coupled to the channel (see *Figure 5*). Traditionally, the coder and the modulator have acted sequentially on the message, *M(t)*. In about 1986, Ungerboeck created a hybrid block which is said to produce trellis code modulation (TCM). TCM is beyond the scope of our present discussion, but the interested reader will find an introduction to TCM in Blahut (1990) or Ungerboeck (1987).

The modulator accepts a sequence of bits as its input and produces a sequence of symbols at its output. A symbol is a complex number. In an unusual sense, the modulator is a digital to analog converter (DAC) in that it converts between number systems. The simplest modulator to imagine is that for binary phase shift keying (BPSK), as in *Figure 5*. The modulator maps each bit at its input to one output symbol by the relation $s = 2b\text{-}1$, where s is the output symbol and b is the

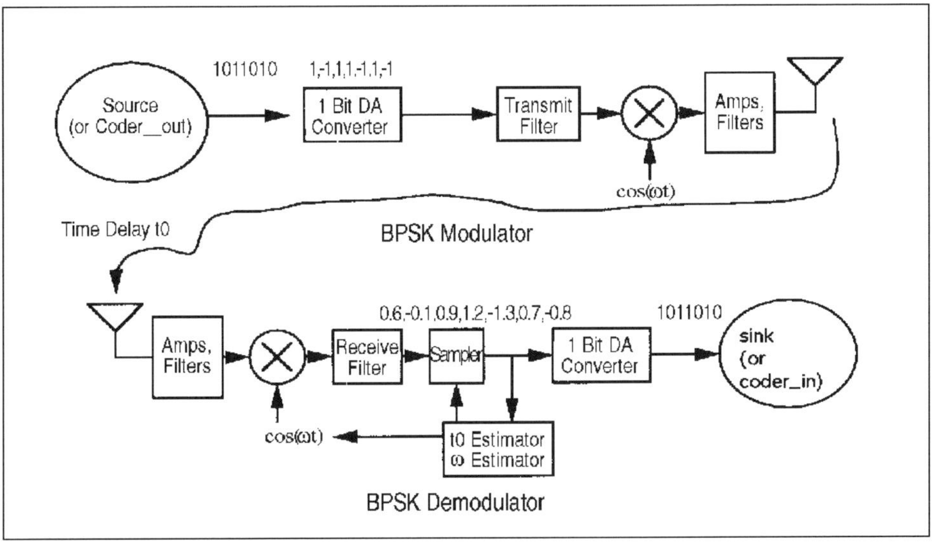

Figure 5 *BPSK modem*

input bit. A binary 1 maps to a real 1 and a binary 0 maps to a real -1. Other modulators map several bits into a single symbol or a single bit into several symbols. Examples are 16 QAM, which maps 4 bits to 1 symbol as in MIRS (Motorola Integrated Radio System), or IS-95 forward link, in which a single coded bit ends up as 64 symbols (each of the 64 is called a chip, which is a fraction of a bit).

The term linear modulation has come into use to distinguish it from modulations which use nonlinear circuits for modulation. In analog frequency modulation, the sine of the integral of the information waveform is the output of the modulator. The output being the sine of the input is clearly the result of the action of a nonlinear circuit. For analog amplitude modulation, the output is a scaled (albeit time-varying) version of the input and is a linear operation. The use of the term "linear" spills over into the mapping of binary sequences into complex symbol sequences (digital modulation). The notion of mapping is not linear in general. For example, suppose one has two sequences, $a(n)$, $b(n)$, and that the digital modulator produces $A(n)$, $B(n)$ when each of these is individually applied. The sum of $a(n)$ and $b(n)$, when modulated, will not be equal to $A(n)+B(n)$.

The goal of the modulator is to map the input sequence into a transmittable sequence which can be detected with satisfactory probability at the receiver, with its finite computing ability, after passing through the channel and suffering from a variety of impairments. An impairment is an obstacle to easy successful detection. Examples of impairments are distortion due to the linear and nonlinear transmitting circuitry; frequency drift of the transmitter, channel, or receiver; additive noise at the receiving antenna or in the receiver circuitry; random amplitude and phase variations due to the channel (fading); more than one version of the received signal arriving at the receiving antenna separated in delay (multipath); shadowing of the transmission path, which leads to slowly changing large-scale variations in received signal strength; timing skews due to timing oscillator drift, deliberate slewing at the transmitter, or dependence on a signal reflected or transmitted from a moving object (Doppler).

The modulator must also produce a signal which fits into the allocated channel bandwidth with the proper timing. In a regulatory sense, a signal fits in its assigned bandwidth if the portion outside the assigned bandwidth is below a prescribed energy limit. The modulator may sin, if it's a small sin. The design of the modulator, which typically operates in low-frequency bands or even in a completely digital baseband mode, is thus tightly linked with the power amplifier which drives the matching stages to the antenna. This is because the power amplifier will amplify the out-of-band portions of the transmit spectrum as well as the in-band portions. Also, it will increase the proportion of out-of-band energy due to the existence of nonlinear terms in the output. For these reasons, a trade-off exists between creating an ideal transmission spectrum from a final receiver-error performance goal and complying with FCC regulations for radiation of power outside the assigned band. Arbitration among the different factors discussed at the beginning of this essay will determine how well the final modem design performs against impairments.

Additional Qualitative Comments

Noise is very important. For a fixed information rate, there is a fixed ratio of signal-to-noise power which must be exceeded for potentially error-free transmission. To even approach this limit, coding is applied.

Coherent modulations operate at a lower signal-to-noise ratio than noncoherent ones. A coherent system is one in which the channel phase is estimated, and consequently so is the absolute phase of the received data symbols. If this can be done, then some noise components can be eliminated before detection. This results in a higher snr (signal-to-noise ratio) on the detection variable and a better probability of error-free performance for the same snr at the antenna. Furthermore, coherent systems are able to combine signal voltages rather than powers from independent arrivals of the noisy message, which yields a higher snr than noncoherent methods. The channel phase must be estimated well or the benefits of coherent modulation will be lost. Noncoherent schemes can tolerate higher frequency drift without correction than can coherent ones. One popular noncoherent scheme is binary frequency shift keying (BFSK) as variations implemented in CT-2, DECT. A technically old but commercially important noncoherent analog modulation scheme is FM, the one used for present-day cellular phone systems like AMPS. The control channel for AMPS uses BFSK, Manchester encoded. IS-54 (USDC) uses differentially encoded QPSK, which does not need (indeed, cannot obtain) absolute phase recovery. GSM and USDC rely on a training sequence, or marker, which is sent with each frame of data to train the receiver as to the instantaneous values of channel parameters. In order to obtain somewhat coherent operation and synchronization, the modeling and measurement of channels such as Rayleigh and lognormal, rural and urban, is beyond the scope of this paper, although it is essential to the design of the modem.

Quantitative Introduction to Coding and Modulation

This portion of the paper will present a few basic results and examples of wireless coding and modulation. The essence of engineering may be seen as the attempt to reconcile methods that have been used in the past with what present theory says is possible and what the customer wants. Coding and modulation designs of commercial wireless systems certainly fit this definition of an engineering task. In general, the customer wants as many bits as possible delivered as fast as possible without errors. There is a theory which tells us, for idealized channels, how many bits per second may be sent with an arbitrarily low probability of error (Shannon, 1948). The theory is tantalizing because it tells how many bits can be sent but not how to do it. As far as design theories are concerned, a matched waveform filter will do if we only want to send 1 bit. On the other hand, engineers (and the manufacturer's factory and the customer representatives) come to the design task equipped with previous solutions to similar problems. Typically, the engineering solution is thus highly correlated with existing systems. One is reminded of attempts to explain the zigs and zags of national economies. What actually has taken place is the ad-hoc evolution of wireless communication systems, with more attention to theory as competition for spectrum has increased.

The Matched Filter

Simple radio communication amounts to the binary detection of a series of bits. Consider the problem of sending a single bit. The successful detection of that bit

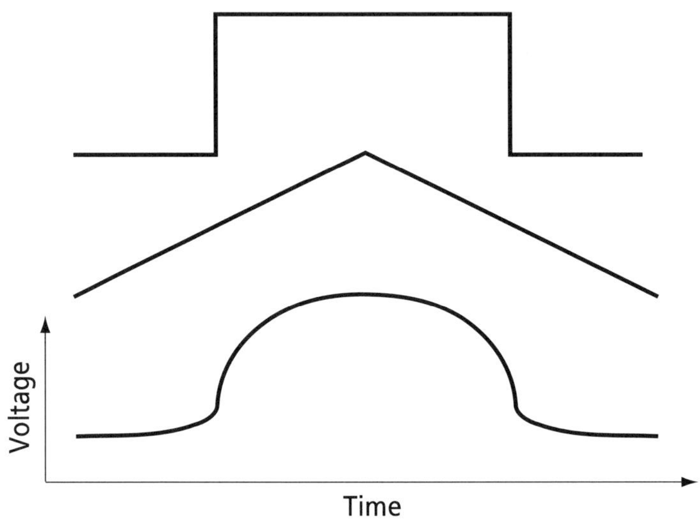

Figure 6 *Matched filter, pulse shape independence*

as a 1 or 0 is only a function of the received energy, assuming only thermal noise as an impairment. Among other things, successful detection is not a function of the waveform shape or bandwidth (see *Figure 6*). Consider the transmission of a single symbol through the communication system shown in *Figure 5*. The transmit filter limits the bandwidth of the transmitted signal. An example of the transmitted pulse shape is shown in *Figure 6*. The pulse in question is delayed by t_0 as it passes through the channel. At the receiver, the pulse and thermal noise will pass through the receive filter and be sampled at time t', producing s'. s' is composed of desired energy and noise energy, with a ratio of S. The receive-filter impulse response is equal to the conjugate of the transmit filter reversed in time:

$$H_{rx}(t) = H_{tx}(-t) \quad \text{A maximum likelihood derivation leads to} \quad (1)$$

$$S\max = E_p/(No/2)$$

where the E_p is the energy per pulse and $No/2$ is the power spectral density of the noise. Again, what this result shows is that S_{max} is not a function of the pulse shape $s(t)$. Any other choice of $H_{rx}(\ )$ will lead to a smaller S. If the error in the timing estimate $t_e = t_0 - t' = 0$, then the actual signal to noise ratio, S, will be equal to S_{max}. If t_e is not zero, then $S<S_{max}$.

Now we can recognize the function of these important system elements. First, we observe bandwidth control. The transmit filter impulse-response shape $H_{tx}(t)$, sets the time domain pulse shape for a digital system. For a system transmitting analog information, H_{tx} is still an impulse response, but an impulse-like signal is never applied, i.e., the exciting signal is already somewhat bandlimited. The bandwidth of H_{tx} sets the bandwidth of the transmitted signal, which is critical in the design of any multiple access system, including CDMA if one considers the idea of bandwidth generally.

Second, we see that timing recovery in coordination with choosing H_{tx} to match H_{tx} will maximize the signal-to-noise ratio for single-pulse transmission. However, the transmission of single pulses is of interest to radar designers and to few others. If the system is to transmit more than one bit of information and to work in a spectrum-efficient manner, then pulse shape is important. Further, if the degradation of S because of difficulty in reducing the variance of t_e is a limitation on error-rate performance, then pulse shape is even more important (usually not the case in first-generation digital wireless systems).

Nyquist (1924) extended the single-pulse case. There is a pulse shape (called the Nyquist pulse shape) which will fit in a specified bandwidth less than or equal to 2W and will allow one to send pulses at rate W without ever overlapping neighboring pulses. Now, if t_e equals 0 always, one may achieve $S = S_{max}$ while trans-

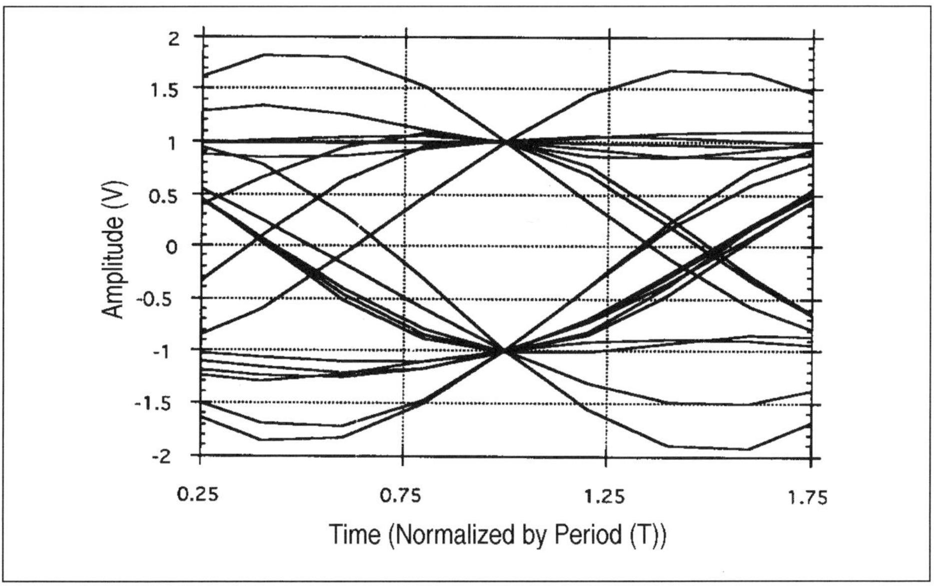

Figure 7 *Eye diagram for sin(x)/x pulse shape*

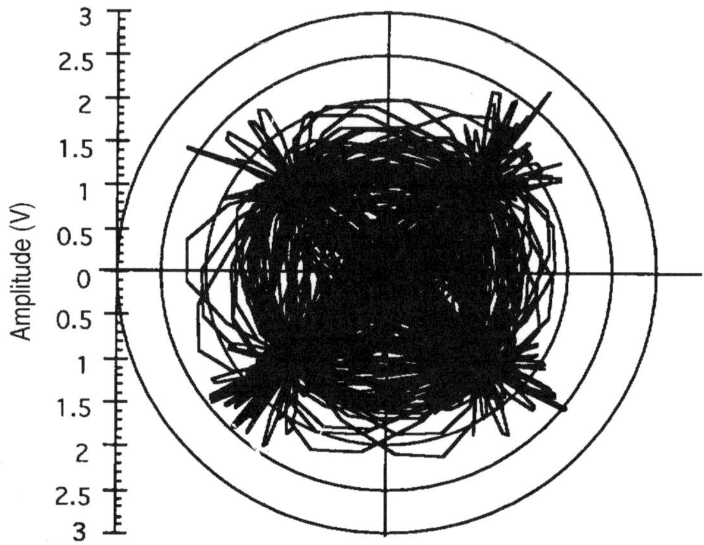

Figure 8 *Polar plot of modem waveform for sin(x)/x pulse shape*

mitting a pulse stream and occupying a finite bandwidth $2W$ or less. Harvesting both of these windfalls involves finding the "square root" of the Nyquist pulse shape and using it as the pulse-shaping filter at the transmitter and the noise-reducing filter at the receiver. The family of pulses which do not overlap at the sampling instant are found by constraining the solution to have a finite value at the origin and passing through zero at +/- kT, k being any nonzero integer, and T, the time between pulse transmissions. A practical radio will exhibit t_e as a random variable (timing jitter). The family of Nyquist pulses known as raised cosine is parameterized by a bandwidth parameter called *rolloff.* For zero rolloff, the pulse shape is $\sin(x)/x$ with tails declining as $1/t$ and the occupied bandwidth is W (see *Figures* 7 and 8).

Figure 7 is a plot of one-and-a-half periods of the real part of a quadrature phase shift keying (QPSK) signal in volts, versus time, similar to what an engineer sees on an oscilloscope. It is called an eye diagram because it looks like an open eye. Additive noise makes the voltage waves shown in the figure change position arbitrarily up or down a little bit gradually, about once per symbol period. If there is enough noise power, the eye closes completely, which causes a simple demodulator to confuse many samples as representing 1 which actually represent 0 and vice versa, thus resulting in bit errors. Or, if there are echoes in the channel (that

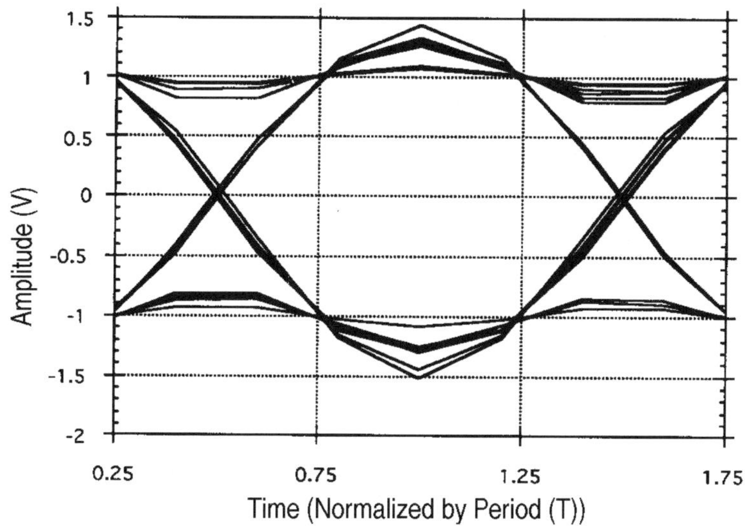

Figure 9 *Eye diagram for 100% rolloff pulse shape (high bandwidth)*

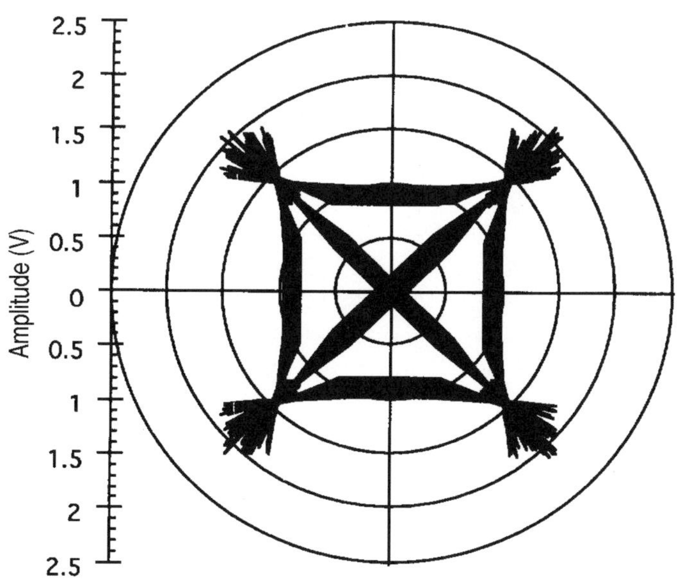

Figure 10 *Polar plot for 100% rolloff pusle shape*

is, delay spread), even if there is no noise, some lines in the eye will shift an arbitrary distance left or right along the time axis, in a random way, potentially closing the eye in a horizontal fashion, again leading to errors. In the case of sufficient delay spread, no matter how high the signal-to-noise power ratio is, errors will occur at a predictable rate called the error floor. *Figure 8* shows the magnitude and phase of a modem waveform plotted at discrete intervals (5 samples per period) over a continuous time span. The trajectory evolution of a 500-symbol QPSK overall $\sin(x)/x$ pulse-filtered sequence is shown. The modulation parameter giving the peak voltage to rms voltage ratio may be estimated from *Figure 8*.

Figures 9 and *10* are for 100% rolloff QPSK pulse shapes. Note that the eye of *Figure 9* is open wider in time and voltage than that of *Figure 8*. Thus, the 100% src pulse shape would be more robust against timing errors and large additive noise power. Clearly, the peak-to-average ratio for *Figure 10* is less than that for *Figure 8*. Peak-to-average is a critical constraint on linear power amplifier design for radio transmitters. Unfortunately 100% src pulses require about twice the bandwidth of 0% src pulses. For a rolloff of 1, the bandwidth is $2W$, but the tails of the pulse shape decline as $1/t^2$.

With a nonzero variance of t_e, the intersymbol interference (ISI) is determined by the number of neighboring pulses that have energy at the actual, suboptimal, sampling moment. Here lies a trade-off. The higher the rolloff, the less significant pulses' tails become and the less stringent the requirement on estimating t correctly. The higher the rolloff, the more bandwidth the signal occupies, and bandwidth is precious. If ISI is not viewed as a significant contributor to the total noise at the receiver, then a low rolloff shape may be selected, or the pulse may not even be of the Nyquist variety.

Systems like GSM and USDC operate at a fairly high symbol-error rate, such as 2% (meaning the noise is high), and bandwidth is an important consideration. The GSM pulse shape is from a large family called OQPSK, in particular Gaussian minimum shift keying. The USDC pulse shape is a 35% square root of raised cosine shape (Nyquist). The IS-95 chip pulse shape is not from the Nyquist family at all.

To understand these terms, we need to introduce the idea of representing a modulated signal in terms of real and imaginary parts. Both parts exist; it is a matter of convenience with Euler's identity

$$e^{jx} = \cos(x) + j\sin(x) \tag{2}$$

that one is called the real part and the other the imaginary. This holds up under linear operations. A nonlinear operation on the composite signal will not be equal to the sum of the nonlinear operations done separately on the real and imaginary parts.

The so-called real part of the signal is the part remaining after correlation with $\cos(w_c t)$, and the imaginary part is what is left (equal to the correlation of the signal with $\sin(w_c t)$). The sine and cosine of a given frequency are orthogonal after lowpassing. The product gives the sine of the sum and difference, the lowpass eliminates the sum, and the difference is sin(0)=0. They are conveniently generated in radio transmitters and receivers, from which we get the terminology of *I* and *Q*, where *I* is in-phase with $\cos(w_c t)$ and *Q* is 90° out of phase (quaternary).

A sequence of bits is placed on the *I* portion of the carrier by successive transmissions of $+P(t)$ or $-P(t)$ for a 1 or zero respectively, and by multiplying this stream by $\cos(w_c t)$. Thus

$$x(t) = \sum (2a_k - 1)P(t - kT)\cos(w_c t) \tag{3}$$

where $x(t)$ is the continuous-time, continuous-amplitude signal launched from the transmitting antenna, a_k is the kth bit transmitted, P is the pulse shape (with an impulse response corresponding to H_{tx}), k, the sum index, ranges over $-\infty \ldots \infty$, and w_c is the carrier frequency. $x(t)$ is a binary phase-shift keyed signal (BPSK). An eye diagram of x would be like *Figure 7* or *Figure 9*, since the real or imaginary part of a simple QPSK signal by itself is a BPSK signal. If we add a similar sum based on an independent information stream b_k and replace the cosine with the sine in this additional term, $x(t)$ now represents a quaternary phase-shift keyed signal (QPSK):

$$x_{QPSK}(t) = \sum (2a_k - 1)P(t - kT)\cos(w_c t) + \sum (2b_k - 1)P(t - kT)\sin(w_c t). \tag{4}$$

If the pulse peaks of the quadrature portion are offset in time from those of the in-phase part, the signal is OQPSK:

$$x_{OQPSK}(t) = \sum (2a_k - 1)P(t - kT)\cos(w_c t) + \sum (2b_k - 1)P(t - kT - T/2)\sin(w_c t). \tag{5}$$

If the QPSK signal has a continuous phase rotation component of π/4 every baud period, (T), it is called π/4 QPSK as in USDC and PDC:

$$x_{p/4}(t) = \sum (2a_k - 1)P(t - kT)\cos(w_c t) + \sum (2b_k - 1)P(t - kT - T/2)\sin(w_c t). \tag{6}$$

If we return to the OQPSK waveform and specify P to be a Gaussian function convolved with a ramp, then we have what is called GMSK (GSM and DECT).

We will now consider modulations which pass through the origin (*Figures 8* and *10*). Consider the $x_{QPSK}(t)$ waveform at time k=0 and suppose $a_0 = b_0 = 1$, for example. If at time $t = T$, (k=1), $a_1 = b_1 = 0$, then $x(T) = -x(0)$. Consider the evolution of x(t) as t progresses from 0 to T. Other than effects from pulses at $k = -1$ and k =2, $x(T/2)$ will be the average of $x(0)$ and $x(T)$; that is, 0 (this is called "passing through the origin" and exhibiting a "high peak-to-average"). This means that the transmitter must faithfully produce signals with an amplitude of $P(0) + jP(0)$ down to 0, a large dynamic range. If the QPSK amplifier does not have the required dynamic range, it will distort the signal-causing creation of energy outside the assigned bandwidth and difficult demodulation at the receiver. The primary reason for selecting OQPSK or π/4 QPSK is that they do not pass through the origin during symbol transitions and do not require an amplifier with as large a dynamic range. Such modulations are sometimes called constant envelope modulations, which is short for "more nearly constant envelope than QPSK."

It may be that although reasonable care has been taken in the design of the modulator and demodulator, the error rate is still too high at the receiver for the specified noise and channel conditions. In this case, coding may be called for in the modem design. Coding is an operation undertaken to lower the received bit error rate for a given information rate and channel bandwidth. The capacity of an idealized channel at a given signal-to-noise ratio has been given by Shannon (1948). Coding is one step in bringing a practical modem closer to the theoretical capacity limit.

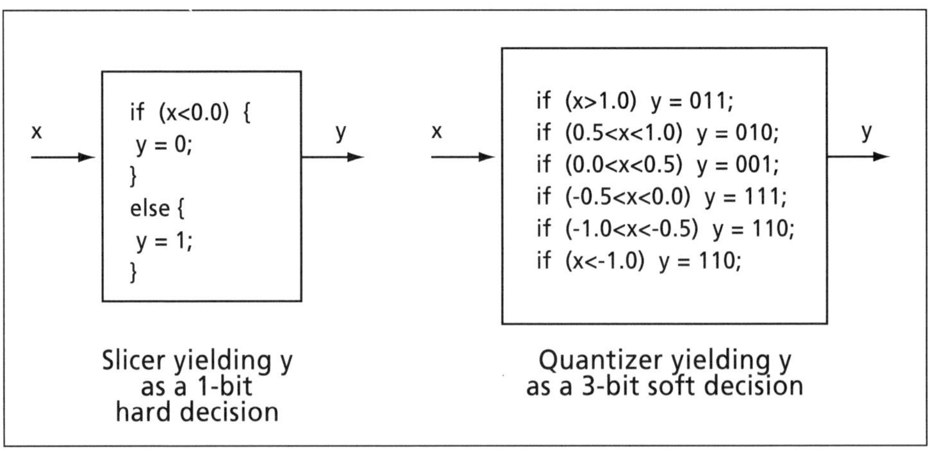

Figure 11 *Mapping of demodulator sample* x *to decoder input* y

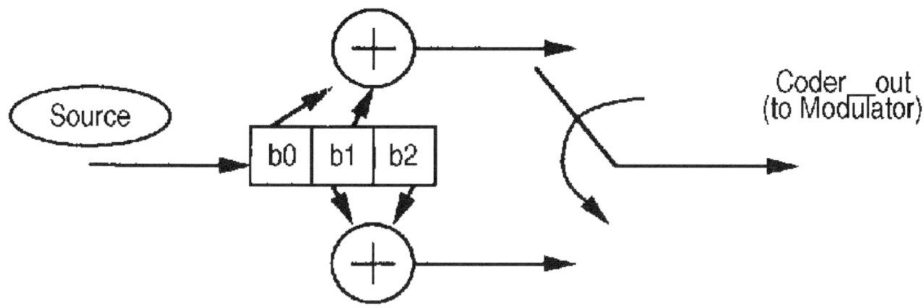

Figure 12 *Rate 1/2 convolutional encoder (1 bit in, 2 out)*

There are two kinds of codes commonly in use, *convolutional* and *block*. Both are linear codes. At least one reason for the use of linear codes is that they are amenable to mathematical design and analysis, as are the corresponding decoders. Block codes accept a block of k bits and produce an n-bit codeword, which is modulated and transmitted over the channel. The receiver removes the carrier terms ($\cos(w_c t)$ and $\sin(w_c t)$) and retains estimates of the transmitted level $I_k = 2a_k - 1$ and $Q_k = 2b_k - 1$. If these estimates pass through a decision device and emerge quantized to the nearest neighbor I_k and Q_k, we consider this a hard-decision input to the decoder. If the quantization retains more bits than just enough to distinguish I_k or Q_k, it is called a soft-decision input (see *Figure 11* for an example). Soft-decision decoders perform better but at higher computational cost than hard-decision decoders.

A contrived convolutional encoder is shown in *Figure 12.* Convolutional encoders are extremely easy to implement in digital hardware, as one can see from the figure. This simplicity of form is not lost as the rate of the code decreases (more bits out for each bit in). The expression "convolutional" comes about from representing the input, code-transfer function and output bit streams as sequences whose elements are drawn from the finite field with two elements, 0 and 1. Denote these functions as x[n], h1[n], h2[n], y1[n] and $y2$[n], where y1[n] is the sequence obtained from the upper output spigot in *Figure 12* and y2[n] is the other output. The orderly selection of samples from y1[] and y2[] comprises the output. $y1$[n] is equal to the convolution of $h1[n]$ with $x[\,]$; similarly for $y2$[] based on h2[]. The multiplication and addition operations used in convolution are those defined for a binary Galois field, GF(2) –, that is, 0• 0=0; 0• 1=0; 1• 1=1; 0+0=0; 0+1=1; and 1+1=0 (Blahut 1983).

The convolutional encoder operates on the input sequence to produce an output

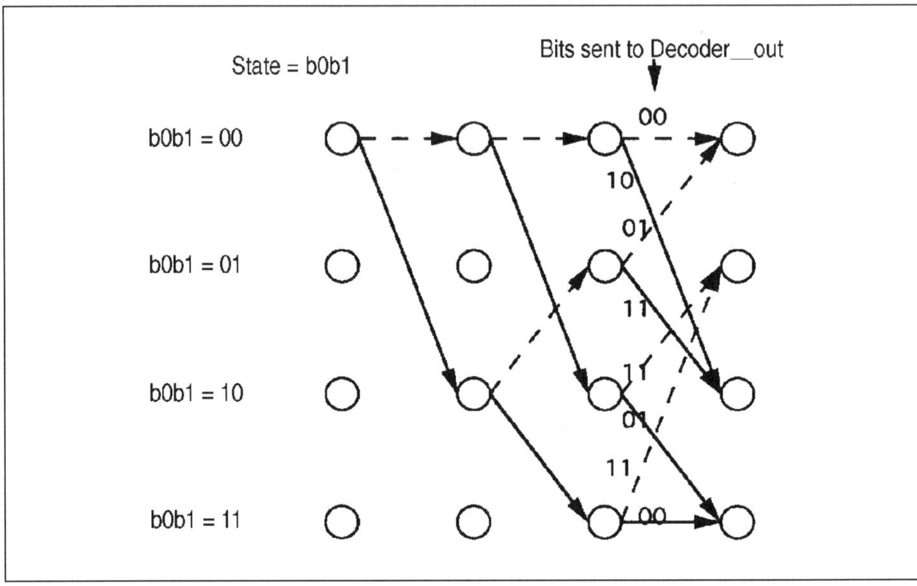

Figure 13 *Trellis corresponding to Figure 12*

sequence. A good encoder will produce significantly different output sequences for two input sequences which are minimally different. This separation of output sequences is called the minimum distance of the code and is very important in determining how good the code is.

Consider the simple encoder shown in *Figure 12*. The encoder generators are {1,1,0} and {0,1,1}. Sometimes the encoder is specified by giving these numbers in octal, in this case 63. This is not a coder that would ever be implemented; it has been chosen for ease of discussion. The constraint length, *L*, is the number of flip-flops in the shift register, in this case 3. The number of output bits, *k*, per input bits, *n* in this case, is 2 output per 1 input. The number of states is 2^{L-1} in this case we have 4 states (see *Figure 13*). The state of the encoder evolves after each *k* input bit has been mapped to *n* output bit. The state is determined by the first *L*-1 bits remaining in the shift register before the new information bit arrives. In *Figure 12*, those bits may be 00, 01, 10, or 11 (or 0, 1, 2, or 3 in decimal).

Now we will consider the use of the trellis shown in *Figure 13* for completely describing the behavior of the encoder circuit in *Figure 12*. The shift register in *Figure 12* is loaded with 0, 0, *X* (the value of *x* can vary) before encoding begins. Hence, the encoder begins encoding with the state being 0 (as opposed to state 1, 2, or 3 in decimal). If the first bit is a 0, the encoder produces the doublet (0,0) at its output and progresses to state 0 at the next input bit time. If the first input bit is a 1, the encoder produces the doublet (1,0) and progresses to state 2 (10 in binary). From state 2, it may evolve to state 1 or 3 depending on whether the input bit is a 0 or a 1.

It is not necessary to construct the trellis at the encoder; the circuit of *Figure 12* is all that is needed. But at the decoder, the trellis is constructed to aid the receiver. Without coding, n bits could have represented one of 2^n different sequences. But when the n bit sequence is produced by the action of a convolutional encoder on k input bits, $k<n$, we realize that the n output bits can only represent one of 2^k sequences. The encoder serves to make these sequences as easy as possible to tell apart at the receiver. In practice, the receiver makes use of the trellis to decide which sequence it is most likely that the received noisy sequence, $R[n]$ (notation of *Figure 4*), represents (thus the term *maximum likelihood* decoder). The method for picking a path through the trellis is called the Viterbi algorithm. Notice that once we decide which codebits were sent (the selected path through the trellis), there is a 1:1 mapping to what the information bits were. This is obvious, but it does require attention in the actual decoding circuit.

Here, then, is the Viterbi algorithm. The receiver has some kind of frame synch, so it knows when the transmitter has begun to encode the information in question. The transmitter always starts in a known state, for example, state 0. Let us call the hardware or software that implements the Viterbi algorithm the *elf.* The elf is in possession of $R[n]$, the trellis, and the fact that the transmitter starts in state 0. The elf knows that at time $v = 0$, states 1, 2, and 3 are unlikely to an extreme, so these are initialized with large negative numbers, like -1000, but the elf sets metric [0]=0 (0>>-1000). The elf will create a measure which is a sufficient statistic for the maximum likelihood function for each state at time v after processing $R[0] \ldots R[v]$. These measures are called the cumulative metric [i], i=0, 1, 2, or 3. The elf examines all possible outcomes. All outcomes must come from one of the four states and end at one of the four states. In this way, the elf at time $v+1$ may consider how a sequence may have come to state 0. It must either come from state 0 or state 1 at time v (see *Figure 13*). In the first case, a 00 was sent, in the second case, a 01 was sent. The elf compares $R[v]$ with 00 and calls the result delta_00, and compares $R[v]$ with 01 and calls the result delta_01. Then, the elf makes a decision about how the data may represent a sequence passing through state 0 at time v+1. The elf compares metric[0]+delta_00 to metric[1]+delta_01; the larger value indicates to the elf what state to label as the ancestor to being in state 0 at time $v+1$. The information concerning the losing state is forgotten, metric[0] takes on the value of the winning sum, and the predecessor state is added to a list which shows how one arrived at state 0 at time v+1 with new metric[0]. Still at time v+1, the elf now "stands" on state 1 and looks back to time v and sees that one may arrive at state 1 via states 2 or 3. The ancestor for state 1 is chosen as either 2 or 3, depending on the outcome of comparing metric[2]+delta_10 to metric[3]+delta_11. States 2 and 3 also have ancestors determined at time v+1, and then the cycle is complete. R[v+2] is processed, starting with finding an ancestor for state 0 at time v+2.

Decoding of the received sequence up to time v-TRACEBACK is done by finding the maximum of metric[i], i=0, 1, 2, or 3 at time v. The list of ancestors attached to this winning state is examined. The information bit corresponding to the arrival at the winning state's ancestor v-TRACEBACK cycles previously is declared the decoded information bit. Hence, there is a delay of v-TRACEBACK in the decoder. TRACEBACK is usually chosen to be something like 5L.

This explanation of the convolutional decoder is original to the best of the author's knowledge. It has been found to get the trellis searching idea across quickly and positively.

An example of a practical convolutional code is the one used in IS-95 (cellular CDMA). For this coder, L = 9 (256 states) and, at the mobile transmitter, it is a rate 1/3 code (1 bit in, 3 bits out).

Conclusion

Coding and modulation for wireless applications is a very broad topic. We have mentioned the elementary idea of trying to pass information through a channel. The varying characteristics of the source have been seen to dominate the entire modem design. The need for understanding the dynamics of conversation versus the costs and benefits of coding has been shown for the very important wireless application in which the source is voice. We have proceeded from the source to the coder and explained the objective and means of coding. Reaching the antenna neighborhood of the modem, we have explained what modulation for digital transmission is. The different purposes of transmit filtering and receive filtering were also considered in the discussion of allocated bandwidth and matched filtering. Finally, a detailed convolutional encoder and decoder example was provided, including the Viterbi algorithm.

We will now point out some references for further reading along with a hint at the difficulty of their use for a newcomer to this area. McGillem and Cooper (1986) is a text which could be used at the undergraduate level and covers AM, FM, digital modulation, and demodulation and includes an introduction to direct-sequence and frequency-hop spread-spectrum. Taub and Schilling (1986), another undergraduate text, presents similar information, along with a chapter on error-control coding. Proakis (1983) is concerned with digital modulation and demodulation, surveys many cases of practical interest, and discusses spread-spectrum at some length as well as coding. Blahut (1990) treats with reasonable detail everything from the matched filter and phase modulation through coding and on to capacity considerations for FDMA, TDMA, and spread-spectrum. Turning to the area of coding specifically, we notice that there are few introductory texts suitable at the undergraduate level. After looking at Taub and Schilling and Proakis, one may turn to Man Young Rhee (1989) and finally Blahut (1984). Two interesting books on radio system design are Lee (1982) – a perennial favorite with engineers – and Hess (1993), which is a bit closer to describing the design decisions that go into selecting and deploying a system.

Works Cited

Blahut, R.E. *Theory and Practice of Error Control Codes.* (Addison Wesley, Reading Massachusetts, 1984).

Blahut, R.E. *Digital Transmission of Information.* (Addison Wesley, Reading Massachusetts, 1990).

Hess, G. *Land Mobile Radio System Engineering.* (Artech House, Boston, 1993).

Lee, W.Y.C. *Mobile Communications Engineering.* (McGraw-Hill, New York, 1982).

McGillem and Cooper. *Modern Communications and Spread Spectrum.* (McGraw-Hill, New York, 1986).

Nyquist, H. "Certain Factors Affecting Telegraph Speed." *Bell System Tech. J.* vol. 3, no. 2, 324 - 346, 1924.

Proakis, J. *Digital Communications.* (McGraw-Hill, New York, 1983).

Rhee, M.Y. *Error Correcting Coding Theory* (McGraw-Hill, New York, 1989).

Shannon, C.E. "A Mathematical Theory of Communication." *Bell System Tech. J.,* vol. 27 (July 1948) 379 - 423.

Sklar, B. *Digital Communications, Fundamentals and Applications,* (Prentice Hall, Englewood Cliffs, New Jersey, 1988).

Taub, H., and D. Schilling. *Principles of Communication Systems.* (McGraw-Hill, New York, 1986).

Ungerboeck, G. "Trellis-Coded Modulation with Redundant Signal Sets." *IEEE Comm. Mag.,* vol. 25 (February 1987), 5 - 21.

Coded Modulation Techniques for Cellular Mobile Systems

Masoud Salehi
Professor, Department of Electrical and Computer Engineering
Northeastern University

John G. Proakis
Chairman, Department of Electrical and Computer Engineering
Northeastern University

Abstract

Trellis-coded modulation is an attractive coding technique for transmission through bandlimited channels. On Gaussian channels this coding scheme achieves high coding gains without expanding the bandwidth. Cellular mobile channels are also severely bandlimited but they can not be modeled as Gaussian channels due to the effects of fading and shadowing. New code design techniques are required for designing good trellis codes for fading channels. In this paper we present some of the techniques for designing good trellis codes for fading channels and also summarize the works that have been published in the technical literature on this topic. We also elaborate on important parameters that effect the performance of trellis codes over fading channels.

Introduction

The ever-increasing demand for mobile radio communications has made analog cellular systems reach their capacity limitation in the late 1980s. It is expected that the market for digital cellular radio telephony will increase dramatically during the 1990s. Today, much research activity is concentrated on various aspects of digital techniques used for information transmission through the mobile radio channel. This body of research comprises diverse areas such as antennas and propagation, channel characterization, data compression, multiaccessing techniques, modulation-demodulation schemes, and coding-decoding methods.

The digital mobile environment is characterized by various types of fading due to terrain configuration, multipath fading from man-made structures, and mountains, which result in a dispersive communication medium and hence cause intersymbol interference (ISI). The move from analog communication techniques to digital systems has a dramatic effect on the noise and interference immunity of the system, especially under the harsh environment of a mobile channel and a limited transmitter power. The main advantage of a digital communication system is that it allows only a finite number of levels (signals) to be transmitted. In a binary system this number is two, and in general in an M-ary system this number is M. In most practical applications, these signals are modulated on a carrier using either quadrature amplitude modulation (QAM) or M-ary phase shift keying (MPSK). Both QAM and MPSK signals consist of two components, one modulated on cos $(2\pi f_0 t)$ (in-phase component) and one modulated on sin $(2\pi f_0 t)$ (quadrature component), where f_0 is the carrier frequency. These signals can be easily visualized by their signal constellation, a representation of each signal as a point in the plane whose components are the in-phase and the quadrature components of the corresponding signals. *Figure 1* shows a 16-QAM and an 8-PSK signal constellation.

In analog systems a range of transmitted values are possible, thus resulting in a continuum of possible values. The conclusion is that in analog systems the possible transmitted signals are very close together, which makes their recognition difficult. However, in digital systems the possible levels (signals) are rather far apart, and therefore are better recognizable when received through a noisy channel. The higher the distance between signal levels, the higher the noise immunity of the system.

The bit rate of an M-ary digital system is given by

$$R_b = R_s \log_2 M \text{ bits per second} \tag{1}$$

where R_s is the signaling rate (the number of signals transmitted per second) and is determined by the channel bandwidth. From the above relation it is seen that

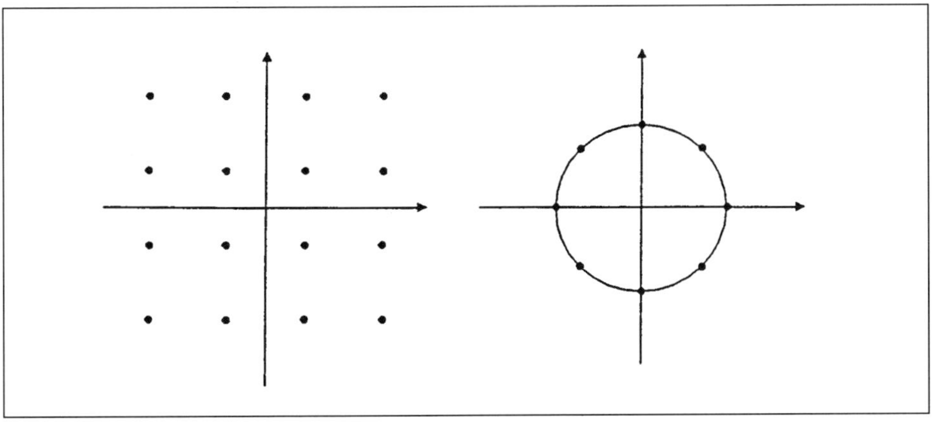

Figure 1 *16-QAM and 8-PSK signal constellations.*

in order to increase the bit rate, one has to either increase R_S (i.e., increase the bandwidth) or increase M (use more signaling levels). At a constant power level, increasing M has the effect of decreasing the distance between adjacent levels and thus reducing the noise immunity of the system. One way to cope with this is to increase the power level of the transmitter simultaneously, such that with increasing M the distance between the adjacent levels remains constant and therefore the error probability (which is a measure of noise immunity) is not decreased. The conclusion of the above discussion is that one cannot increase the bit rate without increasing the bandwidth or the power level of the transmitter.

One way to increase the noise immunity without increasing the power is by *coding*. In the same way that moving from analog to digital results in fewer levels with larger distances between them, by employing coding one transmits sequences of bits (or M-ary levels) that are far apart. In other words from all possible sequences only a fraction are chosen as code words, or legitimate transmitted sequences, thus providing even higher distance between these sequences, and therefore a higher level of noise immunity. The error performance of a system that employs coding is equivalent to the error probability of a system without coding but operating at a higher power level. This difference in power level between a coded system and an uncoded system is called the *coding gain*.

Two general classes of codes have been widely used in communication systems, *block codes* and *convolutional codes*. In block codes, k information bits are mapped into n binary symbols to be transmitted over the channel. The next k bits are mapped independently of the previous mapping. The rate of this code is k/n.

In convolutional codes, at each instant of time, n_0 binary symbols are transmitted over the channel. These binary symbols depend on the most recent $m = Lk_0$ information bits to be transmitted over the channel (the recent k_0 bits as well as the past $(L - 1)\ k_0$ bits), where m is called the constraint length of the code. For convolutional codes the rate of the code is k_0/n_0. A convolutional encoder is implemented by a shift register.

In block codes we are dealing with sequences of length n where not all of them are used, thus providing higher distance between the code words. In convolutional codes not all binary sequences of arbitrary length can be code words, only those which are the outputs of the shift register that generates the codes are valid sequences.

Both block and convolutional codes expand the bandwidth since both transmit longer sequences (n versus k and n_0 versus k_0) and these longer sequences have to be transmitted in the same time duration of the original sequence, which results in narrower pulses and therefore higher bandwidth. Therefore these systems are used in power-limited scenarios where bandwidth is not a premium, such as deep space communication systems. In cases where bandwidth is strictly limited trellis coded modulation (TCM) can provide considerable coding gain without expanding the bandwidth.

Trellis Coded Modulation

Trellis coded modulation is based on the idea of interpreting coding and modulation as a single entity rather than first doing coding and then modulating the encoded sequence. The resulting coding scheme provides sequences of channel symbols that are well separated and thus less vulnerable to errors, without increasing the bandwidth. The fundamental idea in trellis coded modulation is that of set partitioning in which the signal constellation is partitioned into congruent subsets such that the intrapartition distance, defined as the minimum distance between two signal constellation points within a partition, is maximized. A partitioning of a 16-QAM constellation into eight partitions is depicted in *Figure 2,* where the constellation is partitioned into subsets denoted by A, B ,C, D, E, F, G, and H. We have also shown the partitioning of an 8-PSK constellation into four subsets in the same figure.

The TCM Encoder

The encoder for a TCM coding scheme consists of a convolutional encoder and a

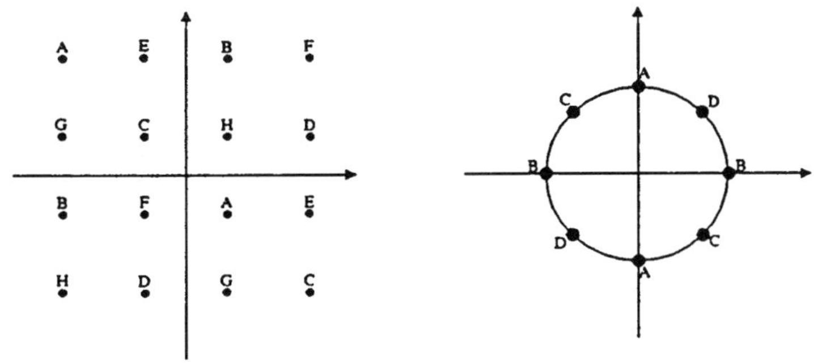

Figure 2 *Set partioning for l6-QAM and 8-PSK signal constellations.*

mapping section. In general k information bits are broken into k_1 and k_2 bits such that $k = k_1 + k_2$. The convolutional encoder is a rate $k_1/(k_1 + 1)$ encoder that takes k_1 bits at the input and generates $k_1 + 1$ binary digits at the output. These $k_1 + 1$ binary digits are used to select one of the 2^{k_1+1} partitions, and the remaining k_2 binary digits are used to choose one of the 2^{k_2} points within the selected partition. This means that the constellation contains $2^{k_1+1} \times 2^{k_2} = 2^{k+1}$ signal points, or twice the number of the binary digits transmitted. A block diagram for a general TCM encoder is shown in *Figure 3*.

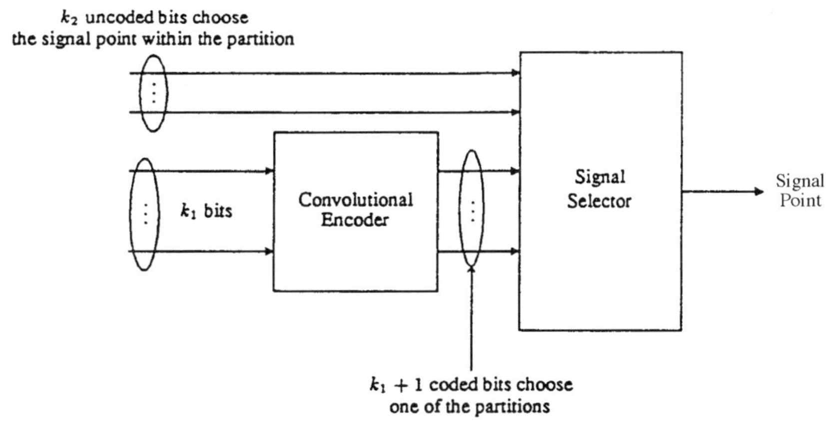

Figure 3 *A TCM modulator.*

The case of k_1=2, k_2=1 can be used in conjunction with a 16-QAM constellation partitioned into eight subsets. The two bits supplied to the convolutional encoder are mapped into three bits, which are used to choose one of the eight partitions A, B, C, D, E, F, G, and H. The remaining encoded bit is used to choose one of the two signal points within the chosen partition. The case of $k_1 = k_2 = 1$ can be used with the 8-PSK constellation partitioned into four subsets in a similar way.

The bandwidth requirement for the system with $k_1 = 2$, $k_2 = 1$ and employing the 16-QAM constellation is exactly the bandwidth needed for an encoded system employing an 8-PSK or 8-QAM signaling scheme, because in both systems for each sequence of three information bits ($k_1 + k_2 = 3$), one signal is transmitted. However in the uncoded system this signal is chosen from a constellation with eight points and in the TCM system it is chosen from a constellation of 16 points. If the transmitted power remains the same, the distance between the constellation points in the 16-QAM system is less and therefore the system is more vulnerable to errors. However, it should be noted that in the TCM system, the convolutional encoder generates some redundancy. In other words, in a TCM system, unlike the 8-PSK or 8-QAM system with no coding, not all sequences of points chosen from the constellation are valid transmitted sequences. Since only certain sequences are valid, the distance between these sequences can be more than the distance between the corresponding sequences in an uncoded system. It turns out that by judicious choice of the convolutional encoder it is possible to achieve sequences with better distance properties compared to uncoded systems. This was shown by Ungerboeck,[1] who found some good codes by exhaustive search and formulated a method for designing good codes.

The TCM Decoder

Assuming that the signal sequence generated by the TCM encoder is transmitted through an additive white Gaussian noise (AWGN) channel, the task of the decoder is to find a sequence of bits such that when this sequence is encoded by the TCM modulator the resulting signal sequence is the closest signal sequence (in Euclidean distance) to the received sequence. This task is performed by the well known Viterbi algorithm.[2]

The Viterbi algorithm is an efficient trellis search technique that finds a path through the trellis such that its Euclidean distance from the received sequence is less than all other paths. Since there exists a one-to-one correspondence between the paths through the trellis and the information bits, finding this path is equivalent to finding the transmitted bits. *Figure 4* shows a four-state trellis used for the

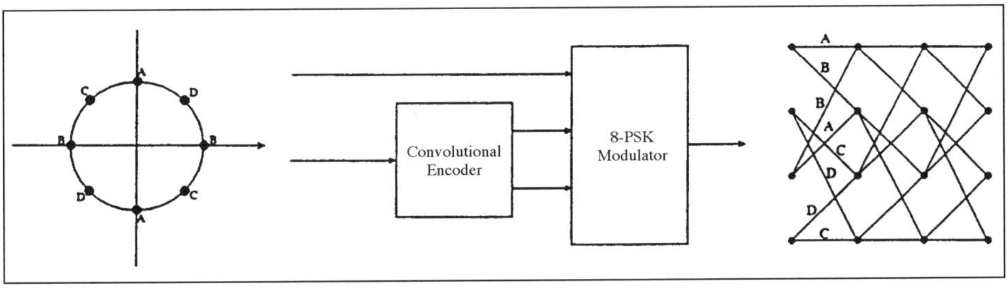

Figure 4 *Partitioned signal set, encoder, and the decoding trellis for a four-state, 8-PSK TCM system.*

decoding of the 8-PSK scheme described previously. The partitioned constellation and the TCM modulator are also show in this figure.

Note that the number of states and the topology of the trellis depend entirely on the particular convolutional encoder employed. Each link in the trellis of *Figure 4* is denoted by A, B, C, and D. As previously discussed, each of the subsets corresponds to two signal points in the 8-PSK signal constellation as shown in the partitioned signal space in *Figure 4*. Therefore, each link in the trellis corresponds to one of the four partitions, or equivalently to two signal points.

In the decoding process first the received sequence is compared to the two signal points corresponding to each link. The signal point closest to it is chosen and the distance between the received signal and this signal point is computed. This distance is subsequently used in the Viterbi algorithm to find the transmitted bits.

Performance of Trellis Coded Modulation

The performance of the trellis coded modulation system depends on the distance properties of the code. To a first approximation this depends on the free Euclidean distance (FED) of the code, which is defined as the minimum Euclidean distance between any two paths through the trellis, and denoted by D_{fed}. If the number of paths at distance D_{fed} from a given path is N, then the error probability is approximated by

$$p_e \cong N Q\left(\sqrt{\frac{D_{\text{fed}}^2}{2N_0}}\right) \tag{2}$$

where

$$Q(x) = \frac{1}{\sqrt{2\pi}} \int_x^{\infty} e^{-t^2/2} dt \tag{3}$$

and $N_0/2$ is the (two-sided) noise power spectral density. It turns out that depending on the complexity of the encoder, TCM codes can provide a coding gain between 3 dB to 6 dB over the uncoded systems.

It is obvious from the relation that gives the error probability that the key issue in designing good TCM schemes is to make D_{fed} as large as possible while keeping N low. However, since Q (·) is essentially an exponential function, the error probability is mainly determined by the FED of the code. Ungerboeck formulated some simple rules for designing trellises with good distance properties.[1] Coding gains of up to 6 dB are typically achieved on the AWGN channel.

TCM for Mobile Channels

The impressive coding gain of trellis coded modulation which is obtained at no increase in bandwidth requirements of the communication channel makes this scheme particularly attractive for transmission in cellular mobile systems. In cellular systems, particularly in large urban areas, the demand for bandwidth is very high and therefore TCM is preferable to other coding alternatives.

Mobile Channel Characterization

Mobile channels are characterized by high attenuation and shadowing, additive Gaussian noise, strong multipath fading, and co-channel interference. The terrain conditions and the low antenna height in mobile transmitters contribute to high attenuation, and man-made structures surrounding the mobile units contribute to multipath fading. The fading characteristics of the mobile channel can be described as Rayleigh fading.

In a Rayleigh fading channel the magnitude of the (time-varying) lowpass equivalent channel impulse response has a Rayleigh density function and its phase is uniformly distributed. The main characteristics of a fading channel are expressed in terms of the multipath spread, denoted by T_m and the Doppler frequency spread, denoted by B_d. The multipath spread is a measure of the time spread of the impulse response of the channel, and the Doppler frequency spread is a measure of how fast the characteristics of the channel vary with time. Two closely related quantities are the coherence time of the channel, denoted by T_c, and the coherence bandwidth of the channel, denoted by B_c. The relation between coherence time and bandwidth and multipath spread and Doppler frequency spread are given by

$$T_c = \frac{1}{2B_d} \tag{4}$$

and

$$B_c = \frac{1}{2T_m} \tag{5}$$

Coherence time is a measure of the time interval over which the channel characteristics will remain essentially constant, and the coherence bandwidth is a measure of the bandwidth over which the channel characteristics are highly correlated.

If the signal bandwidth W is less than the coherence bandwidth of the fading channel, then all frequency components of the transmitted signal are subject to the same attenuation and phase shift. In this case we have frequency nonselective fading. If the signal bandwidth is higher than the coherence bandwidth of the channel, we are dealing with frequency selective fading. If the signaling interval T is much smaller than the coherence time of the channel, the channel characteristics will not change during the transmission of one signal. Under this condition we have a slowly fading channel. Conversely if $T \gg T_c$ the channel is a fast-fading channel. The slow fading condition implies that the channel characteristics vary slowly enough to be measured.[3]

The average multipath spread of local scatterers (surrounding buildings, trees, etc.) is in the 0.5-3, μ-sec range depending on the denseness of the local buildings. There exists yet another type of multipath fading from high rise buildings or from surrounding mountains which is characterized by higher average multipath spread, in the 20-100 μ-sec range.[4] The Doppler frequency spread is given by $B_d = v/\lambda$, where v is the speed of the vehicle and λ is the wavelength of the signal. For a transceiver moving at v = 90 km/hour, assuming a carrier frequency of 900 MHz, we have $\lambda = 3 \times 10^8/9 \times 10^8 = 1/3$ meters, and therefore, B_d = (90000/3600)/(1/3) = 75 HZ.

Coded Modulation Performance over Fading Channels

We have already described the basic principles of trellis coded modulation and gave an approximate expression (which is tight at high SNR's) on the resulting error probability. We observed that the performance of a TCM system is essentially determined by its free Euclidean distance D_{fed}. The essential role played by the

FED (or the Euclidean distance in general soft decision decoding) comes from the basic characteristics of the Gaussian noise. Under the existence of fading, although Euclidean distance still plays an important role, some other properties of the code have a more prominent effect on the performance of the system. In particular it turns out that the codes designed by Ungerboeck, which provide up to 6 dB coding gain in AWGN channels, cannot perform equally well in fading environments.

The performance of trellis coded modulation systems on Rician and Rayleigh fading channels is addressed in the paper by Divsalar and Simon,[5] where it is shown that parameters other than the free Euclidean distance are more important in the overall code performance. In particular it is shown that for Rayleigh fading channels and Rician fading channels with low Rician parameter κ, the *effective code length* (ECL) and the *product distance* (PD) of the code are important factors in code design. The effective code length of a TCM system is the minimum number of branches along which the diverging code sequences differ (or the length of the shortest *error event path*), and the product distance of the code is a measure of how uniformly the Euclidean distance between two diverging paths is distributed among the branches of the two paths. More formally, the product distance of a code is the product of the Euclidean distances of the branches along the shortest error event path. The more uniformly the Euclidean distance is distributed among the branches of the shortest error event path, the higher the PD and the better the performance of the code.

It can be shown that for a Rician fading channel with Rician parameter κ, under the assumption of the availability of channel state information at the decoder (i.e., when the fading state of the channel is known to the decoder),[6] the error probability (under a normalized energy assumption) can be asymptotically approximated by

$$p_e \cong \alpha\left(L, d_p^2(L)\right) \frac{\left(\left(1+\kappa\right)e^{-\kappa}\right)^L}{\left(\frac{1}{4N_0}\right)^L d_p^2(L)} \tag{6}$$

where L is the effective code length, $d^2{}_p(L)$ denotes the product distance of the code and $\alpha[L, d^2_p(L)]$ is the number of code sequences at the effective code length distance from the transmitted sequence. In the above we have assumed that the energy of the transmitted signal is normalized to unity. The product distance of the code is given by

$$d_p^2(L) = \prod_{n:\hat{x}_n \neq x_n} \left|\hat{x}_n - x_n\right|^2 \tag{7}$$

where $\boldsymbol{x}$ is the correct path and $\hat{\boldsymbol{x}}$ is the *shortest error event path* defined as the error event path with the smallest Hamming distance from the correct path. The length of the shortest error event path is of course equal to L, the effective code length. The special case of Rayleigh fading corresponds to $\kappa = 0$, in which case

$$p_e \cong \alpha\left(L, d_P^2(L)\right) \frac{1}{\left(\frac{1}{4N_0}\right)^L d_P^2(L)} \tag{8}$$

In order to decrease the error probability, a code must have a large L and $d^2_p(L)$. It should also be noted that the above results are based on the assumption that perfect interleaving is employed in the transmission of the coded data. Interleaving causes the adjacent bits to be spread over a long time frame such that the characteristics of the channel are independent from one transmission to the next one and therefore the transmission of the signals are almost independent.[7]

Comparing these results with the error probability of TCM on a Gaussian channel it is seen that in a fading channel the error probability varies inversely with the Lth power of the SNR, whereas in a Gaussian channel this dependence is exponential. This is very similar to the error probability of a fading channel when time diversity is employed. In fact, using a trellis code with an effective code length of L is equivalent to introducing L^{th}-order time diversity. This is why the parameter L is sometimes referred to as the *time diversity of the TCM code.* Again it is worth emphasizing that the above results provide upper bounds on the error probability that are exponentially tight at high signal-to-noise ratios, and in the case of Rician channels the upper bounds are only useful for low values of the Rician parameter κ. For larger values of κ the free Euclidean distance has a significant effect on the performance of the code.

The above discussion shows that a good TCM code for the Gaussian channel is not necessarily a good code for the fading channel. It is quite possible that a trellis code has a large Euclidean distance but it has a low effective code length or product distance. In particular, some of the good codes designed by Ungerboeck for the Gaussian channel have parallel branches in their trellises, which is to be avoided for fading channels because the existence of parallel branches severely decreases the effective code length (in fact reduces it to unity).

TCM Code Design for Fading Channels

In order to design a TCM code that performs well on a Rayleigh fading channel,

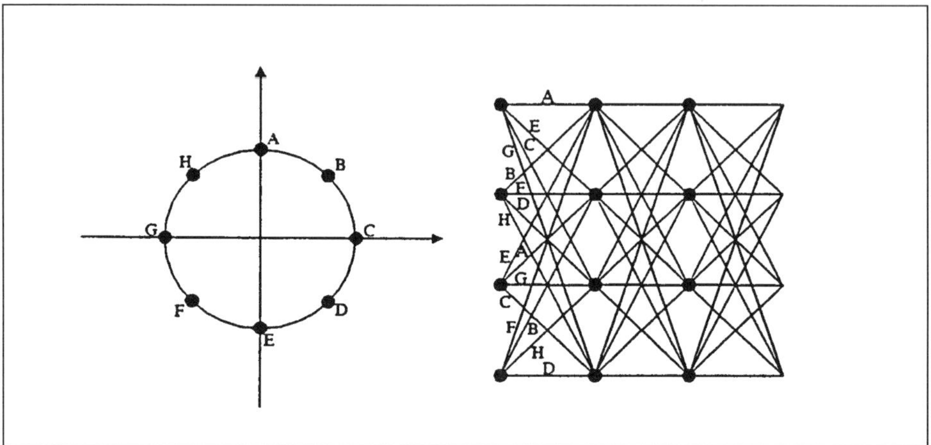

Figure 5 *Wilson and Leung's signal constellation and trellis for the TCM scheme.*

one has to look for codes with high effective code length and product distance. It turns out that the effect of the product distance is more noticeable at lower values of the effective code length. At higher values of the effective code length the dependence of the code performance on the product distance is less noticeable. Using the basic properties of convolutional codes, it can be shown that the maximum effective code length is an almost linearly increasing function of the constraint length of the code and a decreasing function of the parameter k_0, the number of input bits.

The problem of good code design for fading channels has been studied by a number of authors. Wilson and Leung designed a 4-state 8-PSK TCM system that avoids parallel branches.[8] The trellis diagram for this code and the signal constellation are shown in *Figure 5*.

As seen in *Figure 5*, the trellis corresponding to this code is a fully connected trellis and there are no parallel branches on it (i.e., each branch of the trellis corresponds to a single point in the constellation). The effective code length for this trellis is 2, therefore the error probability is inversely proportional to the square of the signal-to-noise-ratio. This means that this code is capable of achieving a time diversity equal to 2. The squared product distance provided by this code is 1.172. Note that the equivalent optimal system for a Gaussian channel contains parallel branches as shown in *Figure 4*. It can be easily verified that the squared free Euclidean distance for this code is $d^2_{\text{fed}} = 2.586$, therefore the coding gain of the TCM scheme in *Figure 5*, when used for transmission over an AWGN channel, is 1.1 dB which is 1.9 dB inferior to the coding gain of the Ungerboeck code of comparable complexity.

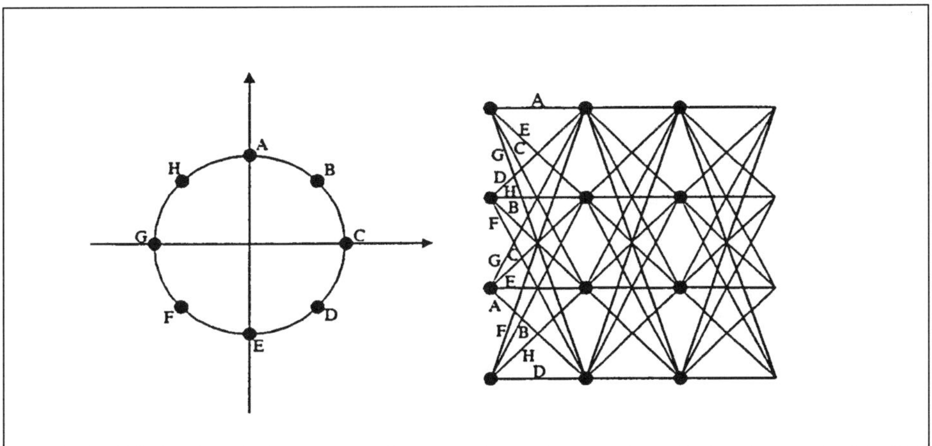

Figure 6 *Signal constellation and trellis for the code designed by applying the rules of Jamali and Le-Ngoc.*

Schlegel and Costello introduce a class of 8-PSK rate 2/3 TCM codes for various constraint lengths.[9] The search for good codes in this work is done among all codes that can be designed by employing a systematic convolutional code followed by mapping to the 8-PSK signal constellation. It turns out that the advantage of this design procedure is more noticeable at higher constraint lengths. In particular, this design approach results in the same codes obtained by Ungerboeck when the constraint length is small.[1] At high constraint lengths these codes are capable of providing both higher effective code lengths and higher product distances compared to the codes designed by Ungerboeck. For example, for a trellis with 1024 states, these codes can provide an effective code length of 5 and a (normalized) product distance of 128. For comparison, the Ungerboeck code with the same complexity can provide an effective code length of 4 and a product distance of 32.

Du and Vucetic employ Gray coding in the mapping from convolutional code output to the signal constellation.[10] An exhaustive search is performed on 8-PSK TCM schemes, and it is shown that particularly at lower constraint lengths these codes have a better performance compared to those designed. by Schlegel and Costello.[9] As the number of states increases, the performance becomes better than the codes designed by Schlegel and Costello. As an example, for a 32-state trellis code, the approach of Du and Vucetic results in an effective code length of 3 and a normalized product distance of 32, whereas the corresponding figures for the code designed by Schlegel and Costello are 3 and 16 respectively.

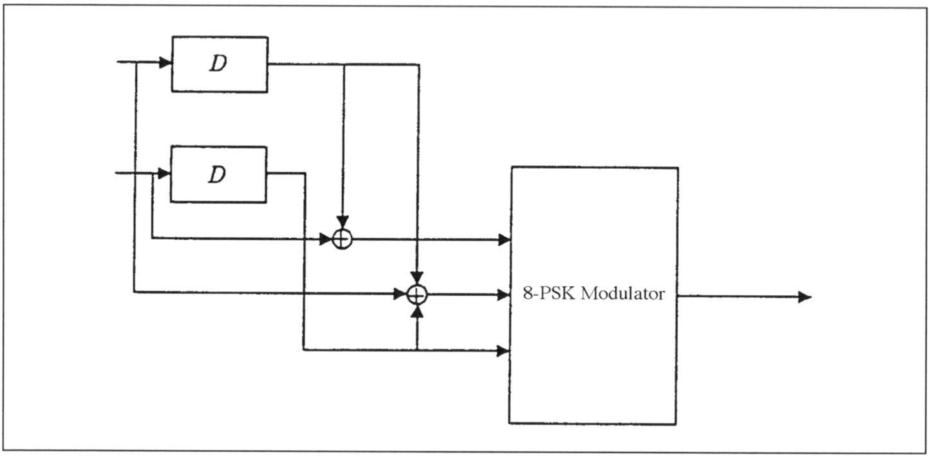

Figure 7 *A realization of the four-state 8-PSK TCM encoder designed by the rules of Jamali and Le-Ngoc.*

Jamali and Le-Ngoc not only address the design of good four-state 8-PSK trellis codes, but also formulate general design rules for the Rayleigh fading channel.[11] These design principles can be viewed as the generalization of the design rules formulated by Ungerboeck in for the Gaussian channel.[1] Application of these rules results in improved performance. As an example, by applying these rules one obtains the signal constellation and the trellis shown in *Figure 6*.

It is easy to verify that the coding gain of this code over an AWGN channel (as expressed by the free Euclidean distance) is 2 dB, which is 0.9 dB superior to the code designed by Wilson and Leung and shown in *Figure 5*, and only 1 dB inferior to the Ungerboeck code with a comparable complexity. It is also easy to see that the product distance of this code is twice the product distance of the code shown in *Figure 5* and therefore the performance of this code over a fading channel is superior to the performance of the code designed by Wilson and Leung.[8] Since the squared product distance of this code can be shown to be twice the squared product distance of the code shown in *Figure 5*, the asymptotic performance improvement of this code compared to the one designed by Wilson and Leung, when used over fading channels, is $10 \log\sqrt{2} = 1.5$ dB. The encoder for this code can be realized by a convolutional encoder followed by a natural mapping to the 8-PSK signal set as shown in *Figure 7*.

Multiple Trellis-Coded Modulation (MTCM)

We have seen that the performance of trellis coded modulation schemes on fading

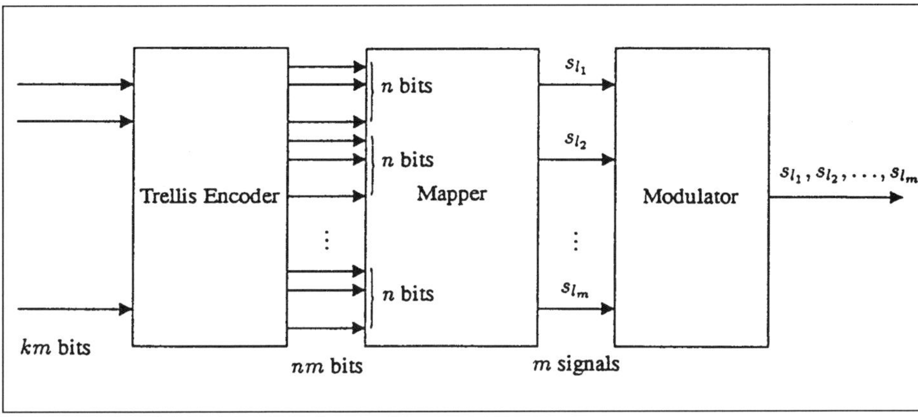

Figure 8 *Block diagram of a multiple trellis-coded modulation scheme (MTCM).*

channels is primarily determined by their effective code length (time diversity) and product distance. In particular we saw that trellises with parallel branches are to be avoided in transmission over fading channels due to their low effective code length. In cases where high bit rates are to be transmitted under severe bandwidth restrictions, the signal constellation consists of many signal points. In such cases in order to avoid parallel paths in the code trellis, the number of trellis states should be very large, resulting in a very complex decoding scheme.

An innovative approach to avoid parallel branches and at the same time to avoid a very large number of states is to employ Multiple Trellis-Coded Modulation (MTCM) as first formulated by Divsalar and Simon.[12] The block diagram for a multiple trellis-coded modulation is shown in *Figure 8*.

In the multiple trellis-coded modulation depicted in *Figure 8*, $K = km$ information bits enter the trellis encoder at each instant of time and are mapped into $N = nm$ bits which correspond to m signals from a signal constellation with a total of 2^n signal points. These m signals are transmitted over the channel. The parameter m is called the multiplicity of the code. The important fact is that unlike standard TCM, here each branch of the trellis is labeled with m signals from the constellation and not only one signal. The existence of more than one signal corresponding to each trellis branch results in higher time diversity and therefore improved performance when used over fading channels. In fact, MTCM schemes can have a relatively small number of states and at the same time avoid a reduced effective code length. The throughput (or spectral bit rate defined as the ratio of the bit rate to the bandwidth) for this system is k which is equivalent to an uncoded (and a conventional TCM) system. In most implementations of MTCM, the value of n is

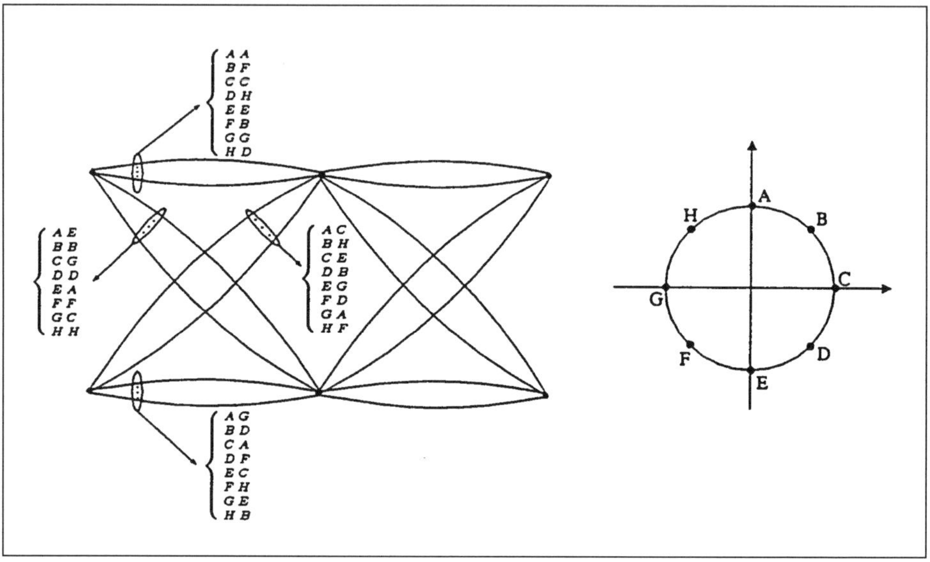

Figure 9 *An example of multiple trellis-coded modulation.*

chosen to be $k + 1$. Note that with this choice, the case $m = 1$ is equivalent to conventional TCM. The rate of the MTCM code is $R = K/N = k/n$.

In the following example, we give a specific MTCM scheme and discuss its performance in a fading environment.[12, 6] The signal constellation and the trellis for this example are shown in *Figure 9*. The multiplicity of the code is assumed to be equal to 2 ($m = 2$) and $k = 2$ and $n = 3$. Therefore, the rate of this code is 2/3 and the trellis chosen for the code is a two-state trellis. At each instant of time $K = km = 4$ information bits enter the encoder. This means that there are $2^K = 16$ branches leaving each state of the trellis. Due to the symmetry in the structure of the trellis, there exist eight parallel branches connecting any two states of the trellis. The difference, however, with conventional trellis-coded modulation is that here we assign two signals in the signal space to each branch of the trellis. In fact, corresponding to the $K = 4$ information bits that enter the encoder, $N = nm = 6$ binary symbols leave the encoder. These six binary symbols are used to choose two signals from the 8-PSK constellation shown in *Figure 9* (each signal requires three binary symbols). The mappings of the branches to the binary symbols are also shown in *Figure 9*. Close examination of the mappings suggested in this figure show that although there exist parallel branches in the trellis for this code, the diversity level provided by this code (the effective code length) is equal to two.

It is obvious from the above example that multiple trellis-coded modulation can

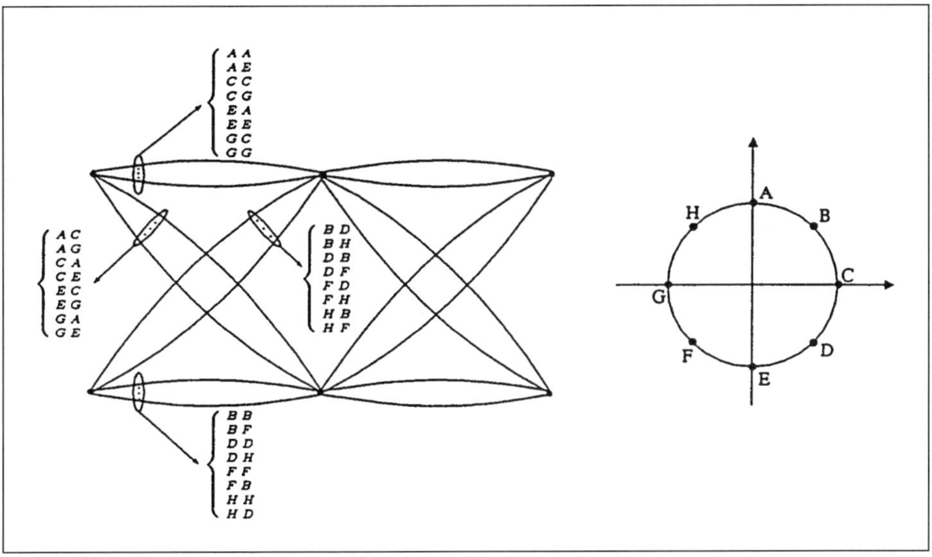

Figure 10 *Signal assignment for an MTCM scheme appropriate for transmission over an AWGN channel.*

achieve good diversity, which is essential for transmission through the fading channel, without requiring complex trellises with a large number of states. It can also be shown that this same technique can provide all the benefits from using the so-called *asymmetric signal sets*, as described by Divsalar, Simon and Yen,[13] without the difficulties encountered with time-jitter and catastrophic trellis codes.[13] Optimum set partitioning rules for multiple trellis-coded modulation schemes are investigated by Divsalar and Simon[14] and Biglieri, et al.[15] It is important to note that the signal set assignment to the trellis branches shown in *Figure 9* are not the best possible signal assignments if this code is to be used over an AWGN channel. In fact the signal set assignment shown in *Figure 10* provides a performance that is 1.315 dB superior to the signal set assignment of *Figure 9* when used over an AWGN channel. However, obviously, the signal assignment of *Figure 10* can only provide a diversity level equal to unity as opposed to the diversity level of two provided by the signal assignment of *Figure 9*. This means that on fading channels the performance of the code shown in *Figure 9* is superior to the performance of the code shown in *Figure 10*.

Multi-Level Codes and Multi-Stage Decoding

An approach to combined coding and modulation, which is somewhat different from Ungerboeck's trellis coded modulation, is the multilevel coding technique introduced by Imai and Hirakawa.[16] In this technique several error correcting codes (block or convolutional) are employed. The information bits are distributed among the encoders and are encoded independently. The outputs of the encoders are used to generate the transmitted signals.

Multi-Level Codes for Gaussian Channels

The block diagram for a multi-level encoding system is shown in *Figure 11*. The information sequence $\boldsymbol{i}$ is distributed into M information sequences $\boldsymbol{i}_1, \boldsymbol{i}_2, \ldots, \boldsymbol{i}_M$. Each of these sequences is the input to the corresponding encoders and the encoded sequences are denoted by $\boldsymbol{c}_1, \boldsymbol{c}_2, .. \boldsymbol{c}_M$. Assuming that the length of the information sequence i is K and this sequence is split into M subsequences of lengths $k_1, k_2, \ldots, k_M$ and the corresponding code words are all of length n, the rate of each code is given by

$$R_i = \frac{k_i}{n} \tag{9}$$

and the overall rate of the multilevel code is

$$R = \frac{K}{n} = \sum_{i=1}^{n} R_i \tag{10}$$

The signal selector transmits n signals each chosen from 2^M signal points. In order to transmit the first signal, the signal selector looks at the first components of each of the M code words $\boldsymbol{c}_1, \boldsymbol{c}_2, \ldots, \boldsymbol{c}_M$ — that is $(c_1^1, c_2^2, \ldots, c_1^M)$ — and depending on them chooses one of the 2^M signal points. The second signal point is chosen by looking at the second set of components of the M code words, and so on. For example, in the case of $M = 3$, the signal points are chosen from a constellation of eight points which can be mapped to an 8-PSK constellation. This sequence of signals $\boldsymbol{s} = (s_1, s_2, \ldots, s_m)$ is then transmitted over the channel and the sequence $\boldsymbol{r}$ is received.

It can be seen that if the minimum Hamming distance of the M codes is denoted by $d_1, d_2, \ldots, d_M$ then the minimum Hamming distance between the transmitted signal sequences is given by

$$d = \min\left(d_1, d_2, \ldots, d_M\right) \tag{11}$$

The optimal decoder for this system is a maximum likelihood decoder, which for the case of convolutional codes can be implemented by a Viterbi-type algorithm. However, implementation of such a decoder is impractical due to the very large

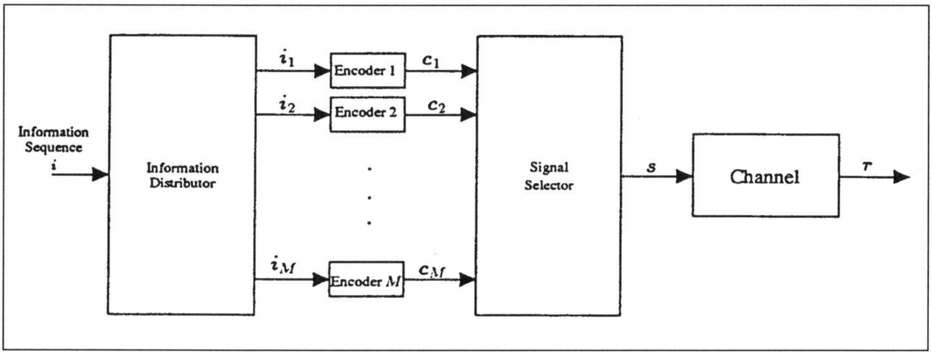

Figure 11 *Block diagram for multi-level encoding.*

number of states required. In fact, the number of required states is the product of the number of states in each individual code. The difficulty of implementation of an optimal decoder has resulted in the development of sub-optimal algorithms. A step-by-step decoding algorithm can be used for decoding of the multilevel code given above, as described by Imai and Hirakawa.[16] In this approach the M codes for constructing the multilevel code are decoded in succession. To be more specific we discuss the case of M = 8 with 8-PSK modulation below.

In *Figure 12*, the structure of a three-level coding scheme with the corresponding set partitioning is shown. We have used the notation of Kofman et al.[17] In this scheme, the information sequence is split into three subsequences which are encoded by the three encoders E_1, E_2, E_3. The outputs of the three encoders are mapped to an 8-PSK constellation as shown in the same figure. Assuming that the three encoded sequences are $\boldsymbol{c}_1 = (c_1^1, c_2^2, \ldots, c_1^M)$, $\boldsymbol{c}_2 = (c_2^1, c_2^2, \ldots, c_2$ and $\boldsymbol{c}_3 = (c_3^1, c_3^2, \ldots, c_3^n)$, n signal points from the constellation are transmitted. The binary representation of the ith signal point is given by $(c_3^i\, c_2^i\, c_1^i)$, where these binary representations are shown in *Figure 12* and are derived by applying Ungerboeck's set-partitioning rules. The resulting n signal points generate the sequence $\boldsymbol{s}$ which is transmitted over the channel. The redundancy of the code is a result of the redundancy of the three encoders. The partitioned signal sets at different stages are denoted by S_0, S_q, and S_b (for octal, quaternary, and binary respectively), and the normalized minimum Euclidean distances for each partition are also shown in *Figure 12*.

The multistage decoding for the code shown in *Figure 12* first uses a Viterbi decoding scheme to decode c_1 and therefore to obtain i_1. Assuming a Gaussian channel, the Viterbi algorithm metric is given by

$$m_1(r, c_1) = -\min_{s \in S_q^{c_1}} |r - s|^2, \quad c_1 = 0, 1 \tag{12}$$

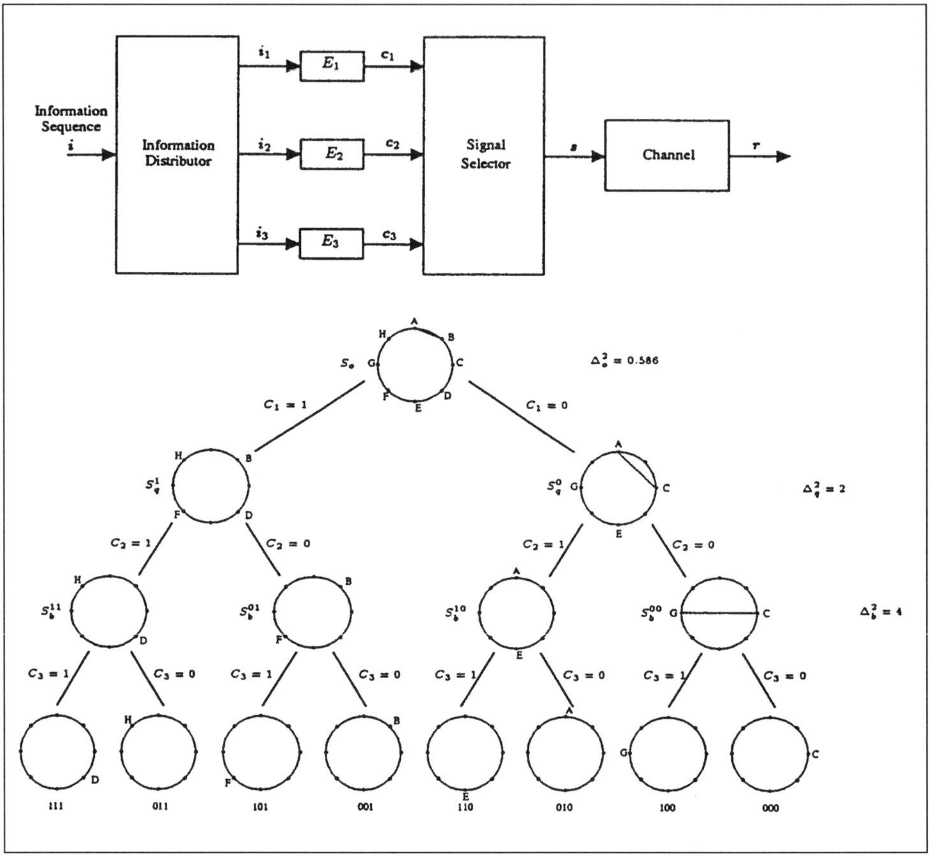

Figure 12 *Encoder and set partitioning for a three-level encoder.*

In the next decoding stage, it is assumed that the result of decoding in the first stage is correct and the Viterbi algorithm for this stage is applied with the following metric

$$m_2\left(r, c_2 \middle| \hat{c}_1\right) = -\min_{s \in S_b^{\hat{c}_2 \hat{c}_1}} \left|r - s\right|^2, \quad c_2, c_1 = 0, 1 \tag{13}$$

where $\hat{c}_1$ denotes the result of decoding in the first stage. Similarly, in the last decoding stage the metric used is

$$m_3\left(r, c_3 \middle| \hat{c}_2, \hat{c}_1\right) = -\left|r - s\right|^2, \quad c_{3,} c_2, c_1 = 0, 1 \tag{14}$$

where $\hat{c}_1$ and $\hat{c}_2$ are the decoder outputs in the first and second stages, respectively. Such a decoder has a complexity which is roughly the sum of the complex-

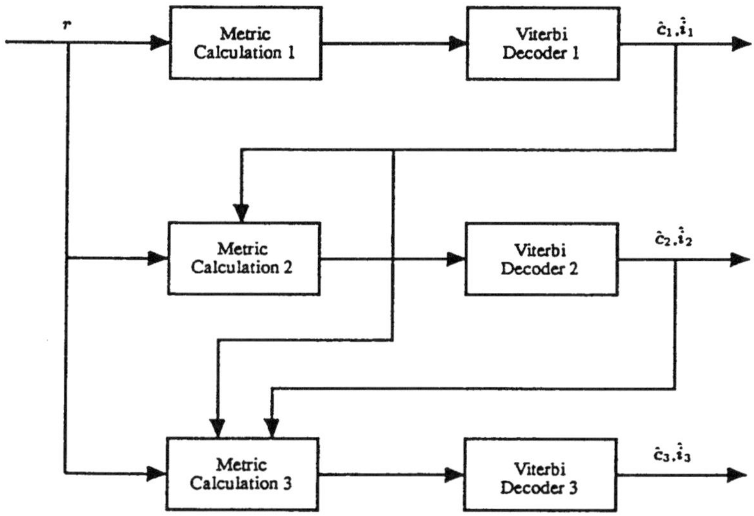

Figure 13 *Multistage decoding for a multi-level code*

ities of individual decoders rather than the product of the complexities. The concept of multilevel codes was introduced by Imai and Hirakawa,[16] and later developed by Calderbank, Potti, Taylor, Sayegh, Yamaguchi, and Imai, among other works.[18-21] A block diagram of a multistage decoder is shown in *Figure 13*, where we have not included the required delays from one decoding stage to another to keep the block diagram simple.

In the description of multistage decoding given above, the Euclidean metric has been employed for the implementation of the Viterbi decoder. This is based on the assumption that the channels for each coding level are both Gaussian and without memory. However, the side information that is passed among the different decoders makes the channels neither memoryless nor Gaussian. Kofman, Zehavi, and Shamai employ interleavers in order to minimize the appearance of error bursts in the component convolutional codes.[17] Such error bursts in the early stages of decoding can significantly affect the performance of the subsequent decoding stages.

The multistage decoding at the second and third stages above have been based on the assumption that the decoding at the previous stages has been done with no errors. It is possible to repeat this entire decoding process after finishing the third stage (last stage in general) of decoding, based on the assumption of correctness of the already decoded codes. In other words we can make tentative decoding in the first round of decoding and use this tentative result in a second decoding round and so forth. This improves the overall performance of the code. For example, in the second round of decoding of the first code, the following metric is used in the Viterbi algorithm

$$m_1^{(2)}\left(r, c_1 \middle| \hat{c}_2, \hat{c}_3\right) = -\left|r - s\right|^2, \quad c_{1,}c_2, c_3 = 0, 1 \tag{15}$$

where $\hat{c}_2$ and $\hat{c}_3$ are the decoded sequences in the first decoding round. Similar metrics for the decoding of the second and the third codes are employed in the second round of decoding. This process of repeated (or iterative) decoding can be carried out many times. However, in practice usually there is not much gain that can be achieved after repeating the decoding process twice.[22]

Multi-Level Codes for Fading Channels

Multilevel trellis codes for fading channels have been studied by Seshadri and Sundberg (see also Sundberg and Seshadri).[22-24] In the paper by Seshadri and Sundberg,[22] an asymptotic comparison is performed among Ungerboeck's codes, the codes designed by Schlegel and Costello and multilevel codes.[22] Coded modulation schemes with 8-PSK constellation and rate 2/3 with almost equal decoding complexity are compared in terms of the minimum Hamming distance (effective code length) that they provide. This comparison shows that multilevel codes outperform Ungerboeck and Schlegel-Costello codes when used for transmission over fading channels. For example, for a code with 256 trellis states, Ungerboeck's codes can provide an effective code length of 3, Schlegel-Costello codes can provide and effective code length of 5 and multilevel codes can have an effective code length of 8. With 1,024 states in the code trellis, the corresponding numbers increase to 4, 5, and 10, respectively. Since the performance of a trellis code over a fading channel is primarily determined by its effective code length (at least at high signal-to-noise-ratios), this comparison clearly shows the advantage of multilevel codes. In the same work, numerous simulations of the performance of multilevel codes over fading channels, including cases employing differential decoding, are presented.

Wu and Lin study both the effective code length and the product distance of multilevel trellis codes and show that a multilevel code C consisting of component codes $C_1, C_2, \ldots, C_M$ with corresponding Hamming distances d_i, $1 \le i \le M$, and squared intraset distances $\Delta_i 1 \le i \le M$, has an overall effective code length L given by:[23]

$$L = \min\left\{d_i, 1 \le i \le M\right\} \tag{16}$$

and a product distance given by

$$d_P^2(L) = \min_{k:d_k=L} \Delta_k^{d_k} \tag{17}$$

These two results serve as guidelines for designing good trellis codes for fading channels which are based on keeping both the effective code length and the product distance high to achieve good performance.

Conclusion

Wireless communication channels are characterized by severe fading and shadowing. The increasing demand for cellular mobile communications over the allocated bandwidth imposes severe bandwidth constraints on these channels and bandwidth-efficient modulation and coding schemes are desirable for communication over wireless channels. The high spectral efficiency of trellis coded modulation makes this coding scheme an attractive technique for communication over these channels. However, trellis codes were originally designed for use over Gaussian channels. Analysis of the performance of trellis codes over fading channels shows that parameters other than the free Euclidean distance are important in determining the performance of these codes. Various techniques for designing good codes for fading channels have been employed in recent years, resulting in considerable error probability improvement and enhanced coding gain.

References

1 G. Ungerboeck, Channel coding with multilevel/phase signals, *IEEE Transactions on Information Theory*, vol. IT-28, pp. 55-67, January 1982.

2 D. G. Forney Jr., "The Viterbi algorithm *Proceedings of the IEEE*, vol. 61, pp.268-278, March 1973.

3 J. G. Proakis and M. Salehi, *Communication Systems Engineering* (Prentice-Hall, 1994).

4 W. C. Y. Lee, Overview of cellular CDMA, *IEEE Transactions on Vehicular Technology*, vol. 40, pp. 291 -302, May 1991.

5 D. Divsalar and M. K. Simon, The design of trellis coded MPSK for fading channels: Performance criteria, *IEEE Transactions on Communications*, vol. 36, pp. 1004-1012, September 1988.

6 S. H. Jamali and T. Le-Ngoc, *Coded Modulation Techniques for Fading Channels* (Kluwer Academic Publishers, 1994).

7 J. G. Proakis, *Digital Communications*, 3rd edition (McGraw-Hill, 1995).

8 S. G. Wilson and Y. S. Leung, Trellis coded phase modulation on Rayleigh fading channels, *Proceedings of the IEEE International Conference on Communications*, June 1987.

9 C. Schlegel and D. J. Costello Jr., Bandwidth efficient coding for fading channels: Code construction and performance analysis, *IEEE Journal on Selected Areas in Communications*, vol. SAC-7, pp. 1356-1368, December 1989.

10 J. Du and B. Vucetic, New MPSK trellis codes for fading channels, *Electronic Letters*, vol. 26, pp. 1267- 1269, August 1990.

11 S. H. Jamali and T. Le-Ngoc, A new four-state 8PSK TCM scheme for fast fading, shadowed mobile radio channels, *IEEE Transactions on Vehicular Technology*, vol. V.C.-40, pp. 216-222, February 1991.

12 D. Divsalar and M. K. Simon, Multiple trellis coded modulation (MTCM), *IEEE Transactions on Communications*, vol. COM-36, pp. 410-419, April 1988.

13 D. Divsalar, M. K. Simon, and J. H. Yen, Trellis coding with asymmetric modulation, *IEEE Transactions on Communications*, vol. COM-35, pp. 130-140, February 1987.

14 D. Divsalar and M. K. Simon, The design of trellis-coded MPSK for fading channels: Set partitioning for optimum code design, *IEEE Transactions on Communications*, vol.36, pp. 11 013- 1021, September 1988.

15 E. Biglieri, D. Divsalar, P. J. McLane, and M. K. Simon, *Introduction to Trellis coded Modulation with Applications.* (New York: Macmillian, 1991).

16 H. Imai and S. Hirakawa, A new multilevel coding method using error-correcting codes, *IEEE Transactions on Information Theory*, vol. IT-23, pp. 371-377, May 1977.

17 Y. Kofman, E. Zehavi, and S. Shamai(Shitz), Performance analysis of a multilevel coded modulation system, *IEEE Transactions on Communications*, vol. 42, pp. 299-312, February/March/April 1994.

18 A. R Calderbank, Multilevel codes and multistage decoding, *IEEE Transactions on Communications*, vol. 37, pp. 222-229, March 1989.

19 G. Potti and D. Taylor, Multilevel codes based on partitioning, *IEEE Transactions on Information Theory*, vol. IT-35, pp. 87-98, January 1989.

20 S. I. Sayegh, A class of optimum block codes in signal space, *IEEE Transactions on Communications*, vol. COM-34, pp. 1043-1045, October 1986.

21 K. Yamaguchi and H. Imai, Highly reliable multilevel channel coding systems using binary convolutional codes, *Electronic Letters*, vol. 23, pp. 939-941, August 1987.

22 N. Seshadri and C.-E. W. Sundberg, Multi-level trellis-coded modulation for Rayleigh fading channels, *IEEE Transactions on Communications*, vol. 41, pp. 1300-1310, September 1993.

23 J. Wu and S. Lin, Multilevel trellis MPSK modulation codes for Rayleigh fading channels, *IEEE Transactions On Communications*, vol. 41, pp. 1311 - 1318, September 1993.

24 C.-E. W. Sundberg and N. Seshadri, Coded modulation for fading channels: An overview, *European Transactions on Telecommunications*, vol. 4, pp. 309-324, May-June 1993.

Introduction to Propagation for Mobile and Portable Radio

Theodore S. Rappaport
Professor, Bradley Department of Electrical Engineering, The Mobile and Portable Radio Research Group
Virginia Polytechnic Institute and State University

Dennis G. Sweeney
Senior Research Associate
Virginia Polytechnic Institute and State University

Introduction

The radio channel places fundamental limitations on the performance of mobile and portable communications systems. The transmission path between transmitter and receiver can vary from a simple direct line-of-sight path to one that is severely obstructed by buildings and foliage. Speed of motion impacts how rapidly the signal level fades as a mobile terminal moves through space. Unlike wired channels that are relatively stationary and predictable, mobile and portable radio channels can be extremely random.

Most mobile and portable radio systems operate in environments where there is no direct line-of-sight path between the transmitter and the receiver. The mechanisms behind electromagnetic wave propagation are diverse, but they can generally be attributed to reflection, diffraction, and scattering. Propagation models that characterize signal strength over large transmitter-receiver (T-R) separation distances (hundreds or thousands of wavelengths) are called *large-scale* models, whereas models that characterize the rapid fluctuations of the received signal strength over short distances (a few wavelengths) or short time durations (on the order of seconds) are called *small-scale* models.

The Free-Space Propagation Model

The free space model is an ideal propagation model that can be accurately applied only to satellite communication systems and short line-of-sight radio links. However, it does demonstrate how the received power decays exponentially as the T-R separation increases. This property is a fundamental characteristic of large-scale modeling. The power received, P_r, from a distant transmitter located d meters away assuming free-space path between the antennas, is given by the Friis free-space equation:

$$P_r = \frac{P_t G_t G_r \lambda^2}{(4\pi)^2 d^2 L} \tag{1}$$

where P_t is the transmitted power, G_t is the transmitter antenna gain, G_r is the receiver antenna gain, d is the T-R separation distance in meters, L is the system losses ($L>1$), and λ is the wavelength in meters. The Friis transmission formula is an inverse square relationship between the received power and the T-R separation distance. This implies the received power decays at a rate of 20 dB/decade with distance.

The path-loss is defined as the difference between the effective transmitted power and the received power, and it may include the effect of the antenna gains. The path-loss (PL) for the free space model, including antenna gain and system losses, is given by

$$PL(\text{dB}) = -10\log\left(\frac{P_r}{P_t}\right) = G_t(\text{dB}) + G_r(\text{dB}) - L(\text{dB}) + 20\log_{10}\left(\frac{\lambda}{4\pi d}\right) \tag{2}$$

The Three Basic Propagation Mechanisms

Reflection occurs when a propagating electromagnetic wave impinges upon an obstruction whose dimensions are very large when compared to the wavelength of the radio wave. Reflections from the surface of the earth and from buildings or walls produce reflected waves which may combine constructively or destructively at the receiver.

Diffraction occurs when the radio path between the transmitter and receiver is obstructed by a surface that has sharp irregularities (edges). The secondary waves resulting from the obstructing surface are present throughout the space and even behind the edge. The waves bend around the obstruction, thus permitting reception of the signal even when a line-of-sight path does not exist. *Scattering* occurs when the medium through which the wave travels consists of objects with dimensions that are small compared to the wavelength and in which the number of obstacles per unit volume is quite large. Scattered waves are also produced by irregularities in the refractive index of the propagation media.

Reflection

When a radio wave propagating in one medium impinges upon another medium having a different dielectric constant, permeability, or conductivity, the wave is partially reflected and partially transmitted. For the case of a plane wave in air incident normally on a perfect conductor, the wave is completely reflected without loss of energy. The electric field intensity of the reflected wave and the incident wave are related through the Fresnel reflection coefficients (Γ). The reflection coefficient is a function of the permittivity of the reflection surface, the wave polarization, the angle of incidence, and the frequency of the propagating wave.

Figure 1 shows the geometry for calculating reflection coefficients. The subscripts v and h refer to the vertical and horizontal polarization, respectively, of the E-field, and the subscripts i and r refer to the incident and reflected E-field, respectively.

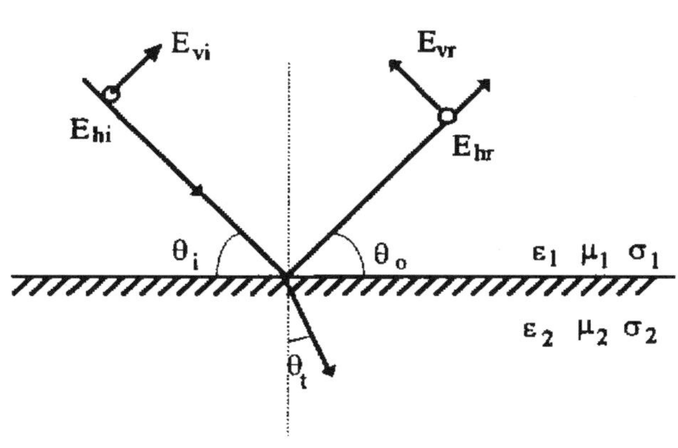

Figure 1 *Geometry for reflection coeffecient*

The permittivity, permeability, and conductance of the two media are $\varepsilon_1,\mu_1,\sigma_1$ and $\varepsilon_2,\mu_2,\sigma_2$, respectively. For the case when the first medium is free space, the reflection coefficients for the two cases of vertical and horizontal polarization are given by

$$\Gamma_v = \frac{E_{vr}}{E_{vi}} = \frac{\varepsilon_r \cos(90-\theta_i) - \sqrt{\varepsilon_r - \sin^2(90-\theta_i)}}{\varepsilon_{vi} \cos(90-\theta_i) + \sqrt{\varepsilon_r - \sin^2(90-\theta_i)}} \text{ (vertical E - field polarization)} \quad (3)$$

and

$$\Gamma_h = \frac{E_{hr}}{E_{hi}} = \frac{\cos(90-\theta_i) - \sqrt{\varepsilon_r - \sin^2(90-\theta_i)}}{\cos(90-\theta_i) + \sqrt{\varepsilon_r - \sin^2(90-\theta_i)}} \text{ (horizontal E - field polarization)} \quad (4)$$

where ε_r is the relative permittivity of the second medium. The incident angle θ_i t which the reflection coefficient Γ_v is equal to zero is called the *Brewster angle* and is given by the value of θ_i which satifies the equation:

$$\sin(90-\theta_i) = \frac{\sqrt{e_r^2 - e_r}}{\sqrt{e_r - 1}} \quad (5)$$

Ground Reflection (Two-ray) Model

In a mobile/portable radio channel, a single direct path between the transmitter and the receiver seldom exists, hence, the free-space propagation model of *Equation 2* is of limited use. The two-ray ground reflection model shown in *Figure 2* is a simple but useful propagation model which illustrates the effect of signals arriving via multiple paths, i.e., multipath. This model assumes that the wavelength λ is much smaller than anything encountered in the channel.

In most mobile communication systems where the separation distance is only a few tens of kilometers, it may be assumed that the terrain is flat. The total received E-field, E_r, is the resultant of the direct line-of-sight component E_{LOS} and a ground reflected component E_g.

Referring to *Figure 2*, h_t is the height of the transmitter and h_r is the height of the receiver. According to laws of reflection,

$$\theta_i = \theta_o \text{ and } E_o = \Gamma E_i \quad (6)$$

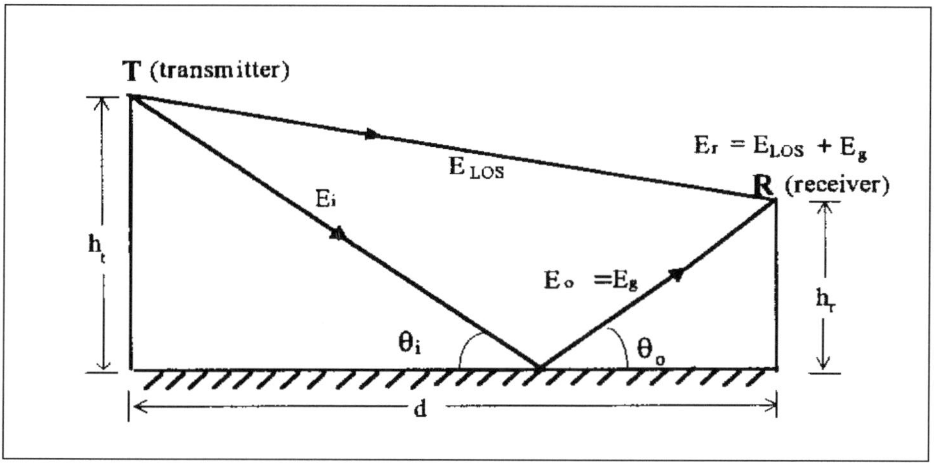

Figure 2 *Two-ray ground reflection model*

where Γ is the reflection coefficient for ground. As θ approaches 0° (i.e., grazing incidence), the reflected wave becomes equal in magnitude and 180° out of phase with the incident wave. The received E-field is the sum of the line-of-sight field, E_{LOS} and the reflected field, E_g. It can be shown that:

$$\left| E_R(d) \right| = 2\, E_{LOS} \sin \frac{\theta_\Delta}{2} \tag{7}$$

where the phase difference θ_Δ is related to the difference in the path's length, Δ, between the direct and ground reflected rays (Parsons 1992). In the mobile environment it is reasonable to assume that $d >> h_t, h_r$, so then:

$$\Delta = \frac{2h_t h_r}{d} \tag{8}$$

and

$$\theta_\Delta = \frac{2\pi\Delta}{\lambda} = \frac{4\pi h_t h_r}{\lambda d} \tag{9}$$

The received power is proportional to $|E_r|^2$ so:

$$\begin{aligned} P_R &= 4\left|E_{LOS}\right|^2 \sin\left(\frac{2\pi h_t h_r}{\pi d}\right) \\ &= P_t\left(\frac{\lambda}{4\pi d}\right)^2 G_t G_r \sin\left(\frac{2\pi h_t h_r}{\lambda d}\right)^2 \end{aligned} \tag{10}$$

Figure 3 shows the path-loss versus distance with single ground reflection.

Notice that the amplitude of the sin^2 term is the Friis transmission equation. This implies that in the absence of a reflected ray, *Equation 10* reduces to the free spacc transmission loss. Noticc also that if d is largc cnough, *Equation 10* reduces to:

$$P_r = P_t G_t G_r \frac{h_t^2 h_r^2}{d^4} \tag{11}$$

At large distances, the received power falls off at a rate of 40 dB/decade, and the received power and the path-loss become independent of frequency. The path-loss for the two-ray model in dB form is given as

$$PL(\text{dB}) = 10\log_{10} G_t + 10\log_{10} G_r + 20\log_{10} h_t + 20\log_{10} h_r - 40\log_{10} d \tag{12}$$

Diffraction

Diffraction around the earth's curvature makes it possible to transmit radio signals beyond the line-of-sight, and it allows an electromagnetic wave to propagate beyond obstructions. Although the field strength decreases rapidly as one moves deeper into the shadowed region, the field is still finite and has sufficient strength to produce a useful signal.

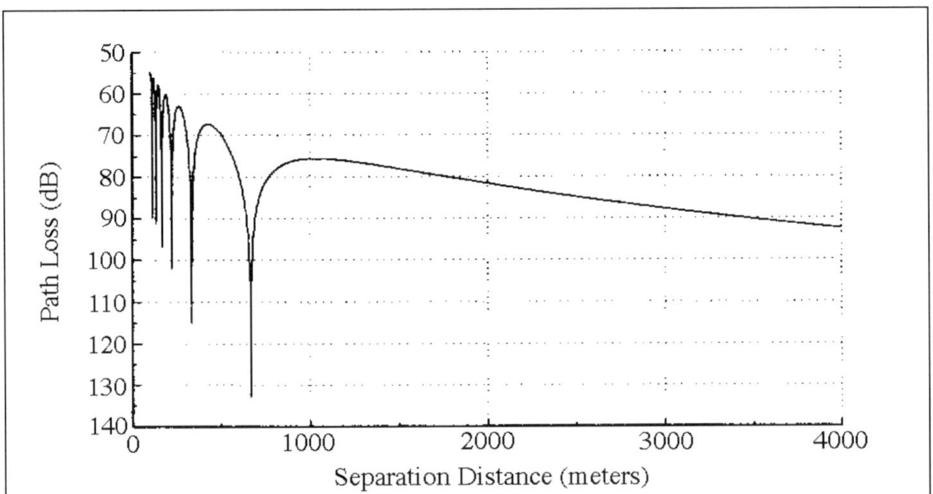

Figure 3 *Path-loss with a single ground reflection. h_t = 100m, hr = 1m, and f = 1000 MHz. Notice that the path-loss increases approximately 12dB from 2000 to 4000 meters, indicating a 40 dB/decade rate.*

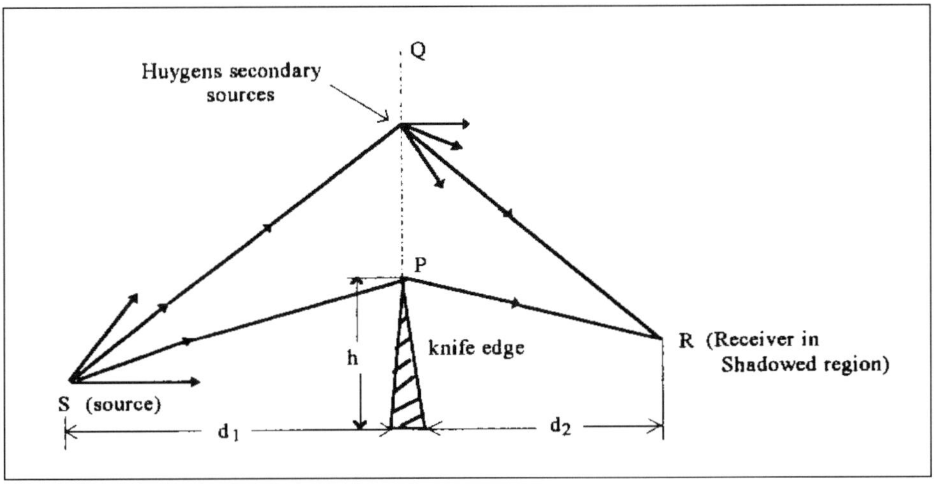

Figure 4 *Knife-edge diffraction*

The phenomenon of diffraction can be easily explained using Huygen's principle, which states that all points on a wavefront can be considered as point sources for the production of secondary wavelets and that these wavelets combine to produce a new wavefront in the direction of propagation. Diffraction is caused by the propagation of secondary wavelets into the shadowed region, and the field in the shadowed region is the vector sum of all the secondary wavelets.

Knife-Edge Diffraction Model

When shadowing is caused by a single object such as a hill, the attenuation due to diffraction over such an object can be estimated by treating the obstruction as a diffracting knife edge. In this simple case, the diffraction loss can be estimated by using the Fresnel solution for the field behind a knife edge or half-plane. *Figure 4* illustrates this approach. The field strength at point R in the shadowed region (also called the diffraction zone) is a vector sum of the fields due to all the secondary Huygen sources in the plane above the knife edge. The field strength E_d of the knife-edge diffracted wave is given by:

$$E_d = E_o F(v) = E_o \frac{(1+j)}{2} \int_v^{\infty} \exp\left(-j\pi t^2 / 2\right) dt \tag{13}$$

where E_o is the free space field strength in the absence of the knife edge, and $F(v)$ is the complex Fresnel integral. The Fresnel integral is a function of the Fresnel-Kirchoff diffraction parameter v, which is defined as

$$v = h\sqrt{\frac{2(d_1 + d_2)}{\lambda\, d_1 d_2}} \tag{14}$$

where h is the height of the knife edge and d_1, d_2 are the distances of the knife edge from the transmitter and the receiver, respectively. If the knife edge protrudes above a line drawn from S to R, h and v are positive. If the obstruction lies below the line-of-sight, then both h and v are negative. The Fresnel integrals are available in tabular form for given values of the diffraction parameter v. The diffraction loss due to the presence of a knife edge is given by

$$L_d(\text{dB}) = 20\log_{10}|F(v)| \tag{15}$$

Figure 5 shows the excess path-loss over free space due to diffraction as a function of the parameter v. Since the expression for diffraction loss involves the Fresnel integral, which is difficult to compute, approximate numerical solutions are often used. An approximate but simple to compute solution is provided by Lee (1985).

In many practical situations, especially in hilly terrain, the propagation path may consist of more than one obstruction, and it may be necessary to calculate the total

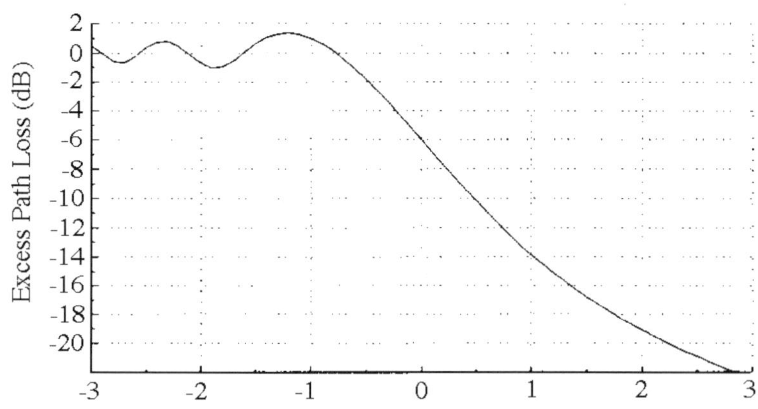

Figure 5 *Path-loss in excess of free space due to diffraction.*

diffraction losses due to all the obstacles. Bullington (1947) suggested that the series of obstacles be replaced by a single equivalent obstacle so that the path-loss could be obtained using single knife-edge diffraction models. In a more rigorous treatment, Millington et al. (1962) gave a wave-theory solution for the field behind two knife edges in series. This solution is very useful and can be applied easily for predicting diffraction losses due to two knife edges. However, extending this to more than two knife edges becomes a formidable mathematical problem. Many models that are mathematically less complicated have been developed to estimate the diffraction losses due to multiple obstructions, such as in Epstein (1953) and Deygout (1966).

Scattering

The path-loss in a mobile radio environment is often less than what is predicted by the reflection and diffraction models alone. This is due to the fact that when a radio wave impinges on a rough surface, the reflected energy is spread out (diffused) in all directions due to scattering. The roughness of a surface is often tested using the Rayleigh criterion which defines a critical height, h_c, of surface protuberances for a given angle of incidence θ_i:

$$h_c = \frac{\lambda}{8\cos\theta_i} \tag{16}$$

A surface is considered smooth if its minimum to maximum protuberance is less than h_c, and it is considered rough if the protuberance is greater than h_c. For rough surfaces, the Fresnel reflection coefficients need to be multiplied by a scattering loss factor (ρ_s) to account for the diminished specularly reflected field. Ament (1953) derived this as

$$r_s = \exp\left[-8\left(\frac{\pi\sigma_h\cos\theta_i}{\lambda}\right)^2\right] \tag{17}$$

where σ_h is the standard deviation of the surface height about the mean surface height. The scattering loss factor derived by Ament was modified by Boithias (1987) to give better agreement with measured results:

$$\rho_s = \exp\left[-8\left(\frac{p\sigma_h\cos\theta_i}{\lambda}\right)^2\right] I_0\left[8\left(\frac{p\sigma_h\cos\theta_i}{\lambda}\right)^2\right] \tag{18}$$

where I_0 is the Bessel function of the first kind and zero order.

Small-Scale Fading

In the mobile/portable radio environment, short-term signal fluctuations caused by signals that are reflected, refracted, and/or scattered are called small-scale fading to distinguish them from the large-scale variation in mean signal level which is dependent on T-R separation. Small-scale fading is caused either by wave interference between two or more multipath components which arrive at the receiver slightly delayed in time or when the mobile travels a short distance (a few wavelengths). These waves combine vectorally at the receiver antenna to give a resultant signal which can vary widely in amplitude over very small distances or observation times depending on the distribution of phases of the waves and the bandwidth of the transmitted signal.

Flat Fading

Small-scale fading is generally classified as being either *flat* or *frequency selective.* If the mobile radio channel has a constant gain and a linear phase response over a bandwidth which is greater than the bandwidth of the transmitted signal, the received signal will undergo *flat fading.* In this type of fading, the spectral characteristics of the transmitted signal are preserved. However, the strength of the received signal will change rapidly with time due to fluctuations in the gain of the channel caused by multipath.

Rayleigh Fading

In a flat fading mobile radio channel, the envelope of the received signal will often have a Rayleigh distribution. The Rayleigh distribution has a probability density function given by:

$$f(r) = \frac{re^{-\left(\frac{r^2}{2\sigma^2}\right)}}{\sigma^2} \qquad 0 \le r \le \infty \tag{19}$$

where σ^2 is the variance of the received signal r.

The *level crossing rate* and the *average fade duration* of a Rayleigh fading signal are two important statistics for determining error-control codes and diversity schemes to be used in a communication system. The level crossing rate (LCR) is defined as the expected rate at which the received signal envelope, normalized to the local mean signal, crosses a specified level R, in a positive direction. The number of level crossings per second is given by

$$N_R = \sqrt{2\pi} f_m \rho e^{-\rho^2} \tag{20}$$

where f_m is the maximum Doppler frequency, and $\rho = R/R_{rms}$ is the value of the specified signal level R, normalized to the local RMS amplitude of the fading envelope. The *average fade duration* is defined as the average period of time for which the received signal is below a specified level ρ. It depends primarily on the speed of the mobile and is given by

$$\bar{\tau} = \frac{e^{\rho^2} - 1}{\rho f_m \sqrt{2\pi}} \tag{21}$$

The duration of a signal fade determines the most likely number of signaling bits that will be lost during a fade. *Figure 6* shows a typical Rayleigh fading signal.

Rician Fading

When there is a dominant signal component, such as a line-of-sight propagation path, the small-scale fading distribution is Rician. In such a situation, random components arriving at different angles are superimposed on a stationary signal. At the output of an envelope detector, this has the effect of adding a dc component to the random multipath. The Rician distribution is

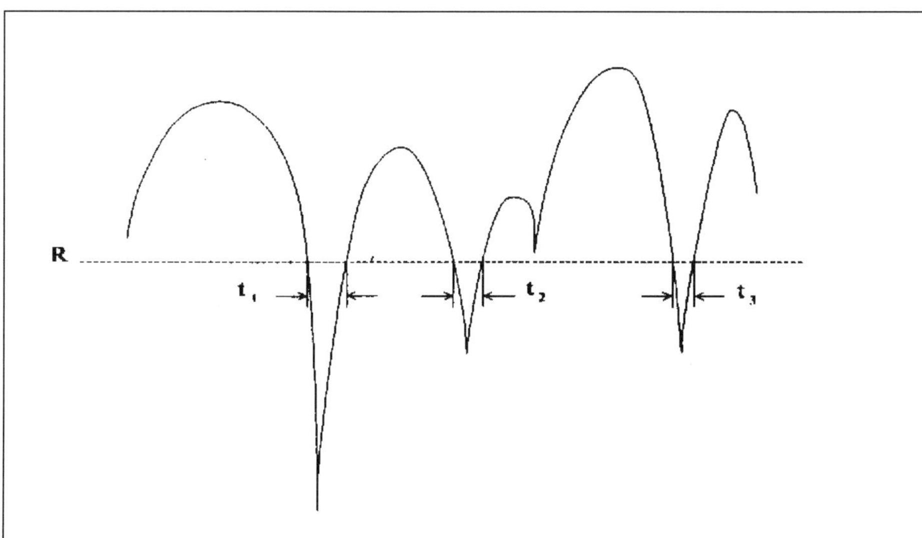

Figure 6 *An example of Rayleigh fading. The level crossing rate (LCR) is the number of times the signal level crosses the mean local level, R, in the positive direction per unit time. The average fade duration is the average of* t_1, t_2, t_3.

given by

$$f(r) = \frac{r}{\sigma^2} \exp - \left[\frac{r^2 + A^2}{2\sigma^2} \right] I_0 \left(\frac{Ar}{\sigma^2} \right), \quad \text{for } \{A \geq 0, r \geq 0\} \tag{22a}$$

$$= 0 \quad \text{for } r < 0. \tag{22b}$$

The Rician distribution is often described in terms of a parameter K which is defined as the ratio between the deterministic signal power and the variance of the multipath:

$$K = 10 \log_{10} \frac{A^2}{2\sigma^2} \text{ dB} \tag{23}$$

The parameter A denotes the peak-to-peak amplitude of the dominant sine wave about the carrier. The parameter K completely specifies the Rician distribution. As $A \to 0$ and $K \to 0$, the dominant path diminishes, and the Rician distribution degenerates to a Rayleigh distribution. *Figure 7* shows a signal undergoing Rician fading.

Frequency-Selective Fading

If the channel has a variable gain and/or non linear phase over the bandwidth of

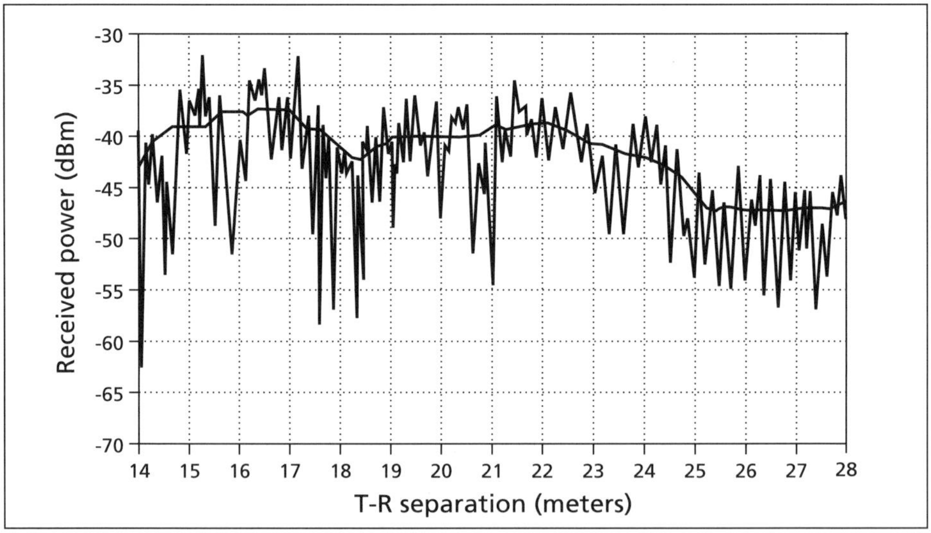

Figure 7 *Signal undergoing Rician small-scale fading, and the running small scale, or local, average*

the transmitted signal, the channel induces frequency-selective fading. Under such conditions, the channel impulse response has a multipath time delay spread which is greater than the time duration of a symbol of the transmitted message waveform. When this occurs, the received signal includes multiple versions of the transmitted waveform which are each fading and delayed in time. Viewed in the frequency domain, this amounts to certain frequency components in the received-signal spectrum having greater gains than others. Frequency-selective fading channels are much more difficult to model than flat fading channels since each multipath signal must be modeled. When analyzing communication systems, statistical impulse-response models such as SIRCIM for indoor channels (Rappaport 1991) and SURP (Turin 1972) and SMRCIM (Rappaport 1993) for outdoor channels can be used to study the effects of frequency-selective, small-scale fading.

The channel can be thought of as a filter with a complex frequency response. Each reflected, diffracted, or scattered signal will arrive at the receiver at a different time depending on the path traveled. A plot of the received signal power versus time delay is called the power-delay profile, and it is the square magnitude of the channel impulse response. *Figure 8* is typical of such a profile. Profiles can be measured either by frequency domain techniques (Howard and Pahlavan 1992), by sending out short pulses (Rappaport 1989) or by noting the time of arrival using correlation techniques (Cox 1975).

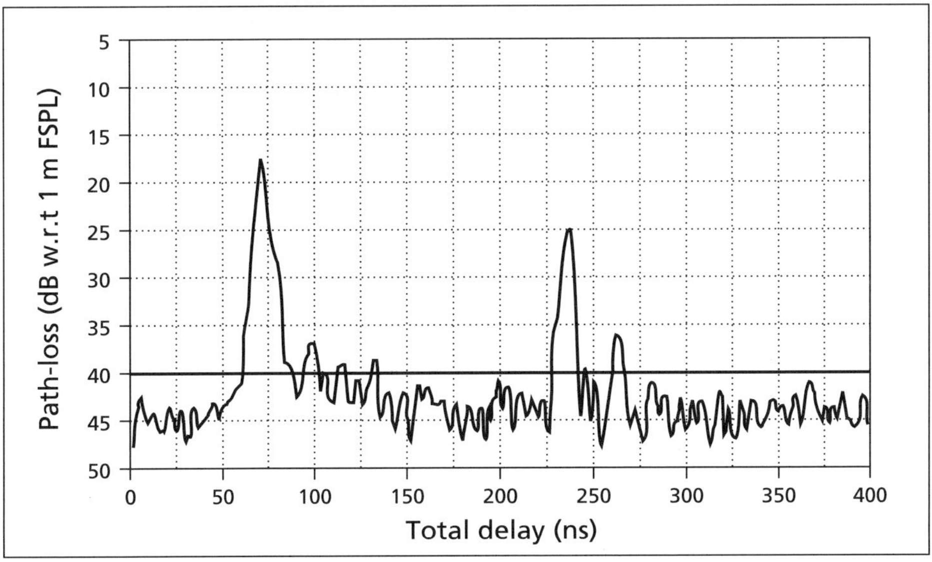

Figure 8 *Typical power-delay profile inside a building. T-R separation: 21.4m; Rms delay spread: 61.5 ns; Threshold: 40.0 dB*

Two parameters are used to describe a power-delay profile: mean excess delay and RMS delay spread (Saleh 1987 and Rappaport 1989). Mean excess delay is the first moment or average of the power-delay profile. It is given by:

$$\bar{\tau} = \frac{\int_0^\infty \tau P(\tau) d\tau}{\int_0^\infty P(\tau) d\tau} \tag{24}$$

where $P(\tau)$ is the received power at delay τ. The RMS delay spread is the second moment or variance of the excess delay given by:

$$\sigma = \sqrt{\frac{\int_0^\infty (\tau - \bar{\tau})^2 P(\tau) d\tau}{\int_0^\infty P(\tau) d\tau}} \tag{25}$$

Practical Propagation Models

Most radio propagation models are derived using a combination of analytical and empirical methods and focus on predicting received power. The empirical approach is based on curve fits or analytical expressions that recreate a set of measured data. This has the advantage of implicitly taking into account all propagation factors, both known and unknown. The limitation of empirical models is that they must be verified by measurement for frequencies and environments other than those originally used to derive the model.

Long Distance Path-loss Model

Both theoretical and measurement-based propagation models show that the average received signal power decreases logarithmetrically with distance. Models which indicate that the mean path-loss increases logarithmetrically with distance have been used extensively in the literature. Here the average path-loss for an arbitrary T-R separation is expressed as a function of distance by using a path-loss exponent:

$$PL(d) \infty \left(\frac{d}{d_o}\right)^n \tag{26}$$

or

$$PL(\mathrm{dB}) = PL(d_o) + 10n\log_{10}\left(\frac{d}{d_o}\right) \tag{27}$$

Environment	path-loss Exponent, *n*
Free space	2
Urban area cellular radio	2.7 to 4
Shadowed urban cellular radio	5 to 6
In-building line-of-sight	1.6 to 1.8
Obstructed in building	4 to 6
Obstructed in factories	2 to 3

Table 1 *Path-loss exponents for different environments. From Rappaport (1996).*

where n is the path-loss exponent, and indicates the rate at which the path-loss increases with distance, d_o is the free space close-in reference distance, and d is the T-R separation distance. When plotted on a log-log scale, the path-loss (in dB) is a straight line with a slope equal to $10n$. The value of n depends on the specific propagation environment. For example, in free space n is equal to 2, and when obstructions are present, it will have a larger value. *Table 1* gives a listing of typical path-loss exponents obtained in various mobile and portable radio environments.

It is important to select a free space reference distance that is appropriate for the propagation environment. In large cellular systems, 1-km and 1-mile reference distances are commonly used (Lee 1985), whereas in microcellular systems, much smaller distances are used. The reference distance should always be in the far-field of the antenna ($d_o > 2D^2/\lambda$, where D is the largest antenna dimension) so that near-field effects do not alter the reference path-loss. The reference path-loss is calculated using the free space path-loss formula given by *Equation 2.*

Measurements have shown that at any value of d, the path-loss $PL(d)$ at a particular location is distributed log-normally (normal in dB) about the mean distance-dependent value. That is, $PL(d) = \overline{PL}(d) + X_\sigma$, where X_σ is a zero-mean log-normally distributed random variable with standard deviation, σ, in dB, and $PL(d)$ is the average path-loss in dB. The log-normal distribution describes the random shadowing effects which occur over a large number of measurement locations having the same T-R separation, which have different levels of clutter on the propagation path. A path-loss exponent n and standard deviation σ will therefore statistically describe the path-loss for an arbitrary location having a specific T-R separation. In practice, the value of n and σ are computed from measured data using linear regression such that the difference between the measured and estimated path-losses is minimized in a mean square error sense.

Outdoor Propagation Models

A number of propagation models are available to predict path-loss over irregular terrain. While all these models aim to predict signal strength at a particular receiving point or a specific small area called a sector, the methods vary widely in their approach, complexity, and accuracy. Most of these models are based on a systematic interpretation of measurement data obtained from the service area.

One of the more popular outdoor propagation models is the Longley-Rice model (Rice 1967). This model is applicable to point-to-point communication systems in the frequency range from 40 MHz to 100 GHz over different kinds of terrain. The median transmission loss is predicted using the path geometry of the terrain profile and the refractivity of the troposphere. Geometric-optics techniques (two-ray ground reflection model) are used to predict signal strengths within the radio horizon. Diffraction losses over isolated obstacles are estimated using the Fresnel-Kirchoff knife-edge models. The Longley-Rice propagation prediction model is also referred to as the ITS irregular terrain model.

The Longley-Rice model is available as a computer program (Longley 1968) to calculate long-term median transmission loss relative to free-space loss over irregular terrain for frequencies between 20 MHz and 10 GHz. For a given transmission path, the program takes as its input the transmission frequency, path length, polarization, antenna heights, surface refractivity, effective radius of the earth, ground conductivity, ground dielectric constant, and climate.

There have been many modifications and corrections to the Longley-Rice model since its original publication. One important modification (Longley 1978) deals with radio propagation in urban areas and this is particularly relevant to mobile radio. This modification introduces an excess term as an allowance for the additional attenuation due to urban clutter near the receiving antenna. This extra term, called the urban factor (UF), has been derived by comparing the predictions by the original Longley-Rice model with those obtained by Okumura (1968).

The major shortcoming of the Longley-Rice model is that it does not provide for environmental factors in the immediate vicinity of the mobile, nor can it account for the effects of buildings and foliage. Further, multipath is not considered.

The Okumura Model

Okumura's model (Okumura 1968) is one of the most widely used models for signal prediction in urban areas. This model is applicable for frequencies in the range 150 MHz-2000 MHz and for distances of 1 km to 100 km. It can be used for base-station effective antenna heights ranging from 30 m to 1000 m.

Okumura developed a set of curves giving the median attenuation relative to free space, A_{mu}, in an urban area over a quasi-smooth terrain with a base-station effective antenna height (h_{te}) of 200 m and a mobile antenna height (h_{re}) of 3 m. These curves are plotted as a function of frequency in the range 100 MHz-3000 MHz and as a function of distance from the base station in the range 1 km-100 km. To determine the path-loss using Okumura's model, the free-space path-loss between the points of interest is first determined, and then the value of $A_{mu}(f,d)$ as read from the curves, is added to the free-space path-loss along with correction factors to account for antennas at other than the reference heights. The model can be expressed as

$$L_{50}\left(\mathrm{dB}\right) = L_F + A_{mu} + G\left(h_{te}\right) + G\left(h_{re}\right) \tag{28}$$

where L_{50} is the median propagation loss, L_F is the free space propagation loss, A_{mu} is the median attenuation relative to free space, $G(h_{te})$ is the base-station antenna-height gain factor and $G(h_{re})$ is the mobile antenna height gain factor.

Various corrections can be applied to Okumura's model. Some of the important terrain-related parameters are the effective base station antenna height (h_{te}), terrain undulation height (Δh), isolated ridge height, average slope of the terrain and the mixed land-sea parameter. Once the terrain-related parameters are calculated, the necessary correction factors can be added or subtracted as required. All these correction factors are also available as Okumura curves.

Okumura's model is wholly based on measured data and does not provide any analytical explanations. For many situations, extrapolations of the derived curves can be made to obtain values outside the measurement range. The validity of such extrapolations depends on the circumstances and the smoothness of the curve in question.

Okumura's model is considered to be among the best in terms of accuracy in path-loss prediction for mature cellular radio systems. It is also a very practical model that has become a standard for system planning in today's land mobile radio systems in Japan. The only major disadvantage with the model is its complexity and slow response to rapid changes in radio path profile. Generally, it is found that the model is particularly good in urban and suburban areas, but not as good in rural areas over irregular terrain.

The Hata Model

The Hata model (Hata 1980) is an empirical formulation of the graphical path-loss information provided by Okumura. Hata presented the urban area propagation loss as the standard formula and supplied correction equations to the standard formula for application to other situations. The standard formula for median path-loss in urban areas is given by

$$\begin{aligned} L_{50}(\mathrm{dB}) &= 69.55 + 26.16\log_{10} f_c - 13.82\log_{10} h_{te} - a(h_{re}) \\ &\quad + (44.9 - 6.55\log_{10} h_{te})\log_{10} d \end{aligned} \tag{29}$$

where f_c is the frequency from 150 MHz to 1500 MHz, h_{te} is the effective transmitter (base station) antenna height ranging from 30 m to 200 m, h_{re} is the effective receiver (mobile) antenna height ranging from 1 m to 10 m, d is the T-R separation distance in km, and $a(h_{re})$ is the correction factor for effective mobile antenna height, which is a function of the size of the service area. For a small- to medium-sized city, the correction factor is given by

$$a(h_{re}) = (1.1\log_{10} f_c - 0.7)h_{re} - (1.56\log_{10} f_c - 0.8)\mathrm{dB} \tag{30}$$

and for a large city, it is given by

$$a(h_{re}) = 8.29(\log_{10} 1.54h_{re})^2 - 1.1\ \mathrm{dB} \quad \text{for } f_c \le 200\ \mathrm{MHz} \tag{31a}$$

$$= 3.2(\log 11.75h_{re})^2 - 4.97\ \mathrm{dB} \quad \text{for } f_c \ge 400\ \mathrm{MHz} \tag{31b}$$

To obtain the path-loss in a suburban area, the formula is modified as

$$L_{50}(\mathrm{dB}) = L_{50}(\mathrm{urban}) - 2[\log_{10}(f_c/28)]^2 - 5.4 \tag{32}$$

and for the path-loss in open areas the formula is modified as

$$L_{50}(\mathrm{dB}) = L_{50}(\mathrm{urban}) - 4.78(\log_{10} f_c)^2 - 18.33\log_{10} f_c - 40.98 \tag{33}$$

Although Hata's formulation does not have any of the path-specific corrections available in Okumura's model, the above expressions have significant practical value. The predictions of the Hata model closely compare with those of the original Okumura model – the difference is generally within 1 dB, so long as d exceeds 1 km.

Indoor Propagation Models

With the advent of personal communication systems (PCS), there is a great deal of interest in characterizing the radio communication channel inside a building. The propagation mechanisms described above apply to in-building radio signals. The dynamic range of fading experienced by a hand-held unit inside a building is smaller than that experienced by a moving vehicle in an urban setting. Due to secondary effects such as motion of people and doors being opened and closed, the channel characteristics change with time, although slowly. It also has been observed that propagation within buildings is strongly influenced by local features such as the layout of the building, the construction materials, and the building type.

Saleh Model

Saleh and Valenzuela (1987) reported the results of indoor propagation measurements between two vertically polarized omni-directional antennas located on the same floor of a medium-sized office building. Measurements were made using 10 ns, 1.5 GHz, radar-like pulses. The method involves averaging the square-law detected received pulse response while sweeping the frequency of the transmitted pulse. Using this method, multipath components within 5 ns were resolvable.

The results obtained by Saleh and Valenzuela show that the indoor channel is quasi-static or very slowly time varying, and the statistics of the channel's impulse response are independent of transmitting and receiving antenna polarization if there is no line-of-sight path between them. They reported a maximum multipath delay spread of 100 ns-200 ns within the rooms of a building, and 300 ns in hallways. The measured RMS delay spread within rooms had a median of 25 ns and a maximum of 50 ns. The signal attenuation with no line-of-sight path was found to vary over a 60-dB range and obey a log-distance power law, *Equations 26* and *27*, with an exponent between 3 and 4. Saleh and Valenzuela developed a simple multipath model for indoor channels based on measurement results.

Sircim Model

Rappaport et al., (1989) reported results of measurements at 1300 MHz in five factory buildings and carried out subsequent measurements in other types of buildings. Multipath delays ranged from 40 ns to 800 ns. Mean multipath delay and RMS delay spread values ranged from 30 ns to 300 ns, with median values of 96 ns in LOS paths and 105 ns in obstructed paths. Delay spreads were found to be uncorrelated with T-R separation, but they were affected by factory inventory, building construction materials, building age, wall locations, and ceiling heights. Measurements in a food processing factory that manufactures dry goods and has considerably less metal inventory than other factories had an RMS delay spread that was half of that observed in factories producing metal products. Newer factories which incorporate steal beams and steel-reinforced concrete in the building structure have stronger multipath signals than older factories which used wood and brick for perimeter walls. The data suggests that radio propagation in buildings may be described by a hybrid geometric/statistical model that accounts for both specular reflections from walls and ceilings and random scattering from inventory and equipment. The authors developed an elaborate empirically derived statistical model and computer code called SIRCIM (simulation of indoor radio channel impulse-response models) that recreates realistic samples of indoor channel measurements (Rappaport 1991) and subsequent work developed models for predicting coverage between floors using building blueprints (Siedel and Rappaport 1992).

Objects and partitions inside a building effect the path-loss exponent. *Table 1* shows that the loss exponent can range from 1.6 to 6 depending on the building. Clearly, accurate loss modeling requires site-specific factors. Seidel and Rappaport (1992) developed a simple site-specific model for their measurements. The standard deviation between measured and predicted path-loss was found to be 5.8 dB. Ray tracing is now being used to estimate an impulse response anywhere in a building. Rays are launched from a transmitter, and the interaction of the rays with partitions, floors, and walls can be modeled using reflection and transmission theory. Either brute force ray tracing (Seidel and Rappaport 1992) or ray tracing based on image theory (Ho et al, 1994, and Valenzuela 1993) can be used to provide accurate predictions. The 3-D ray tracing algorithms of Ho et al., and Velenzuela account for all the transmitted and reflected rays that reach a receiver. In Velenzuela, transmission and reflection coefficients were derived from a multilayer dielectric model for walls, ceilings, and floors while Ho et al. used simple, empirically derived reflection coefficients which are a function of the angle and polarization of the incident ray.

Rays can be tested for intersection at any point on a building floor plan. Ultimately, the accuracy of such techniques relies on the accuracy and detail of the site-specific representation of the propagation medium.

For Further Information

Recently, several books and survey papers on mobile radio propagation have been published. Two particularly well written texts that cover the topic are *The Mobile Radio Propagation Channel,* by Parsons (Wiley 1992) and *Radio Wave Propagation,* by Griffiths (Wiley 1989). Two recent survey papers that cover the subject of indoor radio propagation are those by Molkdar (1991) and Hashemi (1993). The textbook *Wireless Communications: Principles and Practice* by Rappaport (Prentice Hall 1996) provides a thorough treatment of mobile radio propagation in Chapters 3 and 4.

Works Cited

Ament, W. S., "Toward a theory of reflection by a rough surface," *Proceedings of the IRE,* vol. 41, no. 1, January 1953, pp. 142-146.

Boithias, L., *Radio Wave Propagation* (McGraw-Hill Inc., New York, 1987).

Bullington, K., "Radio propagation at frequencies above 30 megacycles," *Proceedings of the IEEE,* vol. 35, no. 10, 1947, pp. 1122-1136.

Cox, D.C. and Leck, R.P., "Distributions of multipath delay spreads and average excess delay for 910 MHz urban mobile radio paths," *IEEE Transactions on Antennas and Propagation,* vol. AP-23, no. 2, March 1975, pp.206-213.

Deygout, J., "Multiple knife-edge diffraction of microwaves," *IEEE Transactions on Antennas and Propagation,* vol. AP-14, no. 4, 1966, pp. 480-489.

Epstein, J. and Peterson, D.W. "An experimental study of wave propagation at 850 MC," *Proceedings of the IRE,* 41, no. 5, 1953, pp. 595-611.

Hashemi, H. "Indoor radio propagation channel," *Proceedings of the IEEE,* vol. 81, no. 7, July 1993, pp. 941-968.

Hata, Masaharu, "Empirical formula for propagation loss in land mobile radio services," *IEEE Transactions on Vehicular Technology,* vol. VT-29, no.3, August 1980, pp. 317-325.

Ho, C. M. P., Rappaport, T. S, and Koushik P., "Antenna effects on indoor obstructed wireless channels and a deterministic image-based wide-band propagation networks," *International Journal of Wireless Information Networks,* vol. 1, no. 1, 1994, pp. 61-67.

Howard, Steven J. and Pahlavan, Kaveh, "Autoregressive modeling for wide-band indoor radio propagation," *IEEE Transactions on Communications,* vol. 40, no 9, September 1992, pp. 1540-1552.

Lee, W. C. Y., *Mobile Communications Engineering* (McGraw Hill Inc., New York, 1985).

Longley, A G. and Rice P. L., "Prediction of tropospheric radio transmission loss over irregular terrain; A computer method-1968," *ESSA Tech. Rep.* ERL 79-ITS 67, 1968.

Longley, A. G., "Radio propagation in urban areas," *Office of Telecommunications*

Report 78-144, April 1978.

Millington, G., Hewitt R., and Immirzi F. S., "Double knife-edge diffraction in field strength predictions," *Proceedings of the IEEE,* vol. 109C, 1967, pp. 419-429.

Molkdar, D. "Review on radio propagation into and within buildings," *IEEE Proceedings H, Microwaves, Antennas, and Propagation,* vol. 138, no. 1, February 1991, pp 61-73.

Okumura, Y., Ohmori E., Kawano T., and Fakuda K., "Field strength and its variability in VHF and UHF land mobile radio service," *Review of the Electronic Communication Laboratory,* vol. 16, no. 9-10, 1968, pp. 825-873.

Parsons, David, *The Mobile Radio Propagation Channel* (John Wiley & Sons, New York, 1992), 2.3.3.

Rappaport, T. S., "Characterization of UHF Multipath Radio Channels in Factory Buildings," *IEEE Transactions on Antennas and Propagation,* vol. 37, no. 8, August 1989, pp. 1058-1069.

Rappaport, T. S., et al., "Statistical channel impulse response models for factory and open plan building radio communication system design," *IEEE Transactions on Communications,* vol. COM-39, no. 5, May 1991, pp. 794-806.

Rappaport, T. S., Huang W., and Feuerstein M.J., "Performance of decision feedback equalizers in simulated urban and indoor radio channels," *IEICETransactions on Communications (Japan),* vol. E76-B, no. 2, February 1993, pp. 78-89.

Rappaport, T. S., *Wireless Communications, Principles, and Practice,* Prentice Hall, 1996.

Rice, P. L., Longley A. G., Norton K. A., and Barsis A. P., "Transmission loss predictions for tropospheric communication circuits," NBS Tech Note 101, issued May 7, 1965; revised May 1, 1966 and January 1967. Saleh, A. M., and Valenzula, R. A., "A statistical model for indoor multipath propagation," *IEEE Journal on Selected Areas in Communications,* vol. SAC-5, no. 2, February 1987, pp. 128-137.

Seidel, S. Y. and Rappaport T. S., "914 MHz Path-loss prediction models for indoor wireless communications in multifloored buildings," *IEEE Transactions on Antenna and Propagation,* vol. AP-40, no. 2, February 1992, pp. 1-11.

Seidel, S.Y. and Rappaport T. S., "A ray tracing technique to predict Path-loss and delay spread inside buildings," *IEEE Globcom 92,* December 7, 1992, pp. 1826-1829.

Turin, L. George, et al., "A statistical model of urban multipath propagation," *IEEE Transactions on Vehicular Technology,* vol. VT-21, no. 1, February 1972. pp 1-9.

Valenzuela, R.A., "A ray tracing approach to predicting indoor wireless Transmission," *IEEE Vehicular Technology Conference,* 1993, pp. 214-218.

Instantaneous Wideband Channel Sounding for 1.7 GHz PCS

Robert Castle
Roberto Corvaja
Michael Lawton
Matteo Vocale
Hewlett-Packard Laboratories

Abstract

This paper presents the technique for channel sounding experiments carried out at the Hewlett-Packard Laboratories in Bristol. The channel sounding measurements were taken at 1.7225 GHz, a frequency similar to those used by DCS 1800 and DECT, and to those which will be employed by other emerging PCS standards. The technique presented is based on a software correlation between the input and the output of the channel, and the dynamic performance achieved is better than that obtained with conventional techniques. In particular, unlike conventional correlation sounders, this technique does not suffer from the trade-off between dynamic range and maximum resolvable Doppler shift.

Introduction

The aim of the experiments carried out at the Hewlett-Packard Laboratories in Bristol was to obtain a characterization of a radio channel around 1.7 GHz (close to the frequencies used by typical emerging PCS systems) by means of its impulse response. Due to the time-varying characteristics of these radio channels, it is important to reflect the variations of the channel impulse response in the experiments, since the average impulse response gives only limited information for the analysis of radio communication systems. Therefore, the channel sounding is performed in such a way that a sequence of almost instantaneous "snapshots" is obtained, representing the channel impulse response at different times, without the need for averaging over relatively long intervals. The time between two subsequent snapshots, in this technique, is chosen by necessity to be greater than the maximum expected delay in the multipath impulse response, but it must not be too long, otherwise the information on the time variations of the channel is lost: in these experiments the time resolution between different snapshots is almost two orders of magnitude better than in conventional correlation sounders.

Sounding Technique

The impulse response of a radio channel can be obtained by different and almost equivalent methods. In the first method, called pulse sounding, the impulse response is obtained by sending a sequence of narrow-time power impulses through the channel and collecting the output. The disadvantages of this method are the small dynamic range and the difficulty in the generation of the narrow power peaks, constrained by the limitations of the transmitters.

A second method is the pulse compression technique, where a wideband signal is transmitted through the channel and the impulse response is obtained by the correlation between the input and the output. In particular, the usual way in which the correlation is performed is the sliding correlation technique: the transmitter and the receiver use a clock with a slight frequency offset to evaluate the impulse response for different values of the delay. This technique suffers from a trade-off between the dynamic range and the maximum measurable Doppler shift (i.e., the maximum frequency in the variation of the channel) for a fixed number of values per second in the representation of the impulse response. If instead the dynamic range is kept fixed, then the trade-off is between the maximum Doppler shift and the number of values per second obtained in each representation of the impulse response (this number sets a limit to the bandwidth of the sounding). Using a software correlation technique, the above trade-off can be overcome, since both a sufficient dynamic range and a high maximum Doppler frequency can be reached at the same time, allowing one to measure fast variations of the radio channel.

In this case, the transmitted signal $x(t)$ is a pseudo-random binary sequence (PRBS), and the correlation between the input $x(t)$ and the output $y(t)$, say $rxy(t)$, yields an estimation of the channel impulse response as a sequence of snapshots that is "almost periodic" with the same period Tp as the input PRBS. More specifically, due to the variability of the channel, one can denote with $h(T, t)$ the impulse response at time t (T represents the delay), and the correlation gives):

$$r_{xy}(t) = \sum_{n=0}^{M} h(t - nT_p, nT_p) \tag{1}$$

where M is the number of periods observed (each of duration Tp) and it is assumed that the impulse response is not varying within each PRBS period (a lighter assumption than that needed when the period of computation of the impulse response is longer).

Numerical Algorithms

As explained in the previous section, the channel sounding requires the computation of the correlation between the input PRBS and the output of the channel. This correlation can be performed in software once the output signal is available in digitized format; the PRBS used at the transmitter is known and can be generated by a simple software routine, and as the channel is band-limited to a bandwidth B, using a sampling rate greater than $2B$ guarantees ideally no loss of information in a discrete-time representation of the received wave forms.

From the software point of view, the PRBS signal is a sequence $x(nTc)$, where Tc is the chip time and is periodic with period $Tp = NTc$. The channel output is a signal $y(kT)$, where T is the sampling period of the received data. The relations that must hold between the limit Tc, B, and T are the following:

- $1/T_c >> B$, to neglect in the computation of the impulse response the effects of the nonideality of the PRBS autocorrelation function.

- $1/T \geq 2B$, to satisfy the requirements of the sampling theorem.

To perform the correlation between $x(nTc)$ and $y(kT)$, the two signals must be defined in the same time domain. The choice of the authors is to adopt $Z(T)$ as this time domain. To this end, the PRBS signal must be low-pass filtered to remove the frequency components greater than $1/(2T)$ and avoid aliasing, and then the signal can be sampled (with sampling rate $1/T$) to the desired time domain. The design of the digital low-pass filter is not a critical task; the passband is required to be equal to the channel bandwidth and the stopband to start from $1/(2T)$. In the experiments, the margin between B and $1/(2T)$ is always quite wide to allow various design techniques to be used. For the sounding algorithms, both a windows-based approach and an equiripple filter obtained by the Remez

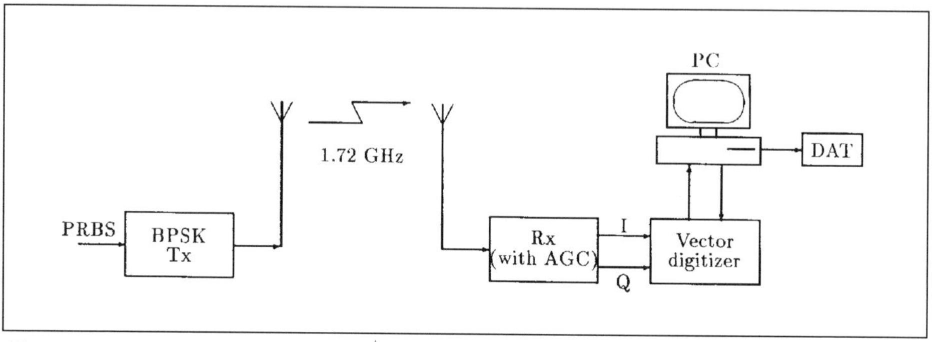

Figure 1 *Experimental setup for the channel sounding*

exchange algorithm have been used. Using these design techniques, a slight difference has been experienced in the time resolution of different rays in a multipath profile (given the same bandwidth of the filter). The different resolution can be related to the fact that the filter impulse response obtained with the window approach is narrower (in the time domain) than the corresponding one obtained by the Remez algorithm, even if its spectral properties are somewhat poorer.

Once the two signals are defined on the same temporal domain, the correlation can be computed straightforwardly or via fast fourier transform (to reduce the complexity of the calculation for large values of the PRBS period $Np = Tp/T$.

Experimental Setup

A schematic of the experimental setup is presented in *Figure 1.* The key elements for the channel sounding experiments are a conventional BPSK transmitter, a superetherodyne receiver, and a high-speed, deep memory digitizer.

The PRBS generated by the transmitter is upconverted to 1.7225 GHz and amplified prior to transmission. At the opposite end of the channel, the radio signal is downconverted by the receiver and the baseband in-phase (I) and quadrature (Q) components are sampled by the digitizer.

In these experiments, 8-bit ADCs have been used for the quantization of signals. To maintain a good system dynamic range when dealing with low signal levels, a closed-loop automatic gain control (AGC) has been introduced before the quantization, to ensure that the dynamic of the signal would constantly fill the range of the ADCs. The voltage controlling the gain is sampled and stored together with the I and Q channels so that the amplitude information can be reapplied to the impulse response after having performed the correlation on the stored I and Q. The capture process is controlled by a PC, and the data are then transferred to a DAT archive or to a Unix workstation for the off-line analysis and presentation of the results.

System Parameters and Performance

The main parameters of the system described above are the following:

- Chipping rate of the PRBS $Fc = 1/Tc = 40$ MHz
- Length of the PRBS $N = 213 - 1$
- Bandwidth of the filter at the receiver $B = 2.5$ MHz
- Sampling frequency for I and Q components $F = 1/T = 8$ Msamples/s
- Quantization with 8-bit A/D converters
- Maximum amount of data stored for each measurement =120 Mb, or 7.5*s*

The bandwidth B of the filters on I and Q at the receiver has been chosen to be suitable for the analysis of typical PCS systems, and the chipping rate has been chosen great enough to guarantee a white spectrum of the PRBS over this bandwidth.

With regards to the performance of the channel sounding system characterized by these parameters, the benchmarks to be observed are:

- Time resolution between different echoes (i.e., bandwidth of the sounding technique)
- Dynamic range, defined as the power ratio between the largest and smallest echoes that can be perceived
- Maximum measurable Doppler shift
- Minimum measurable Doppler shift

The maximum time resolution, namely the minimum delay that allows two different paths to be clearly distinguished, is limited by the bandwidth B at the receiver filter. In this case it is about 400ns.

The dynamic range for a conventional pulse compression system is given by 20 log N (N is the length of the PRBS used). In this case, the dynamic range is only limited by the number of bits used to quantize the received signal, since the length

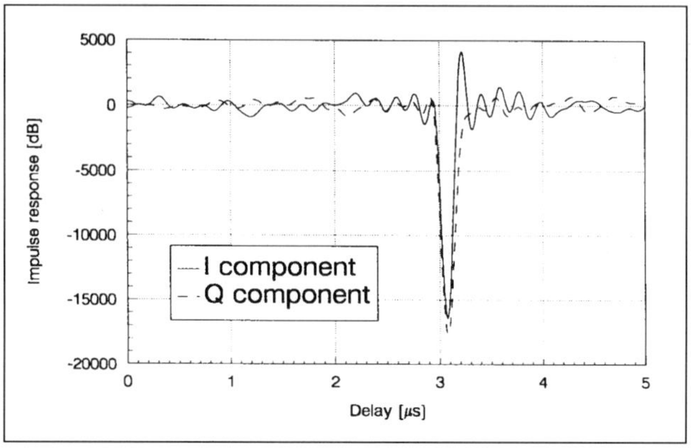

Figure 2 *Back-to-back impulse response (I and Q components)*

of the PRBS can be chosen to be great enough (213 – 1) to guarantee a much wider dynamic range. With 8-bit A/D converters, the dynamic range is about 48 dB. Therefore, as long as $N >$ 28 – 1, the limit is only determined by the quantization error (this bottleneck for system dynamic range can be eased by simply upgrading the ADCs). Parameter N can then be chosen only on the basis of Doppler considerations.

The maximum measurable Doppler shift is limited by the minimum time between two snapshots, which for this system is $Tp = NT_c = 0.2$ ms, allowing us to measure Doppler shifts up to 2500 Hz, and channel variations that are almost two orders of magnitude faster than with conventional sliding correlation sounders.

Finally, the minimum measurable Doppler shift is limited by the time the I and Q channel outputs can be monitored. Given the sampling rate $F = 8$ Msamples/s, and given that the digitizer can store up to 120 Mb of data, this is equivalent to an observation time of about 7 seconds, so that Doppler shifts below 1 Hz can be measured reliably.

Although the main characteristic of this method is the off-line processing, allowing a greater flexibility in the choice of the parameters, this does not prevent the use of the equipment for measurements in the field. In fact, in a possible application in the field, a time of less than one minute allows a detailed characterization of the radio channel to be obtained. For example, a sequence of 20 snapshots each spaced 10 ms apart would require a computational time of about 40 s, mainly determined by the time required to access the large files of stored data.

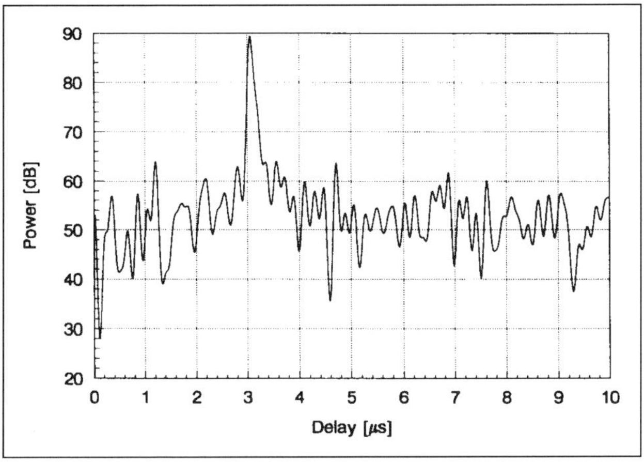

Figure 3 *Back-to-back impulse response (power profile)*

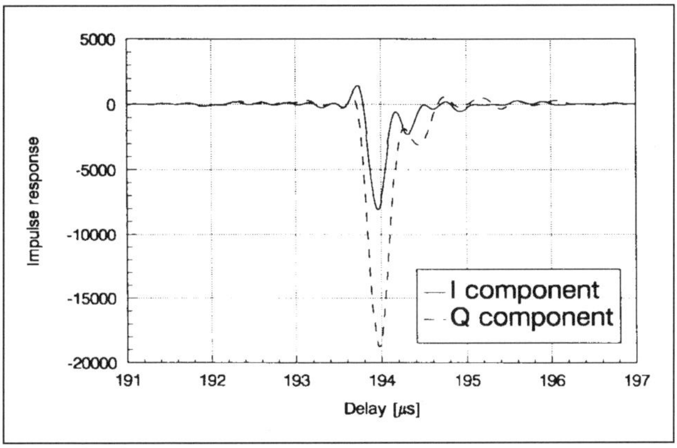

Figure 4 *Complex impulse reponse (I and Q components) for an indoor channel*

Results

First of all, the back-to-back impulse response is presented in Figures 2-3. This corresponds to an ideal channel, obtained by connecting the receiver and the transmitter with a coaxial cable. *Figure 2* shows the I and Q components, while the power profile is presented in *Figure 3*.

Then, some results are shown for both indoor and outdoor channels as an example of the kind of observations allowed by these measurements. For the indoor channel, *Figure 4* shows the real and imaginary parts of the complex baseband

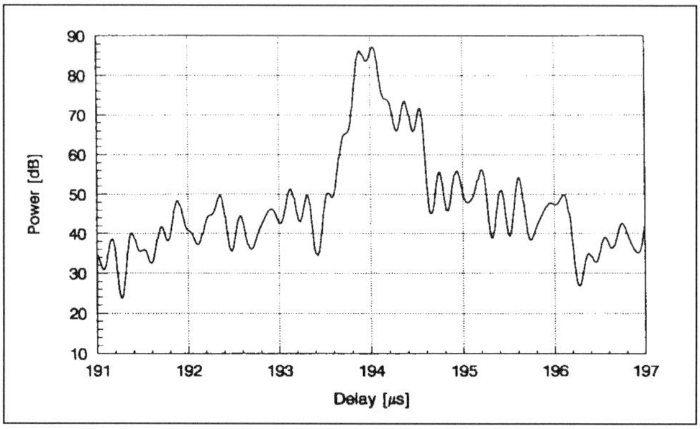

Figure 5 *Power profile for an indoor channel*

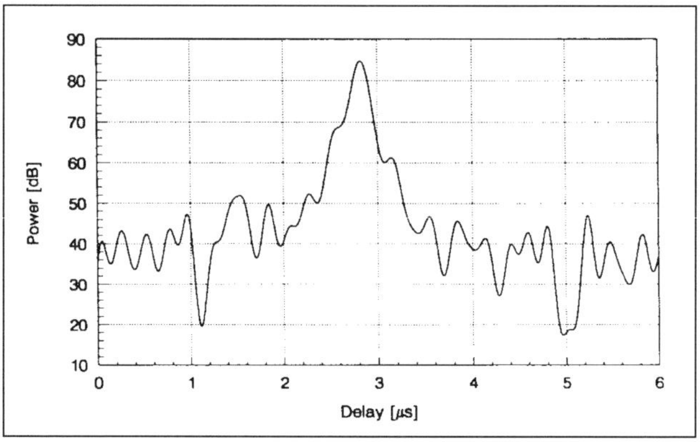

Figure 6 *Power profile for an indoor channel*

impulse response for a channel without a line-of-sight (LOS) path. In *Figure 5,* the corresponding power profile is presented, showing that the impulse response is almost confined within 1ms, which, incidentally, is a time period very close to the bit period of DECT and DCS 1800. The channel can be considered almost static, since both the transmitter and the receiver were not moving during the capture.

With regards to the outdoor channel, the transmitter was fitted in a car, while the receiver was fixed in an area where nearby buildings and land irregularities provided reflections and the lack of a direct line of sight. *Figure 6* shows the power profile at a particular time for this channel, while *Figures 7* and *8* show the evolution with time of the power profile. In the first, the separation between successive power profiles is the minimum allowed by the system, so that the power profiles shown in the figure correspond to successive snapshots. On the other hand, in Figure 8, the separation between two power profiles is chosen to

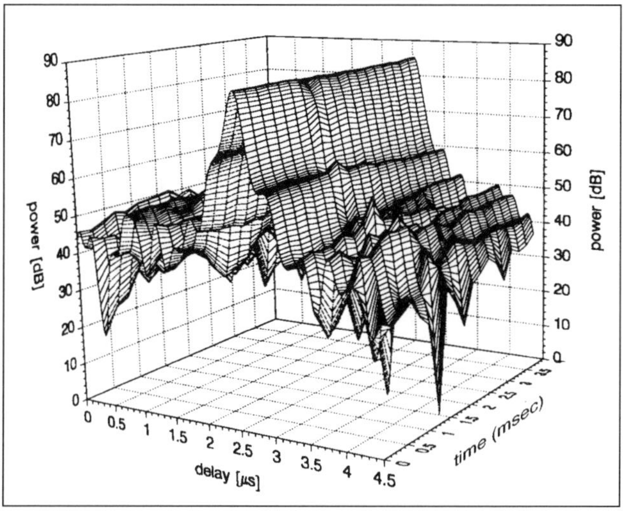

Figure 7 *Time evolution of the power profile for the outdoor channel*

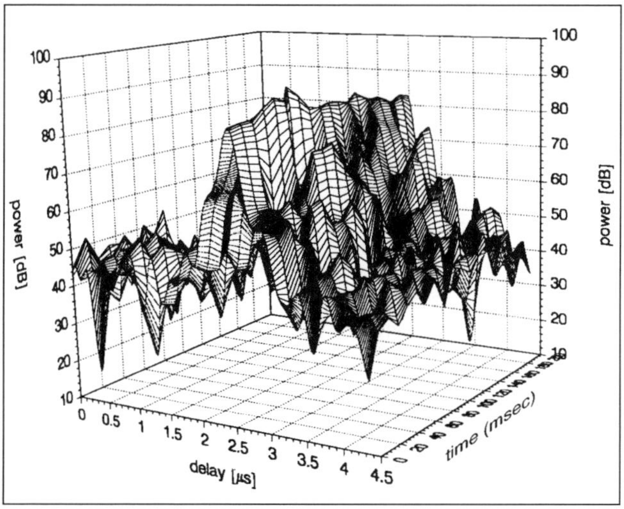

Figure 8 *Time evolution of the power profile for the outdoor channel*

be of 50 snapshots, corresponding to a time of about 10 ms, and a larger variation of the power profile can be noticed. The Doppler spectrum for this channel has been evaluated for the delay corresponding to the first peak of the power profile, and for a delay of 0.5/1ms from it, and is presented in *Figures 9* and *10:* the car was moving at a speed of approximately 30 mph, so that the maximum Doppler

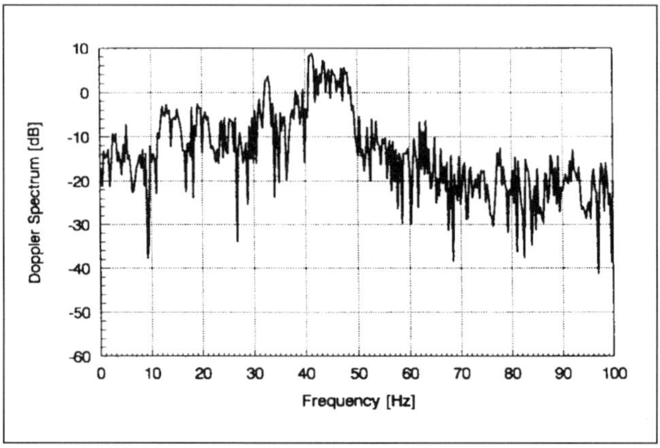

Figure 9 *Doppler power spectrum for an outdoor channel evaluated for the delay corresponding to the first path*

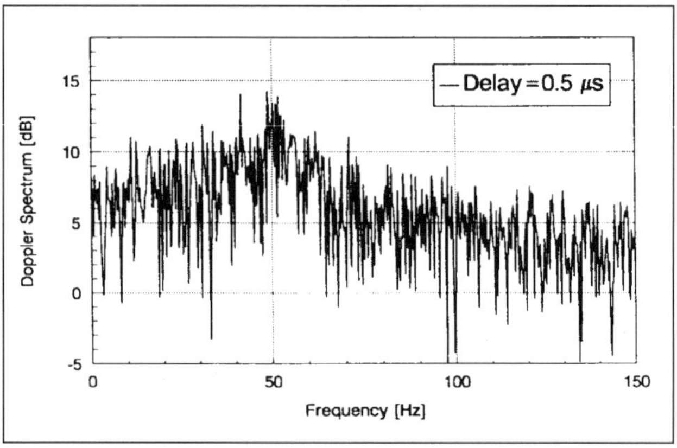

Figure 10 *Doppler power spectrum for an outdoor channel evaluated for a delay of 0.5 μs from the first path*

frequency expected (due only to the car speed) is about 75 Hz. From the results shown in *Figures 7* and *8,* and from the Doppler spectra, the variation of the channel between two subsequent snapshots (i.e., spaced 0.2 ms apart) does not seem noticeable, but a variation of some dB in the power profile can be experienced between different impulse responses taken some ms apart. Indeed, the Doppler spectrum evaluated for 0.5 ms from the first path shows some frequency components (with a power within 10 dB from the maximum) even at relatively high frequencies (about 150 Hz). This fact suggests that to follow the variation of the channel used in this experiment with sufficient accuracy, a reasonable value for the time between two different snapshots should be smaller than 5ms.

Conclusions

This paper presented a sounding technique based on software correlation, which achieves a performance comparable to, or even better than, the commonly used techniques, without the disadvantages of hardware correlation techniques and without the trade-off between dynamic range and Doppler shift.

The interval between subsequent snapshots of the channel impulse response is much less than with sliding correlation, and the dynamic range is only limited by the quantization step chosen. Indeed, simply upgrading the ADCs used in these experiments from 8 to 10 bits would lead to a 12-dB increase in dynamic range. The first results elaborated so far offer already a good insight, for a few channels, on how this technique can be used to measure the extent to which multipath effects are contained within the bit period of a given transmission standard, and the rapidity with which multipath echoes change their delay and intensity.

Bibliography

Cox, D.C.. "Delay Doppler characteristics of multipath propagation at 910 MHz in a suburban mobile radio environment", *IEEE Trans. on Antennas and Propagation,* vol. 20, No. 5, pp. 625-635, Sep. 1972.

Parks, T.W.; J. H. McClellan, "A program for the design of linear phase finite impulse response digital filters", *IEEE Trans. Audio and Electroacoustics,* vol. AU-20, pp. 195-199, Aug. 1972.

Parsons, J.D.; D. A. Demery, A. M. D. Turkmani, "Sounding techniques for wideband mobile radio channels: a review", *IEE Proceedings I,* vol. 138, No. 5, pp. 437-445, Oct. 1991.

Proakis, J.G.; D. G. Manolakis, *Introduction to Digital Signal Processing.* New York: Macmillan, 1988.

Rappaport, T.S.; S. Y. Seidel, K. Takamizawa, "Statistical channel impulse response models for factory and open plan building radio communication system design", *IEEE Trans. on Comm.,* vol. 39, No. 5, pp. 794-807, May 1991.

Intelligent Antennas: Spatial Division Multiple Access

Martin Cooper
Chairman and Chief Executive Officer
ArrayComm, Inc.

Marc Goldburg, Ph.D
Research Team Leader
ArrayComm, Inc.

Introduction

Antennas collect radio frequency energy from space for reception purposes and distribute energy into space for transmission. To date, they have been the most neglected of all the components in personal communications systems. Yet the manner in which radio frequency energy is distributed into and collected from space has a profound influence upon the efficient use of spectrum, the cost of establishing new personal communications networks, and the service quality provided by those networks. The combination of the enormous increase in low-cost computing capacity and the development of new algorithms for processing signals from arrays of simple antennas make possible "intelligent antennas" for cellular-like communications systems, providing greater coverage area for each cell site, the rejection of interference, and substantial capacity improvements.

There are many methods for utilizing data from antenna arrays in wireless communications systems, the most advanced of which is called spatial division multiple access (SDMA). At the base station, SDMA continually adapts to the radio environment, providing each user with uplink and downlink signals of the highest possible quality. At the network level, this improved base station performance can be exploited to increase base station range – thereby reducing network cost – or to increase system capacity and, hence, spectral efficiency. SDMA is compatible with any practical modulation method or frequency band. Practical SDMA intelligent antenna systems have been constructed and demonstrated. Such systems will appear in commercial communications networks by the end of 1995. This essay describes the principles of SDMA, explains the benefits of SDMA, and compares SDMA with other smart antennas.

Antenna Principles

Radio antennas couple, more or less efficiently, electromagnetic energy from one medium, space, for example, to another, such as wire, coaxial cable, or waveguide. The simplest antennas radiate and receive equally well in all directions.[1] Antennas can also be constructed to have certain fixed, preferential directions (such antennas are used in so-called sectorized systems). One might reasonably ask, "Why use anything more than a single omnidirectional (no preferential direction) antenna at a base station?". The goal of the next several sections is to answer this question by describing, in order of increasing benefits, the principal schemes for employing multiple antennas at a base station.

Sector Antennas

Antennas can be constructed to cover a fixed arc or sector of, say, 120 degrees. Three antennas could therefore cover all directions. Everything else being equal, these sector antennas provide increased gain over a restricted range of azimuths as compared to an omnidirectional antenna. In a cellular radiotelephone system, each sector would typically be treated as a different cell whose range is greater than in the omnidirectional case. Sector antennas increase the reuse possible in such systems by reducing potential interference, and they are widely used for this purpose. As many as six sectors per cell have been used in practical service. This results in a single antenna per (effective) base station and, therefore, does not really qualify as "multiple antennas at a base station" in our context. In the sequel, a sector will be considered to be the same as an independent base station.

Diversity Systems

Small displacements in an antenna's location can have a significant effect on the signal amplitude that appears at its output. This effect results from a propagation phenomenon known as fading. A class of antenna system called "diversity antennas" employs a group of physically separated antennas with the assumption that at least one of the antennas will be in a favorable location at any given time. So-called "switched diversity" systems continually switch among the antennas so as always to use the antenna with the largest output. Other diversity systems, such as maximal ratio combining systems, for example, combine the outputs of all the antennas to maximize the ratio of combined received signal energy to noise. These diversity systems can be useful in environments where fading is the dominant mechanism for signal degradation. In environments with significant interference, however, the simple strategies of locking onto the strongest signal or extracting maximum signal power from the antennas are clearly inappropriate and can result in crystal-clear reception of an interferer at the expense of the desired signal.

Principles of Antenna Arrays

By combining the outputs of the individual antennas in an array, a single effective antenna can be created with gain and directional characteristics that are very different from those of the individual elements comprising the array. For example, consider a row of, m, simple and identical, antenna elements (called a linear array) with elements spaced one-half wavelength apart. Suppose that one simply adds the outputs of all the elements. Signals arriving at the array from broadside, or perpendicular to the axis of the array, arrive simultaneously at each element of the array, and their sum will therefore be m times as large (m^2 times as powerful) as the signal received by a single antenna. Simply adding the outputs of the m element array therefore results in an amplitude gain of m for signals arriving from broadside.

Figure 1 depicts the effective antenna pattern resulting from the above strategy for linear arrays consisting of a single antenna, two antennas, and six antennas (omnidirectional element patterns and half-wavelength spacing in all cases). The radius of the pattern in *Figure 1* is proportional to the gain or strength of the signal at the output of the array. The increased gain in the broadside direction and the increases in gain with the number of array elements are evident in the figure. It is also evident that there are certain directions in which the effective antenna has reduced sensitivity, or nulls.

Classical phased-array antenna systems have gain patterns or beam patterns that look very much like the one in the figure, except that the direction of maximum gain might not be the broadside direction. More advanced systems have patterns that are optimized to enhance a particular user's signal while simultaneously

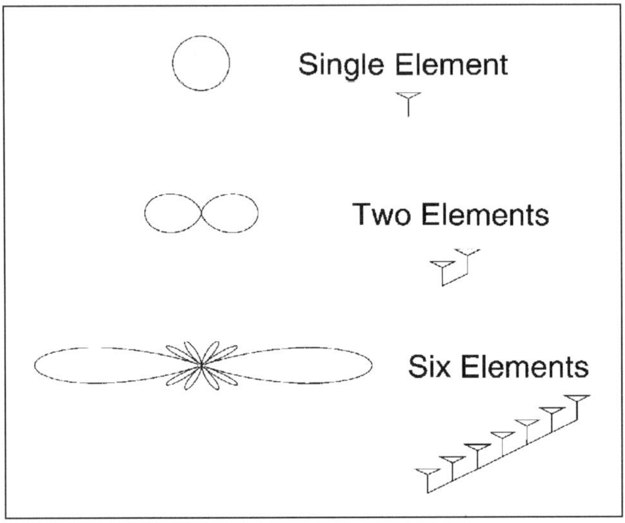

Figure 1 *Broadside antenna pattems for arrays with different numbers of elements.*

rejecting interferers. Another differentiator of antenna array systems is whether or not the beam patterns can change with time. Antenna arrays can be built with combining strategies that are fixed. From an operational standpoint, this sort of array is no different from a conventional antenna with the same directional sensitivity. Alternatively, arrays can be equipped with combining hardware and software that make it possible for the pattern to be changed over time and adapted to the current operational scenario.

Because radio reception and transmission are reciprocal, directive transmission with gain is also possible from an array. Any directivity pattern achievable for reception is also achievable for transmission.

Switched-Beam Antennas

It is possible, using array antennas, to create a group of overlapping beams that together result in omnidirectional coverage. For example, one might design overlapping beam patterns (combining strategies) similar to the ones in *Figure 1* but pointing in slightly different directions so that the patterns' main lobes are adjacent like the petals of a flower. Every so often, the system scans the outputs of each beam and selects the beam with the largest output power. If the signal/user moves from one beam to another, the antenna switches to the new beam. Only a single beam pattern is employed at any given time; on average its gain will be slightly less than the factor of m described above.

Switched-beam antennas are normally used only for reception of signals since there can be ambiguity in the system's perception of the location of the received signal. The consequences of transmitting in the wrong beam are evident. Switched-beam antennas offer range extension as their primary feature but also suppress interference arriving from directions away from the active beam's center.

Adaptive Array Antenna Systems

Adaptive array antenna systems continually monitor their coverage areas, attempting to adapt to their changing radio environment, which consists of (often mobile) users and interferers. In the simplest scenario – that of a single user and no interferers – the system adapts to the user's motion by providing an effective antenna pattern that follows the user, always providing maximum gain in the user's direction. The base station component of spatial division multiple access (SDMA) is an advanced adaptive array system, described below in greater detail. The principle of SDMA is quite different from the beam-forming approaches described above. In fact, SDMA's operation is analogous to that of human hearing.

Consider the following analogy: if you close your eyes and listen to a sound source, you can identify the direction from which the sound is coming with remarkable accuracy. You accomplish this by using your ears as an array of acoustic sensors, not unlike antennas that are sensing radio energy. Each ear receives the sound at a different time depending upon the direction from which the sound is coming. Your brain processes the information from both ears (now not sound at all but rather some form of neural signal) and computes the direction from which the sound is coming. Your brain then combines the sounds, coming from the direction you selected, constructively. Sound from other directions adds together incoherently. The net effect is that you can hear the sound you have decided to listen to twice as loud as sounds from other directions, and you know what direction the sound is coming from.

Of course, the radio frequency array is not limited to two "ears" (array elements) and can thus "hear" with even greater gain and selectivity. Further, the radio frequency array can respond with multiple "mouths" (array elements) and achieve the reciprocal advantages.

The following sequence of events occurs many times each second in an SDMA system:

1) A "snapshot," or sample, is taken of the signals coming from all of the antenna elements, converted into digital form, and stored in memory.

2) A computer, the SDMA processor, then analyzes the sample to obtain an estimate of the radio environment, identifying users and interferers and their locations.

3) The processor calculates the combining strategy for the antenna signals that optimally recovers the users signals. With this strategy, each user's signal is received with as much gain as possible and with the other users' and interferers' signals rejected as much as possible.

4) An analogous calculation is done to allow spatially selective transmission from the array. Each user's signal is now effectively delivered through a separate communications channel – a spatial channel.

5) The system now has the ability to both transmit and receive information on each of the spatial channels making them two-way channels.

The net result of the above process is that the SDMA system can create a number of two-way spatial channels on a single conventional channel, be it frequency, time, or code. Each of these spatial channels enjoys the full gain and interference rejection capabilities of the array. In theory, an array with m elements can support m spatial channels per conventional channel. In practice, the number is somewhat less and depends on the environment.

The benefits of an SDMA system include the following:

- Range extension – the coverage area of the array is greater than that of any element as a result of the gain provided by the array. When a system is constructed using SDMA, the number of cells required to cover a given area can be substantially reduced. A ten-element array offers a gain of ten, which typically doubles the range of the cell and thereby quadruples the coverage area.

- Interference from other systems and from users in other cells is significantly reduced. In "noisy" areas where range is limited by interference, spatially selective transmission and reception result in range extension.

- The destructive effects of multipath signals – copies of the desired signal that have arrived at the antenna after bouncing from objects between the signal source and the antenna – can often be mitigated. In certain cases, the multipath can actually be used to reinforce the desired signal.

- The capacity of the system is increased in two distinct ways:

  a) Channel reuse patterns in cellular systems can be significantly tighter because the average interference resulting from co-channel signals in other cells is markedly reduced (e.g., moving from a 7-cell to a 4-cell reuse pattern nearly doubles capacity).

  b) Separate spatial channels can be created in each cell on the same conventional channel. In other words, intra-cell reuse of conventional channels is possible.

- Because SDMA employs spatially selective transmission, an SDMA base station radiates much less total power than a conventional base station. One result is a reduction in network-wide RF pollution. Another is a reduction in power amplifier size. First, the power is divided among the elements, and then the power to each element is reduced because the energy is being delivered directionally. With a ten-element array, the amplifiers at each element need only transmit one-hundredth the power that would be transmitted from the corresponding single-antenna system.

- The direction of each spatial channel is known and can be used to accurately establish the position of the signal source, a prerequisite for location-based services.

- SDMA is compatible with almost any modulation method, bandwidth, or frequency band including AMPS, GSM, PHP, DECT, IS-54, IS-95, and other formats. SDMA can be implemented with a broad range of array geometries and antenna types.

Designers of a communications system will avail themselves of the benefits described above in different ways and in different mixtures as each system matures. When a communications system is initially constructed, costs can be minimized and roll-out can occur more rapidly by using the range extensions available from SDMA. Because the cells are large, better channel reuse is possible than would be with small cells. It is unnecessary to add sites and reduce cell size until required by subscriber demand and then only after the benefits of spatial channels have been absorbed. The major benefit of SDMA is in the flexibility offered to the system implementer.

Comparison of Switched Beam and SDMA Antenna Systems

In clean environments with little or no interference, switched beam antennas can provide range extension. Because the effective pattern of a switched-beam antenna is scalloped (*Figure 2*) and because the peak gain provided by the switched-beam antenna is generally less than that provided by SDMA, the range extension provided by switched-beam systems will be less than that provided by SDMA, assuming comparable antenna arrays. Switched-beam antennas do not offer significant capacity increase since they are customarily used only in the receive direction for range extension. In high-interference areas, switched-beam antennas are further limited since their pattern is fixed and they lack the ability to adaptively reject interference.

Conclusion

It is unlikely that new personal wireless systems will be built a few years from now without the use of some form of smart antenna. Smart antennas, properly applied, can substantially reduce the initial cost of such systems. Additionally, intelligent antenna systems such as SDMA can improve signal quality and increase system capacity. Most importantly, SDMA technology offers system designers the tools to create communications systems that are continually and automatically optimized.

End Note

1 More precisely, when we speak of "direction" we are generally referring to azimuth. It is impossible to build a truly omnidirectional antenna and, for a number of good reasons that are beyond the scope of the present discussion, base-station antennas are typically designed to have a small window of elevation angles over which they operate efficiently.

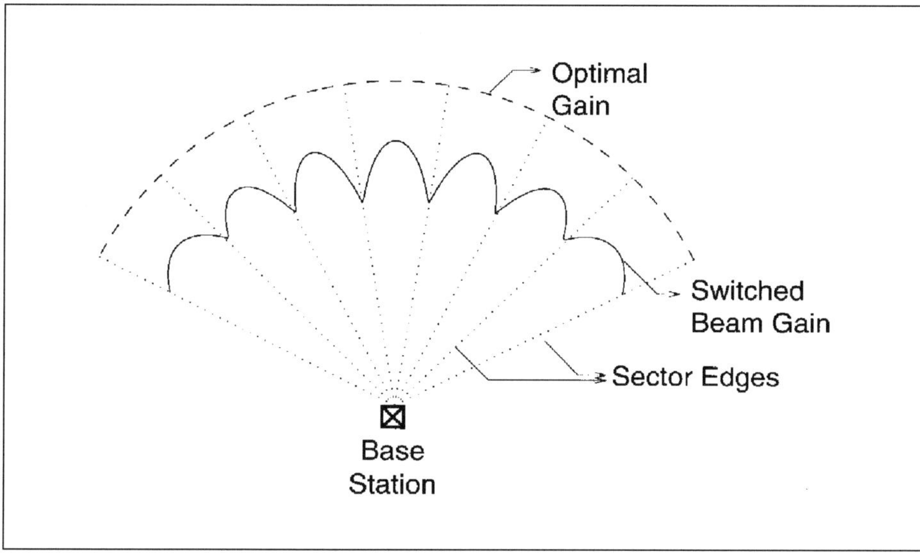

Figure 2 *Gain patterns for switched-beam and ideal systems*

Biological Effects of Radio Frequency Fields

Frank S. Barnes
Professor, Department of Electrical and Computer Engineering
University of Colorado at Boulder

Viviana E. Irizarry-Sepulveda
Northern Telecom, BNR

Abstract

If the extensive use of hand-held transmitters for cellular phones should prove to result in health effects, it could have both significant legal and financial implications for the industry. In this paper, we begin by reviewing some of the problems in setting safety standards. Next, we consider the coupling of radio frequencies or radio frequency waves into the body in order to estimate the power levels that are likely to occur. These levels are then discussed in terms of possible mechanisms for the interaction of the fields with biological systems. A review of some of the literature on biological effects of low levels of radio frequency power is presented. This is followed by a brief discussion of product liability history for other devices that might be presumed to be similar to cellular phones and some speculation on how the public is likely to react to charges that the use of cellular phones causes cancer or other health problems.

Introduction

The question of whether exposing biological systems to low levels of radio frequency fields causes health effects has been controversial since at least the late 1940s and remains unresolved. This is at least partly because of the complexity of the problem. To begin with, there is a rather complicated problem of determining how much energy from a given source is coupled into the biological system and how it is distributed. Second, there is the problem of how these fields interact with the biological material. Finally, there is the complex problem of how these changes affect the functioning of the biological system. The size scales for these effects range from atoms through cells to organs and the whole body. The time scales range from 10^{-12} seconds to generations. Building a logical chain of events that leads us from basic interactions, such as the force exerted by an electric field on an ion or molecule through changes in chemical reaction rates or the transport of ions across a membrane to a health effect, is a long and complex process. Thus, we should not be surprised that it is taking time and that our understanding improves slowly.

In the meantime, we have a public that wants answers and is being aroused by the press and by lawsuits. There have been several legal cases where the plaintiff has claimed that his or her health problems were the result of exposure to radio or microwaves; for one of these, the source of the exposure was a cellular phone.[1-3] As a result of these lawsuits, there is substantial public interest in the possibility that radio waves can cause cancer or increase its growth rate. Given the current state of the legal system in the United States, it is highly likely that a significant number of additional cases will be filed as the use of cellular phones grows, independent of whether or not a cause-and-effect relationship is found.

In this essay, we will review a number of studies on the coupling of radio frequency energy from the antenna of a cellular phone into a human. Assuming these energy distributions, we will look at possible mechanisms of interaction between the radio waves and biological materials. This will be followed by a review of some of the literature on the effects of radio waves on biological systems. Finally, we will briefly review some of the legal history on product liability and make predictions for the future of cellular phones in the health effects arena.

Before we begin, however, it is useful to review briefly how we define what is safe and how we define safety standards. Typically, when a new technology is developed (such as X-rays or microwaves), a number of useful applications drive the development, which in turn leads to the exposure of a significant number of people. It is then discovered that high levels of exposure result in biological damage or health problems. As a result, standards bodies such as ANSI and the IEEE attempt to define safe levels of exposure. The way in which these levels are set depends greatly upon *philosophy*. In the past, the United States has tried to set standards based on the highest level of exposure *not known to be dangerous*. This

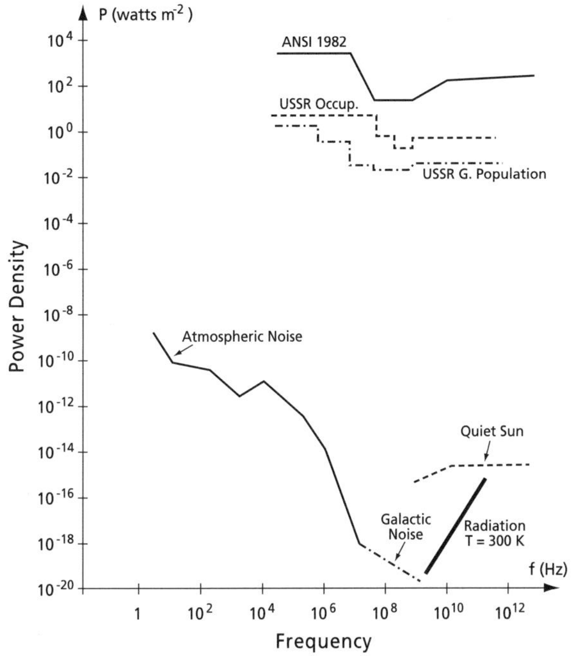

Figure 1 *The average atmospheric noise power density incident on the Earth (Smith, 1982) is compared with the 1982 ANSI and U.S.S.R. exposure standards. Additionally, some low-level signal strengths are indicated for measured biological perturbations and the average signal level for 15 U.S. cities. Median population exposure level is 5 x 10^{-5} W m^{-2} for 15 U.S. cities (U.S. Environmental Protection Agency, 1978).*

is done with inputs from the manufacturers, and an attempt is made to provide the maximum flexibility for the application of technology. For example, *Figure 1* shows allowed power levels in the current safety standards for the continuous exposure of humans to plane waves as a function of frequency. These power levels range from 0.2 mW/cm^2 to 10 mW/cm^2. In contrast, Soviet standards were set by estimating the *lowest level* at which biological effects had been observed and then setting the standard well below that. This resulted in values near 10 μW/cm^2 and a standard which has a much different meaning. Soviet citizens did not sue their government, and major violations of the standard were known to occur.

At the present time, we have a significant element of the U.S. population demanding not to be exposed to anything unless its safety can be proven. We contend that we cannot prove anything is safe against possible effects we have not thought of. Additionally, we should expect that, as our understanding of the interactions and the biology improves over time, we will find radio frequency fields at lower levels, which may or may not lead to health effects. Thus, the overall problem reduces to what risks we are willing to take given the incomplete state of our knowledge. In general, people are willing to take greater risks when they feel they understand them and have some control over them. For example, people are willing to take the risk of driving their cars or skiing but not the risks of living next to a nuclear power plant.

The Coupling of Radio Frequency Energy to the Head from Cellular Phones

In general, the problem of calculating or measuring the energy deposited in a biological object is a complex function of frequency and geometry. Biological objects are irregular in shape and constructed of materials with varying dielectric constants and conductivities which are also a function of frequency. In the past, first-order approximations of power absorbed by the body have been made by assuming that the body could be modeled by a sphere or an ellipsoid with the average dielectric and conductivity of tissue. These models were improved upon by making computer block models as well as models with materials having the appropriate complex dielectric properties.[4] Recently, finite-difference time domain models (FDTD) have been made for portions of the body which take into account a significant portion of the anatomical structure.

The coupling of energy from the antenna of a hand-held cellular phone to the head has been studied by several authors.[5-10] Jensen and Rahmat-Samii have used the FDTD approach to model the fields radiated from a number of antenna configurations *(Figure 2)* and anatomical models for the head which take into account the dielectric constant and loss tangent for the various materials which make up the head and the ear. This information was used to compute values for the peak and average specific absorption rates (SAR), as shown in *Table 1*.[5] Similar results have been obtained by Chen and Wang.[6] Both groups show peak SARs in the range from 1 W/kg/W to 3 W/kg/W of radiated power. The calculated SARs for the radiation of 0.6 W from the antenna show a strong dependence on the distance of the antenna from the head or ear. The power densities decrease rapidly with distance and are less likely to be significant in other parts of the body. New antennas are being designed for reduced power absorption in the head and the ear which can lead to reductions of as much as a factor of 10 in the peak SARs. It is also important to note that hand-held sets may generate a wide

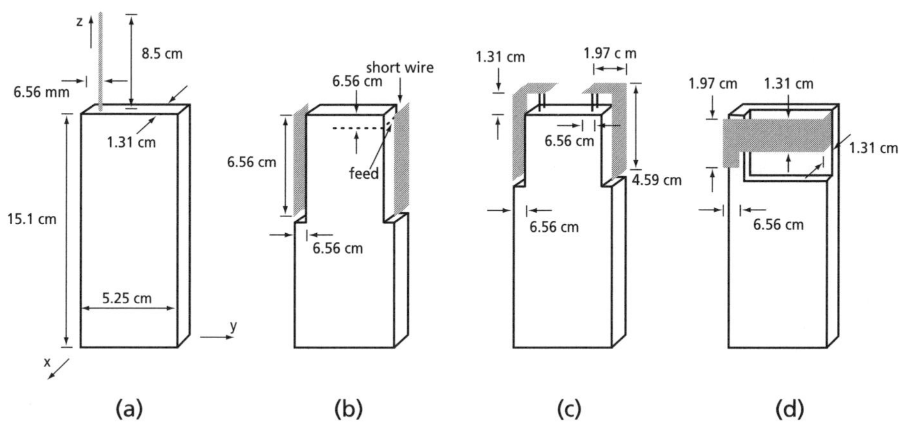

Figure 2 *Antenna geometries and dimensions for hand-held transceiver units: (a) monopole; (b) side-mounted dual PIFA; (c) top-mounted dual BIFA; (d) back-mounted PIFA. These chassis dimensions (105 cm³) shown in (a) apply to all four configurations.*

Configuration						
			Handset Rotated 60° with Respect to Head			
Monopole	0.359	0.169	0.472	1.97	2.29	0.0875
Side-mounted PIFA	0.382	0.301	0.317	3.81	4.54	0.0931
Side-mounted PIFA*	0.212	0.318	0.470	3.14	5.42	0.1003
Back-mounted PIFA	0.260	0.222	0.518	1.32	3.58	0.0634
			Handset Upright with Respect to Head			
Monopole	0.351	0.184	0.465	2.06	2.43	0.086
Side-mounted PIFA	0.332	0.324	0.344	2.07	4.91	0.0809
Back-mounted PIFA	0.258	0.225	0.517	0.90	3.53	0.0629

* Homogenous spherical head (εr = 58.5, σ = 1.21)

Table 1 *Computed normalized power absorption and peak SAR (mW/g/W) in the head and hand. All data is computed at 915 MHz.*

range of frequencies down to as low as 2 Hz via various modulation formats and station-keeping activities. Additionally, low frequency magnetic fields up to 8μT have been measured. These fields are generated by currents in the power supply. The fields around base stations have been studied by Petersen and Testagrossa.[11] The results of their measurements on a variety of cellular base-station antennas show that the peak-power densities that can be expected for 46-m towers at 2 m above the base for distances greater than 1 m from the tower are less than 30 mw/m^2 (3 μm/cm^2) for a 96-system channel radiating 100 w per channel. These power densities are far bellow the ANSI standards.

Mechanisms of Interaction for Radio Waves and Biological Materials

It is useful to begin at the lowest level of interaction among electric and magnetic fields and biological materials with the force, F, which is generated on a charged particle.[12] This is given by:

$$F = q(E + v \times B) \tag{1}$$

where q is the charge, E is the electric field, v is the velocity, and B is the magnetic field. The application of these forces results in current flows. In most cases, the second term is small, and we can describe the induced current density, J, as given by:

$$J = \Sigma q_i \mu_i n_i E = \sigma E \tag{2}$$

where q_i is the charge on a given ion, μ_i is the mobility, n_i is the density of particles, and s is the conductivity of the material. If we go to the slightly more complicated case of a dipole, then

$$F = \alpha \nabla E \quad or \quad F = m \nabla B \tag{3}$$

where α is the electric dipole moment and m is the magnetic dipole moment. ∇E is the gradient of the electric field, and ∇B is the gradient of the magnetic field. Again the magnetic forces are usually small. A possible exception to this may be the recently discovered particles of magnetite, Fe_3O_4, which have been found in the human brain and elements of the immune system.[13] It is convenient to treat the effect of the electric fields on the dipole moments in terms of a complex polarization, p, where p is given by:

$$p_i = \varepsilon_o \Sigma \chi_{ij} E_j \tag{4}$$

where p_i is the component of the polarization in the i-th direction, ε_o is the dielectric constant of free space, χ_{ij} is the electric susceptibility, and E_j is the electric field in the j-th direction.

At radio frequencies, we are not usually concerned with the direct force effects of the electric or magnetic fields; the rate of oscillation is so rapid that the net displacement of a charged particle is small. However, the absorbed energy or the power delivered to the material is often important, and it is possible that radio frequency fields could change the distribution of states for atoms or molecules which are not at thermal equilibrium.[14] The latter possibility is being studied in the case of free radicals and is used as a diagnostic tool in nuclear magnetic resonance imaging; however, we will not discuss it further in this essay.[15]

Depending on the situation, it is convenient to express the absorbed power, P_L, per unit volume, or specific absorption rate (SAR), in the form

$$P_L = \sigma E^2 \quad \text{or} \quad P_L = \omega \varepsilon_o \chi'' E^2 / 2 \tag{5}$$

where w is the angular frequency. The first expression is most convenient when the dominant loss is due to ionic conductivity, and the second expression is most useful when the dominant loss is associated with the imaginary part of the electric susceptibility χ'' and dipole moments.

The most easily observed effect of the absorbed power is heating, and an increase in the temperature of the biological material is

$$\frac{\partial T}{\partial t} = \frac{P_L}{\rho C} - \frac{P_D}{\rho C} \tag{6}$$

where T is the temperature, C is the specific heat, ρ is the density, and P_D is power dissipated by conduction, convection, and radiation. For short exposures, the heat dissipation is dominated by conduction. In this case, our equation takes the form

$$\frac{\partial T}{\partial t} = \frac{P_L}{\rho C} - \frac{T - T_0}{\tau} \tag{7}$$

where τ is the thermal time constant. For a homogeneous sphere, the thermal time constant is given by

$$\tau = r^2 / 4K' \tag{8}$$

where K' is the thermal diffusivity. For water, a sphere with a radius r of one millimeter yields $\tau = 1.8$ seconds. If an applied current pulse is short compared to τ, the maximum temperature change is given by

$$\Delta T_{\max} = \left(\frac{3}{2\pi e}\right)^{3/2} \frac{\overline{H}}{\rho' C_p a^2} \tag{9}$$

where $\overline{H}$ is the total input energy in calories, e is the base of natural logarithms, and a is the radius of the heated sphere.

Changes in temperature are important not only because they change transport properties such as viscosity, η, mobility, μ, and the diffusion coefficient, D but also because they change chemical reaction rates. Typical biochemical reactions can be described by an equation of the form

$$\frac{dS}{dt} = -K'S \tag{10}$$

where S is the fraction of the material which has undergone the chemical reaction, t is the time, and K' is the reaction rate. K' is often given by

$$K' = \frac{kT}{h} \exp\left(\frac{+\Delta H' - T\Delta S'}{R'T}\right) \tag{11}$$

where k is the Boltzmann constant, T is the absolute temperature, H' is the free energy, S' is the entropy, h is Planck's constant, and S' is the gas constant. The significant feature is that the reaction rate K' varies exponentially with temperature, and $\Delta H'$ and $T\Delta S'$ are large numbers. Thus, very small changes in temperature can lead to big changes in chemical reaction rates. It is interesting to note that amplitude-modulated signals are effectively demodulated by the conversion of an AM-modulated field to temperature changes which are proportional to the field squared. This in turn can lead to amplitude-modulated chemical signals.[15]

In addition to chemical reaction rate changes, there may be changes in the binding of proteins to cell membranes which lead to a shedding of proteins and a small increase in temperature. An exponential temperature dependence of the binding to membrane receptors is to be expected just as it is for chemical reactions.

A rule of thumb which the author uses to estimate whether or not significant biological changes are likely is to see if ΔT is >10° C for 10^{-6} sec, or 5° C for 1 sec, or 2° C for hours. If the ΔTs are larger, then they can be expected to lead to important changes in the biological system. Typical mammalian temperature regulatory systems will hold the internal body temperature constant to better than 0.5° C. However, the biological effects of small periodic changes have not been studied. In addition to the magnitude of the temperature change, it can be shown that the rate of temperature rise, dT/dt, is important and can induce current to flow across membranes. Changes in the firing rate of pacemaker cells from the ganglia of *Aplysia* have been induced by total temperature changes of as little as 1/10° C when the rates of change are about 1° C/sec.[16] This change of the firing rate corresponds to the injection of approximately 1 nA to the cell. By taking the time derivative of the Nernst equation, which describes the passive equilibrium potential across a membrane for a single ion, it can be shown that a current proportional to the temperature derivative is to be expected, or

$$I = -qV_1C_1\left(\frac{\phi}{\phi_T}\right)\left(\frac{\dot{\phi}}{\phi} - \frac{\dot{T}}{T}\right) \qquad (12)$$

where q is the charge on the ion, V_1 is the volume of the cell, C_1 is the concentration of ions inside the cell, ϕ is the resting potential, ϕ_T is given by $\phi_T = \frac{kT}{q}$, $\dot{\phi}$ is the derivative of the membrane potential with respect to time, T is the temperature, and $\dot{T}$ is the temperature derivative with respect to time.[16]

Some Biological Effects of Radio and Microwaves

The best understood effects of radio and microwaves are associated with changes in temperature and the rate of change of temperature. These include changes in heart rate and the firing rate of nerve cells, and they occur at relatively high power levels or peak power levels. For example, brachycardia, or slowing of the heart, has been observed under CW irradiation at 960 MHz (Lords 1973), and we have observed tachycardia in turtle hearts with microwave heating when the rate of temperature rise was approximately 1°C per minute.[17] With pacemaker cells from the ganglion of an *Aplysia,* we observe a decrease in firing rate when the rate of temperature rise is on the order of degrees per second; for slow heating, the firing rate increases.

At lower exposure levels, the mechanisms by which the observed biological effects are caused are not understood in sufficient detail to show how we go from the physics to the observed biological changes. However, there are a sufficient number of results, so these effects need to be taken into account because the SARs on these experiments are at levels which correspond to those calculated to exist in the head during cellular phone conversations. Some examples are as follows:

Cleary et al. have shown that the exposure of human lymphocytes for two hours at 27 MHz and 2.45 GHz in a constant-temperature bath changes the uptake of ^{3}H thymidine.[18] These results indicate changes in cell proliferation and activity which may be either positive or negative, depending on the exposure level in the range from 5 W/kg to 50 W/kg. In turn, these changes effect the behavior of an important part of the human immune system. Bolshakov et al. (1982) and Liburdy (1992) have observed shifts in membrane permeabilities for SARs of 0.1 W/kg at 2.45 GHz after 15 minutes.[19,20] Increases in O^2 free radical penetration of liposomes, and the release of vitamin E has also been observed. Chou, Guy, et al. performed an extensive study of long-term, low-level exposures of 200 rats to SARs of 0.4 W/kg at 2.45 GHz at 800 PPS and pulse widths of 10 ms.[21] Measurements on 155 parameters were negative, and just a statistically significant increase in primary malignancies was observed.

Even more puzzling results have been observed for modulated waves. Bawin and Adey have observed an increase in efflux of calcium from a cat brain when it was irradiated at 450 MHz at 1 mW/cm^2 and modulated in the range from 0 Hz to 60 Hz.[22] The maximum effect occurred at approximately 15 Hz, and no changes were observed at 10 Hz and 20 Hz. Byus has observed increases in ornithine decarboxylase activity for hepatome cells after 1-hour exposures at 450 MHz and power levels of 1 mW/cm2 when the signal is AM-modulated at 16 Hz. The increased activity persists for several hours.[23] Dutta, Ghosh, and Blackman have observed changes in acetylcholine esterase activity in neuroblastoma cells upon exposure to 915 MHz AM-modulated at 16 Hz.[24] Thirty-minute exposures to SARs of 0.02 W/kg increased the activity by 30 percent, and SARs of 0.01 W/Kg led to a decrease in activity of 10 percent. There was no response to 16 Hz modulation. Kolomytkin, Yuniska, et al. have shown an increase in the binding of 3H-glutamate and a decrease in 3H-muscimole to synaptic membranes from rat brains for five- minute exposures to 915-MHz pulses at 1 GHz for power densities as low as 10μW/cm^2.[25] This is a very detailed study which includes effects on both rat brains and black lipid membranes with careful temperature controls and dose response curves. The source of the changes in receptor binding seems to be associated with increases in the intracellular calcium concentration. The net result is extremely convincing evidence that neural signal transmission can be effected by modulated radio waves in the region around 16 Hz at power levels ranging from 10 to 1,000μW/cm^2. Additional modifications have been seen for exposures of cerebral cortex slices to modulated signals at 915 MHz.[26] There is also a report of the suppression of the ability of rats to remember a maze after exposure to microwaves.

In isolated chick hearts, Caddeni et al. have observed injection locking at 1 Hz for 2.45-GHz waves modulated at 1 Hz at power levels of 3 mW/cm^2 and SARs of 0.95 W/kg.[27] Perhaps the most direct indication that modulated signals at power levels in this range are biologically active are the results of Rosch on the use of 27 MHz at power levels less than 100 mW to treat chronic insomnia.[28] By inserting a coil in the mouth and modulating at frequencies less than 100 Hz for 20 minutes three to four times a week, the time required for these patients to fall asleep was reduced by an average of 52 minutes, and the average sleep cycle was lengthened by 123 minutes. These results are better than those obtained with any drug.

In summary, a variety of biological effects have been observed at SARs that may be generated by hand-held cellular phones and related wireless systems. However, to the authors knowledge it should also be stated that adverse health effects in humans have not been demonstrated from the use of cellular phones as of July 1995.

Product Liability and Industry Risks

Product liability laws vary from state to state both in the law and in the way it is interpreted. The general feeling of the public bears upon the way a potential hazard is viewed, and this is at least partially determined by the press. It is also determined in part by the degree to which the public feels the individual involved has control over the risk to which he or she is exposed. The greater the degree of individual control, the higher the risk that people will assume and the less they will hold the manufacturers, distributors, and so forth, responsible. The perception of risk is more important than statistical data from experts, particularly from those who may be presumed to represent the industry which profits from the sale of the product. In general, for a plaintiff to win a product liability case it is necessary for the plaintiff to show at least one of the following: (1) defective construction or materials, (2) failure to comply with codes, (3) failure to properly warn the user of hazards, (4) failure of the product to perform as advertised, (5) improper design, (6) failure with a fault of two or more manufacturers, or (7) failure to research. In the case of the cellular industry, items three, five, and seven are the most likely to be important.

A possible plaintiff's case might be constructed as follows: the biological effects described above and others which are likely to be observed over the next few years should have been interpreted as health hazards. Additionally, potential users should have been warned of the possible hazards by the manufacturer, retailer, cellular phone operator, and so on, and these groups should have sponsored more research on the potential hazards. Finally, they should have developed antenna systems that radiated less power into the head. Counters to such a case might include the assertions that the biological effects listed above are not proven health hazards, the ANSI standards have been complied with, and the antenna systems being sold are state-of-the-art. The outcome of such a case is hard to predict except that it is likely to be expensive. Recent cases have been decided in both directions. In response to the legal environment, the industry is funding a major epidemiological study through the Wireless Technology Research Corporation and a number of smaller studies. The first results from the epidemiological study, which will cover about 7,000,000 cell phone users, are several years away. Some of the most important studies may take as long as 20 years to gather the data.

It is interesting to speculate on how lawsuits and future research, which may or may not show that there are significant health effects from exposure to radio waves, may affect the cellular industry. A worst-case scenario would be along the lines of what has happened to the use of asbestos or in the nuclear power industry. In this case, the public has developed the opinion that the benefits of these technologies are not worth the risk, and as a result these industries have been virtually shut down. In spite of the fact that nuclear power is probably the safest source of electric power we have (in terms of the number of lives lost per megawatt hour of power delivered), the public perception is that it presents to people living in the area a risk of a nuclear explosion and death from radiation. (Recent data that low levels of radiation are healthy is not likely to change this perception in the near future.) It is also viewed as a risk people cannot control but is imposed upon them from the outside. The opposite situation occurs with cigarettes and automobiles, which are estimated to cause about 400,000 and 50,000 deaths per year, respectively. In these cases, the benefits are viewed as being worth the risks.

It is our view that the cellular industry is going to find itself treated more like cigarettes than nuclear power, at least with respect to the hand-held phone. The benefits of convenience and safety of hand-held phones are likely to be perceived by the public as outweighing the potential risk that the fields may promote brain tumors, cancer, and so forth. On the other hand, the cellular industry is likely to find itself in the role of the nuclear power industry with respect to the location of base stations even though the fields to which people are exposed are much smaller. Here, the potential hazard of radio waves may be viewed as outside the control of the individuals being exposed and their direct effects considered minimal. Thus, it may be hard to obtain approval for base station sites from zoning boards. Additionally, owners and operators of these sites need to consider the possibility that they may be sued for causing cancer, interfering with plant growth, and decreasing property values.

The overall situation is one in which the number of biological effects being observed as a result of exposures to power levels from hand-held cellular phones will increase with time. Adverse health effects may or may not be found; however, there are likely to be a significant number of lawsuits, and the industry needs to build this into their planning.

References

1 Yannon vs. New York Telephone Company, 0714-2308 and 075200-3602 (New York State Workers' Compensation Board).

2 Robert Pool, "Flying Blind: The Making of EMF Policy," *Science,* Oct. 5, 1990, vol. 250, no. 4977, p. 23.

3 Robert Kane vs. Motorola Inc., Illinois Circuit Court for Cook County, Na 93 L 15226.

4 James C. Lin, "Computer Methods Tor Yield Intensity Predictions," *Handbook of Biological Effects of Electromagnetic Fields,* eds. Charles Polk and Elliot Postow, CRC Press, 1986.

5 M. A. Jensen and Y. Rahmat-Samii, "EM Interaction of Handset Antennas and a Human in Personal Communications," *Proc. IEEE,* vol. 83, no. 1, Jan. 1995, pp. 7- 17.

6 Hsing-Yi Chen and Hou-Hwa Wang, "Current and SAR Induced in a Human Head Model by the Electromagnetic Fields Irradiated from a Cellular Phone," *IEEE Trans. MTT,* vol. 42., no. 12, Dec. 1994, pp. 2249 - 2254.

7 Robert F. Cleveland, Jr. and Whit Athey, "Specific Absorption Rate (SAR) in Models of the Human Head Exposed to Hand-Held UHF Portable Radios," *Bioelectromagnetics,* vol. 10, 1989, pp. 173 -186.

8 Arthur W. Guy and Chung-Kang Chou, "Specific Absorption Rates of Energy in Man Models Exposed to Cellular UHF Mobile-Antenna Fields," *IEEE Trans. MTT* vol. 34, no.6, June 1986, pp. 671-680.

9 Q. Balzano, O. Garay, and F.R. Steel, "A Comparison Between the Energy Deposition in Portable Radio Operators at 900 MHz and 450 MHz," Proceedings of the 28th *IEEE Veh. Tech. Conf.,* 1977, pp.46-49.

10 Mark Fischetti, "The Cellular Phone Scare," *IEEE Spectrum,* June 1993, p. 4347.

11 R. C. Petersen and P.A. Testagrossa, "Radiofrequency Electromagnetic Fields Associated with Cellular Radio Cell-Site Antennas," *Biolelectromagnetics,* vol. 13, no. 6, 1992, pp. 527-542.

12 F.S. Barnes, "Radio-Microwave Interactions with Biological Materials," *Health Physics,* vol. 56, no. 5, May 1989, pp. 759-766.

13 J.L. Kirschvink, A.K. Kirschvink, J.C. Diaz-Ricci, and S.J. Kirschvink, "Magnetite in human tissue: A mechanism for biological effects of weak ELF magnetic fields," *BEMS Supplement 1,* 1992, pp. 101-114.

14 F.S. Barnes and C.L. Hu, "Nonlinear Interactions of Electromagnetic Waves with Biological Materials, in *Nonlinear Electromagnetics,* ed. P.L.L. (New York: Uslenghi, Academic Press, 1980), pp. 391-426.

15 F. S. Barnes, "The Effects of ELF on Chemical Reaction Rates in Biological Systems," to be published by Plenum Press, 1995.

16 F.S. Barnes, "Cell Membrane Temperature Rate Sensitivity Predicted from the Nernst Equation," *Bioelectromagnetics,* vol. 5, 1964, p. 113.

17 S.J. Lords, C.H. Aurney, A.M. Borg, and C. Tinney, "Rate effects in isolated hearts induced by microwave irradiation," *IEEE Trans. on Microwave Theory and Technology,* vol. 21, pp. 834-836.

18 S.F. Cleary, L.M. Liu, and R.E. Merchant, "In vitro lymphocyte proliferation induced by radio-frequency electromagnetic radiation under isothermal conditions," *Bioelectromagnetics* vol. 11, no. 1, 1991, pp. 47-56.

19 M.A. Bol'shakov and S.I. Alekseyev, "Change in the Electrical Activity of the Pacemaker Neurons of L. Stagnalis with the Rate of Their Heating," *Biophysics,* vol. 31, 1986, pp. 569-571.

20 R.P. Liburdy and A. Penn, "Microwave Bioeffects in Erythrocyte are Temperature and pO-dependent: Cation Permeability and Protein Shedding Occur at the Membrane Phase Transition," *Bioelectromagnetics* vol. 5, 1984, pp. 283-291.

21 C.K. Chou, A.W. Guy, L.L. Kuna, R.B. Johnson, J.J. Crowley, and J.H. Krupp, "Long-term, low-level microwave irradiation of rats," *Bioelectromagnetics,* vol. 13, no. 6, 1992, pp. 469-496.

22 S.M. Bawin and W.R. Adey, "Sensitivity of calcium binding in cerebral tissue to weak environmental electric fields oscillating at low frequency," *Proc. Nat. Acad. Sci. USA,* vol. 73, no. 6, pp. 1999-2003.

23 Craig V. Byus, K. Kartun, S. Pieper, and W.R. Adey, "Increased ornithine decarboxylase activity in cultured cells exposed to low energy modulated microwave fields and phorbol ester tumor promoters," *Cancer Research,* vol. 48, August 1, 1988, pp. 4222-4226.

24 S.K. Dutta, K. Das, B. Ghosh, and C.F. Blackman, "Dose dependence of acetycholinesterose activity in neuroblastoma cells exposed to modulated radiofrequency electromagnetic radiation," *Bioelectromagnetics,* vol. 13, no. 4, pp. 317-322, 1992.

25 O. Kolomytkin, M. Yuvinska, S. Zharikov, V. Kuznetsov, A. Zharikova, "Response of brain receptor systems to microwave energy exposure," in *On The Nature of Electromagnetic Field Interactions with Biological Systems,* Allen H. Frey, ed., (Austin, TX: R.G. Landus Co.), pp. 195-206.

26 N.M. Zakharova, S.I. Alexseyev, and M.N. Zhadin, "Effect of ultra-high frequency radiation on the spontaneous impulse activity of surviving cerebral cortex slices," *Biophysics* vol. 88, no. 3, pp. 521-525, 1993.

27 A. Caddemi, C.C. Tamburello, L. Zanforlin, and M.V. Torregrossa, "Microwave effects on isolated chick embryo hearts," *Bioelectromagnetics,* vol. 7, no. 4, 1986, p. 359.

28 P. Rosh, private communication, 1995.

High-Efficiency DC-DC Converters for Battery-Operated Systems with Energy Management

Robert W. Erickson
Department of Electrical and Computer Engineering
University of Colorado at Boulder

Dragan Maksimovic
Department of Electrical and Computer Engineering
University of Colorado at Boulder

Abstract

Dc-dc converters suitable for use in battery-powered electronic equipment with energy management are discussed. To realize the benefits of energy management, these converters must be capable of maintaining regulation of their output voltages at no load while maintaining high efficiency.

Variable-frequency operation is shown to give an improvement in efficient operating range of approximately two orders of magnitude. Proportionality of the battery and load currents can be maintained over approximately five orders of magnitude, with minimum battery currents of less than 100μA and maximum load currents approaching 10A.

Steady-state and small-signal ac characteristics of converters operating with variable-frequency control are tabulated, and an ac two-port model is described.

Introduction

Energy management in battery-operated systems can allow for greater capabilities in energy-limited systems. Elements with high instantaneous power consumption can be used, such as transmitters, microprocessors, backlit displays, and flash memory, provided that they are switched to a low-power standby mode when not needed. The use of low-voltage power supplies (3.3 V or less) to reduce power consumption can also allow greater capability at the expense of lower noise margins and a requirement for better voltage regulation. In the standby mode, some payloads are switched off while others operate with reduced capability and at very low power consumption, listening for a wake-up signal.

Battery leakage currents are typically a fraction of a microampere; this could be taken as an ultimate goal for the standby mode current. Standby currents of one to two orders of magnitude greater than this figure, in the 10-100μ range, would yield good performance for many applications. Meeting these goals requires improvements in both payload and dc-dc converter technologies.

To fully realize a battery-powered energy management system, dc-dc converters are needed which meet new requirements. Simply turning off the converter in the standby mode is insufficient since part of the payload must continue to operate. Hence, the converter must maintain regulation of the load voltage with zero load current. High converter efficiency is required over many orders of magnitude variation in the load power, typically from tens of microwatts to tens of watts. Since system volume and weight are dominated by the batteries rather than the

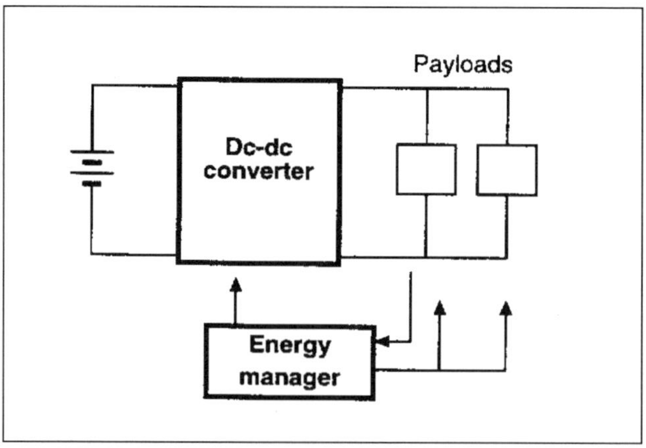

Figure 1 *A simple energy management system for a battery-powered mobile system.*

converter, high efficiency is more important than converter size and weight; nonetheless, minimization of converter volume and weight is also a goal.

Conventional fixed-frequency dc-dc converters cannot meet these requirements. A conventional 0 to 25 W, 4.8 V input, 3.3 V output, fixed-frequency design is examined in Section 2 and is shown to exhibit high efficiency for 25W ≥ P_{out} ≥ 0.2W. The current drawn from the battery at no load (standby current) is 5 mA. In addition to exhibiting high fixed (load-independent) losses, this approach is unable to regulate its output at zero load.

As illustrated in *Figure 2(a)*, a conventional dc-dc converter contains substantial fixed losses which are independent of the load current. These fixed losses lead to a significant battery current at no load. As discussed in Section 2, the total

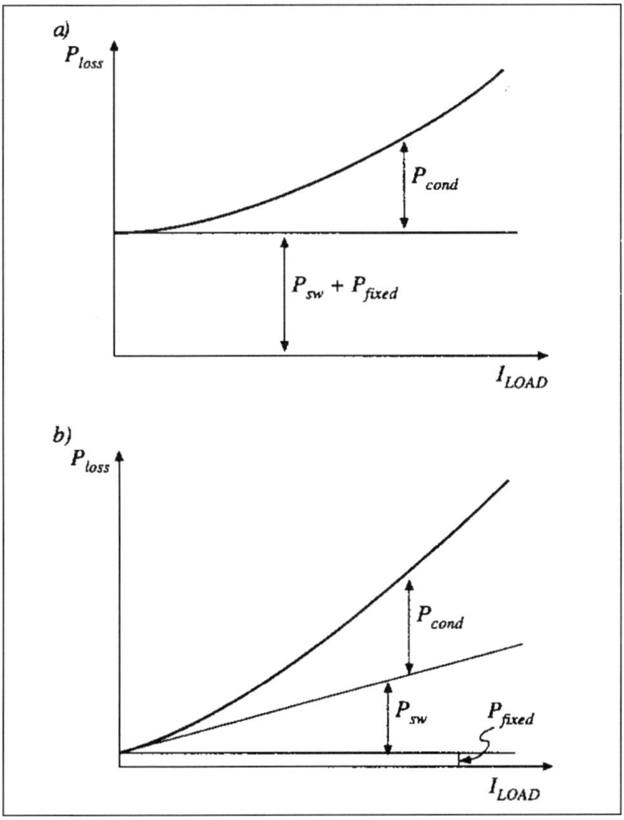

Figure 2 *Variation of converter power losses with load current: (a) in a conventional fixed-frequency converter; (b) with switching frequency proportional to load current.*

converter power losses P_{loss} can be expressed as

$$P_{loss} = P_{cond}\left(I_{load}\right) + W_{sw} f_{sw} + P_{fixed} \tag{1}$$

where P_{cond} is the converter conduction loss, directly dependent on load current, W_{sw} is the energy lost during the transistor switch-on and switch-off transitions (depending on the switching frequency f_{sw}), and P_{fixed} is the fixed loss, which is dependent on neither load current nor switching frequency. The switching loss $W_{sw} f_{sw}$ accounts for nearly all of the fixed loss. This suggests that one should reduce the switching frequency f_{sw} as the load current decreases, causing the frequency-dependent losses to vary with load, as in *Figure 2(b)*. Decrease of the converter standby current by one to two orders of magnitude is feasible using such a scheme. The performance, design, and optimization of variable-frequency converters for this application are summarized in Section 3. With this approach, converter-fixed losses can be reduced by nearly two orders of magnitude, with a no-load battery current of less than 100μA.[3,4] With the variable-frequency approach, standby performance is limited by semiconductor device leakage currents, power controller chip standby current, and payload standby current.

The characteristics of converters operating in discontinuous conduction mode with variable-frequency control are tabulated in Section 4. A small-signal equivalent circuit model is also given, which is solved to yield the small-signal control transfer functions of these converters. The results are summarized in Section 5, and some directions for future R&D are suggested.

Performance of the Conventional Fixed-Frequency Approach

Several basic dc-dc converter circuits are shown in *Figure 3*. The buck converter, *Figure 3(a)*, reduces the dc voltage magnitude and is used as an example here to reduce a four-cell battery voltage (4.8 V minimum) to a regulated 3.3V dc. The boost converter, *Figure 3(b)*, is commonly used to increase the voltage magnitude and can interface a one- or two-cell battery pack to a 5V dc load. The buck-boost converter, *Figure 3(c)*, inverts the voltage polarity. All three converters exhibit similar loss behavior.

Typical MOSFET voltage and current waveforms are given in *Figure 4*. These waveforms lead to several types of power loss listed in *Table 1*. Conduction losses are directly dependent on the load current and are nearly independent of switch-

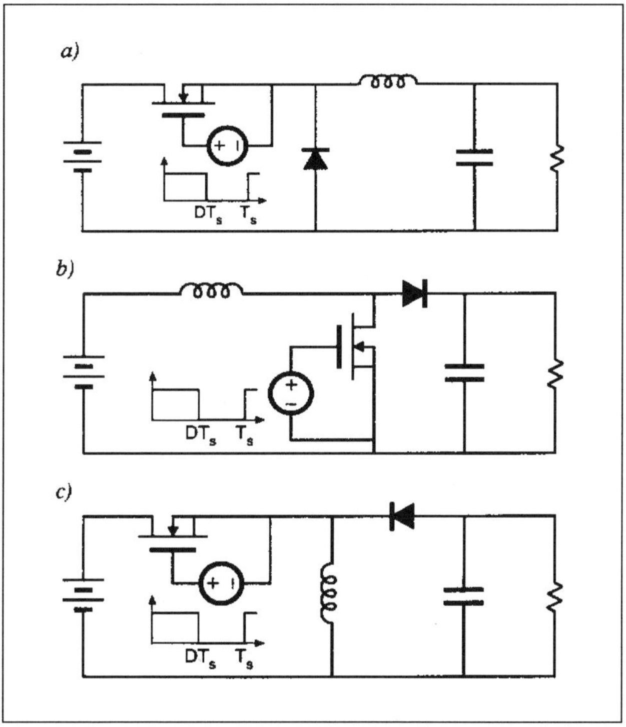

Figure 3 *Basic dc-dc switching converter circuits: (a) buck, (b) boost, (c) buck-boost.*

ing frequency. An example is the power loss due to current flowing through the MOSFET on-resistance and the diode-forward voltage drop. Conduction losses dominate the other types of loss at maximum load power but have little effect at light load or under standby conditions.

Switching losses are nearly independent of the value of the load current but depend directly on the converter switching frequency. An example is the loss due to the energy stored in the MOSFET output (drain-to-source) capacitance. When the MOSFET is switched on, this capacitance is shorted out, and the energy stored in it is lost. The total switching power loss is equal to the total energy lost during the turn-on and turn-off switching transitions multiplied by the switching frequency f_{sw}. The various sources of switching loss in a conventional dc-dc switching converter are now well documented in the literature.[2,5] Depending upon the value of the switching frequency, switching losses may or may not be significant at full load. However, at light load and under standby conditions, the switching losses are a serious source of power loss in a fixed-frequency converter.

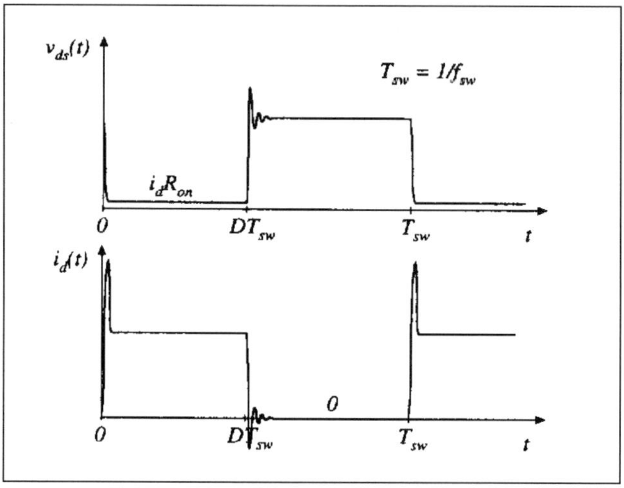

Figure 4 *Typical MOSFET voltage and current waveforms in a conventional dc-dc converter operating in continuous conduction mode.*

Other sources of fixed loss, independent of both switching frequency and load current, include the power supply controller standby current and the leakage currents of the transistors, diode, and filter capacitors. These losses remain even when the switching converter is shut off. They represent the minimum power which can be drawn from the battery and should be minimized.

At the low voltage levels typical of battery-operated equipment, synchronous rectification is often used. As shown in *Figure 5*, the diode is replaced by a MOSFET. A control circuit gates the MOSFET on and off at the times when the diode would have switched. If the MOSFET is sufficiently large, the MOSFET forward-voltage drop is less than the diode forward-voltage drop would have been, and the conduction losses are reduced. Synchronous rectification is of significant benefit when the battery or load voltage is comparable to the diode forward-voltage drop, as is the case in most battery-operated equipment. An added benefit is the reduction of leakage currents since the MOSFET leakage current (on the order of one microampere at room temperature) is much lower than the leakage current in a low-V_F schottky diode (in the milliampere range). The gate drive control circuit should be carefully designed to ensure that the transistor is correctly switched on and off in the discontinuous conduction mode.

Efficiency and battery current are plotted versus load power in *Figures 6* and *7*, for a well-designed conventional fixed-frequency switching converter in a typical

Load-Dependent (Conduction) Losses
• MOSFET on resistance
• Diode forward voltage drop
• Inductor winding resistance
Frequency-Dependent (Switching) Losses
• MOSFET output capacitance
• MOSFET gate capacitance
• Diode capacitance
• Diode stored minority charge
• Inductor and transformer core loss
• Snubber loss
• Gate driver losses
Other Fixed Losses
• Controller standby current
• MOSFET, diode, and capacitor leakage currents

Table 1 *Origins of loss in a conventional fixed-frequency converter*

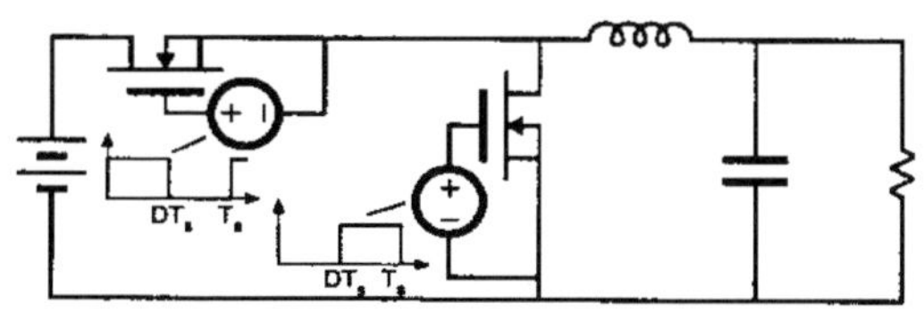

Figure 5 *Buck converter with synchronous rectification*

application. A buck converter, shown in *Figure 3(a)*, is used to supply a regulated 3.3V to a 0-25W load, from a four-cell battery pack having a minimum voltage of 4.8V. The switching frequency is a fixed 100 kHz. A logic-level MOSFET (MTP15N05EL) and schottky diode (80SQ) are assumed. The ferrite inductor has the value 10μH with a winding resistance of 10mΩ. The controller current is assumed to be a fixed 75μA, plus 20nC*f_{sw} . All losses listed in *Table 1* are modeled, and it is assumed that the converter regulates the output voltage to 3.3V over the entire output power range of 0-25W.

It can be seen from *Figure 6* that the predicted converter efficiency is above 80 percent for load powers in the range of 0.2W - 25W. However, the efficiency is poor at lower power levels. More illuminating is *Figure 7*, in which the battery

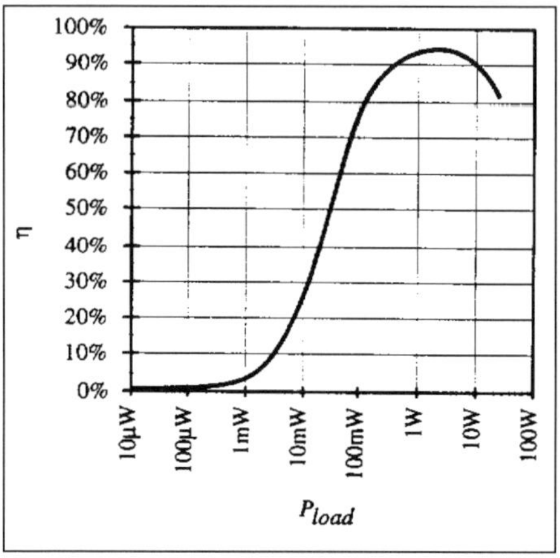

Figure 6 *Predicted converter efficiency vs. output power, conventional fixed-frequency converter.*

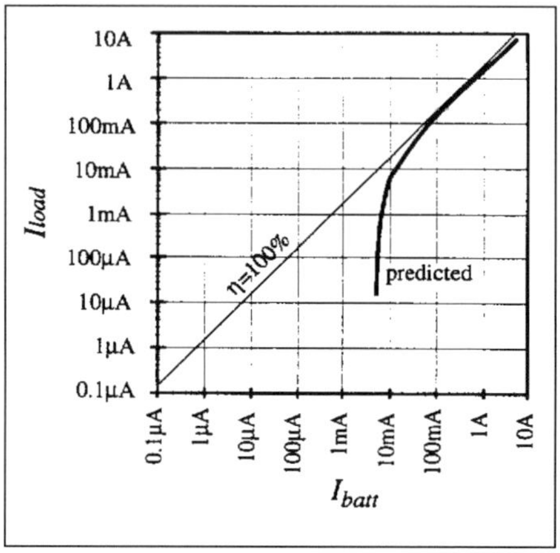

Figure 7 *Predicted converter input current vs. output current, conventional fixed frequency converter.*

current is plotted vs. load current. The minimum battery current is approximately 5mA regardless of the load current. Even with the load disconnected, the converter requires 5mA of battery current to maintain regulation of the converter output voltage. A well-designed dc-dc converter using the conventional fixed-frequency approach is able to maintain high efficiency over less than three orders of magnitude of load powers. Apparently, reducing the load current to the microampere range does not yield corresponding reductions in battery current; hence, the goals of energy management are not realized.

Performance of the Variable-Frequency Approach

According to *Equation 1*, the switching loss $W_{sw}\, f_{sw}$ is a fixed loss when the switching frequency f_{sw} is constant as in the conventional approach. However, if the switching frequency varies with load current, then the switching loss will also vary directly with load. This suggests that the standby-mode efficiency and battery-current draw could be improved using a variable-frequency approach.

At light load and low switching frequency, the converter circuits of *Figure 3* operate in the discontinuous conduction mode (DCM), with the inductor-current waveform shown in *Figure 8*. Each switching period of length T_{sw} contains three subintervals. During the first, the MOSFET conducts and energy is transferred from the battery to the inductor. During the second subinterval, the diode conducts, and

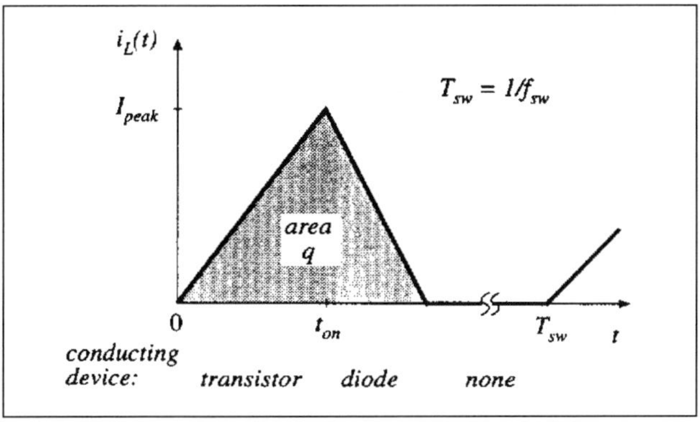

Figure 8 *Inductor current waveform* $i_L(t)$ *in the discontinuous conduction mode*

the inductor stored-energy is transferred to the load. At the end of the second subinterval, the diode becomes reverse-biased when the inductor current reaches zero. Both transistor and diode are off during the third subinterval, and the inductor current is zero for the remainder of the switching period. During the third subinterval, no conduction losses occur, and the only losses are due to leakage currents.

For a given transistor conduction time, t_{on}, a fixed amount of energy per switching period is transferred to the load, given by

$$W_{load} = qV \tag{2}$$

where for the buck converter, q is the integral of the inductor current waveform, and V is the dc load voltage. The output power is then given by

$$P_{load} = W_{load} f_{sw} \tag{3}$$

In the process of transferring this energy to the load during the first and second subintervals, some amount of energy, W_{loss}, is lost as conduction and switching losses. To the extent that the fixed losses arising in the controller chip due to leakage currents can be neglected, the efficiency is given by

$$\eta = \frac{W_{load}}{W_{load} + W_{loss}} \tag{4}$$

Note that if the quantities W_{load} and W_{loss} can be made independent of the load current, the efficiency also becomes essentially independent of load current. There are several methods for accomplishing this, including constant on-time variable-frequency control and constant peak-current programming with variable-frequency control.

In the constant on-time variable-frequency control scheme, the transistor on-time t_{on} is fixed by the control circuit. As a result, the energy transferred to the load during one switching period, W_{load} as well as the loss W_{loss} incurred are also fixed. The efficiency given by *Equation 4* is therefore independent of load current, and high efficiency occurs at light load. The output voltage is regulated by variation of the switching frequency f_{sw}.

Current programming is a popular means of controlling conventional dc-dc converters and can also be used to control variable-frequency DCM converters for battery-powered applications.[6] Operation is similar to the constant on-time control scheme. Rather than operating the transistor with a constant on-time t_{on}, the controller monitors the transistor current and switches the transistor off when the transistor current reaches a predetermined level I_{pk}. The value of I_{pk} is constant, and the output voltage is regulated by variation of the switching frequency. For given values of the inductance L and the battery and load voltages, there is a direct relation between t_{on} and I_{pk}.

How should t_{on} or I_{pk} be chosen? If these quantities are increased to large values, high-peak currents will occur, leading to large conduction losses. The converter will transfer a large amount of energy each switching period, and the converter will operate with low-switching frequency. Alternatively, if t_{on} or I_{pk} are decreased to small values, low-peak currents will occur with low conduction losses. However, a small amount of energy will be transferred each switching period, which necessitates a high-switching frequency with high-switching losses. Hence, there is an optimum value of t_{on} or I_{pk} which leads to maximum efficiency.

Equation 4 was used to predict the efficiency of a dc-dc buck converter controlled by variable-frequency current programming. The same power stage elements assumed in Section 2 to generate *Figures 6* and *7* were used. Efficiency is plotted as a function of peak transistor current in *Figure 9*, with controller standby current and power-stage leakage currents ignored. Depending on the choice of operating point, the converter may enter the continuous conduction mode at low values of I_{pk}; discontinuous conduction mode is assumed in the generation of the figure. The efficiency is a mild function of the battery current; however, the optimum efficiency occurs at approximately the same choice of I_{pk} for a typical range of battery voltages. The optimum efficiency for this example occurs with a peak current of

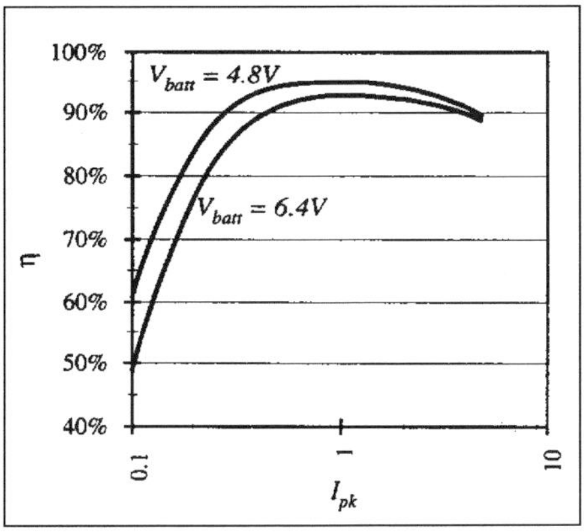

Figure 9 *Efficiency vs. peak transistor current, for constant-current-programmed DCM buck converter example with variable-frequency control*

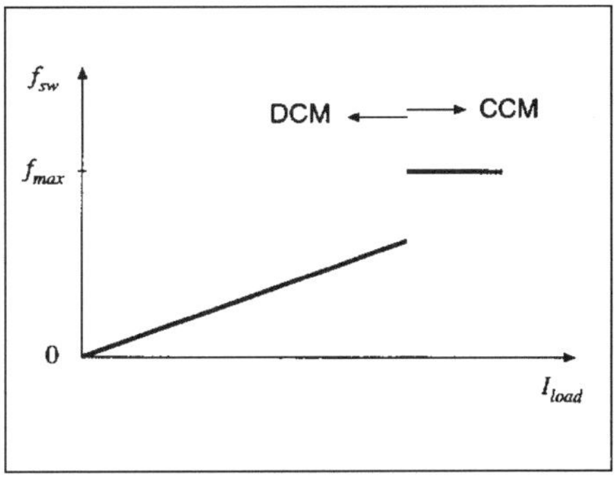

Figure 10 *The operating range can be further extended by use of the continuous-conduction mode near full load and variable-frequency discontinuous-conduction mode at light load.*

approximately one ampere. Experimental results presented by Arbetter et al., confirm this behavior.[1]

It is possible to extend the useful range of load currents further by allowing the converter to operate in the continuous-conduction mode at high current. The switching frequency is increased so that the third subinterval is eliminated. A sketch of switching frequency vs. load current is given in *Figure 10.* Full-load efficiencies greater than the value predicted by *Equation 4* are possible using this scheme.

Efficiency and battery current are plotted vs. load power in *Figures 11* and *12* for a well-designed variable-frequency switching converter. The same buck converter power-stage elements used to generate *Figures 6* and *7* are assumed. The converter supplies the same regulated 3.3V to a 0-25W load from a four-cell battery pack having a minimum voltage of 4.8V. All sources of loss are modeled, including conduction and switching loss, leakage currents, and controller standby current. For output powers greater than 2W, the converter operates in the continuous-conduction mode with a fixed frequency of 100 kHz. For output powers less than 2W, the converter operates in the discontinuous-conduction mode with variable-frequency current-programmed control. The peak current is then held constant at 1A.

By comparing *Figures 6* and *11,* it can be seen that the range of high efficiency (η>80%) is increased by approximately two orders of magnitude in load power.

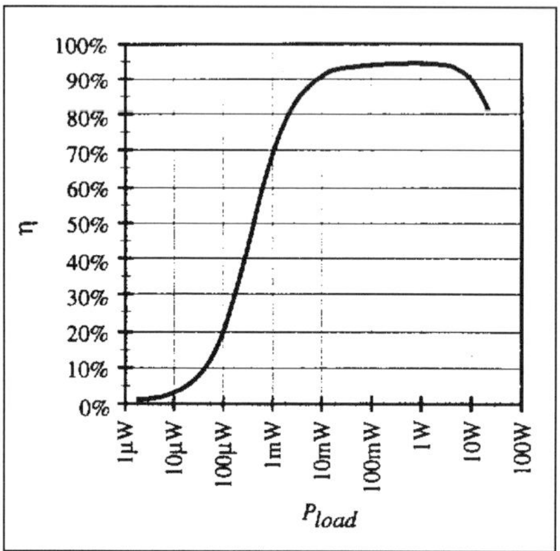

Figure 11 *Predicted converter efficiency vs. output power, variable-frequency converter.*

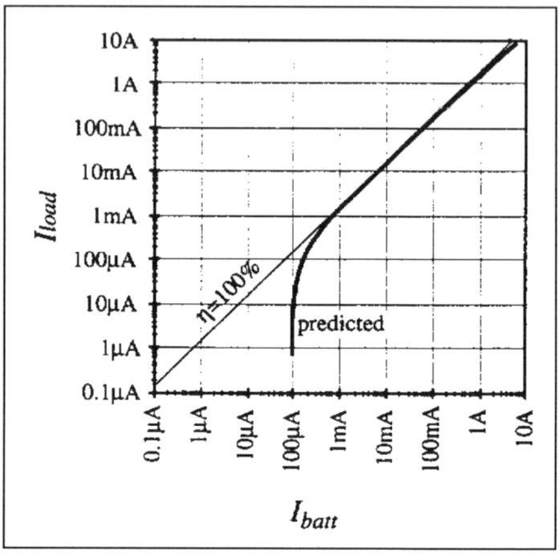

Figure 12 *Predicted converter input current vs. output current, variable-frequency converter.*

Efficiency greater than 80 percent is obtained for all load powers in the range 2.5mW - 25W. The no-load battery current is also decreased by nearly two orders of magnitude, from approximately 5mA in the conventional converter to less than 100μA with the variable-frequency converter.

Characteristics of Variable-Frequency Converters

The steady-state and small-signal ac behaviors of variable-frequency DCM converters are briefly summarized here, for both constant on-time and peak-current control.

For variable-frequency constant-on-time control, the converter voltage conversion ratio M is given in *Table 2* for the basic buck, boost, and buck-boost converters. The voltage conversion ratio M is defined as

$$M = V / V_{batt} \tag{4}$$

The discontinuous conduction mode (DCM) occurs when the load current I_{load} is less than the critical value, I_{crit}, where

$$I_{crit} = \left(V_{batt} / R_e\right)\left(1 - D\right) / D \tag{5}$$

for all three converters. Otherwise, the converters operate in continuous conduc-

Converter	M (CCM)	M (DCM)
Buck	D	$\frac{2}{1+\sqrt{1+4R_e/R}}$
Boost	$\frac{1}{1-D}$	$\frac{1+\sqrt{1+4R_e/R}}{2}$
Buck-Boost	$-\frac{D}{1-D}$	$-\sqrt{\frac{R}{R_e}}$

Duty cycle: $D=t_{on}/T_{sw}$
Effective resistance: $R_e=T_{sw}2L/t_{on}^2$

Table 2 *Summary of steady-state characteristics, variable-frequency constant on-time control*

tion mode (CCM). R is the load resistance, defined as V/I_{load}.

The converter characteristics in steady-state for current-programmed, variable-frequency control are summarized in *Table 3*. Current programming tends to cause the converter output characteristic to approach that of a current source. The dc output current I_{load} is listed in *Table 3* as a function of the battery and load voltages, as well as the programmed peak, current I_{pk}, and the effective power P defined as

$$P = \frac{1}{2} L I_{pk}^2 f_{sw} \tag{6}$$

It is assumed that the current-mode controller does not contain an artificial ramp. The voltage-conversion ratio M and the critical current I_{crit} are also listed; the boundary between continuous- and discontinuous-conduction modes is again given by $I_{load} = I_{crit}$.

Design of the output-voltage-regulating feedback loop requires knowledge of the small-signal control-to-output transfer function (i.e., the transfer function $G_c(s)$ illustrated in the block diagram of *Figure 13*). The output voltage is expressed as the desired steady-state quantity V plus a perturbation, $\hat{v}(t)$. The other signals are represented in a similar manner. The controller contains a voltage-controlled oscillator or equivalent device, whose small-signal gain is

$$K_{vco} = \frac{\hat{f}_{sw}}{\hat{v}_c} \tag{7}$$

Converter	I_{out} (CCM)	I_{out} (DCM)	M (DCM)	I_{crit}
Buck	$I_{pk} - \frac{V_{out}(1 - V_{out}/V_{batt})I_{pk}^2}{4P}$	$\frac{P}{V_{out}(1 - V_{out}/V_{batt})}$	$\frac{P_{out} - P}{P_{out}}$	$\frac{1}{2} I_{pk}$
Boost	$\frac{I_{pk}}{M}\left(1 - \frac{V_{out}(V_{out}/V_{batt} - 1)I_{pk}}{4P}\right)$	$\frac{P}{V_{batt} - V_{out}}$	$\frac{P_{out}}{P_{out} - P}$	$\frac{1}{2} \frac{I_{pk}}{M}$
Buck-Boost	$\frac{I_{pk}}{1 - M}\left(1 + \frac{V_{batt} I_{pk}}{4P} \frac{M}{1 - M}\right)$	$\frac{P}{V_{out}}$	Depends on load characteristic : $P_{out} = P$	$\frac{1}{2} \frac{I_{pk}}{1 - M}$

Voltage conversion ratio: $M = V/V_{batt}$ Load power: $P_{load} = VI_{load}$

Table 3 *Summary of steady-state characteristics, variable-frequency current mode control*

where $\hat{f}_{sw}$ is the perturbation in the switching frequency caused by a control-signal perturbation v_c. By perturbation and linearization of the nonlinear averaged converter models, the two-port equivalent circuit model of *Figure 14* can be derived. General expressions for the two-port parameters are summarized in *Tables 4* and *5* for DCM variable-frequency constant-on-time control and for DCM variable-frequency current-programmed control, respectively.

By solving the equivalent circuit model of *Figure 14*, one can derive the small-signal control-to-output transfer functions $G_c(s)$ of the various converters with the two control methods. In each case, the transfer function is simple and contains a single pole and a dc gain of the form

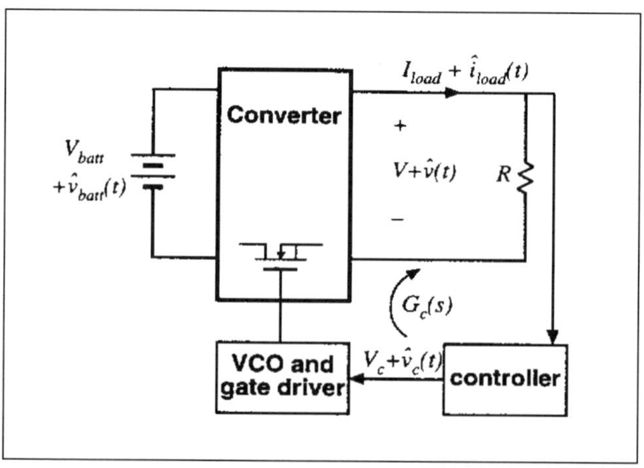

Figure 13 *Output voltage regulation system block diagram*

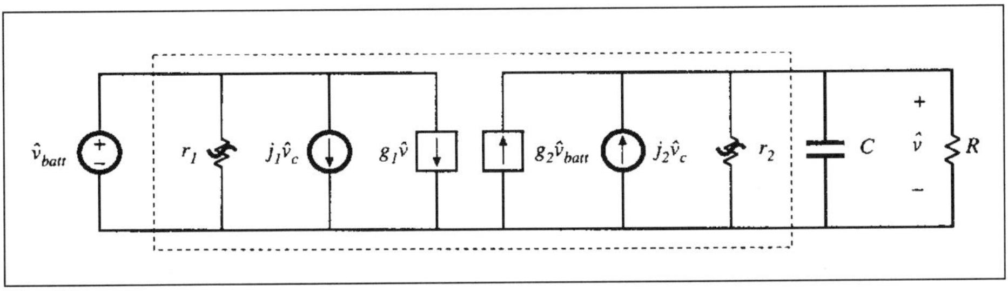

Figure 14 *Equivalent circuit model which predicts the small-signal transfer functions of variable-frequency DCM converters.*

Converter	g_1	j_1	r_1	g_2	j_2	r_2
Buck	$\frac{-1}{R_e}$	$\frac{V_{batt}K_{vco}}{R_e f_{sw}}(1-M)$	R_e	$\frac{(2-M)}{M R_e}$	$\frac{V_{batt}K_{vco}}{R_e f_{sw}}\frac{(1-M)}{M}$	$M^2 R_e$
Boost	$\frac{-1}{(M-1)R_e}$	$\frac{V_{batt}K_{vco}}{R_e f_{sw}}\frac{M}{M-1}$	$\frac{(M-1)^2}{M}R_e$	$\frac{(2M-1)}{(M-1)^2 R_e}$	$\frac{V_{batt}K_{vco}}{R_e f_{sw}}\frac{1}{M-1}$	$(M-1)^2 R_e$
Buck-Boost	0	$\frac{V_{batt}K_{vco}}{R_e f_{sw}}$	R_e	$\frac{2M}{R_e}$	$\frac{V_{batt}K_{vco}}{R_e f_{sw}}\frac{1}{M}$	$M^2 R_e$

Table 4 *Small-signal model parameters, constant-on-time variable-frequency DCM*

Converter	g_1	j_1	r_1	g_2	j_2	r_2
Buck	$\frac{\frac{1}{2}LI_{pk}^2 f_{sw}}{(V_{batt}-V)^2}$	$\frac{\frac{1}{2}LI_{pk}^2 K_{vco}}{V_{batt}-V}$	$-\frac{(V_{batt}-V)^2}{\frac{1}{2}LI_{pk}^2 f_{sw}}$	$-\frac{\frac{1}{2}LI_{pk}^2 f_{sw}}{(V_{batt}-V)^2}$	$\frac{\frac{1}{2}LI_{pk}^2 V_{batt}K_{vco}}{V(V_{batt}-V)}$	$\frac{V^2(V_{batt}-V)^2}{\frac{1}{2}LI_{pk}^2 f_{sw}B_{batt}(B_{batt}-2V)}$
Boost	$\frac{\frac{1}{2}LI_{pk}^2 f_{sw}}{(V_{batt}-V)^2}$	$\frac{\frac{1}{2}LI_{pk}^2 K_{vco}}{V_{batt}(V-V_{batt})}$	$\frac{V_{batt}^2(V_{batt}-V)^2}{\frac{1}{2}LI_{pk}^2 f_{sw}V(2V_{batt}-V)}$	$\frac{\frac{1}{2}LI_{pk}^2 f_{sw}}{(V_{batt}-V)^2}$	$\frac{\frac{1}{2}LI_{pk}^2 K_{vco}}{V-V_{batt}}$	$\frac{(V_{batt}-V)^2}{\frac{1}{2}LI_{pk}^2 f_{sw}}$
Buck-Boost	0	$\frac{\frac{1}{2}LI_{pk}^2 K_{vco}}{V_{batt}}$	$\frac{V_{batt}^2}{\frac{1}{2}LI_{pk}^2 f_{sw}}$	0	$\frac{\frac{1}{2}LI_{pk}^2 K_{vco}}{-V}$	$\frac{V^2}{\frac{1}{2}LI_{pk}^2 f_{sw}}$

Table 4 *Small-signal model parameters, variable frequency current programmed DCM*

$$G_c(s) = \frac{G_0}{\left(1 + \frac{s}{\omega_p}\right)} \tag{8}$$

Expressions for G_0 and ωp are collected in *Tables 6* and *7*. The quantity K_{vco} is the controller gain in Hz/volt.

Converter	DC gain G_o	Pole freq. ωp
Buck	$\frac{VK_{vco}}{f_{sw}}\frac{(1-M)}{(2-M)}$	$\frac{(2-M)}{RC(1-M)}$
Boost	$\frac{VK_{vco}}{f_{sw}}\frac{(M-1)}{(2M-1)}$	$\frac{(2M-1)}{RC(M-1)}$
Buck-Boost	$\frac{VK_{vco}}{f_{sw}}$	$\frac{2}{RC}$

Table 6 *Small-signal transfer function parameters, DCM constant on-time variable-frequency control*

Converter	DC gain G_o	Pole freq. ωp
Buck	$\frac{VK_{vco}}{f_{sw}}\frac{(1-M)}{(2-3M)}$	$\frac{(2-3M)}{RC(1-M)}$
Boost	$\frac{VK_{vco}}{f_{sw}}\frac{(M-1)}{(2M-1)}$	$\frac{(2M-1)}{RC(M-1)}$
Buck-Boost	$\frac{2VK_{vco}}{f_{sw}}$	$\frac{2}{RC}$

Table 7 *Small-signal transfer function parameters, DCM peak-current-programmed variable frequency control*

The current-programmed buck converter operating in discontinuous-conduction mode exhibits instability for voltage conversion ratios greater than two-thirds. This instability was observed experimentally by Arbetter, Erickson, and Maksimovic.[1] Nonetheless, simple output voltage feedback can stabilize the system, resulting in good output-voltage regulation. No instabilities occur in the other converters operating in discontinuous-conduction mode.

Conclusions and Future Directions

Energy management allows increased capabilities in mobile battery-powered equipment through the use of higher-powered components placed in a low-current standby mode when not needed. To realize the benefits of energy management, dc-dc converters must be capable of maintaining regulation of their output voltages at no load while maintaining high efficiency. It is now feasible to build such converters by supplying a variety of load voltages from a single battery.

Variable frequency operation is shown to give an improvement in efficient operating range of approximately two orders of magnitude compared with the conventional constant-frequency approach. Proportionality of the battery and load currents can be maintained over approximately five orders of magnitude, with minimum battery currents of less than 100μA and maximum load currents approaching 10A.

Two variable-frequency control schemes are discussed here: constant-on-time and current-programming. Steady-state behavior and small-signal ac transfer functions are given as well as an ac two-port converter model.

At room temperature, the minimum battery current is essentially equal to the controller chip standby current. Additional work is needed to further reduce this current and to implement improved control algorithms which allow for optimized CCM and DCM operation. Efforts continue to integrate both controller and power semiconductor devices on the same chip, which reduces size and cost.

Additional work in converter modeling and optimization could allow further improvements in size, efficiency, and usable power range. Other control schemes are also possible. Improvements in converter efficiency and noise generation may be possible, using a resonant or other converter approach.

References

1 B. Arbetter, R. Erickson, and D. Maksimovic, "DC-DC Converter Design for Battery-Operated Systems," *IEEE Power Electronics Specialists Conference,*1995 Record, June 1995.

2 M. Schlecht and L. Casey, "Comparison of the Square-Wave and Quasi-Resonant Topologies," *IEEE Applied Power Electronics Conference,*1987 Record, March 1987, pp. 124-134.

3 B. Huffman and R. Flatness, "Power Conversion from Milliamps to Amps at Ultra-High Efficiency," *Linear Technology Application Note 54,* March 1993.

4 "Battery Management and Dc-Dc Converter Circuit Collection – A Power Supply Application Guide for Portable Equipment," MAXIM Integrated Products, 1994.

5 S. Cuk, "Modeling, Analysis, and Design of Switching Converters," Ph.D. thesis, California Institute of Technology, November 1976.

6 R. Redl and N. Sokal, "Current-Mode Control, Five Different Types, Used with the Three Basic Classes of Power Converters: Small-Signal Ac and Large-Signal Dc Characterization, Stability Requirements, and Implementation of Practical Circuits," *IEEE Power Electronics Specialists Conference, 1985 Record,* June 1985, pp. 771-785.

7 A. Dauhjare and R. D. Middlebrook, "A Simple PWM-FM Control for an Independently Regulated Dual Output Converter," *Proceedings of POWERCON 10,* March 1983.

Interoperability of Satellite and Terrestrial Networks

D. M. Chitre, Associate Director of Operations, COMSAT Laboratories

T. Henderson, Member of Technical Staff, COMSAT Laboratories

Abstract

Satellite communications networks can be an integral part of the newly emerging national and global information infrastructures if the issues of interoperability of satellite and terrestrial networks are resolved satisfactorily. This essay discusses the role of future communication satellites in specific newly emerging market segments. The obstacles to the seamless integration of the terrestrial and satellite components of these future telecommunications networks are described. Specific examples and solutions to achieve interoperability are presented within the context of narrowband ISDN, data communications, and asynchronous transfer mode (ATM).

Introduction

We are in the midst of a global information revolution brought about by the dawn of the information age. This revolution is being spawned by digital technology. The enormous strides taking place in digital technology and its applications have fundamentally changed the way in which information is generated, processed, stored, and transported. The resulting information infrastructure will play a critical role in bringing about profound changes in all walks of life, such as health care, education, crisis management, environmental monitoring, electronic commerce, and so on. Telecommunications networks will be the arteries of this infrastructure. Terrestrial telecommunications networks are evolving rapidly in response to the challenge of the complete and stringent requirements of the new environment. Satellite communications can play a multifaceted role in these evolving telecommunications networks, in addition to its well-established niche markets of international communication, broadcast TV distribution, mobile communication, and interactive VSAT systems. The role of future communication satellites within the context of the national and global information infrastructures will be addressed in section two.

There are major hurdles that need to be overcome if satellite communication is going to play a significant role in the general-purpose telecommunications networks being established as foundations for the new information infrastructure. Section three discusses in detail the barriers against the seamless integration of terrestrial and satellite networks. Section four presents specific examples chosen from the three major areas, narrowband ISDN, data communications, and asynchronous transfer mode (ATM), to describe the interoperability problems and possible solutions for achieving seamless integration.

The Future Role of Communications Satellites

The rapidly evolving information infrastructure will play a critical role in realizing the vision of the world as a "global village." Satellite communications systems are essential in realizing the vision of the global information infrastructure (GII). Satellites connect the United States with the rest of the world and are the only means of providing telecommunications services in a number of places. The strengths of satellite systems include coverage to even the most remote areas of the world, with rapid deployment, distance insensitivity, bandwidth on demand, and coverage of developing countries at affordable rates.

Satellites can play a multifaceted role in this global marketplace. The future role of communication satellites is addressed below in three specific newly emerging market segments: the communications infrastructure of developing countries, multimedia communication, and the information superhighway.

Developing Countries

In many developing countries, infrastructure for information transport is either minimal or nonexistent except perhaps in major urban centers. However, most of these countries have realized the importance of communications infrastructure for their economic development.

In some of these countries, computers, work stations, and local area networks have been deployed in many business locations. However, the wide area networking (WAN) necessary to interconnect these locations is woefully inadequate. The modernization of the WANs by wireline facilities is proceeding on a limited basis in certain regions. Satellite communications can prove to be a much more affordable solution for interconnecting widely dispersed areas. The capacity of N times 64 kbps (with low values for N) could be sufficient for information transport in the initial stages in some areas.

There is a growing market for interactive VSAT systems in many developing countries, with an installed base of around 20,000 VSATs. The second generation mesh VSAT networks can indeed provide the much needed higher capacity. Innovative network architectures for both the ground and space segment can lead to an affordable and large-scale deployment of such networks. Thus, with the deployment of digital satellite networks, the developing countries can leap-frog into the twenty-first century.

Multimedia Applications

Some of the major applications of the NII are going to require multimedia services. The ubiquitous public-switched telephone networks in the U.S.A. are transitioning in many parts of the country in response to the high bandwidth requirements of multimedia applications. However, the high-speed (fractional Tl to multimegabits per second) wireline networks provide inadequate coverage in remote, rural, or low-density areas. Satellite communications can provide accessibility to locations that are not connected by wireline telecommunications networks. The current cellular networks are totally inadequate to provide the bandwidth required for the broadband (multimegabits per second) multimedia services.

The concept of hybrid terrestrial and satellite network architectures arises out of the realistic assessment that the terrestrial information infrastructure is likely to be deployed in urban areas of the United States and in other industrialized nations along with a digital "crystal grid" interconnecting these areas with high-speed digital superhighways. Satellites can indeed play an important role in allowing the benefits of the multimedia services to be extended to a much larger segment of the population. Satellite networks will be most cost-effective if they are targeted to the user community that does not have an affordable terrestrial access (if any) to this system. Thus, the hybrid architecture comprises many low-cost broadband earth terminals, which provide access to the terrestrial backbone via a suitable satellite system and a few hub earth terminals.

The satellite network architecture envisaged here will provide universal access to interactive multimedia broadband capabilities. It will provide a wide range of multimedia services in different fields. One such field is health care. With universal access to interactive broadband bandwidth on-demand via a low-cost VSAT, doctors and health care givers in remote clinics can consult with specialists thousands of miles away. It will be possible to share medical records such as X-rays, MRIs, and CAT scans. The interactive video consultation with primary care physicians in rural hospitals will alleviate the shortage of specialists in these areas.

Another area to benefit is education. The Clinton Administration has adopted as a goal the interconnection of all classrooms to the NII. It is a bold and laudable vision that can be realized at an affordable cost only through the broadband VSAT network discussed here. With the advent of digital libraries, satellite networks will allow for remote access to collections of electronic books, periodicals, and other publications. The interactions between users and digital libraries will be somewhat asymmetric. Numerous small messages embodying requests will originate from the users and be sent to the digital library. In response, the library will send messages containing relatively large amounts of data. Information stored in digital libraries could include books, journals, documents, bibliographies, newspapers, and various other documents that provide public information and scientific data.

The asymmetric nature of information transfer requires a high transfer bandwidth from a digital library to the users and is ideally matched with the satellite network architecture described above. In this system, the digital libraries will be connected to the terrestrial high-speed backbone. Low-cost earth terminals located on school premises should be capable of transmitting at low speeds (for small queries) and receiving information at multimegabits per second.

Health care and distance learning are just two of many multimedia services that will be available in all corners of the United States through satellite access to NII. Electronic commerce, advance manufacturing, remote publishing, environmental monitoring, and disaster recovery are some of the other important multimedia applications. Some of these applications are most suitable to be carried via satellite. For example, publishing involves the creation and distribution of electronic books, periodicals, music, film, and other forms of art. These electronic publications can be distributed to many different types of destinations, including private homes, businesses, and government offices. They may also be sent to digital library systems where they can be accessed and retrieved by users. Remote publishing could support applications such as electronic advertising; electronic publication and distribution of books, journals, literature, arts, and music; on-line newspaper subscriptions; and entertainment presentations like pay-per-view. Most of the above multipoint distribution applications are ideally suited for delivery by a satellite system with a hub and VSAT earth stations with appropriate multicast protocol and varying fidelity requirements.

However, the provision of a broadband connectivity via satellite is only the first step. The significant differences in the fundamental characteristics of the terrestrial and satellite networks lead to some formidable problems for their interoperability, which are discussed next.

The Information Superhighway

There is a paradigm shift in the fundamental characteristics of the new telecommunications networks. Terrestrial telecommunications networks are evolving rapidly in response to complex and stringent requirements of the new environment. ATM is the agreed-upon technology for the design and development of the information superhighway's underlying networks. A satellite communications system can play a multifaceted role in these evolving telecommunications networks and vastly strengthen the capabilities of the NII and the global information infrastructure (GII) in a number of ways. For example, the infrastructure would be able to reach remote, inaccessible areas, providing connectivity in regions where the terrestrial infrastructure has been damaged. This would allow for cost-effective broadcast/multipoint services and for global connectivity anywhere, anytime.

More importantly, satellite communications networks, if designed properly, can be a seamlessly interoperable part of global telecommunications networks. This is because the major characteristics of the new networks (namely, the ATM virtual networks) provide variable bandwidth when needed and are designed for multimedia services to be carried uniformly as fixed cells. Hence, they are ideally matched with the unique strengths of satellite communications. Satellite communications systems have a global reach and can have very flexible bandwidth-on-demand capabilities – from the agility of the spot beams to the accessibility of the satellite channel from any earth station in the coverage area. These features, coupled with the inherent multipoint/broadcast nature of satellite links when suitably harnessed, allow for the satellite communications system to play an important role in the information superhighway.

Hurdles Against Interoperability

There are major hurdles that need to be overcome if satellite communications are going to play a significant role in the general-purpose telecommunications networks being established. The barriers against the seamless integration of terrestrial and satellite networks fall into different categories. The concept of seamless integration will be based on two basic tenets: quality of service (QOS) and cost. First, the QOS that the end user expects from a fully terrestrial network should not be impacted (or degraded) because of delay or specific bit error distribution when the end-to-end connection of the user includes a satellite link. Second, the services provided via a hybrid (satellite/terrestrial) network must be cost competitive. The barriers to this fall into two categories – administrative and technical. These barriers, the challenges arising from them, and the innovative solutions to overcome these barriers are described below.

There are two types of administrative hurdles. The first one arises out of nonaccessibility to terrestrial signaling (e.g., Signaling System No. 7). As a result of this, the satellite resources cannot be used in a switched manner to provide new information-related services like Nx64 kbps. The cost penalty of providing satellite links as permanently assigned bit pipes is considerable. The second type of administrative hurdle is due to a lack of awareness, in the vendor and terrestrial carrier community regarding the appropriate parameters and protocol procedures needed for efficient operation over satellite links.

The technical barriers also have two major components. First, many important national and international standards (e.g., X.25/X.75, TCP/IP, or ATM) are designed for terrestrial networks and perform poorly over typical satellite links. Second, the developers of the next generation digital transport architecture (ATM/SDH) have considered bandwidth efficiency as a low priority item. A fundamental decision was made to base the SDH and ATM on the characteristics of fiber optic media. Many characteristics of SDH and ATM, such as the relatively high percentage of overhead and various protocol designs) are not bandwidth efficient. As a result, the onus is on the operator of a bandwidth-limited network to either live with the bandwidth excesses of these technologies or design and deploy special signal processing equipment that reduces the overhead across a bandwidth-constrained network path.

A series of examples is presented in the next section to illustrate the above administrative and technical hurdles and the proposed solutions for seamless integration. The solutions range from the development of new standards, protocols, and satellite interface units to the development of new satellite network architectures.

Solutions for Seamless Integration

Some specific examples of the operation of N-ISDN, data communication, and ATM via satellite are presented here. The issues, challenges, and their resolutions are detailed. Examples from both the administrative and technical hurdles are discussed.

N-ISDN

The development and deployment of the Integrated Services Digital Network (ISDN) has progressed over the last two decades.[1] The ISDN is based on the principles of delivering an array of services to the customer through one of a few standardized network interfaces, using the same digital transmission network to support all services. In order to support the ISDN, common channel Signaling System No. 7 (SS7) has been developed to exchange messages related to call setup, supervision, teardown, and network management.[2] Any active network elements in the international ISDN must participate in the SS7 signaling procedures. However, for administrative reasons, the satellite network may not be assigned a signaling point code in the international network, thereby making it transparent to the international signaling traffic. This situation will consequently preclude the satellite system from using bandwidth-on-demand techniques, or demand-assigned multiple access (DAMA), to use satellite capacity efficiently. Recently, however, a new signaling interface between an international gateway exchange and an ISDN satellite subnetwork has been developed and standardized. This interface specification, known as ITU-T Recommendation Q.768,[3] will enable satellite networks to incorporate bandwidth-on-demand techniques even if they are not fully integrated with the international SS7 network.

The basic function of Q.768 signaling is to inform an earth station when a trunk circuit is being put into use; upon such a request, the satellite system seizes unused bandwidth and interconnects the correspondent trunk pairs. The architecture offers a higher system utilization at the expense of the probability of blocking, but the latter can be engineered to a desired low level. Q.768 allows interaction between the gateway international switching center (ISC) and the satellite connection manager (SCM) in a satellite traffic terminal. Q.768 is not part of SS7, nor does it require any protocol changes to the existing signaling network. The satellite network performs no switching; all inter-ISC correspondent trunk pairs are fixed. *Figure 1* illustrates a satellite system based on Q.768 from a signaling standpoint.

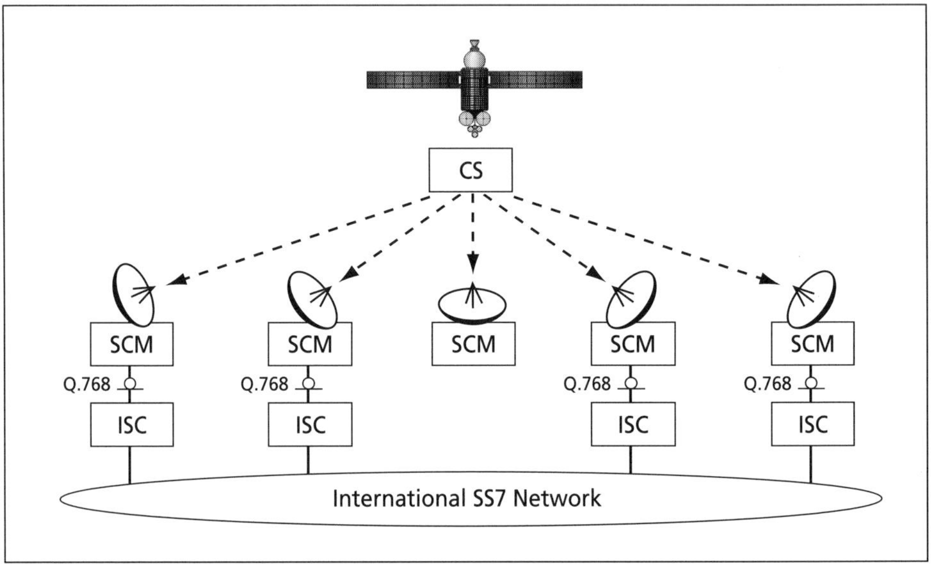

Figure 1 *Signaling relationships between network elements based on Q.768*
CS=communications satellite; SCM=satellite connection manager; ISC=international switching center

The signaling protocols used across the Q.768 interface are a subset of those used in SS7. The call control at an ISC communicates with the call control of the satellite network in order to activate circuits dynamically. *Figure 2* provides one example of how Q.768 can be used to design the signaling information flows of an on-demand satellite network. SS7 system elements and messages are shown with dashed lines, and Q.768 signaling and network elements, along with internal signaling, are shown with solid lines. In this example, bandwidth assignment of the satellite network is centrally controlled via a network control center (NCC). At an outbound ISC, the receipt of an SS7 initial address message (IAM) starts the circuit selection process within the ISC. If the ISC selects a satellite route served by Q.768, it sends a Q.768 setup message to the SCM. When the SCM receives a Q.768 setup message, it relays an assign message to the NCC via an internal signaling system. In this case, the NCC is able to assign suitable network capacity to the call, and the NCC forwards circuit assignment to the inbound SCM (via the assign message) and the outbound SCM (via the assign acknowledge message).

If the circuit assignment had been unsuccessful, the NCC would have sent a release message to the outbound SCM, which would have followed with a Q.768 release message, indicating no network capacity. The ISC would then have had to consider an alternate route for the call. Upon receipt of the Q.768 setup acknowl-

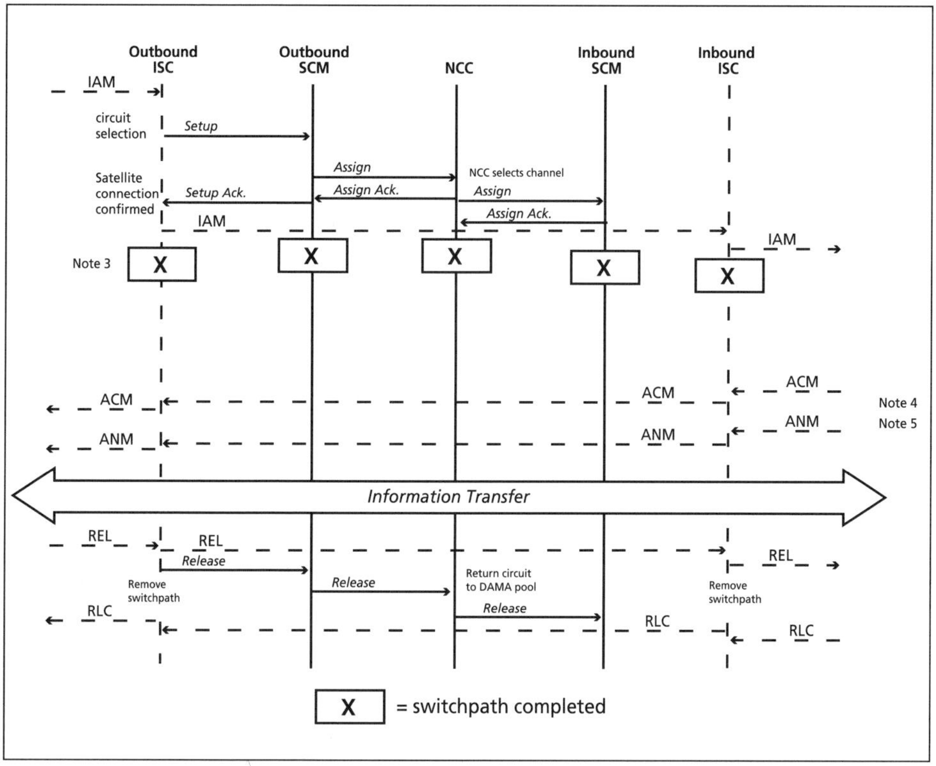

Figure 2 *An example of an on-demand signaling system based on Q.768*
IAM=initial address message; ACM=address complete message; ANM=answer message; NCC=network control center; REL=release message; RLC=release complete message

edge, the ISC forwards the SS7 IAM to the next international exchange. After information transfer has been completed, the ISC will receive an SS7 release message. It will forward a Q.768 release message to the SCM, and the satellite network will return the circuit to the DAMA pool.

Advance Notification

International switching centers are not presently accustomed to holding an IAM message while querying a satellite network for bandwidth availability. Although such a wait (nominally under 1 second) will not impact any SS7 protocol timers, it will add to the post-dialing delay of the user. Therefore, some administrations may want to obtain an earlier acknowledgment of the call setup from the satellite network. A satellite network can be designed to offer advance notification to the ISC, based on circuit availability information disseminated from the NCC.

In *Figure 2*, if the SCM is given the responsibility of either accepting or rejecting the call request, the signaling flow would be modified by placing the Q.768 setup acknowledge above the internal IAM signaling message. Two drawbacks to this approach are that it introduces the possibility of erroneous decisions at the SCM due to the latency of the distributed information, and it incurs an extra internal signaling load. A trade-off exists between the granularity of the control information (e.g., how often it is updated, how much information is conveyed) and the signaling load of the internal signaling system. However, by strategically determining when to provide advance notification (e.g., when the system load is light) and when to forward the request to the NCC (e.g., when the load is heavy), the probability of an incorrect decision can be lowered to acceptable levels.

Deployment of Q.768

International traffic has historically been dominated by voice and voiceband data (e.g., facsimile). In terrestrial networks, a digital hierarchy has evolved based on 64 kbps time division multiplexing, and voice (although nominally only a 4-kHz baseband signal) has been encoded into 64 kbps time slots. Over the past 25 years, a number of techniques have been developed to reduce the bit rate of voice calls. These relate to the techniques of digital speech interpolation and low rate encoding.[4,5] Digital speech interpolation takes advantage of the fact that the average activity level in a conversation is only 30 percent to 40 percent in each direction,[6] while low-rate encoding uses a more optimized quantization technique and memory in the quantizers to achieve a lower bit rate, sometimes down to a few kilobits per second. In addition, compression in voiceband data is achieved through demodulation and remodulation techniques, although as modems become faster these compression ratios will decrease. Advances in digital signal processing continue to advance these technologies.

The highest speech gains can be achieved if these techniques are used on an aggregate number of circuits; hence, a type of equipment known as digital circuit multiplication equipment (DCME) has been developed and widely deployed. An overview of this technology is provided by Onufry.[7] In DCME, a number of transmission channels are compressed into a reduced number of bearers. DCME circuit compression ratios of four- or five-to-one are now achievable. This technique has the added advantage of spreading out the effects of overload on random channels in such a manner that service degradation is forestalled. DCME should continue to be quite attractive for telephony for some time. However, it has limited applicability to data traffic, and it requires the gateway switch to groom voice traffic to the DCME and to seek alternate routes for other calls. For example, DCME are not

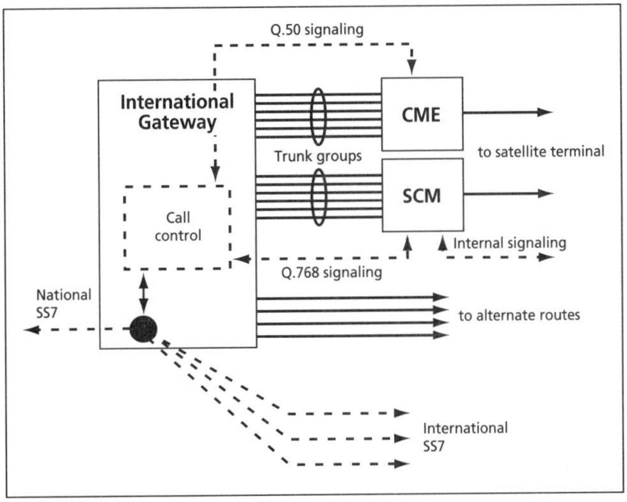

Figure 3 *Independence of CME and SCM*
CME=circuit multiplication equipment
SCM=satellite connection manager

designed to handle 384-kbps ISDN calls. In response, manufacturers have developed packetized circuit multiplication equipment (PCME) that uses packetized voice protocols to combine voice with data traffic.[8] Again, as with DCME, PCME input trunks must be provisioned (and balanced) by the network operator, and the signaling designed to support such equipment is not designed for on-demand circuit assignment for data calls.

However, equipment that provides DAMA based on Q.768 may eventually evolve to encompass circuit multiplication techniques as well, or next-generation DCME or PCME could include a Q.768 interface. Initially, as *Figure 3* illustrates, the services supported by the SCM have been separated from CME traffic in order not to disturb existing implementations of CME. In the future, Q.768 will be robust enough to support the DCME signaling traffic conveyed via signaling based on ITU-T Recommendation Q.50,[9] thereby reducing the number of signaling interfaces that must be supported at the international exchange. *Figure 4* illustrates the potential evolution of satellite transmission equipment to incorporate both circuit multiplication techniques for voice and packet data calls, with the added efficiency of on-demand circuit assignment to account for variations in overall traffic.

Data Communications

Certain data communications protocols can suffer from throughput and delay

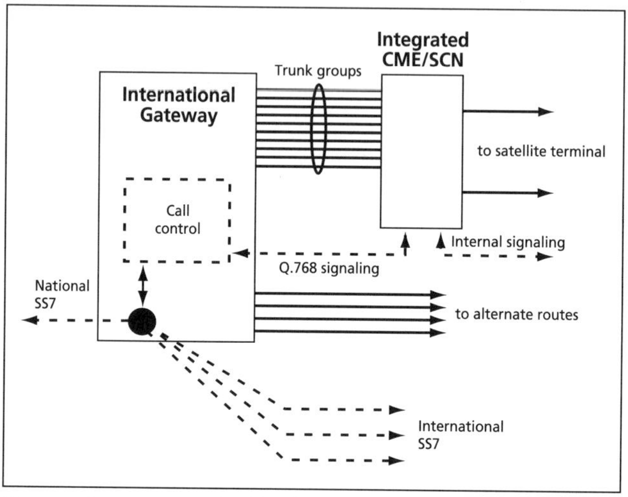

Figure 4 *Circuit multiplication integrated with Q.768 signaling*

degradation when operated over the satellite link. This degradation is a result of both the protocol semantics, which may incorporate procedures that do not work well over long round-trip delays, and the protocol syntax, which may not have enough space in the protocol fields to accommodate the coding of satellite-efficient parameters. Two notable examples are ITU-T Recommendations X.25 and X.75. [10,11]

X.25 is a well-established international standard used to connect data terminals to a packet-switched network. X.25-based packet-switched networks are deployed extensively worldwide. In order to interconnect X.25 networks, the X.75 specification was developed. The X.75 specification is divided into three layers: the physical layer, the data link layer, and the network (or packet) layer. Unfortunately, link and packet layer protocols were not designed to operate over links with geosynchronous satellite delays (up to 270 ms one way). In particular, X.75 specifies the use of basic sequence numbering (modulo 8) or optionally extended sequence numbering (modulo 128) at the link and packet layers. However, a majority of X.75 switching equipment manufacturers do not implement the extended sequence numbering since it is specified as an option. Subsequently only the modulo 8 sequence numbering option is available, which results in satellite links being significantly underutilized at link rates as low as 19.2 Kbps. Even when extended sequence numbering is available, the largest window size possible (127) cannot compensate for satellite links beyond link speeds of 256 Kbps. This results in a scenario in which the link speed may be as high as 2.048 Mbps even though

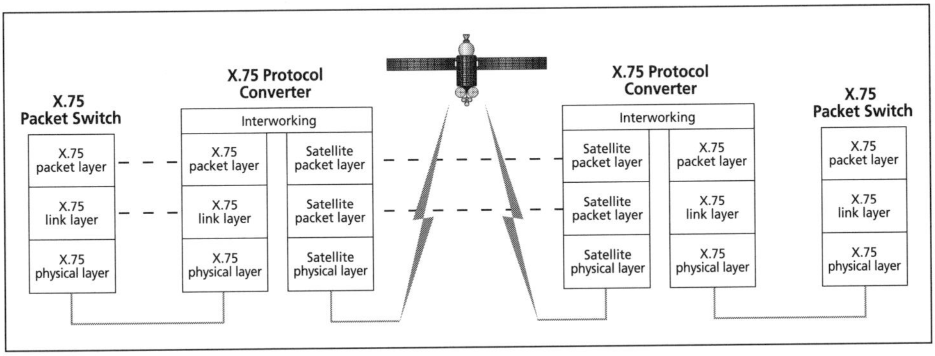

Figure 5 *Protocol conversion at the link and packet layers*

the maximum throughput that can be obtained is only 256 Kbps. Furthermore, the go-back-N error recovery protocol incorporated into X.75 Layer 2 does not operate with adequate performance over satellite links that offer reasonable bit error ratios (e.g., a BER of lE-7), and it provides extremely poor throughput efficiency when the satellite link operates with a somewhat degraded BER (e.g., lE-6).

One solution to this problem is to deploy protocol converters across the satellite network. If the satellite network knows the protocols used by the end terminals, it can insert protocol converters into the circuit in order to improve the user-perceived throughput and delay. This protocol conversion can be made in a manner transparent to the end user; the only result perceived by the user is that the throughput performance is comparable to that of a terrestrial link. *Figure 5* illustrates the protocol architecture of a protocol converter operating at the link and packet layers.

In the case, of X.75, two modifications are necessary. First, extended sequence numbering (modulo 32768) is needed in order to allow more data to be outstanding on the link. Second, a change in the retransmission logic, from go-back-N error recovery to selective retransmission, improves the performance as the BER degrades. *Figure 6* illustrates some of the throughput gains that can be achieved by substituting a satellite-efficient selective retransmission protocol for a go-back-N protocol across the satellite link. [12]

X.25 networks are typically interconnected via leased line. For such arrangements, an X.75 protocol converter can be statically assigned to a circuit group. For other types of ISDN data calls (such as facsimile), calls are switched and not assigned permanently. Fortunately, in ISDN the SS7 signaling can carry identifiers of the particular in-band protocols in use across the link. The setup message of the Q.768

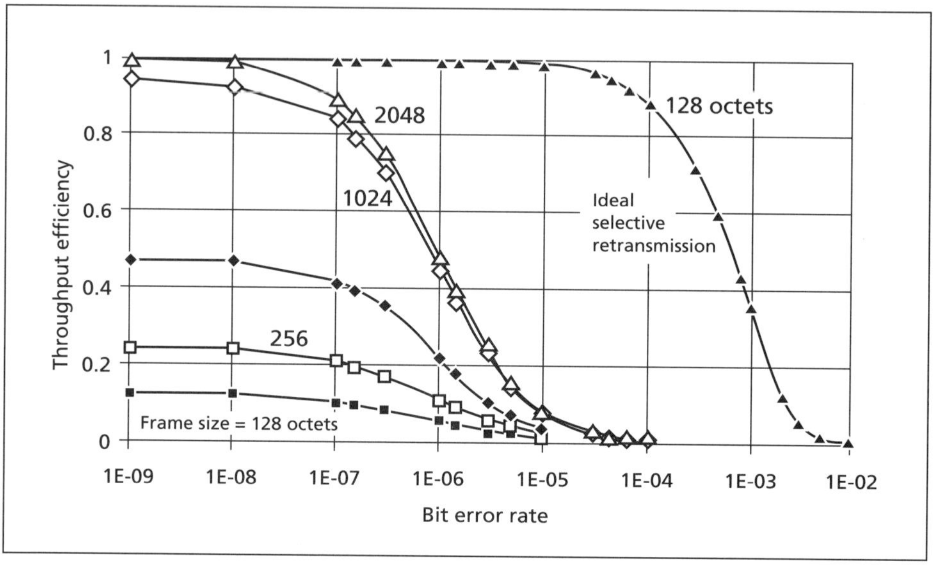

Figure 6 *Throughput efficiency of X.75 LAP-B for various frame sizes at various BERs* link rate=2.048 Mbps; link delay=270 ms; window size=127

signaling, discussed above, is defined to carry such protocol identifiers to the SCM so that protocol conversion may be performed if appropriate. The international gateway switch simply extracts the relevant SS7 signaling information and sends it to the SCM, where it is interpreted. If the protocol being used is one that could benefit from protocol conversion (such as X.75 or Group 4 Facsimile[13]) and if the SCM has an appropriate protocol converter available, the SCM attempts to setup the call with protocol conversion. Coordination of the insertion and removal of protocol conversion across the satellite network is performed via internal signaling.

The current trend in data communications is to push the error recovery procedures from link-to-link procedures to end-to-end procedures, such as in frame relay or TCP/IP. This will make it much more difficult to perform protocol conversion across satellite links because the protocols that need to be converted are not as accessible at the satellite interface. Consequently, it is even more important that protocol designs take into account the possibility of a long delay in the path. Some progress has recently been made in standardizing such protocols, but much more remains to be done before the need for protocol conversion is removed.

ATM

The basic transport mechanism that has been defined for the broadband integrated services digital network (BISDN) era is asynchronous transfer mode (ATM), which uses fixed-size data units known as cells. The ATM cell format is made up of 53 bytes; 48 of these contain data, and 5 contain overhead. The data bytes correspond to the ATM adaptation layer (AAL), and the overhead bytes define the ATM cell header, which includes cell routing information and an error detection/correction mechanism to protect routing information. The AAL also includes an error detection mechanism to mitigate transmission errors in the AAL data.

In principle, ATM cells may be transported using any digital transmission format. However, the recently approved synchronous digital hierarchy (SDH) and plesiochronous digital hierarchy (PDH) signal formats are of interest from the viewpoint of national and international standardization. An important feature in SDH transmission is the inclusion of pointer bytes, which are used to indicate the location of the first byte in the payload of the SDH frame and also to avoid slips and their associated data loss due to slight phase or frequency variations between the SDH payload and the frame. ATM cell transmission via SDH further incorporates a cell delineation mechanism for the acquisition and synchronization of ATM cells on the receiver side of the network.

Transmission considerations and performance criteria for both ATM and its use in the SDH and PDH frames are currently based on the assumption that transmission bit errors are randomly distributed. This assumption is generally valid for most terrestrial microwave and fiber optic-based transmission systems. However, because of the necessity of including channel coding into the satellite channel to reduce the station size of the earth receiver and thereby minimize ground segment cost, transmission bit errors in a satellite channel environment are likely to occur in bursts. The burst errors on a coded satellite link impact the operation of both ATM and AAL protocols and their transportation in the SDH/PDH frame.

ATM cell performance parameters include cell acquisition time, cell in-sync time, and cell discard probability. ATM performance depends upon the bit error ratio (BER) of the transmission channel and also on the nature of transmission bit errors. In the case of a satellite transmission channel, ATM cell performance parameters can only be quantified if the nature of transmission bit errors is first characterized. Here, bit errors are likely to occur in bursts due to the presence of encoding and decoding. The major performance objective for BISDN links is specified in terms of acceptable ATM cell discard probability (defined as the ratio of the number of ATM cells that are discarded because of uncorrectable errors to the total number of cells received). The stringent performance requirements being defined are driven by the characteristics of optical fiber, which can provide for

BERs that are better than 1 x 10^{-10}. Moreover, the single-bit correcting ATM header error correction (HEC) code is capable of correcting most errors encountered, given the random distribution of errors over fiber links.

Satellite links that operate at high rates (BISDN at 155 Mbps, as well as the 45 Mbps) employ error correction schemes for providing acceptable link BER (1 x 10^{-7} or better). The burst error characteristics generated as a result of using these error correction schemes have been studied extensively. The burst errors cannot be corrected by the ATM HEC since it is capable of correcting only single-bit errors. The ATM cell discard probability over such links is therefore orders of magnitude higher than over links with random errors.

To make ATM service feasible over satellites at the same quality level as fiber, COMSAT has developed a proprietary ATM link enhancement (ALE) technique that is capable of providing significant improvements in cell discard probability.[14] The ALE incorporates a selective interleaving technique that does not impose additional overheads (in terms of additional synchronization octets), which allows it to be transparently introduced into the satellite transmission link. The selective interleaving also takes into account the constraints imposed by the error-detection mode of operation (which precludes the correction of single-bit errors in consecutive cells), specified in ITU-T Recommendation I.432. Hardware that implements the ALE functionality can be transparently introduced into a satellite transmission path. The ALE, which includes an interleaver and deinterleaver module, is inserted in the transmit and receiver data path between the ATM switch and the satellite modem.

Figure 7 shows the improvement in the ATM cell discard probability when ALE is used to transport ATM over a 45-Mbps satellite link. A comparison between the curves labeled “No interleaving” and “Full bit interleaving” in *Figure 7* clearly shows that the ATM cell discard probability with the receiver in the “correction mode” is considerably degraded only when transmission is through a satellite system that introduces burst errors. In particular, there is a high probability that an ATM cell header will be in error by two or more bits every time it is ‘hit’ by an error burst. The single-bit error-correction capability of the FEC in the ATM cell header is ineffective in this situation, and the ATM cell is highly likely to be discarded every time its header is hit by an error burst.

Full-bit interleaving of ATM cell headers, on the other hand, tends to distribute or spread the bit errors among ATM cell headers after deinterleaving on the receiver side. The probability of an ATM cell header being in error by two or more bits is

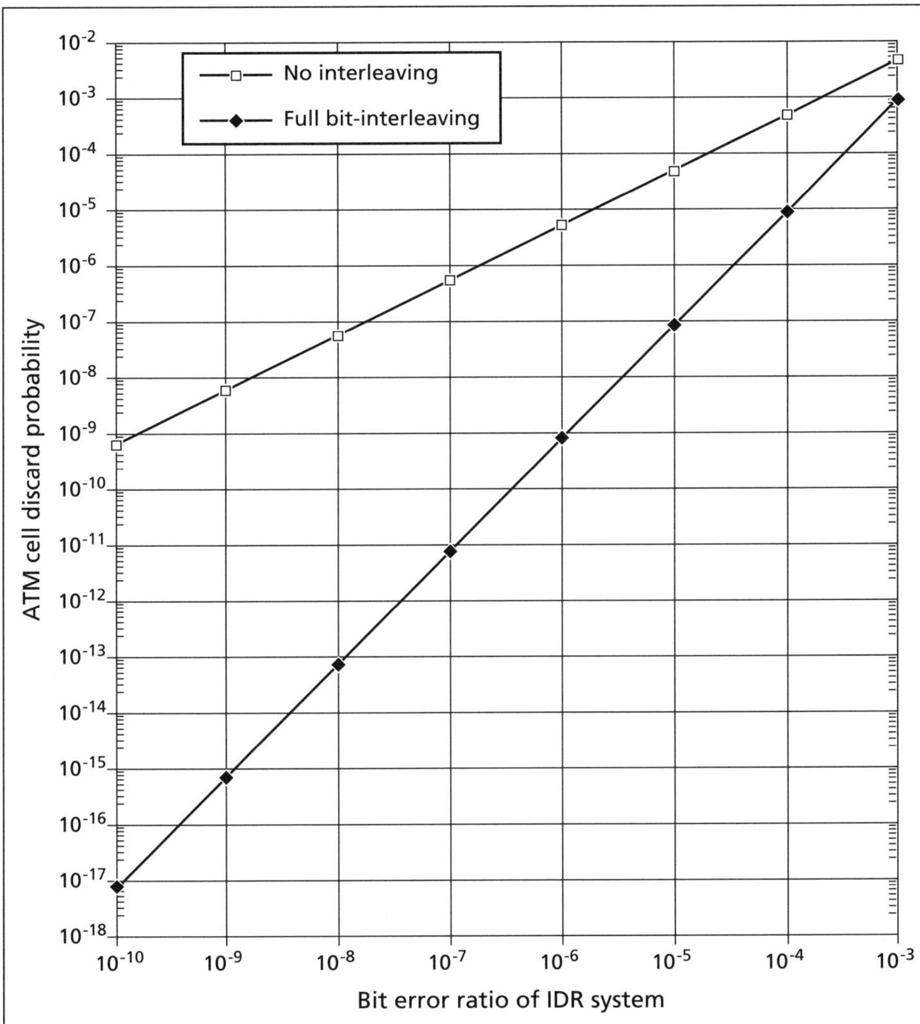

Figure 7 *ATM cell discard probability vs. BER*

reduced. Furthermore, eroded cell headers having single-bit errors are more likely, and these can be corrected by the cyclic code in the cell header.

A series of next generation ALEs is being developed by COMSAT to reduce the bit errors in the ATM cell payload. These products are targeted for both ATM via DS3 and ATM at lower rates (a few megabits per second) and use cell based outer coding to improve the performance.

Conclusion

Satellites are a key element in both the communications infrastructure of developing countries, and the multimedia communication and information superhighway of the more technically advanced countries. Satellite systems provide many unique advantages, including remote coverage with rapid deployment, distance insensitivity, bandwidth on demand, and affordable rates, which are particularly attractive to developing countries. Thus, along with the terrestrial network, satellite communications can play an important role in the rapidly evolving information infrastructure. However, there are major technical and non-technical hurdles which need to be overcome to achieve seamless integration of terrestrial and satellite networks. Specific barriers were identified for N-ISDN, data communications, and ATM, and suitable solutions to overcome these barriers were presented. These solutions ranged form the development of new standards for N-ISDN and protocol conversion functions for data communications to intelligent satellite communications interfaces for ATM.

References

1 ITU-T Recommendation I.120, "Integrated Services Digital Networks (ISDNs)," Geneva, Switzerland, latest version published 1993.

2 A. Modarressi and R. Skoog, "Signaling System No. 7: A Tutorial," *IEEE Communications Magazine*, July 1990, pp. 19-35.

3 Draft ITU-T Recommendation Q.768, "Signaling Interface between an International Switching Center (ISC) and an ISDN Satellite Subnetwork," COM 11-R X, Geneva, Switzerland, 1994.

4 S. Campanella, "Digital Speech Interpolation," *COMSAT Technical Review*, vol. 6, no. 1, spring, 1976, pp. 127-158.

5 CCITT Recommendation G.726 "40, 32, 24, 16 kbit/s Adaptive Differential Pulse Code Modulation (ADPCM)," Geneva, Switzerland, 1991.

6 ITU-T Recommendation G.763, "Digital Circuit Multiplication Equipment Using ADPCM (Recommendation G.726) and Digital Speech Interpolation," Geneva, Switzerland, latest version published 1994.

7 M. Onufry, "DCME in the INTELSAT VI Era," *COMSAT Technical Review*, vol. 22, no. 1, pp. 19-56.

8 ITU-T Recommendation G.765, "Packet Circuit Multiplication Equipment," Geneva, Switzerland, 1993.

9 ITU-T Recommendation Q.50, "Signaling between Circuit Multiplication Equipments (CME) and International Switching Centres (ISC)," Geneva, Switzerland, latest

version published 1994.

10 ITU-T Recommendation X.25, "Interface between Data Terminal Equipment (DTE) and Data Circuit-Terminating Equipment (DCE) for Terminals Operating in the Packet Mode and Connected to Public Data Networks by Dedicated Circuit," Geneva, Switzerland, latest version published 1993.

11 ITU-T Recommendation X.75, "Packet-Switched Signaling System between Public Networks Providing Data Transmission Services," Geneva, Switzerland, latest version published 1993.

12 D. M. Chitre, "Analysis of Throughput Efficiency and Delay in ARQ Systems," *COMSAT Technical Review*, vol. 11, no. 2, fall 1981, pp. 345-368.

13 CCITT Recommendation T.90, "Characteristics and Protocols for Terminals for Telematic Services in ISDN," Geneva, Switzerland, latest version published 1992.

14 D. M. Chitre, et al., "Asynchronous Transfer Mode (ATM) Operation Via Satellite: Issues, Challenges and Resolutions," *International Journal of Satellite Communications*, vol. 12, 2/94, pp. 211-222.

A Comparative Analysis of Near Earth Satellite Communications Systems

Joseph N. Pelton
Director, Center for Advanced Research in Telecom
University of Colorado at Boulder

Abstract

For 30 years, there has been continuous technological innovation within the field of geosynchronous communications satellites. During the last few years, however, new system concepts and new service needs have served to create a new era of change and have refocused the entire field of satellite communications. The forces of competition *vis à vis* fiber optic cable plus new technological innovations in space communications are giving rise to new types of architecture, orbits, and innovative systems design.

The characteristics of these new systems are driven by such factors as low latency echo cancellation, lower path-loss, improved elevation angles, high latitude coverage, and higher levels of frequency reuse. In many ways, this process is the active pursuit of lower costs, enhanced user mobility, and improved quality of service.

To date, these new systems as formally proposed can be grouped as follows: little low earth orbitals (LEOs), operating at 800 MHz or below; big LEOs and medium earth orbitals (MEOs), operating at 1.6 GHz or above; and mega-LEOs, which could operate at various frequencies but have so far been slated for 30/20 GHz. Finally, the most recent concepts have been for new types of telecommunications platforms that could maintain geosynchronous operation in what has been called proto-space. These are called high-altitude long-endurance (HALE) platforms and can presumably offer a broad range of services in a variety of frequency bands.

The purpose of this article is to describe the basic characteristics of the various types of new satellite systems that have been proposed and to contrast and compare their operational, technical, service, and financial characteristics. The final analysis will seek to compare the new systems with broadband fiber optic cable systems to determine their technical and financial competitiveness with the best in terrestrial wireline systems.

Introduction

To understand the evolution of satellite system designs over the last 30 years, it is first necessary to explore how geosynchronous satellite systems first evolved and then developed increased performance over a number of years. The next key stage of evolution occurred when new requirements for mobile satellite services with high-look angles and small mobile transceivers suggested totally new satellite architectures. This directly led to new forms of low and medium earth orbit systems.

The system designs that have emerged in the last few years attempt to overcome certain liabilities of traditional geosynchronous satellite systems. This can be best demonstrated by reviewing the technical development of the Intelsat satellite series up to the present. Intelsat represents the longest continuous history of design evolution for geosynchronous satellites. It also reflects the general patterns of technical development that are generally representative of the overall geosynchronous industry. In short, this technological heritage is important in helping to define patterns of technical growth and expansion and in showing the important departures from this heritage that modern low and medium earth orbit systems now define.

Since no major commercial near earth orbit systems now exist, an analytic framework will be developed that shows the many trade-offs that various types of systems can achieve in terms of coverage, look-angles, frequency reuse, path-loss, operational performance, net operating costs, and system costs. This will allow a type of comparative algorithm to be developed to present not only the individual characteristics of all proposed systems but also to present a comparative figure of merit for all types of systems including geosynchronous, medium earth orbit, little and big LEOs, mega-LEOs, and even HALE-type platforms.

The key to applying this comparative algorithm is the relative weighting that one applies to the various factors used to make the evaluation. These include such factors as latency, coverage of higher latitudes, levels of frequency reuse, cost, size, portability of ground transceivers, and so forth. This comparative analysis attempts to apply reasonable and fair assessment factors, but its conclusions can be altered by placing different priorities on which service characteristics are considered most important.

Ultimately, the demand for services and the marketplace will decide which systems are funded, built, deployed, and operated successfully. In the meantime, this analysis attempts to present in a comprehensive manner all systems now envisioned and the comparative strengths and weaknesses of these systems. Clearly, satellite systems designed for fixed-satellite service, VSAT service, personal communications services, or direct broadcast service will have different figures of merit. Thus, the comparative index developed within this article can be best applied in comparing systems designed for a particular service rather than systems with totally disparate service objectives. Despite these limitations, it is hoped that the comparative methodology presented here provides at least a better way of understanding how the proposed new systems both differ and compare to traditional systems.

Figures 1 and *2* show the evolution of the Intelsat satellite system over time. The first figure shows the traditional measures of satellite mass, size, communications capacity, available frequency, power, and lifetime. The second figure shows the performance in terms of direct financial costs on a per-MHz basis. It is interesting that despite claims by critics to the contrary, Intelsat satellites in objective terms exhibited dramatic cost reductions both before and after they entered a directly competitive market.

At the risk of oversimplifying the process, the areas of fixed-satellite technological advancement that have proven most important in terms of producing cost reductions have included increased on-board power; increased frequency reuse through multi-beam antennas and cross- polarization; constantly focused spot beam antennas; digital modulation, encoding, and compression techniques; beam switching and interconnection; and lifetime extension techniques. Clearly other factors provided some advantages as well. The simplification of earth station antenna design, for instance, accompanied the evolution of more powerful and higher-gain satellites in orbit. These factors have all combined to make geosynchronous satellite systems over a thousand-fold more cost-effective and have allowed them to offer a higher quality signal as well.

In most industries, a technology which achieved a thousand-fold increase in

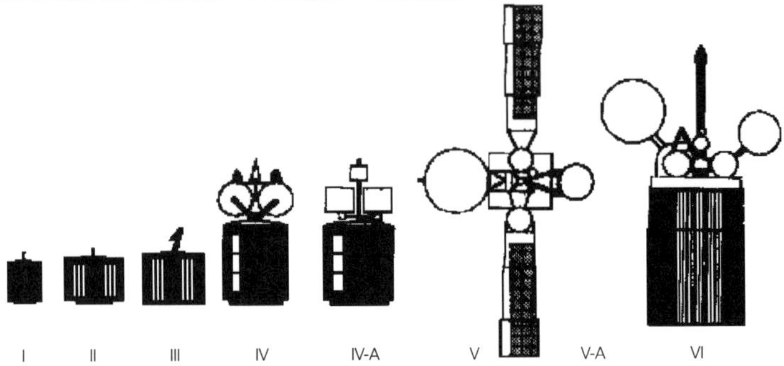

INTELSAT Designation	I	II	III	IV	IV-A	V	V-A	VI
Year of First Launch	1965	1966	1968	1971	1975	1980	1983	1986
Prime Contractor	Hughes	Hughes	TRW	Hughes	Hughes	Ford Aerospace	Ford Aerospace	Hughes
Dimensions (undeployed)								
Width, m.	0.7	1.4	1.4	2.4	2.4	2.0	2.0	3.6
Height	0.8	0.7	1.0	6.3	6.8	6.4	6.4	6.4
Launch Vehicles	Thor Delta			Atlas Centaur	Atlas Centaur	Atlas Centaur or Ariane 1,2	Titan 34D	Ariane 4
Spacecraft Transfer Orbit Mass, kg.	68	182	293	1,385	1,489	1,946	2,140	12,100/3,720
Communications								
Payload Mass, kg.	13	36	58	185	190	235	280	800
Power EOL Equinox, watts	40	75	134	480	800	1,270	1,270	2,200
Design Lifetime, yrs.	1.5	3	5	7	7	7	7	14
Rated Voice Channel								
Capacity in AOR	480	480	2,400	8,000	12,000	25,000	30,000	80,000
Bandwidth, MHz.	50	130	300	500	800	2,300	2,180	3,680

Figure 1 *Evolution of INTELSAT Satellites*

productivity or cost-efficiency over a 30-year period would be considered truly phenomenal. In the fields of telecommunications and computers, this rate of innovation in terms of the so-called Moore's law is considered normal or expected. The parallel development of fiber optic cable systems and digital processing has indeed applied competitive pressure to the satellite field. The industry has begun to evolve in new directions as the potential of new markets and services have become better understood.

Satellite System Characteristics and Key Trade-offs

The possible trade-offs in a satellite system design are nearly infinite. The principal options, however, represent some 12 parameters, which although large is still a practical number to be comparatively evaluated. These factors are as follows:

- connectivity

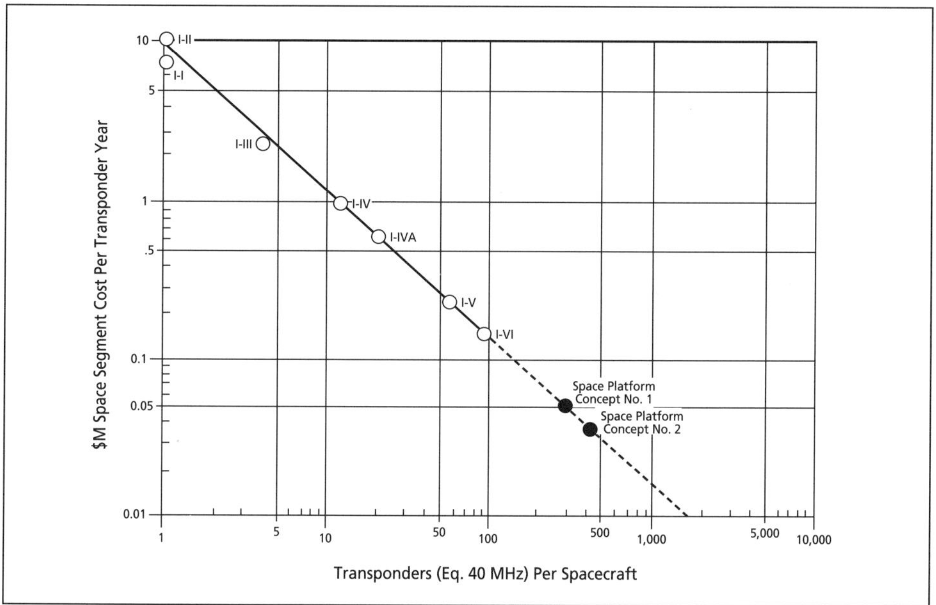

Figure 2 *INTELSAT series demonstrates economy of scale in communications spacecraft, past and present.*

- cost and complexity of ground transceivers
- cost and complexity of satellites
- ease or difficulty of achieving frequency reuse
- efficiency of coverage of prime areas
- ground antenna tracking and gain requirements
- launch cost and launch options
- lifetime
- look-angle to satellite
- number of satellites needed to achieve coverage
- operational requirements for TTC&M and autonomous control
- path-loss
- propagation time, transmission delay, or latency
- requirements for an inter-satellite link (ISL)
- sparing strategies and concepts

The relative importance of these factors is dependent upon many other factors such as the type of service to be provided, the frequency band to be used, and the patterns and configurations of intended users. This is why the factors are listed in alphabetic order. The most likely way of determining a relative figure of merit

System Type	System Cost ($U.S.)	Coverage (sq. km.)	Beam Throughput	Beam Performance Index
GEO (3 Sats) 8 Beams (7 yr. life)	$1.2 B	12.6×10^6	200 Mbs	$2,834/Mbs/Km² per year
MEO (12 Sats) 20 Beams (5 yr. life)	$2 B	3.1×10^6	250 Mbs	$6,451/Mbs/Km² per year
LEO (50 Sats) 40 Beams (5 yr. life)	$4 B	1.5×10^6	50 Mbs	$8680/Mbs/Km² per year
Mega-LEO 800Sats- 50 Beams (5 yr. life)	$12 B	1.5×10^6	100 Mbs	$6000/Mbs/Km² per year
HALE 12 Cells (10 yr. life)	$10 M	0.18×10^6	200 Mbs	$1852/Mbs/Km² per year
Terrestrial 30 Cells (20 yr. life)	$50M	$.008 \times 10^6$	90 Mbs	$82,000/Mbs/Km² per year

Sample calculation:

GEO SYSTEM: $\dfrac{\text{Costs (\$1.2 Billion)}}{\text{(200Mbs)xBeam size(12.6 M.Km) x Sats(3)xbeams(8)xlife(7 yrs)x(eff)100\%}}$

Resulting Beam Performance Index: $2834/Mbs/Sq. Km/Yr

NOTE: Efficiency index for MEOs, LEOs, and MEOs is 30% because of ocean and artic coverage 70% of the time.

NOTE: The HALE platform cost assumes two platforms of $5 million, each continunously providing service on a rotational cycle each lasting 4 months duration.

Table 1 *Comparing satellite system capabilities*

is a calculation of cost of service that is somehow figured on a parallel or common basis. *Table 1* essentially seeks to calculate the cost of comparable satellite service delivery (i.e., megabits per second) over the service area covered. This is not a perfect measure of a figure of merit. Certainly it gives more advantage to those systems with higher cost and more elaborate ground systems and less advantage to those with lower cost ground systems. Nevertheless, it is still an interesting and useful measure.

The trade-offs among the various types of systems presented in *Table 2* suggest that geosynchronous satellites are highly effective for large coverage areas and conventional fixed point-to-point satellite services with concentrated traffic and a more limited number of earth station facilities. If one adds a large number of users

Type of System	Number of Satellites	Frequency Reuse Opportunity	Transmit Delay	Path Loss	Look Angle	Ground Segment
GEOSYN	few (3)	low (4-6 times)	high (0.25 sec)	high	low to high	typically expen. medium
MEO	medium (12 to 18)	medium (8-12 times)	medium (0.1 sec)	medium to high	medium	medium expen. to low
LEO	high (50 up)	high (30-100 times)	low (>0.05 sec)	low	high	low
MEGA LEO	very high (hundreds)	high (30-100 times)	low (>0.05 sec)	low	high	ranges low to medium
Equator Circular Orbit	typically med.(8-15)	medium (8-12 times)	medium (>0.10)	medium to high	medium	ranges low to medium
HALE* Platform	one per region	medium (8-12times)	almost none (>0.001 sec)	none	none	low

* High-altitude long endurance platforms operate at altitudes of 18 to 21 kilometers and are in what has been characterized as proto-space, or the 18- to 110-kilometer region of aviation space. They typically can cover a region of 180,000 sq. kilometers. This is included for providing a wider range of comparisons.

Table 2 *Comparison of different types of satellite systems capable of real-time services – voice and above*

to the need for lower cost and mobile terminals, the various LEO and MEO options will start to look increasingly better. Perhaps the biggest surprise of all is the new concept of a high-altitude long-endurance platform. These systems seem to hold the advantage of coverage, throughput, look-angle, and cost-per-area-served without sacrificing the ability to provide low-cost earth segment facilities as well.

At this stage, these comparisons are clearly highly theoretical since geosynchronous earth orbit (GEO) systems exist while big LEOs, mega-LEOs, MEOs and HALE platforms are only proposed. One of several key unknowns in all this is the importance of latency and transmission delay. Many believe that in the twenty-first century satellite technology must reduce delay significantly if it is to stay competitive with fiber optic technology. If this is indeed true, low earth orbit satellites, whether they are global constellations or equatorial circular orbits or some other configuration, will be increasingly important. Likewise, HALE technology will increase in importance and market share.

The complexity of satellite typology has greatly increased in the last few years, at least in terms of planned satellite systems. The actual or true advantages of such new LEO and MEO satellite systems will be demonstrated as systems are deployed and tested in the real-world market. In the interim, there will be some type of comparative chart that indicates the relative strengths and weaknesses of the various approaches. Such an attempt is provided in *Table 2*. This chart does not attempt to develop a comparative single metric as that presented in *Table 1* but rather seeks to show within a single chart the strengths and weaknesses of various types of satellite systems as well as that of the HALE platform that operate not in outer space but in so-called proto-space.

Finally, some review of fiber optic cable systems in contrast to satellite technology is necessary. It is important to note that satellite and fiber optic technology are perhaps as often complementary as they are competitive. The satellite telecommunications services that are most often considered to be naturally defined or at least best suited to the media are as follows:

- mobile services of all types – aeronautical, maritime, or land
- broadcast or single point to multi-point
- large mesh-type networks
- rural and remote services
- quickly deployed services such as for emergency and relief efforts
- services to vulnerable areas such as those subject to earthquake, flooding, and so on.

In contrast, fiber optic cable is a service often thought best suited to high-volume trunk lines between high-density traffic nodes, urban networks, and corporate campus operations that require high-quality, high-volume, and low-latency transmission. A detailed review of most telecommunications currently in place would show that satellite services and fiber optic systems are, generally speaking, not in direct competition. The only real exception is with regard to international telecommunications services. Today, satellites are increasingly handling the international links for more isolated developing countries while fiber mostly is claiming the high-density routes.

Table 3 shows a comparison of today's fiber technology with current satellite technology. It can be seen that the primary advantage of fiber optic systems is largely with regard to quality of service (i.e., bit error rates) and to speed of connection (i.e., lack of significant latency). These are, of course, not of minor importance. MEO, LEO, and HALE technologies can overcome some of these liabilities.

The most obvious point is with regard to latency. Clearly MEO, LEO, and HALE systems offer major improvements. MEO systems are about six times faster than GEO systems. LEO systems can be 40 times faster, and HALE can be well over one thou-

Parameter	Intelsat System		Submarine Cables
Capacity	Equivalent Voice Circuits		Equivalent Voice Circuits
64 kbit/s	VA	VI	TAT-8
32 kbit/ss/DSI	15,000	24,000	7,500
		120,000	37,800
Performance (Error Rate)	Full ISDN compatibility		Low error rates
Reliability			Frequently restored via satellites
Networking	Multiple access strongest attribute of satellites		Essentially point-to-point, hence dependent on terrestrial talls
Economics	Competitive, distance-independent		Heavy upfront investment, distance dependent

Table 3 *Comparison of fiber technology with current satellite technology.*

sand times faster. Since light travels through space about 40 percent faster than light through cable and because space systems can skip over switches, LEO and HALE systems can actually contain less latency or delay than a terrestrial fiber system.

The issue of quality is more complex. If one considers the basic design features and frequency use patterns of many of the proposed little or big LEO systems, it is clear that the quality of these systems is not only less than fiber optic networks but also less than conventional geosynchronous satellite systems. However, if one considers the potential of future broadband networks with even more sophisticated cellular antenna designs and with on-board, totally complete signal regeneration (i.e., bit-by-bit processing), it is conceivable that quality fully comparable to fiber can be achieved even in the millimeter waveband. This must also assume technologies such as those promised by the Teledesic satellite system in terms of providing a "Cray in a Shoe Box."

Conclusions

The field of satellite communications is currently in a period of major transition. This transition is being driven by the development of a wide range of new services

System Name	System	# of Satellites Type	Basic Rationale
ECO-8	ECO	8	Use of a series of low to medium-earth orbit, equatorial circular orbit, equatorial zone satellites to provide continuous global coverage of the equatorial zone. Developed by INPE, the Brazilian Space Agency, to cover not only Brazil but other equatorial countries as well. Relatively low cost, but it will be difficult for a nationally-based system to sell service to other equatorial countries. **Key Constraints:** Raising capital needed to complete project. Limited frequency. Essentially limited to voice and data. Limited spectrum.
Ellipso	Elliptical	N/A	Use of a number of elliptical orbit satellites to concentrate coverage over North America. This medium-earth orbit elliptical satellite system thus places maximum capacity over key market areas. This is a rather straightforward bent-pipe design that will be able to provide satellite cellular service in the $0.50-per-min. and up range. **Key Constraints:** Raising capital for a start-up enterprise. Competing with other satellite cellular systems which are more well known plus competing with terrestrial cellular as well. Limited spectrum at 2 GHz.
Globalstar	LEO	50	Use of a LEO constellation to provide worldwide cellular telephone and data services. Use of a simpler bent-pipe design without ISL to simplify the system and make it much lower cost. Concept is to deliver a basic cellular telephone service at about $0.50 per minute or at one-eighth the cost of Iridium. Also uses CDMA spread-spectrum techniques to boost capacity since Qualcomm, one of the partners in Globalstar, holds a leading position in this technology. Also Globalstar has international backing as a U.S.-French venture with major investment by ALCATEL through the Loral Corporation. **Key Constraints:** Limited spectrum at 2 GHz. Market appears to have too many entrants. Limited value-added service offerings.
INMARSAT PROJECT 21	MEO	N/A	Use of medium-earth orbit constellation in conjunction with existing INMARSAT geosynchronous constellation.Existing traffic base in aeronautical and especially maritime mobile service gives major head start. New Project 21 will, however, be semi-autonomous. Limited number of satellites, perhaps 12 or less, will reduce capital investment in comparison to Iridium system. **Key Constraints:** Medium-earth orbit will have more delay, more path-loss, and some lower look angles. Satellites will need to be more complex and more powerful to compensate for this. Limited spectrum at 2 GHz. Price must be competitive with market yet not undercut existing maritime and aeronautical service.
Iridium	LEO	66	Use of a LEO constellation with ISL to achieve the following: (a) global coverage; (b) mobile communications with hand-held transceiver; (c) 48-fold frequency reuse; (d) executive mobility land service anywhere; and e) bypass of terrestrial network when needed. The system's bid to be first into the market with pioneer's preference was denied. Key profitability to be in hand-held units. Global consortium to finance $4 billion space segment. **Key Constraints:** High cost of space segment and high cost of service (i.e., $4 per minute).

System Name	System	# of Satellites Type	Basic Rationale
Odyssey	LEO	12	Use of a medium-earth orbit system to provide global cellular voice and data services. Reduce the number of satellites required by going to a higher orbit. This reduces opportunity for frequency reuse, increases path-loss and alters look angles, but overall it allows for a low-cost service, projected to be under $ 0.50 per minute. **Key Constraints:** Despite very good technical design, there has been difficulty raising the capital for this project by its key proponent, TRW Inc.
Spaceway	GEO	8	Use of highly efficient cellular spot beam technology and the Ka-band frequencies to attempt to provide a global system with the range of service offerings comparable to a Teledesic mega-LEO system. This system suggests that if GEO systems push the technology as far as possible, they can provide a cost-effective and competitive service for mobile hand-held transceivers, village-based rural telecommunications systems, and even high-speed broadband services. Seeks to deliver service at under $0.25 per minute, **Key Constraints:** Reliance on Ka-band technology which has yet to be commercially proven despite experiments (e.g., ACTS, ETS-6). Latency associated with geosynchronous transmission. Very large capital investment in unproven markets.
Teledesic	Mega	910	Use of a huge number of mass produced, one-ton but very high performance satellites with more than 50 fold frequency reuse in the Ka-band. Unique technologies with up to 8-fold ISL connections, fixed or painted spot beams, supercomputer in a shoe box , etc. Based on conversion of U. S. military Brilliant Pebbles program. Also envisions commercial use of Ka-band. Seeks to provide service at under $0.25 per minute. Projected cost of system estimated at $12 billion or more, making it the most expensive and largest capacity of any system. **Key Constraints:** Even with the backing of Microsoft's Bill Gates and Greg McCaw, the huge capital needed for such an exotic and unproved design would seem very difficult. Marketing and sales for such a huge system on a global scale would also seem to be a major challenge.

Table 4 *Review of new satellite systems providing voice and above service*

and applications that includes direct broadcast satellite services, all forms of land, air, and sea mobile communications, plus navigational and radio determination services. This is in addition to the growth of conventional geosynchronous fixed satellite services. Likewise, the field of satellite communications is being strongly influenced by a range of new technologies, system architectures, and satellite orbital configurations. Among these technologies are on-board switching and processing, including signal regeneration, advanced cellular antenna designs, and enhanced reuse and beam interconnection. Finally, on top of these innovations, which are specific to the field of satellite communications, is a series of technical innovations that apply to all telecommunications. These include advanced modulation and encoding techniques, digital compression techniques, and high-performance and low-cost digital coder/decoders (codecs).

The cumulative effect is that the satellite communications field is in a period of rapid and highly dynamic change in almost every regard. These changes are mostly driven by the use of new orbital configurations to achieve better cost or technical performance. This is closely followed in importance by the use of more technology to make satellites smarter – less like bent pipes and more like a signaling, switching, and processing centers. Finally, digital processing, modulation, coding, and compression techniques are further defining and refining the field of satellite communications today.

In light of the above analysis, is it possible to indicate which satellite systems of the next decade will succeed and which will fail? Obviously, the answer is no. The key to success in the satellite field is first and foremost rooted in market factors. Those with extremely effective market ties and with well-established systems such as INTELSAT and INMARSAT have a great advantage that transcends technological and operational issues. Second, the various approaches to deploying satellite systems have widely different technical, operational, and market factors behind their designs.

Table 4 examines the rationale behind the various MEO, LEO, mega-LEO, and ECO systems that have been filed in the United States or internationally. This analysis assesses the logic behind the systems, using the comparative methodologies found in *Tables 2* and *3*. Because there are many subjective or non-comparable factors involved in this analysis, *Table 4* cannot be considered scientific or even quantitative. Nevertheless, the differences among the systems can be shown in clear relief and the strengths and weaknesses made more identifiable.

Propagation Effects in Satellite Systems

Ernest K. Smith
Professor, Electrical and Computer Engineering Dept.
University of Colorado at Boulder

Abstract

Propagation aspects in satellite systems normally include three areas: ionospheric effects, tropospheric effects, and antenna noise.

Ionospheric effects are dominant for transmission frequencies less than 3 GHz, and in most cases, do not need to be considered for systems operating above 10 GHz. In recent years there has been increased recognition of the effect of ionospheric scintillation at low latitudes in the evening hours during periods of high solar activity. Ionospheric scintillation is particularly troublesome at L-band and longer wavelengths (lower frequencies).

Tropospheric effects dominate for frequencies above 10 GHz and are usually negligible for frequencies below 3 GHz. Rain constitutes the principal limitation among the tropospheric effects. Attenuation (in decibels) by rain increases roughly as the square of the frequency. However, for service 99% of the time or less, rain can usually be ignored. Extraterrestrial noise is normally the primary antenna noise source for systems operating at frequencies below 1 GHz, but with the exception of the sun become negligible above 2 GHz. Above 2 GHz noise emitted by the atmospheric gases in proportion to their absorption is the limiting source, but antenna noise temperatures do not exceed the ambient temperature of the medium, say 285° K.

Propagation Effects in Satellite Systems

Propagation, as treated here, has to do with satellite communications – that is, communications between earth-orbiting satellites and stations on the surface of the earth. Readers interested in deep-space propagation are referred to Yuen (1982). Following an introductory section, earth-space propagation is treated under three headings: ionospheric effects; tropospheric effects; and terrain, foliage, and building effects. In the case of propagation from fixed-earth stations, only the first two effects need to be considered. The third category becomes important for the mobile satellite and broadcast satellite services. Texts on satellite communications which also treat propagation are Gagliardi (1991), Miya (1975), Morgan and Gordon (1989), Pratt and Bostian (1986), and Roddy (1989). Texts which specialize in the propagation aspects of satellite communications are Allnutt (1989), Flock (1987), and Ippolito (1986, 1989). The International Radio Consutative Committee (CCIR) of the International Telecommunications Union has provided an international forum on telecommunication questions. Propagation effects have been historically treated under Study Groups V and VI. The CCIR (1990) treats satellite communications in several of its Green Books (e.g., Volumes II, IV, VII, X, and XI) but most notably in Volume IV. Propagation aspects are found in Volume V (non-ionized media), most notably in Report 5644; and in Volume VI (ionized media), most notably in Report 263-7. The CCIR has ceased to exist in name, but the work still goes on under the heading ITU-R (International Telecommunications Union - Radio). The two propagation study groups have been consolidated into a single group – Study Group III.

Introduction

The decibel (dB) is the preferred unit for expressing relative signal strength whether voltage or power. A ratio in decibels is 20 times the logarithm to the base ten of current or voltage ratios and 10 times the logarithm to the base ten for power ratios. A typical example is signal carrier, *C*, to external noise, *N*, ratio *C/N* as will be shown below.

Propagation in Free Space

The range equation relates received power P_r to transmitted power P_t, as

$$P_r = \frac{P_t G_r G_t \lambda^2}{(4\pi R)^2} \text{ watts} \tag{1}$$

where:

G_r is the receiving antenna gain above an isotropic antenna in the direction of the transmitter

G_t is the transmitting antenna gain similarly defined

λ is the wavelength in free space, *m*

R is distance between antennas, *m*

P_t is the transmitter power, watts.

path-loss is given by the following equation. It follows from (1) if G_r and G_t are isotropic antennas. L_{fs} is also called "free space loss."

$$L_{fs} = \left(\frac{4\pi R}{\lambda}\right)^2 = \frac{P_t}{P_r} \tag{2}$$

Another expression of (2) in terms of carrier power C and receiving antenna aperture is

$$C = \frac{P_t G_t A_r}{4\pi R^2} = \frac{(EIRP) A_r}{4\pi R^2} \tag{3}$$

where A_r is the effective area of the receiving antenna and is related to gain above an isotropic radiator by

$$A_r = \frac{G\lambda^2}{4\pi} = G A_i \tag{4}$$

where:

$$A_i = \frac{\lambda^2}{4\pi} \tag{5}$$

is the effective aperture for an isotropic radiator. EIRP stands for effective isotropic radiated power ($=P_t\ G_t$).

Carrier-to-Noise Ratio (C/N)

C/N is basic to microwave transmission. If the noise can be expressed in terms of thermal noise, then

$$\frac{C}{N} = \frac{P_t G_t G_r}{L_{fs} k T_{sys} B} = \frac{(\mathrm{EIRP}) G_r}{L_{fs} k T_{sys} B} \tag{6}$$

where:

T_{sys} is the system noise temperature in kelvin

$k = 1.38062 \times 10^{-23}$ Joules/K is Boltzmann's constant

B is bandwidth in Hertz

L_{fs} is transmission loss in free space
G_t, G_r are the transmitting and receiving antenna gains relative to an isotropic radiator.

It is common practice to express relation (1) - (5) in terms of decibels. For example, expressions for (2) and (6) are (log = logarithm to base 10).

$$(P_r)\mathrm{dB}w = (P_t)\mathrm{dB}w + (G_r)\mathrm{dB}i + (G_t)\mathrm{dB}i + 20\log\lambda - 20\log R - 21.98 \quad (7)$$

$$(L_{fs})\mathrm{dB} = 20\log R - 20\log\lambda + 21.98 \quad (8)$$

Example: Assume R = 38,000 km and λ=3 cm

$$(L_{fs}) = 20\log(38\times10^6) - 20\log 0.03 + 21.98 = 151.60 + 30.46 + 21.98 = 204.04\mathrm{dB}$$

$$(C)\mathrm{dB}w = (P_t)\mathrm{dB}w + (G_t)\mathrm{dB}i + 10\log A_r - 20\log R - 10.99$$

$$10\log A_r = (G)\mathrm{dB}i + 20\log\lambda - 10.99\mathrm{dB}\ \text{above 1 meter}^2$$

$$10\log A_i = 20\log\lambda - 10.99\mathrm{dB}\ \text{above 1 meter}^2$$

$$(C/N)^{\mathrm{dB}} = (\mathrm{EIRP})^{\mathrm{dB}w} + (G_r)_{\mathrm{dB}i} - (L_{fs})_{\mathrm{dB}} + 228.60 - 10\log T_s - 10\log B$$

Ionospheric Effects

Ionospheric effects in satellite communications can be divided into two categories – those resulting from the total number of electrons in the transmission path, and those caused by electron density irregularities (ionospheric scintillations). Both types are strongly influenced by time of day, season of the year, and phase of the solar cycle. A good modern reference on ionospheric behavior is Davies (1989).

Earth's Magnetic Field

The earth's magnetic field is quite complex near the surface (Davies 1989, section 2.4). However, as altitude increases, the field tends to approach that of an earth centered dipole inclined about 11 degrees from the axis of rotation and penetrating the earth's surface at 79.1 degrees north and 289.1 degrees east of Greenwich in the Northern hemisphere. The radial or vertical component Z of the field, considered positive if directed inward, is given by

$$Z = -2H_o(a/r)^3\cos\phi \quad (9)$$

The horizontal component H of the field is given by

$$H = H_o(a/r)^3 \cos\phi \tag{10}$$

and the total field F is given by

$$F = H_o(a/r)^3(1 + 3\sin^2\phi)^{1/2} \tag{11}$$

where:

H_o = 0.31 gauss = 0.31 × 10^{-4} tesla (the SI unit)
a = 6371 km is the mean radius of the earth
r is the distance to the center of the earth in km
ϕ is the geomagnetic latitude in degrees (i.e., the latitude oriented on the magnetic dipole axis instead of the rotational axis)

The geomagnetic dip I is given by

$$\tan I = Z/H = 2\tan(\phi) \tag{12}$$

Example: What is the total field F at 6.6 earth radii (geostationary orbit) at a longitude where the geographic and geomagnetic equators cross?

Solution: We take the value of H_o as 0.31 gauss or 0.31 × 10^{-4} tesla. Then

$$\begin{aligned} F &= \left\{0.31/(6.6)^3\right\}\left\{1 + 3\sin^2(0)\right\}^{1/2} \\ &= 1.08 \times 10^{-3} \text{ gauss} = 1.08 \times 10^{-7} \text{ tesla} \end{aligned} \tag{13}$$

Faraday Rotation

Faraday rotation refers to the twisting of the polarization axis of the radio wave as it passes through the ionosphere (Davies, 1989; Budden, 1985). At frequencies in the vicinity of or below the peak plasma frequency (usually between 1 MHz and 10 MHz) of the ionosphere encountered by the radio wave, the full Appleton Hartree expression (Davies 1989) for refractive index in a magneto-ionic medium must be used. At microwave frequencies, for all considerations except Faraday rotation, the refractive index simplifies to that of a non-magnetic medium. In the case of Faraday rotation the quasi longitudinal (QL) approximation to the Appleton-Hartree equation is used (this is good over 99 percent of the earth's surface at frequencies above 1 GHz). The QL approximation for the refractive index, n, is

$$n^2 = 1 - X/\left(1 \pm Y_L\right) \tag{14}$$

where:

X = $(f_c/f)^2$
R = ±1
Y_L = Y cos = (e/m) F cos $\theta/(2\pi f)$
f_c = 9 $(N_e)^{1/2}$ = plasma frequency, Hz
f = operating frequency
R = polarization (+l = right hand circular, -1 = left hand)
e/m is the charge to mass ratio of electrons
N_e is the electron density in electrons/m^3.

The plus and minus signs in the denominator of (14) indicate the signal divides into two counter-rotating circularly polarized components that travel at slightly different velocities upon entering the ionosphere and then recombine on exiting. The different velocities cause the changed polarization. One might expect that a signal rotated one direction in going from space to earth would unwind on the return trip. In fact the rotation is additive (Budden, 1985, #13.7).

The equation for Faraday rotation is

$$\psi = \left(2.36 \times 10^4 / f^2_{\text{Hertz}}\right)\int N_e B_L \, dl \text{ radians} \tag{15}$$

Example: Assume the ionosphere can be represented by a uniform slab of electrons 400-km thick centered at a height of 400 km and having a density of 10^{12} electrons m^{-3}. Assume B_L is given as 0.3 gauss and the operating frequency is 1 GHz. In this scenario,

$$\Psi = (2.36\ 10^4 / 10^{18})(4 \times 10^5)(3 \times 10^{-5})(10^{12})$$
$$= 0.283 \text{ radians} = 16.2 \text{ degrees.}$$

Excess Group Delay ΔR

The group range delay or excess group delay refers to the excess delay over that in free space and is made up of two parts (Allnutt, 1989; Davies, 1989), one due to the ionosphere and one part attributable to the troposphere (Flock, Slobin, and Smith, 1982). The ionosphere part of the range delay is given by

$$\Delta R = \frac{40.3}{f^2_{\text{Hz}}}\int N_e dl = \frac{40.3}{f^2_{\text{Hz}}}\left(\text{TEC}\right) \text{ meters} \tag{16}$$

where: $\int N_e \, dl$ = the total electron content (TEC) of the column between the space station and the earth station (or vice versa).

Example: Let f =1 GHz and TEC = 10^{18} electron/m^2. The excess range delay is given by

$$\Delta R = \frac{40.3}{f_{\text{Hz}}^2}\left(10^{18}\right) = 40.3 \text{ m} \tag{17}$$

Thus the equivalent free space path is 40.3 m longer than the actual path due to ionosphere.

TEC may vary by an order of magnitude diurnally. The troposphere delay is significant but is much less variable. The variable part is due largely to water vapor and may vary diurnally only 10-20 cm. However, the two-frequency technique used to assess the ionospheric term cannot be used for tropospheric excess range delay because the tropospheric refractive index is essentially non-dispersive.

Excess Phase Delay $\Delta\phi$

Phase change due to the ionosphere is negative while that due to the troposphere is positive. This is a consequence of the ionospheric refractive index being less than one while that of the troposphere is greater than one. The ionospheric reduction in phase path is given by

$$\Delta\Phi = \frac{8.44 \times 10^{-7}}{f_{\text{Hz}}} \times \text{ TEC radians} \tag{18}$$

Ionospheric Scintillation

Ionospheric scintillations are caused by irregularities in electron density of a few meters to a few kilometers. Scintillations affect amplitude, phase, and angle-of-arrival of radio waves (Davies, 1989; Basu et al. 1988). A useful index to compare scintillation data is the scintillation index, S_4, which is defined as the standard deviation of received power, $\overline{P}$, divided by the average value of received power, $\overline{P}$, (Briggs and Parkin, 1963). Thus:

$$S_4 = \frac{\left[\Sigma\left(P - \overline{P}\right)^2\right]^{\frac{1}{2}}}{\overline{P}} \tag{19}$$

For simplicity of scaling, a scintillation index (SI) has been defined (Whitney et al., 1969).

$$SI = Af^{-n} \tag{20}$$

where A is a proportionality constant and n varies between about 1.6 for low values of S_4 to 0.8 for high values.

Temporal and spacial signal impairment due to ionospheric scintillation is illustrated in *Figure 1* for solar maximum and minimum (Basu et al. 1988).

Tropospheric Effects

The troposphere is the atmospheric layer that extends from the earth's surface to a height of 12 km to 20 km in which weather phenomena take place. The three most important effects are refractive index effects (e.g., excess range), attenuation (absorption and scattering of the signal) and cross-polarization (e.g., frequency reuse).

Atmospheric Refractive Index

The most widely used expressions are as follows: (Smith and Weintraub, 1953; CCIR, 1990a; Flock, 1987):

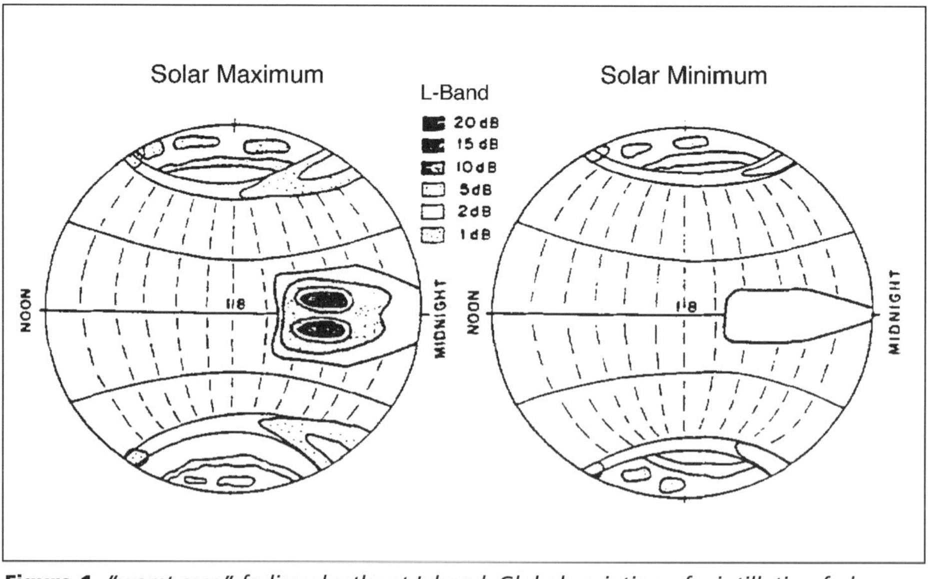

Figure 1 *"worst case" fading depths at L-band. Global variation of scintillation fades during solar maximum and solar minimum.*

Total $$N = (n-1) \times 10^6 = \frac{77.6}{T}\left(p + \frac{4810e}{T}\right) \quad (21)$$

Dry air $$N = 77.6\, P_d / T \quad (22)$$

Water vapor $$N = \frac{72e}{T} + 3.75 \times 10^5 \frac{e}{T^2} \quad (23)$$

where:

n is the index of retraction
N = $(n\text{-}1) \times 10^6$ is refractivity
T = is the temperature in kelvin
p = is the total atmospheric pressure in millibars (mb)
P_d = is the partial pressure of dry air in mb
e = is the partial pressure of water vapor in mb.

In a recent survey for his Ph.D. thesis, Solheim (1993) recommends the coefficients published by Bevis et al. (1993) which are not in conflict with (21). Other determinations of the coefficients have been made by Thayer (1974) and Hill et al. (1982). For a treatment of the complex refractive index, see Liebe (1989).

Attenuation Due to the Gaseous Atmosphere

Two major atmospheric gases produce significant absorption at frequencies between 1 GHz and 300 GHz. These are molecular oxygen and water vapor.

Oxygen

Oxygen (0_2) has a magnetic dipole moment. There is modest oxygen absorption at UHF/EHF as is illustrated in *Figure 2*. These merge into a single broad line at sea level but separate as the presssure decreases with altitude as can be seen in *Figure 3* (Liebe, 1985).

Water vapor

For water vapor, the specific attenuation, γ_ω, is shown in *Figure 2* for the absorption lines at 22.2 GHz, 183.3 GHz and 32S.4 GHz. An expression, including the quadratic dependence on water vapor density is given by the following: (Gibbons 1986; CCIR, 1990a):

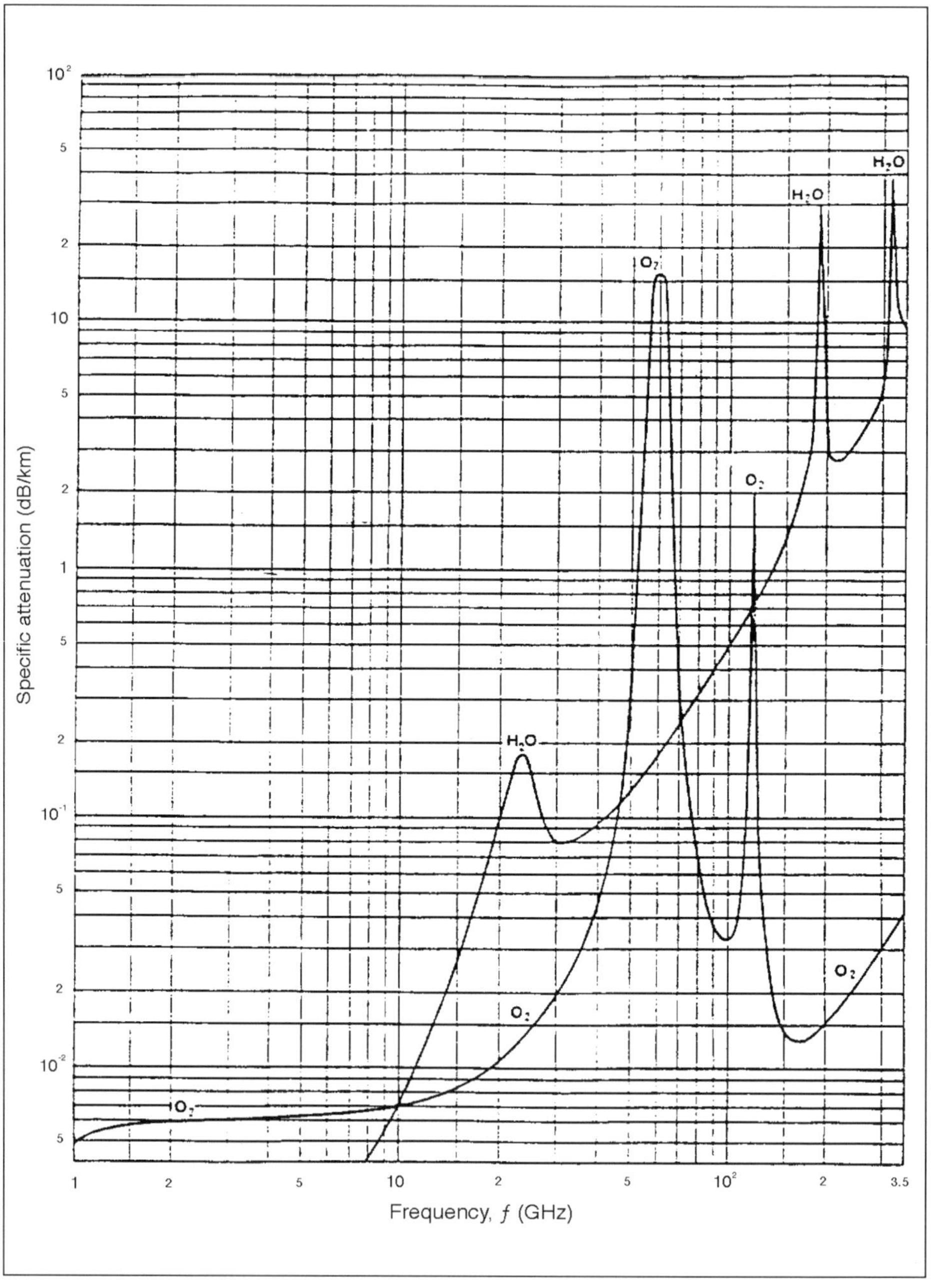

Figure 2 *Specific attenuation due to atmospheric gases. Pressure: 1013 mb; Temperature: 15°C; Water Vapor: 7.5 g/m*^{3}

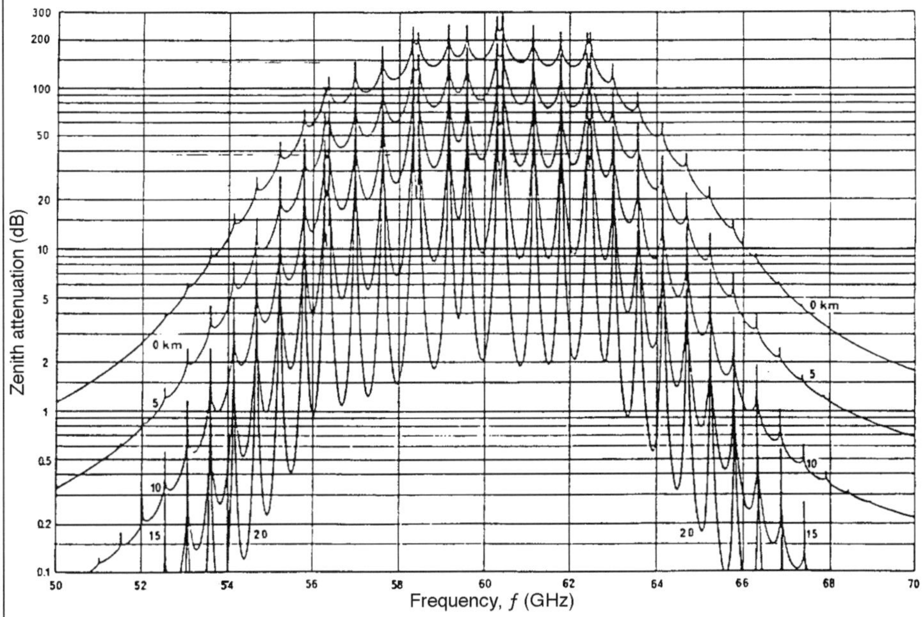

Figure 3 *Zenith oxygen absorption for some initial heights.*

$$\gamma_{\omega} = \left\{0.050 + 0.0012\rho + \frac{3.6}{(f-22.2)^2 + 8.5} + \frac{10.6}{(f-183.3)^2 + 9.0} + \frac{8.9}{(f-325.4)^2 + 26.3}\right\} f^2 10^{-4} \text{db/km} \quad (24)$$

where:

f is frequency in GHz

ρ is water vapor density in g/m^3

Conversion of water vapor partial pressure ***e*** in millibars (mb) into density ρ is given by $\rho = 216.5\, e/T$.

Conversion of relative humidity, H_{ω}, into water vapor density ρ can be obtained from:

$$H_{\omega} = 100 \frac{e}{e_s} \quad (25)$$

where e_s is the saturation water pressure over liquid water in millibars (Buck, 1981).

$$e_s = 6.1121\ \exp\left\{\frac{17.502\,t}{t + 240.97}\right\} \tag{26}$$

where: ***t*** is °C = T- 263.15.

$$\rho = \frac{2.167\,H_\omega e_s}{T} \tag{27}$$

Example: What is the specific attenuation at 30 GHz of water vapor at sea level if the relative humidity is 50% and the temperature is 75°F?

Solution:

$$t(°\mathrm{C}) = \frac{75 - 32}{1.8} = 23.9°$$

from (26)

$$e_s = 6.1121\ \exp\left\{\frac{17.502(23.9)}{23.9 + 240.97}\right\} = 6.1121\ \exp(1.5786)$$
$$= 29.63\ \mathrm{mb}$$

from (27)

$$\rho = \frac{2.167(50)(29.63)}{292.78} = 10.97\ \mathrm{g/m^2}$$

from (23)

$$\gamma_\omega = \left\{0.050 + 0.0021(10.97) + \frac{3.6}{(30 - 22.2)^2 + 8.5} + \frac{10.6}{(30 - 183.3)^2 + 9.0}\right.$$
$$\left. + \frac{8.9}{(30 - 325.4)^2 + 26.3}\right\}(30)^2(10.97)10^{-4}\ \mathrm{dB/km}$$
$$= \left\{0.050 + 0.023 + 0.052 + 4.5 \times 10^{-4} + 10^{-4}\right\}(0.9873)$$
$$= 0.123\ \mathrm{dB/km}$$

Note: *Figure 4* shows 0.08 dB/km for ρ = 7.5 g/m^3 and t= 15°C. As γ_ω is roughly linear with ρ to 12 g/m^3, extrapolating the value in *Figure 4* would yield a γ_ω of 0.117 dB/km.

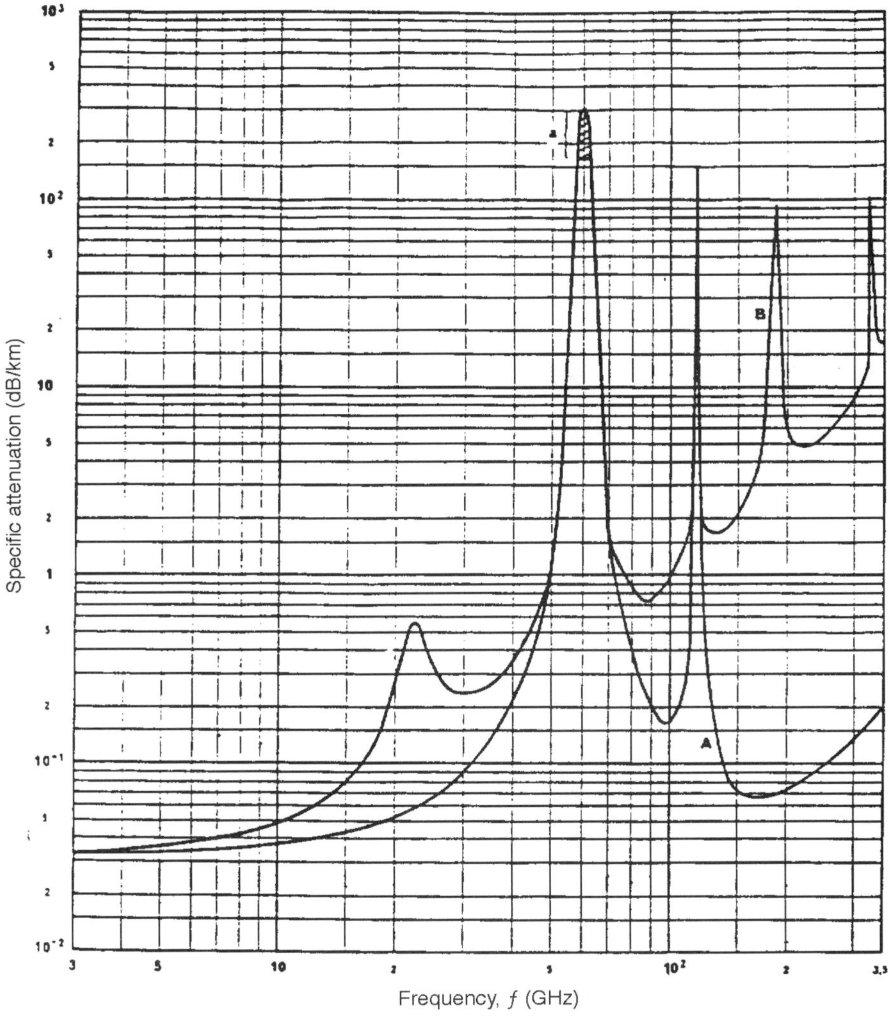

Figure 4 *Total zenithal attenuation at ground level. Pressure: 1013 mb; Temperature: 15°C; a: range of values (see Figure 3); A: for a dry atmosphere; B: with an exponential water vapor atmosphere of 7.5 g/m*^{3} *at ground level, and a scale height of 2 km.*

Attenuation Due to Rain

Rain consists of water droplets with diameters ranging from about 0.2 mm (drizzle) to 7 mm. Small raindrops (1mm) are spherical but large drops become flattened and acquire a dimple in the bottom-side. The best known studies of drop-size distribution as a function of rain-rate are Laws and Parsons (1943), Marshall and Palmer (1948), Joss et al. (1968), and Prupacher and Pitter (1971). Below 30 GHz attenuation is due almost solely to absorption. Above 30 GHz scattering becomes increasingly important. A distinction is then made between the absorbed (coherent) signal and the scattered (incoherent) component (Oguchi, 1986).

Predicting attenuation due to rain is an important but complex problem and a variety of models have been developed (Crane, 1985; Ippolito, 1989). The principal ones are the Rice-Holmberg Model (Rice and Holmberg, 1973) for percent of year in which rainfall exceeds a given rate, the Dutton-Dougherty model (Dutton and Dougherty, 1973; Dutton, 1977; Dutton, Kobayashi, and Dougherty, 1982), the global model (Crane, 1980), the two-component model (Crane, 1982), the Lin model (Lin, 1980; Ippolito, 1986), the simple attenuation model (SAM) (Stutzman and Dishman, 1982, 1984) and the CCIR model (CCIR, 1990b). The CCIR method will be outlined here. Specific attenuation due to rain is equally applicable to terrestrial and earth-space (slant) paths.

Rainfall Statistics

Specific attenuation due to rain is a function of frequency, rainfall rate, drop-size distribution, fall velocity and path orientation. Good reviews of the various aspects may be found in Allnutt [1989], Boithias [1987], CCIR [1990b], Flock [1987], and Ippolito [1986, 1989].

The CCIR proposes two methods for regions in which local rainfall statistics are not readily available (CCIR, 1990b). One method divides the world into 15 rain climate zones and specifies the rainfall intensity exceeded for 0.001 % to 1.0% of the year for each zone. The disadvantage of the zonal method is that boundary discontinuities of a factor of two may exist. The second so-called contour method is reproduced here. The land areas of the world are contoured for rain-rate exceeded 0.01 % of the year (52.56 minutes). The rainfall rates at other percentages of the year are then derived from this single value. This method has the advantage of continuity but does not allow for the variation of rainfall distribution with climate. For example, in the Midwest United States the rainfall is largely convective (thunderstorm related) while on the west coast the rain is largely stratiform and the cumulative distributions are very different, although the 0.01% rain rate may be the same. The CCIR rainfall contours for 0.01 % of the time are given in *Figures 5, 6,* and *7.* The curve parameter is rain rate in millimeters of rain per hour with an integration time of 1 minute. The contour maps in these figures come from Report 563-4 (CCIR, 1990b).

Step 1. Obtain the rain rate $R_{0.01}$ exceeded for 0.01% of the time from *Figures 5, 6,* and *7.*

Step 2. Compute the specific attenuation, γ_{ω} (dB/km) from the nomogram in *Figure 5.*

Step 3. Determine the specific attenuation A_p for the rainfall rate exceeded for the percent of time P desired by the following relation.

$$\frac{\gamma_P}{\gamma_{0.01}} = 0.12\ P^{-(0.546+0.043\log P)} \tag{28}$$

This formula gives factors of 0.12, 0.39, 1, and 2.14 for 1%, 0.1%, 0.01% and 0.001% of the time.

Example: What is the specific attenuation for rain exceeded 1% of the time for Boulder, Colorado, for horizontal polarization at 30 GHz?

Method:

Step 1. Obtain rain-rate from *Figure 4a* for 40°N, 105°W longitude. This is found to be 32 mm/h.

Step 2. From the nomogram in *Figure 8,* we obtain a value of 6.4 dB/km for 0.01 % of the time.

Step 3. From Equation (35) we deduce that for 1% of the time the factor is 0.12 of the attenuation for 0.01% of the time.

Hence: $\gamma_{1.0}$= 0.12(6.4) = 0.77dB/km

Path Attenuation

Heavy rain usually occurs in cells of a few kilometers in extent. Thus, the attenuation applicable for 0.01% of the time at a location should not be multiplied

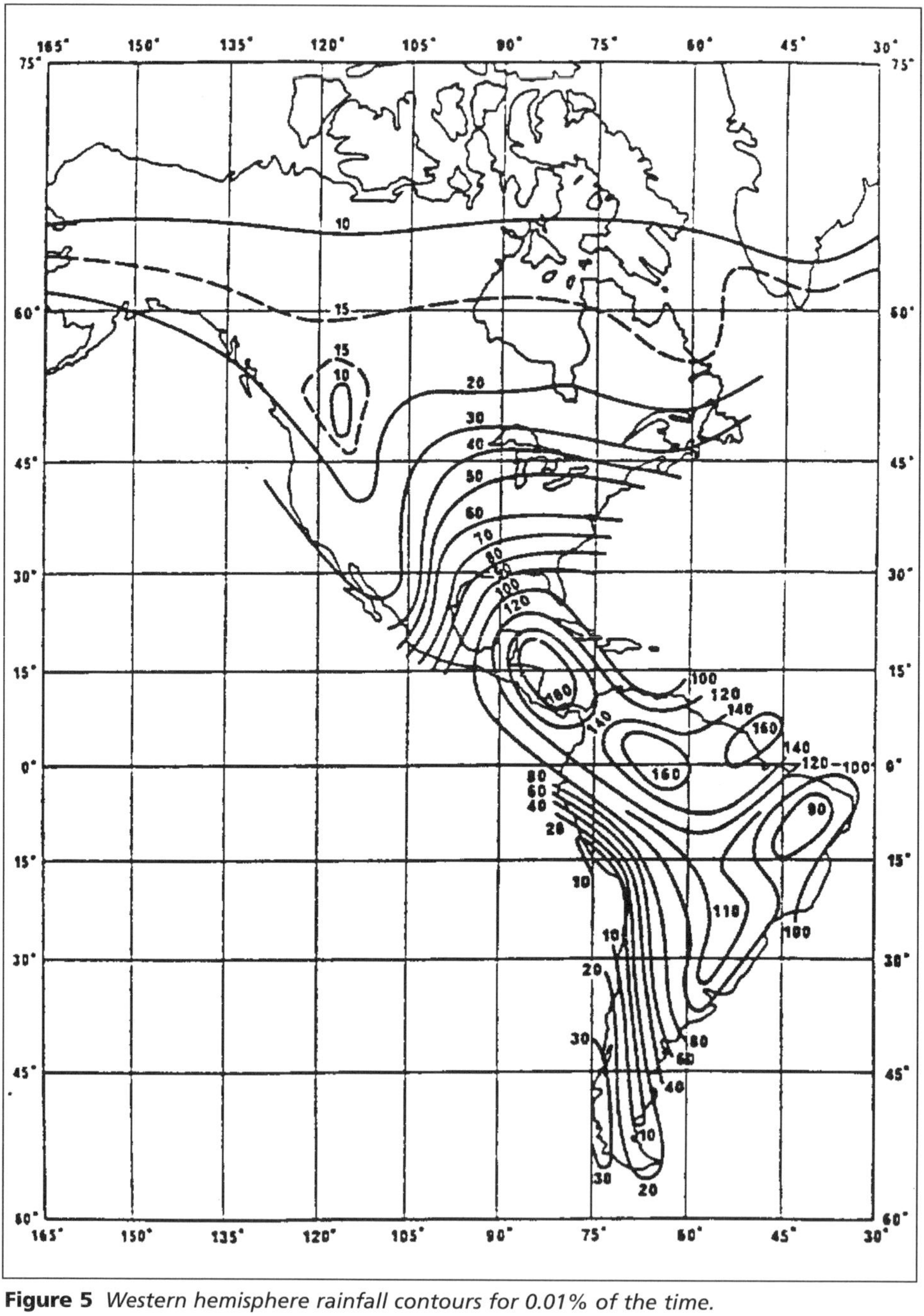

Figure 5 *Western hemisphere rainfall contours for 0.01% of the time.*

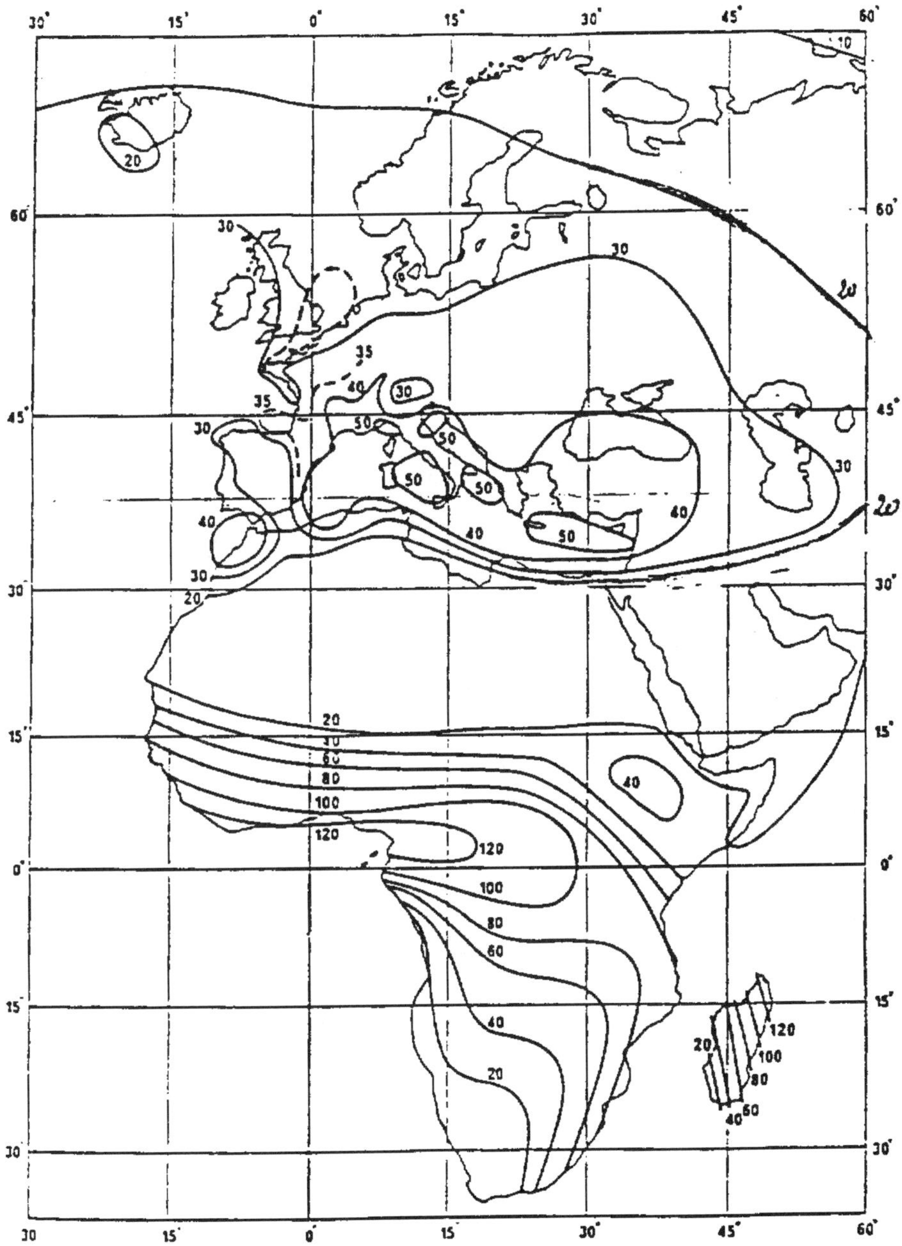

Figure 6 *Rainfall contours for 0.01% of the time in Africa, Europe, and the Middle East.*

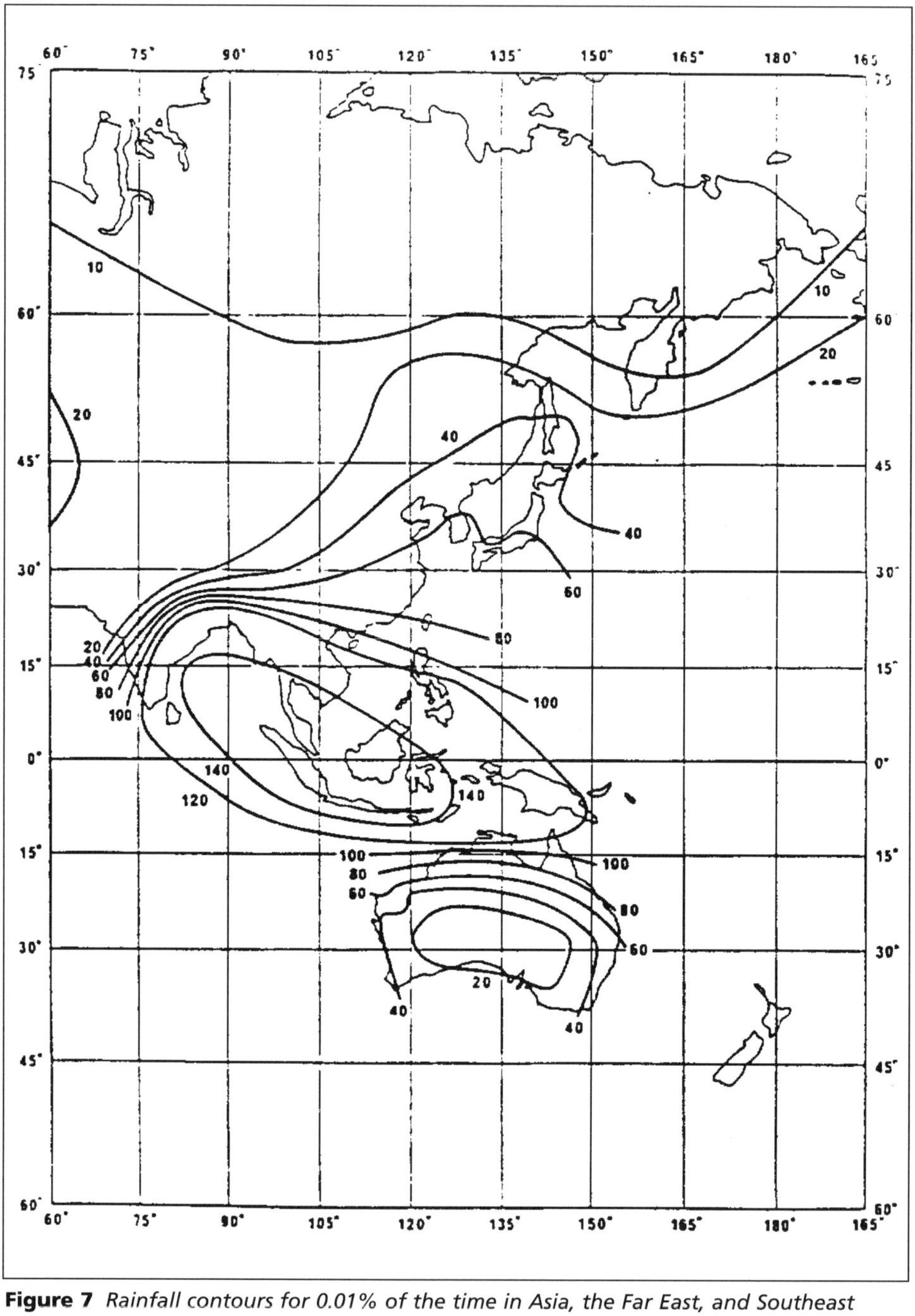

Figure 7 *Rainfall contours for 0.01% of the time in Asia, the Far East, and Southeast Asia.*

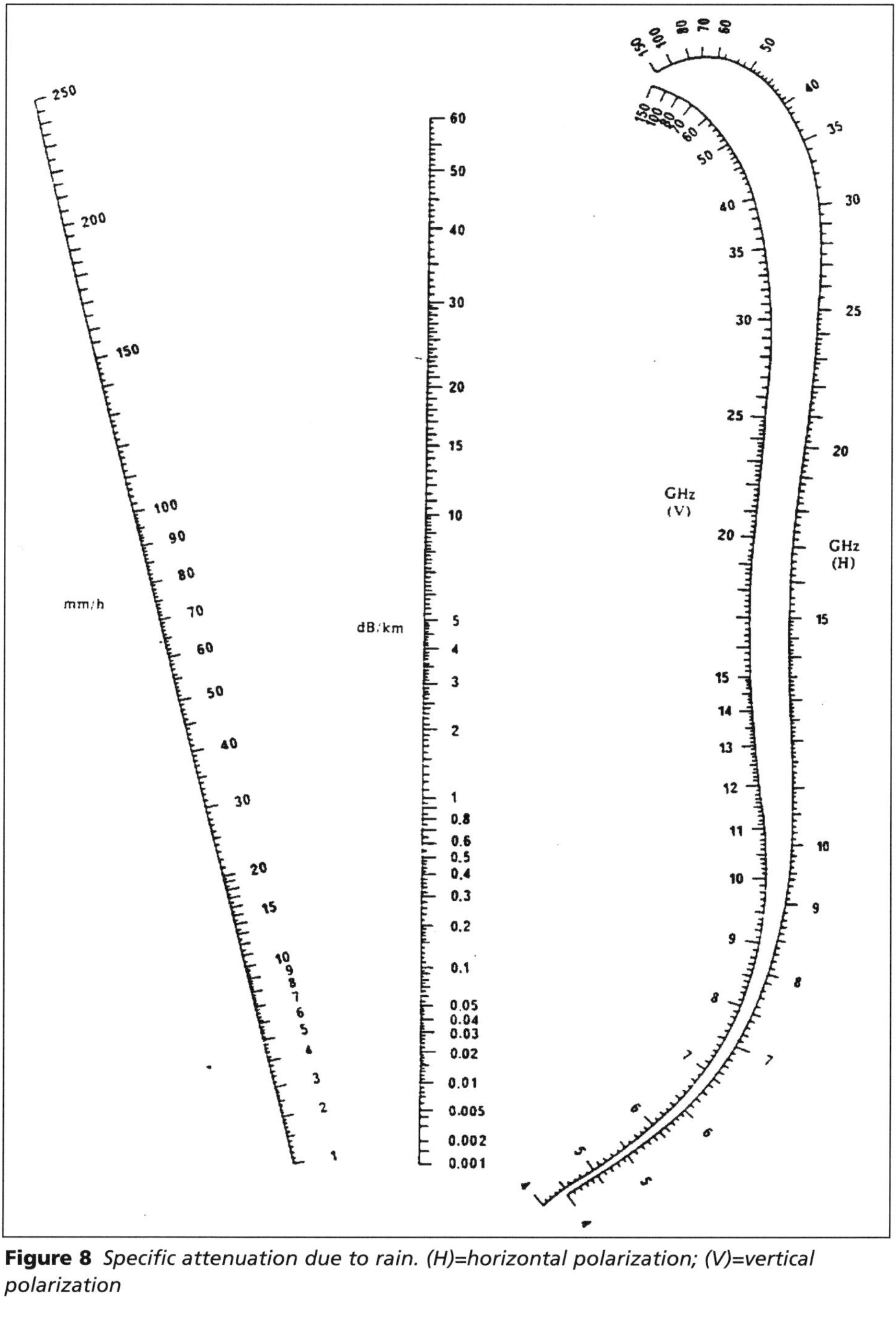

Figure 8 *Specific attenuation due to rain. (H)=horizontal polarization; (V)=vertical polarization*

directly by the path length but by a shorter length to account for lighter rain over some parts. The effective path length L_{eff} is obtained by multiplying the actual path length L by a path reduction factor r.

$$r = \frac{1}{1 + L/L_o}$$

where: $L_o = 35 \exp(-0.015 R_{0.01})$

Earth-Space(Slant) Paths

The procedure for earth-space paths for elevation angle $\theta > 5°$ is (CCIR 1990b, Report 564-4).

Step 1. Calculate the effective rain height, h_R for the station latitude ϕ from

$$h_R(\text{km}) = \begin{array}{ll} 30 + 0.028\phi & 0 \le \phi < 36° \\ 4.0 - 0.075\left(\phi - 36\right) & \phi \le 36° \end{array} \tag{29}$$

Step 2. Calculate the slant path length L_s in the rain

$$L_s = \frac{\left(h_R - h_s\right)}{\sin\theta}\ \text{km} \tag{30}$$

where h_s is the station altitude above sea level

Step 3. Calculate the horizontal projection L_G of L_S

$$L_G = L_S \cos\theta \tag{31}$$

Step 4. Calculate the path reduction factor $\boldsymbol{r}$ as before but substituting L_G for L.

$$r_{0.01} = \frac{1}{1 + L_G/L_o} \tag{32}$$

where $L_o = 35 \exp(0.015\ R_{0.01})$

Step 5. Calculate the path attenuation $A_{0.01}$ for 0.01% of the time

$$A_{0.01} = \gamma_R L_s\ r_{0.01} \tag{33}$$

Example: (Slant Path) What is the attenuation due to rain exceeded 0.01% of the time for an earth station in Boulder receiving a satellite transmission at a 30° elevation angle and a frequency of 30 GHz?

Method: From previous examples the specific attenuation due to $R_{0.01}$ at 30 GHz = 6.4 dB/km.

Step 1. $h_R = 4.0 - 0.075\ (40\text{-}36) = 3.7$ km

Step 2. Station altitude $h_s = 5400\text{ft} = 1.658$ km

$$L_s = \frac{(3.7 - 1.66)}{\sin 30°} = 4.08 \text{ km}$$

Step 3. $L_G = 4.08 \cos 30° = 3.53$ km

Step 4. $L_o = 35 \exp\ (0.015 \times 32) = 21.66$ km

$$r_{0.01} = \frac{1}{1 + 3.53/21.66} = 0.86$$

Step 5. $A_{0.01} = (6.4)\ (4.08)\ (0.86) = 22.46$ dB

Attenuation Due to Fog and Cloud

Both fog and cloud involve water (or ice) particles too small to precipitate (smaller than 0.17 mm). A relation proposed by Altshuler (1984) appropriate from about 30 GHz to 100 GHz is

$$A = -1.347 + 0.0372\lambda + \frac{18}{\lambda} - 0.022t \tag{34}$$

where:

A is attenuation in dB/km/g/m^3
λ is wavelength in mm
t is temperature in degrees C

Gerace and Smith (1990) have compared several cloud models.

Attenuation Due to Snow and Ice Crystals

Snow, as a consequence of its different complex refractive index, produces much less effect on attenuation than rain. An exception is at the melting layer (close to the 0° C isotherm) where snow (and ice crystals) becomes covered with a layer of water. Ice crystals (in ice clouds) can have a significant effect on depolarization of the signal however (Bostian and Allnutt, 1979).

Attenuation Due to Sand and Dust

Sand and dust particles in the air, particularly in the case of sand and dust storms in desert areas, are generally not significant at frequencies below 30 GHz. Theoretical work by Chu (1979) who related radio attenuation to optical visibility, measurements by Ghobrial (1980) in the Sahara and a review by Goldhirsh (1982) should be accessed by the interested reader.

Depolarization in Satellite Communications

Depolarization due to the ionosphere is treated under Faraday Rotation. Tropospheric depolarization can occur because of rain, ice crystals, or multipath. Good treatments may be found in Allnutt (1989), Ippolito (1986), Pratt and Bostian (1986) and others. The two terms most commonly used to express depolarization consequences are cross polarization isolation, *xpi*, and cross polarization discrimination, *xpd*.

Consider orthogonal transmitted fields "a" and "b". At a receiving antenna copolarized components a_c and b_c are received, but if there is depolarization in the propagation medium, some energy from "a" will show up on the "b" polarization as b_X. We define (Pratt and Bostian, 1986).

$$xpi = 20\log_{10}\left|\frac{a_c}{b_x}\right| \text{dB} \tag{35}$$

$$xpd = 20\log_{10}\left|\frac{a_c}{a_x}\right| \text{dB} \tag{36}$$

In most instances, the two terms *xpi* and *xpd* will be the same. Rain depolarization is caused by non-spherical raindrops which induce differential attenuation and phase shift in the orthogonal components of radio waves passing through them.

Tropospheric Scintillation

Tropospheric scintillation tends to be a significant problem only for elevation angles at the earth station of less than 5°. Thus, as most satellite circuits operate at higher elevation angles, tropospheric scintillation is rarely a problem.

Measurements have been made by Crane (1976) and theory developed by Tatarski (1967). Calculation methods for elevation angles above and below 4° are given in CCIR Report 564-4 (CCIR, 1990b).

Mobile Satellite Propagation Effects

There are special problems in the mobile-satellite services that are unique to them. For the land-mobile-satellite service, these have to do with the effects of foliage, terrain, and man-made structures. For the maritime mobile-satellite service, reflection and scattering from the ocean surface is the problem. In the aeronautical-mobile-satellite service, it is long delayed echoes from the earth's surface.

Land Mobile Satellite Service

Empirical results taken over a variety of terrain and foliage condition for elevation angles from 10° to 50° most notably by Vogel and Goldhirsh (Goldhirsh and Vogel, 1987; Goldhirsh and Vogel, 1988; Vogel and Hong, 1988) at UHF and L-hand are the most significant data in this area. The CCIR treatment is found in Report 1009-1 (CCIR, 1990d). Theoretical treatment is primarily based on Beckmann and Spizzichino (1963). At frequencies around 1.5 GHz, single-tree attenuations range from 10 dB to 20 dB, and 99% service typically requires a 10 dB attenuation margin. Ionospheric and tropospheric effects are as in other satellite services.

Maritime Mobile Satellite Service

The maritime situation is more amenable to analysis than the land case in that the direct signal is not obstructed and the ocean surface is easier to model. A two-step process is required (CCIR, 1990e). The amplitude of the coherently reflected component, E_r, relative to the direct signal is given by (Miller et al., 1984).

$$E_r = |R| G_r \left[\frac{-u^2}{2}\right] I_o\left(\frac{-u^2}{2}\right) \tag{37}$$

$$u = \frac{4\pi h_o \sin\theta_i}{\lambda} \tag{38}$$

where:

R is the specular reflection coefficient of the sea
G_r is the antenna gain toward the specular point relative to boresight gain G_o
I_o is the modified Bessel function of the first kind and of zero order
h_o is r.m.s height of the sea waves measured from the mean level
λ is the wavelength of the radio signal
θ is the grazing angle (taken as equal to the elevation angle)

Aeronautical Mobile Satellite Service

For the aeronautical service, the altitude of the aircraft may be above most tropospheric attenuation. Ionosphere effects will be the same as for other services. Important factors are high Doppler shifts due to high aircraft speeds, small antennas, and problems of visibility while banking. The analysis for sea-surface reflection for maritime satellite systems is appropriate for aeronautical satellite systems for over water flights. Measurements have been made by Sutton et al. (1973), and Zaks and Anderson (1986) and others; and simulations have been made by Davarian (1988). The subject is treated in CCIR Report 1148 (CCIR, 1990f).

Indoor Propagation

Propagation into buildings is of interest at microwave frequencies because of current satellite broadcasting plans for possible operation in the 1 Ghz to 3 GHz band and for portable or personal communication for intra-building communication. Penetration into buildings at 850 MHz is described by Walker (1983). Ground-floor penetration losses averaged 18.0 dB with a standard deviation of 7.7 dB for urban samples, and 13.1 dB with a standard deviation of 9.5 dB for suburban samples. CCIR treatments are found in Report 567-4, section 6 and 10 (CCIR, 1990g) which treats buildings, and Report 880-2 which concentrates on propagation in tunnels and mines (CCIR 1990h).

Noise

Noise at microwave frequencies is primarily of natural origin. Extraterrestrial noise of galactic and solar origin dominates the UHF part of the spectrum while noise of atmospheric origin dominates above 10 GHz (Smith, 1982). For a medium in local thermodynamic equilibrium, Kirchoff's Law states that emission is equal to absorption at all frequencies (Rybicki and Lightman, 1979). Report 720-2 (CCIR 1990e) treats natural noise at microwave frequencies.

Works Cited

J. E. Allnutt. "Satellite-to-Ground Radiowave Propagation - Theory, Practice and System Impact of Frequencies Above 1 GHz," *IEEE.* Peter Peregrinus, Ltd., 1989.

E. E. Altshuler. "A simple expression for estimating attenuation by fog at millimeter wavelengths," *IEEE Transactions Antennas and Propagation,* vol. 32, 1984, pp. 757-758.

S. Basu, E. MacKenzie, and S. Basu. "Ionospheric constraints on VHF/UHF communication links during solar maximum and minimum periods," *Radio Science,* vol. 23, 1988, pp. 363-378.

B.R. Bean, and E.J. Dutton. *Radio Meteorology,* Dover, 1968.

P. Beckmann and A. Spizzichino. *The Scattering of Electromagnetic Waves from Rough Surfaces,* Pergamon Press, The MacMillan Co., New York 1963.

M. Bevis, S. Businger, T.A. Herring, C. Rocken, R.A. Anthes, and R.H. Ware. "GPS meteorology: mapping zenith delays onto precipitable water," *Journal of Applied Meteorology,* vol. 33, 1993, pp. 379-386.

L. Boithias. *Radiowave Propagation.* McGraw Hill, 1987.

C.W. Bostian and J. E. Allnutt. "Ice-crystal depolarization on satellite-earth radio paths," *Proceedings of the IEEE,* vol. 126, 1979, pp. 951-960.

A.L. Buck. "New equations for computing vapor pressure and enhancement factor," *Journal of Applied Meteorology,* vol. 20, 1981, pp. 1527-1532.

K. G. Budden. *The Propagation of Radio Waves.* Cambridge University Press, 1985.

CCIR (1990). The International Radio Consultative Committee (CCIR) of the International Telecommunication Union (ITU) meets every four years in Plenary Session. Its conclusion in the form of *Recommendations and Repons* is issued about one year later. The XVIIth Plenary Assembly of the CCIR met in Dusseldorf in May 1990. Available through: ITU Geneva or National Technical Information Service, Springfield, VA 22161.

CCIR (1990a). *Reports of the CCIR* - Annex to Volume V - "Propagation in Non-Ionized Media" XVII Plenary Assembly of the CCIR (International Consultative Committee), Dusseldorf, May 21-June 1, 1990). Available through NTIS.

CCIR (1990b). Rain attenuation is drawn from Reports 563-4, 721-3, 338-6, and 564-4 of *Reports of the CCIR,* Annex to Vol. V, "Propagation in Non-Ionized Media," ITU Geneva.

CCIR (1990c). "Radio emission from natural sources in the radio frequency range above about 50 MHz," *Reports of the CCIR 1990,* Annex to Volume V, Report 720-2, pp. 205-225.

CCIR (1990d). "Propagation data for land-mobile satellite systems for frequencies above 100 MHz," *Reports of the CCIR 1990,* Annex to Volume V, Report 1009-1, pp. 536-519.

CCIR (1990e). "Propagation data for maritime mobile-satellite systems for frequencies above 100 MHz," *Reports of the CCIR,* 1990, Annex to Volume V, Report 884-2, pp. 515-535.

CCIR (1990f). "Propagation data for aeronautical mobile-satellite systems for frequencies above 100 MHz," *Reports of the CCIR,* 1990, Annex to Volume V, Report 1148, pp. 550-564.

CCIR (199Og). "Propagation data and prediction methods for the terrestrial land-mobile service using the frequency range 30 MHz to 3 GHz," *Reports of the CCIR,* Annex to Volume V, Report 5674, pp. 310-341.

CCIR (1990h). Short distance radio wave propagation in special environments - buildings, tunnels and mines, etc.," *Reports of the CCIR,* 1990, Annex to Volume V, Report 880-2, pp. 350-351.

T. S. Chu. "The effect of sand storms on microwave propagation," *BSTJ,* vol. 58, 1979, pp. 549-555.

R. K. Crane. "Prediction of attenuation by rain," *IEEE Trans. Comm.,* vol. COM-28. 1980, pp. 1717-1733.

R. K. Crane. "A two-component rain model for the prediction of attenuation statistics," *Radio Science,* vol. 17, 1982, pp. 1371-1387.

R. K. Crane. "Comparative evaluation of several rain attenuation prediction models," *Radio Science,* vol. 20, 1985, pp. 843-863.

F. Davarian. "Irreducible error rate in aeronautical satellite channels," *Electron. Lett.,* vol. 24, no. 21, 1988, pp. 1332-1333.

K. Davies. *Ionospheric Radio.* Peter Peregrinus, Ltd., 1989.

E. J. Dutton and H. T. Dougheny. "Modeling the effects of clouds and rain upon satellite-to-ground system performance," *NTIS* No. COM-75-10950, Springfield, VA, 1973.

E. J. Dutton. "Earth-space attenuation prediction procedures at 4 and 16 GHz," *Off. Telecom. Report,* (1977).pp. 77-123.

E. J. Dutton, H. K. Kobayashi, and H. T. Dougherty. "An improved model for Earth-space microwave attenuation distribution prediction," *Radio Science,* vol. 17, 1982, pp. 1360- 1370.

W. L. Flock. "Propagation Effects on Satellite Systems at Frequencies Below 10 GHz - A Handbook for Satellite System Design," 2nd ed., NASA Reference Publication 1108 (02) 1987.

W. L. Flock, S. D. Slobin, and E K. Smith. "Propagation effects on radio range and noise in earth-space telecommunications," *Radio Science,* vol. 17, Nov-Dec. 1982, pp. 1411-1424.

R. M. Gagliardi. *Satellite Communications.* Van Nostrand Reinhold, 1991.

G. Gerace, and E. K. Smith, "A comparison of cloud models," *IEEE Antennas and Propagation Magazine,* vol. 32, no. 5, Oct., 1990, pp. 32-38.

S.F. Ghobrial. "Effect of sandstorms on microwave propagation," *Proc. Nat'l Telecommunications Conference,* Houston, TX, vol. 2, 1980, pp 43.51-4.

C. J. Gibbons. "Improved algorithms for the determination of specific attenuation at sea level by dry air and water vapor, in the frequency range 1-350 GHz," *Radio Science,* vol. 21, 1986, pp. 945-954.

J. Goldhirsh. "A parameter review and assessment of attenuation and backscatter properties associated with dust storms over desert regions in the frequency range of 1 to 10 GHz, " *IEEE Trans. Ant. Prop.,* vol. AP-30, 1982, pp. 1121-1127.

J. Goldhirsh and W. J. Vogel. "Roadside tree attenuation measurements at UHF for land-mobile satellite systems," *IEEE Trans. Ant. Prop.,* vol. AP-35, 1987, pp.589-596.

J. Goldhirsh and W. J. Vogel. "Fade measurements at L-band and UHF in mountainous terrain for land-mobile satellite systems," *IEEE Trans. Ant. Prop.,* vol. AP-36, 1988, pp. 104- 113.

L. J. Ippolito. *Radio Wave Propagation in Satellite Communications.* Van Nostrand Reinhold, 1986.

L. J. Ippolito. "Propagation Effects Handbook for Satellite Systems Design - A summary of Propagation Impairments on 10 to 100 GHz Satellite Links with Techniques for System Design," 4th ed., NASA Reference Publication 1082 (04) 1989.

J. Joss, J. C. Thams, and A. Waldvogel. "The valiation of rain drop size distribution at Locarno," *Proc. Int'l Conf. on Cloud Physics,* 1968, pp. 369-373. Toronto.

Y. Karasawa, and T. Shiokawa. "Characteristics of L-band multipath fading due to sea surface reflection," *IEEE Trans. Ant. Prop.,* vol. AP-35, no. 8, 1984, pp. 956-966.

J. O. Laws, and D. A. Parsons. "The relation of rain drop size to intensity," *Trans. Amer. Geophys. Union,* vol. 24, 1943, pp. 43-460.

H. J. Liebe. "An updated model for millimeter wave propagation in moist air," *Radio Science,* vol. 20, no. 5, 1985, pp. 1069-1089.

H. J. Liebe. "MPM - An atmospheric millimeter-wave propagation model," *Int'l J. of Infrared and Mm Waves,* vol 10, no. 6, 1989, pp. 631-650.

S. H. Lin, H. J. Bergmann,and M. V. Pursley. "Rain attenuation on Earth-space paths - Summary of 10-year experiments and studies," *BSTJ,* vol. 59, 1980, pp. 183-228.

J. S. Marshall and W. M. Palmer. "The distribution of raindrops with size," *J. Meteorology,* vol. 5, 1918, pp. 165- 166.

A. R. Miller, R. M. Brown, and E. Vegh. "New derivation for the rough surface reflection coefficient and for the distribution of sea-wave elevations," *Proc. IEEE,* vol. 131, Part H, no. 2, 1984, pp. 114-116.

K. Miya, ed. *Satellite Communications Engineering*, Lattice Co., Tokyo, 1975.

W. A Morgan, and G. O. Gordon. *Communications Satellite Handbook.* Wiley Interscience, 1989.

T. Oguchi. "Effects of incoherent scattering on attenuation and depolarization of millimeter and optical waves due to hydrometeors," *Radio Science,* vol. 21, 1986, pp. 717.

R. L Olsen, T. Tjelta, L. Manin, and J. E. Doble. "Towards a more accurate method of predicting the distribution of multipath fading on terrestrial microwave links," *Electron Lett.,* vol. 22, Aug. 1986, pp. 902-903.

F. C. Owen, C. D. Pudney. "Radio propagation for digital cordless telephones," *Elec. Lett.,* vol. 25, No. 1, Jan. 1989.

J. D. Parsons. "Propagation effects on mobile services," IEE Electromagnetic Wave Series 30, Chap. 30, 1989, pp. 262-278. Peter Peregrinus, Ltd.

T. Pratt, C. W. Bostian. *Satellite Communications.* John Wiley and Sons, Inc., 1986.

H. R. Prupacher, R. L. Pitter. "A Semi-empirical determination of the shape of drops," *J. Atmos. Science,* vol. 28, 1971, pp. 86-94.

P. L. Rice, N. R. Holmberg. "Cumulative time statistics of surface point rainfall rates," *IEEE Trans. Com.*, vol. Com-21, 1973, pp. 1131-1136.

D. Roddy. *Satellite Communications*. Prentis Hall, 1989.

G. W. Rybicki, A. P Lightman. *Radiative Processes in Astrophysics*, Wiley Interscience, 1979.

E. K. Smith and S. Weintraub. "The constants in the equation for atmospheric refractive index at radio frequencies," *Proc. IRE*, vol. 41, 1953, pp. 1035-1037.

E. K Smith. "Centimeter and millimeter wave attenuation and brightness temperature due to atmospheric oxygen and water vapor," *Radio Science*, vol. 17, No. 6, 1982, pp. 1455-1464.

F.S. Solheim. (1993). "Use of pointed water vapor radiometer observations to improve vertical GPS surveying accuracy," Ph.D. thesis, Department of Physics, Univ. of Colorado. Boulder. CO 80309

A. D. Spaulding, G. H. Hagn, and J M. Goodwin, ed. "Worldwide minimum environmental radio noise levels (0.1 Hz to 100 GHz)," Proceedings of the Symposium on the Effects of the Ionosphere on Space and Terrestrial Systems, ONR/NRL, Jan. 1978, pp. 177- 182.

W. L. Stutzman, and W.K. Dishman. "A simple model for the estimation of rain-induced attenuation along Earth-space paths at millimeter wavelengths," *Radio Science*, vol. 17, 1982, pp. 1465- 1476.

W. L. Stutzman and Dishman. "Correction to 'A simple model..."' *Radio Science*, vol. 19, 1984, pp. 946.

R. W. Sutton, E. H. Schroeder, A. D. Thompson, and S. G Wilson. "Satellite aircraft multipath and ranging experiment results at L-band," *IEEE Trans. Comm.*, vol. COM-21, no. 5, 1973, pp. 639-647.

G. D. Thayer. "An improved equation for the radio refractive index of air," *Radio Science*, vol. 8, no. 10, Oct. 1974, pp. 803-809.

W. J. Vogel, U. S Hong. "Measurement and modelling of land-mobile satellite propagation at UHF and L-band," *IEEE Trans. Ant. Prop.*, vol. AP-36, 1988, pp. 707-719.

E. H. Walker. "Penetration of radio signals into buildings in the cellular radio environment," *BSTJ*, vol. 62, no. 9, 1983, pp. 2719-2734.

H. E. Whitney, J. Aarons, and C. Malik. "A proposed index for measuring ionospheric scintillations," *Planet Space Sci.*, vol. 17, 1969, pp. 1069.

J. H. Yuen, ed. *Deep Space Telecommunications Engineering*, JPL, NASA, 1982.

C. Zaks, and S. Anderson. "Aeronautical Satellite data link measurements over the North Atlantic: Test Results," Proceedings 7th Int'l. Conf. on Digital Satellite Communications, Munich; FRG, May, 1986, pp. 557-563.

Section III

Wireless Outside the United States

Wireless Local Loop for Developing Countries – A Technology Perspective

Arogyaswami Paulraj
Professor, Information Systems Laboratory
Stanford University

Abstract

Wireless local loop (WLL) is emerging as an important technology to meet the rapidly growing demand for telephone services in developing countries in Asia and Latin America. This essay surveys some technology choices for WLL in such applications.

Introduction

The world's telecommunication system is by far its most complex technological machine. In less than 120 years, since Alexander Graham Bell invented the telephone in 1876, the telephone network has expanded to 800 million phones worldwide and carries more than 600 billion calls per year. Major advances in satellite and fiber optic transmission have dramatically reduced the cost of telephone service.

Despite these advances, large segments of the world's population, more than three billion people, living in several parts of Africa, Asia, and Latin America have comparatively little access to telecommunications.

India for example has seven access lines per 1000 people as compared to a world average of 255 access lines and the U.S. level of over 550 access lines. ITU has set a teledensity target of 200 access lines per 1000. Therefore, there is a tremendous gap in teledensity that needs to be bridged quickly. Countries such as India, China, and others in the Asia-Pacific region have rapidly growing economies and urgently need expansion of telecommunications, especially of basic telephony services. Many of these countries, notably India, have now begun to permit competition even in basic services and have begun to attract a flood of foreign and local investments, hopefully correcting decades of neglect of this vital sector.

Wireless local loop (WLL) refers to the use of wireless access to provide the basic telephone service. WLL provides a radio link from the subscriber terminal to base station, which in turn is connected to the PSTN network through a switch. WLL is attracting considerable attention in developing countries since it appears to offer several advantages over the traditional wired telephone access. The standard wire line local loop consists of distribution cables, splice boxes, and feeder plant or drop lines to connect the local switch to the customers. This copper-based technology has essentially remained unchanged for over 100 years, since there was no effective alternative technology to copper. Use of digital loop carriers in the wireline plant is about the only modern technology to be introduced in recent years. WLL finally appears to be challenging this situation by providing a completely new alternative to copper. Wireless access still remains a small fraction of the total telephone business. Although there are 38 million cellular users today, nearly 96 percent of the current telecom revenues remains in wired services. However, studies suggest that wireless revenue is likely to reach 25 percent of total telecom revenue within the next decade. The new wireless services include mobile cellular (service to mobiles), lower tier PCS (indoor and outdoor service to pedestrians), and WLL (service to static phones). In this essay, we review the types of WLL markets and their needs in developing countries and discuss the technology options available.

Telephony Markets in Developing Countries

The telephone market in developing countries spans several segments. It includes: (a) increasing basic telephone services penetration in large area urban population centers, including competition with existing wired loop service, (b) local coverage for high density, small area telehubs, such as shopping complexes and train/bus stations, (c) providing rapid roll out telephone services in medium size towns and growth centers, and (d) finally expanding the coverage to widely dispersed villages that have yet to get their first telephone. These market scenarios cover a variety of subscriber density, traffic levels, and area coverage, necessitating specific approaches for each market. *Figure 1* illustrates these emerging telephone markets in terms of these three parameters. We should note that the rural telephone market in developing countries is different from the rural market in the developed world

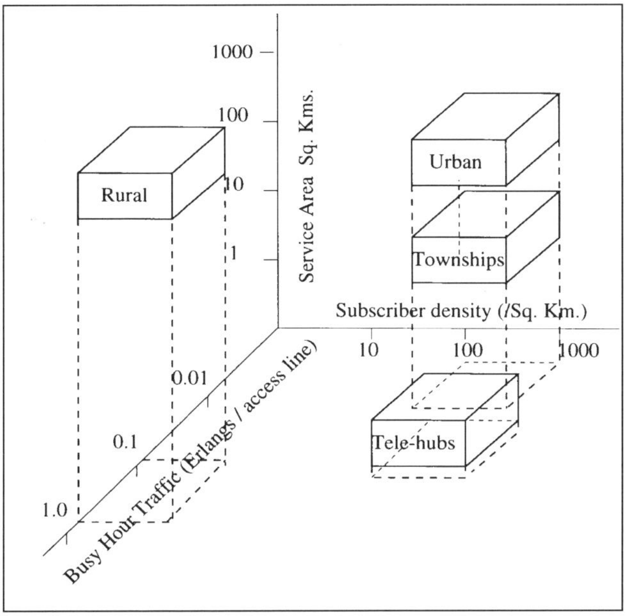

Figure 1 *Wireless local loop market structure*

since the subscriber density is much lower, with usually only one phone per village. This often means high traffic density (0.3 to 0.7 Erlangs) and low area density (0.1 to 1 phone per sq. km.) coupled with large area coverage (hundreds of sq. km.). The urban telephone market, on the other hand, is similar to the standard pattern with high subscriber density (100 to 1000 per sq. km.), large area coverage (1000 sq. km.), and low traffic density (0.05 to 0.08 Ehrlangs). Townships and growth centers are similar to urban networks, except that traffic density and area coverage can be somewhat smaller and with subscriber density at about the same level. Telehubs are needed to serve extremely high density locations, such as crowded markets and public places, with customer base similar to the current CT-2 markets in Hong Kong and Singapore. The traffic density for shared phones can be quite high (0.3 Ehrlang), subscriber density is likely to be very high (greater than 1000 per sq. km.), and area coverage needed is about 1-2 sq. km.

Advantages of WLL Technology in Developing Countries

WLL can play a significant role in developing countries, since it has several promising features:

Rapid Deployment

WLL enables deployment of a network in weeks as opposed to the years needed to install a wired loop plant. Many countries have realized the enormous potential of the telephone to stimulate economic growth and are anxious to hasten telephone penetration. Also, in many developing countries, there is inadequate city planning records and archival data, which means that laying down a wired plant is cumbersome and prone to unexpected delays.

Lower Lifecycle Costs

Lifecycle costs include the initial capital cost and maintenance costs. The per-line capital cost for wired loop runs about $500-1500 on the average, while costs in low density areas can run two to three times higher. The per-line costs are fairly sensitive to the subscriber density and the coverage diameter. WLL costs are estimated in the region of $650-1000 per line installed, excluding any customs duty. Also, the per-line WLL costs are not as sensitive to subscriber density and cell sizes as in wired loops. Precise WLL costs will depend on the WLL technology employed.

A point to note about WLL capital costs is that they are mainly embedded in the electronics and therefore can be expected to drop with time. For example, the per-line costs in cellular systems have dropped between 40 and 50 percent over the past 10 years and we can expect this to continue. On the other hand, costs in the wired loop plant may not show such a strong downward trend since much of the costs associated with the wired loop plant are labor costs, which may in fact rise over time. However, the increasing use of digital loop carrier and distributed switching technologies in the wired loop can lower prices over time.

Another factor in favor of WLL is that the operating costs are expected to be lower in WLL due to lower potential for trouble reports. In countries such as India, the trouble reports per access line are high, with areas reaching as many as 20 calls per year. These problems arise from inadequate maintenance of the loop plant, low quality of engineering and cables, and the high rate of accidental or water damage to cables. WLL can be expected to be free of these problems, therefore WLL may have significantly lower operating costs. However, the higher skill levels needed for radio maintenance personnel as compared to the wireline personnel may offset some of this advantage.

Yet another factor that might give a cost advantage to WLL is the lower investment cost granularity versus higher granularity required for wired loop plants. Therefore, larger financial exposure becomes likely in wired loop plants if demand drops below forecast. In contrast, WLL's smaller granularity allows investment to better match true demand.

In summary, WLL appears to have many potential advantages over the wireline loop in lifecycle costs and speed of deployment. Fair comparative assessments should take into account the network architecture, type of technology used, and market growth projections.

Service Quality

Service quality is measured in terms of voice quality, setup delays, call blocking probability, and access reliability. WLL service quality should be close, or equal to good wireline service levels. Initially, customers may be willing to accept a slightly lower service quality due to the fact that many have been without any service whatsoever for several years, and the wired service, if any, has been substandard. Eventually, however, demand for better quality will follow, and in a competitive access environment, high service quality is essential for customer loyalty.

While achieving wireline (toll) quality may be easier in microcell-based technologies, the macrocell systems currently used in mobile applications have generally poor voice quality. The exception may be CDMA-based systems that have much greater immunity to fading. Service quality requirements need special attention if macrocell technologies are adapted for WLL.

Service Breadth

Though the initial demand in WLL systems is for basic voice telephony, later fax service will also be of high priority. Data service is less important, except in some metro markets. Therefore, WLL should offer voice and fax universally, and perhaps limit inband data and/or enhanced ISDN services to large metro markets. In systems using voice coders, auto bypass of fax and data signals will be needed for successful operation. This issue needs further study if GSM, DAMPS, and IS-95 are to be adapted for WLL. Finally, paging and short message service should be incorporated into the WLL system for service to static phones and mobile paging units.

Interworking and Use of In-Place Infrastructure

Since the WLL market is highly cost sensitive, reducing overall system costs is important. Therefore, maximum use of the in-place network infrastructure should be made and dedicated backhaul networks and switches should be minimized. This requirement again implies some adaptation to the network interfaces of current cellular systems.

Spectrum Efficiency

In high-density urban areas, busy hour traffic may reach 200-500 Erlangs per sq. km. This can stretch even the most generous of spectrum allocations, and hence high spectrum efficiency is critical. Moreover, the usage in urban business areas can double as economic activity expands. While higher traffic densities can be achieved by using smaller cells, the high cost sensitivity in WLL makes this unattractive. Therefore, maximizing the traffic supported per base station per unit bandwidth is critical. Moreover, in many developing countries, due to the highly diversified purchases of military communication equipment, the military spectrum allocation is often fairly dispersed and fragmented, further reducing the spectrum availability and contiguous bandwidths for civilian applications. Also, the fixed cost in a base station is about 30 percent of the fully loaded cost. Therefore, high spectrum efficiency implies lower per line costs for the WLL customer. Higher spectrum efficiency of course drives the WLL towards higher technology options and therefore perhaps higher costs.

While techniques, such as voice coding, spread spectrum multiple access (CDMA), and smart antennas can all significantly improve spectrum efficiency, they may also drive up per unit base station costs. However, the per subscriber cost is usually better under full loading, and lifecycle costs reduce, thus improving the cumulative debt balance.

Multiposition Service

Again due to the cost sensitivity of many emerging WLL markets, significant sections of subscribers, i.e., small business persons, might be interested in using a single phone to support access from the home, office, and even while on the move. WLL architectures that allow such flexibility will have a competitive advantage. Also, such temporal reuse cuts down overall infrastructure costs, since the same infrastructure is reused several times a day, unlike the current wired infrastructure where the network resources (switches, local loop, and cross connects) are duplicated or triplicated to support communications phases at home, office, and on the move. Also, if wireline infrastructure is eventually laid in place, it will be desirable to have the WLL service transition into a primarily mobile service, thus savings cost for new infrastructure.

Security and Authentication

Early cellular systems, such as AMPS and TACS, did not provide voice security and this precluded its use for sensitive business discussions, which had to be postponed until a wireline connection could be established. Since WLL phones are the primary business communication vehicle, it is essential that WLL provides acceptable voice security. Emerging cellular standards have improved security levels and further enhancements are planned. WLL technology can benefit from these developments.

Authentication to prevent fraud is another major problem in cellular systems. These problems remain equally significant in WLL systems and need to be properly addressed.

Technology Comparison

There are several competitive technologies for WLL. We will limit our discussions to terrestrial options. The alternative architectures include: adaptation of macrocellular-based systems originally designed for service to mobiles, microcellular-based systems originally designed for service to pedestrians (both indoor and outdoor), and trunked radio systems. These options are briefly discussed below.

Macrocellular Systems

Macrocellular systems (DAMPS, AMPS, IS-95, and GSM), originally designed for service to mobiles, feature large cells with cell radii up to 35 km. Several adaptations are needed in these systems to make them useful for WLL. Mobile features, such as hand-off and roaming (HLR and VLR support), are not needed in a strictly static WLL network. This implies simpler protocols and switch interfaces. The switch itself will resemble more a local office switch rather than the more complex MTSO. Also, the subscriber terminals can use directive antennas (Yagi arrays with gain of 9-12 dB are typical). Furthermore, subscriber terminals can also use antenna diversity on the rooftop antenna. These features can of course improve terminal range or reduce transmit power on both up and down links. Even more significant is the reduced interference potential of the subscriber terminal into neighboring base stations, therefore improving frequency reuse factors and spectrum efficiency. In CDMA systems, directional antennas increase capacity by about 20 percent. Another WLL advantage is that higher antenna heights (10 to 200 feet) can result in lower path-loss and therefore lower transmit power. *Figure 2* shows

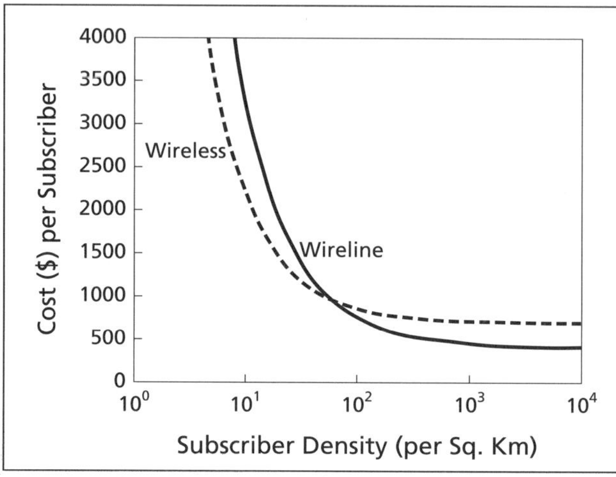

Figure 2 *Cost comparison for wireline and wireless – large cells*

a comparison of per-line capital costs for typical macrocell WLL systems and wired local loops. The cost model used is fairly typical, but variations must be expected depending on the network, technology, and market specifics. Examining *Figure 2* shows that WLL has cost advantages in large macrocells for subscriber densities below 80 subscribers per sq. km. For higher densities, wireline has an increasing cost advantage. This makes WLL advantageous for rural and low-density urban (business only) markets.

In macrocellular systems, the transmission design features adaptive equalization and forward error control to manage fading and high speed mobility. The higher link margins possible in WLL and the availability of antenna diversity at the subscriber unit (in addition to antenna diversity at the base station) reduces the need for these more expensive transmission methods.

Finally, a big advantage of adapting macrocellular systems for WLL is the accumulated experience in deploying these cellular systems. Also, high production volume also makes for lower costs.

Microcellular Systems

Microcellular systems (DECT, PHP, WACS, or CT-2+) were designed for small cell indoor and outdoor coverage with cell radius not exceeding 500 meters. Larger cell sizes (through directive antennas and higher transmit power) will not be

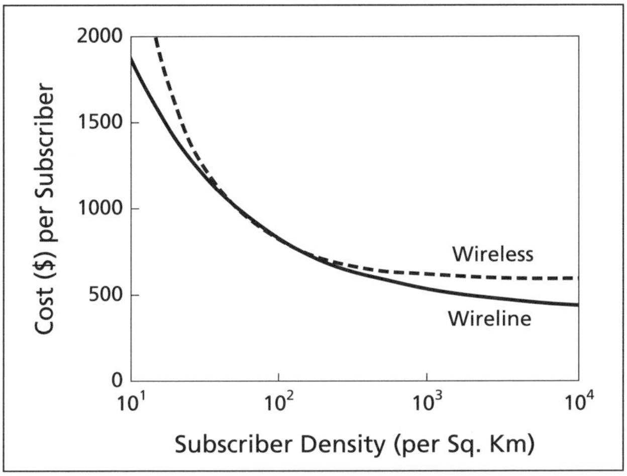

Figure 3 *Cost comparison for wireline and wireless – small cells*

possible beyond one kilometer due to the lack of equalization in the transmission. The voice quality in these networks is typically good and they are designed for distributed interfaces to the PSTN network – both desirable WLL features. Microcellular WLL therefore has some market niches. However, their limited coverage makes them unsuitable for many large urban and rural markets. High density telehubs and small town applications may be a better fit for this technology. While several DECT-based systems and even CT-2 + systems are currently being evaluated for WLL applications, there are new technologies emerging in the U.S., such as PACS (hybrid PHP and WACS), Omnipoint 2000, and the Ultraphone that may also offer WLL potential.

Figure 3 shows a comparison of per-line capital costs for a typical microcell-based WLL system with the wired local loop. We notice that in small cells, wireless does not have a cost advantage over wireline networks and perhaps can only be justified if the advantage of mobility is included.

Trunked Radio Systems

A number of low-capacity trunked point-to-multipoint systems (MARR, BETRS) have been fielded over the years, mainly for rural service. These systems (usually FDMA/FM) have excellent range but poor spectrum efficiency and also cannot support any local mobility. They appear to hold promise if they can be hybridized with other microcell systems, such as DECT, to provide local mobility.

Multiple Access and New Technologies

Choice of multiple access (MA) method for WLL is a contentious problem just as it has been in the cellular arena. While it is probably true that no MA approach has any fundamental theoretical advantage in spectrum efficiency, the key issue is which MA technique offers advantage in building spectrum-efficient systems that are also low cost. Though currently CDMA appears to have an advantage here, it is possible that future TDMA designs may have comparable cost-effectiveness.

The two new technologies that are expected to make substantial impact on WLL network performance and/or economics are the wideband transreceivers and smart antennas. Wideband transreceivers allow digitization further up the receiver chain, allowing most if not all of the channelization to be done digitally, thus driving down costs. Once most of the processing at the base station is digital, the base station can become largely integrated onto silicon and be made software definable, initiating economies of scale and of learning curves typical in silicon-intensive systems.

Smart antennas are another emerging technology. The technique exploits multiple antenna arrays at the base station. Smart antennas at the mobile is not considered cost-effective, at least for now. In a smart antenna-enhanced base station, the signals emerging out of the antenna array or going into the antenna array are processed using sophisticated algorithms. The basic goals of smart antennas are to improve one or more of the following performance indices: coverage, capacity, and voice quality. The improvement in coverage and capacity trade into better network economics. Smart antenna technology involves adaptive transmission and reception, where directive beams are formed in both forward and reverse links. The problem of beam forming is complicated by the difficult propagation environment that introduces a variety of fading and spreading in the channel, requiring sophisticated signal processing.

Figure 4 shows the projected improvement in network economics from the use of smart antennas in large- and small-cell WLL systems. Use of antennas appear to promise about a factor of two advantages in per-subscriber costs in large cells, with a much smaller advantage in small cells.

Conclusion

WLL is at the threshold of tremendous growth worldwide. It offers many cost advantages over wired local loop and it will likely be the technology of choice for

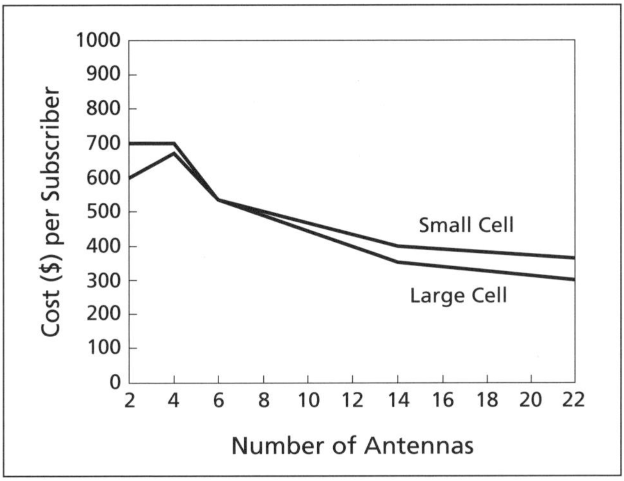

Figure 4 *Network economics using smart antennas*

basic telephone service in many developing countries. Network economics is a key element for WLL acceptance, and it is likely that there will be many different WLL solutions depending on the exact nature of the WLL market. In the short run, adaptation of upper and lower tier versions of cellular mobile and PCS services may dominate WLL technology. However, a ground up design of WLL technology may emerge soon to optimize cost and performance.

Bibliography

C. Clarkson-Short, "Economics of the local loop competition by radio access," *IEE Conference on Communications,* 11-14 October 1993, Edinburgh 1993.

Federal Communications Commission Docket 90-314, Notice of Enquiry on PCS, Oct. 1991.

G. Hawley, "Historical perspectives on the US telephone loop," *IEEE Comm.,* March, 1991.

Liang and A. Paulraj, "Coverage extension using smart antennas," Submitted to IEEE Dual-Use Technologies and Applications Conference, 1995.

R.J. Saunders, J.J. Warford, and B . Willenius, "Telecommunications and Economic Development," A World Bank Report.

Wireless Communications in Russia

Olga Mnatsakanova
CIS Analyst
Innova International Corporation

Wireless technology always seemed to be a good solution for Russia, with its large territory, severe weather conditions, and the present impaired state of telecommunications infrastructure development.

The total area of Russia is 17,026,280 square kilometers (6,592,800 square miles), which makes it the largest country in the world by area. Russia is the sixth largest country in the world by population after China, India, the United States, Indonesia, and Brazil. Of the approximately 149 million Russians, 74 percent are urban inhabitants and 26 percent are rural inhabitants.[1]

Compared to other eastern European countries' telephone penetration rates, Russia's is one of the lowest in the region. Russia has 11 telephones per 100 persons in urban areas and 8.78 phones in rural regions.[2] There are 47.5 million phones installed and 50 million phones urgently needed. The average waiting time for a line is 30 years. More than 150,000 villages and 36,000 summer settlements in Russia do not have any phones.[3] Only 67 percent of enterprises and 85 percent of hospitals in rural areas have telephones.

Quality of service is very low because approximately 57 percent of Russia's telephone lines were laid more than 20 years ago. Local calls can take 10 minutes to connect, and interurban calls generally need the intervention of an operator.[4] International gateways were built only in Moscow to enable the government to control information flow and communications with the West. International calls from all republics were purposely routed through one location with only one international switch located in Moscow. There was no direct-dialing capability, and all international calls had to be scheduled with an operator in advance. All calls were monitored by the authorities. Even though there are more international gateways at present at various locations around Russia, it is still a limited capability.

Switching equipment is inadequate for both present and emerging telecommunications needs. This presents a serious problem since solving the penetration problem without increasing the amount and quality of switching equipment is an irrelevant goal. At present, manual switches account for 1 percent of the switches in use, step-by-step account for 22 percent, crossbar 75 percent, and digital switches only 2 percent.

In the 1980s, Russia invested only 0.3 percent of its GNP in telecommunications, which is less than half the amount invested by western Europe. Investment per capita in eastern Europe in the 1970s and 1980s was estimated to be at one-sixth that of western Europe.[5] In terms of revenue per subscriber, Russia and other CIS states fall behind most of eastern Europe and the Baltic states.[6] For 1993, revenue per subscriber was approximately $3 per month.

Implementing wireless networks in Russia can help solve all these technical problems quicker than would building a wireline network. Since Russia went to the method of charging the real cost of line installation for new telephone lines, wireless is competitive with wireline even in terms of cost. The real cost of a new phone line can be as high as several hundred dollars. Those customers that can afford the telephone line installation can spend a little extra and get all the advantages of a mobile phone. It is also important to remember that the total size of the Russian telephone network is the third largest in the world after the United States and Japan.[7] This implies that Russia has one of the highest numbers of potential customers.

Frequency Allocations

It is important to understand that investment in telecommunications systems in Russia had little to do with economics; rather, it was a political decision. During the socialist years, the Russian government defined all service industries as inferior to real industries such as manufacturing and production. Since communications is a service industry, it suffered from this definition and consequently suffered from severe underinvestment. The GosPlan, the operations tool of the Communist Party (one of the largest organizations in the Soviet Union, which performed planning for five-year periods), directed and controlled investment and development activities in most sectors of the economy. Investments were focused on serving organizations of high political priority such as the Communist Party, the Ministry of Defense, the KGB, the Ministry of Internal Affairs, and such "real" industries' organizations as the Ministry of Oil and Gas, the Ministry of Transport, and so forth. In general, communications networks were developed as a means for different departments of the government to coordinate their planning efforts and had little to do with real demand for services.

The extinction of GosPlan resulted in every ministry or organization once governed by GosPlan pulling the blanket their way. This put the frequency allocation and control process in great chaos. The Ministry of Communications had the second most important role next to GosPlan in telecommunications development and now wants to maintain its role in the licensing process. The military had control over its own frequencies and now is not willing to give up this privilege. Various governmental agencies started their own telecommunications networks for internal purposes. Since many former republics and regions of the USSR received their independence, they are not willing to honor licenses issued by any federal bodies and want to manage their frequency spectra locally. It is a major challenge for Russia to put some order into the combination of licenses issued by the Ministry of Communications, the military, and numerous governmental agencies for their own use on a regional and federal level.

Standards

There are several cellular standards in use in Russia now. The American cellular standard – AMPS, or advanced mobile phone service – is quite widely used in Russia. There are AMPS networks in Moscow, St. Petersburg, Nizhni Novgorod, and the Russian Far East. Nordic mobile telephone, NMT 450, is also in use. This is an old analog mobile standard that is used throughout Scandinavia and other parts of the world. It was implemented in Russia because NMT can be easily modified to operate on whatever frequencies are available in the 450-MHz range. Licenses for commercial operation in this part of the spectra in Russia are easier to acquire because this frequency range was not heavily used in the country previously (compared to the GSM 800-900 MHz spectrum that was occupied by the Russian military). A third standard in use is the British analog mobile standard, or TACS. This standard works at high frequencies and is very similar to the AMPS standard in its characteristics. The global system for mobile communications (GSM) – the European digital standard – is being implemented by Deutsche Bundespost Telecom in Moscow and by U S WEST in several cities across Russia. By the end of 1995, 12 percent of Russia's cellular service will use the NMT 450 and 900 standard, 11 percent the GSM, 49 percent the AMPS, and 24 percent TACS.

A large part of the 800-MHz to 900-MHz spectrum that GSM systems use is already occupied by the Russian military. This creates technical and regulatory problems and makes obtaining licenses very difficult. Professionals in the industry hope that the Russian Ministry of Telecommunications will take another look at the 800-MHz to 900-MHz spectrum allocations and will take away some spectrum from the military forces "as the need for security decreases due to the ending of the Cold War."[8]

The collapse of the Soviet Union influenced the standardization process across the country. Federal organizations are attempting to set standards, but regional organizations, which have been under their thumb for 70 years, are resisting those standards. Regional organizations in Russia need to understand that playing democracy by refusing to accept federal standards costs the telecommunications industry a lot of potential customers and revenue. Democracy is characterized by the ability of individuals to make responsible and reasonable choices; democracy is not the rejection of everything that is suggested as if it were an order. Democracy is impossible in chaos.

Cellular

Mobile telecommunications are not new in Russia. Party officials and military figures had mobile telephone sets in the 1960s. The network they used internally in Moscow is now the Altai commercial wireless network with 3,000 customers and is booked to its full capacity.[9] The single-node Altai system covers a range of 40 km to 70 km with a central antenna located on the Ostankino television tower. Providing one of the least expensive mobile services in Moscow, Altai is popular with many Muscovites and is fully booked at present. However, the system is slowly becoming obsolete as its technical limitations prevent it from meeting customers' needs.

There are at least five cities in Russia that have operational cellular networks: Moscow, St. Petersburg, Tver, Nizhni Novgorod, and Rostov-on-Don. More networks are being put in place every day. The companies involved in Russia's mobile market fall into two categories: the network operators and the manufacturers together with their dealers. The main network operators are U S WEST, Rostelecom, United Telecom, Millicom, Italtel, Deutsche Bundespost Telekom, Telecom Finland, and Moscow Cellular. Major equipment suppliers are Nokia, Motorola, AT&T, and

Siemens. It becomes easier and easier, however, to confuse the two categories, as many manufacturers are starting their own wireless operations.

Corruption is causing a serious problem for cellular operators. Tests performed by the military in 1994 determined that approximately 30 percent of cellular phone users on Moscow networks were illegal. Illegal phones are being offered for sale by young telecom entrepreneurs all over the city. This problem is causing cellular customers to get upset when they receive somebody else's phone bills.

Pagers

Many simple communications needs were solved in Russia with pagers, and there are at least several operational pager networks in Moscow; all have a good customer base. By mid-1995, a total of 21 licenses were granted by the Russian Ministry of Communications. Although more affordable than cellular telephones, the prices on pagers are still much higher than in other countries. There is usually a charge of $300 to $450 for the pager and between $25 and $45 per month rent. In comparison, pagers in big U.S. cities can cost as little as $15 per month, and pager units are distributed free of charge.

Satellites

The Soviet Union had the world's most extensive satellite industry and its largest satellite production and launching facilities.[10] Russia's space program is ranked second only to that of the United States in terms of resources dedicated to it, and Russian scientists, engineers, and technicians were well trained. The former Soviet Union's space program had a different approach to spacecraft manufacturing and launching than that of Western countries. Instead of working longer to employ the most advanced instrumentation and electronic technologies, Russian scientists concentrated on launching as many satellites as soon as possible. This resulted in the overall life of Russian satellites being shorter than that of Western satellites.[11] Nevertheless, the Soviet space program managed to establish an efficient, inexpensive production line of spacecraft, accompanied by an extensive launch vehicle capacity that resulted in an effective operational space industry. At least one TV program carrying news from the Moscow Central TV station was transmitted to almost every household in Russia. Satellite links also connected most of the socialist block countries. Even distant Cuba could communicate with Moscow via Russian satellites.

After the socialist block fell apart, Russia was left with many satellites in orbit but designed for the wrong applications, covering the wrong areas (in terms of potential customers), and with ground stations in poor condition. Most satellites were designed and built to accommodate military purposes and applications. Of those few designed for telecommunications, the majority were oriented toward TV broadcasting rather than communications. Of the few that were designed for communications purposes, most were pointing at eastern Europe and Cuba rather than at western Europe or the United States, where most of the new demand is to be found. After the end of the Soviet army's presence in Europe, many base stations formerly used and operated by the Russian military were not maintained by the local authorities and simply stopped working. Some links to Germany, Czech Republic, and Hungary have ceased to function.[12]

Even if coverage could be extended to the more important locations in Europe and North America, Russian satellite signals are incompatible with Western earth-station receiving equipment because Russian and Western satellite industries developed secretly from each other. Most Russian satellites were based on C-band technology developed for satellites in the late 1960s. Western satellites operate in the Ku band, and their earth stations have smaller, less expensive antennas. Another major investment will be needed to overcome the problems of different frequencies used for transmission and reception and the problem of incompatible equipment parameters.

The Russian military owns a large network operating in the more modern Ku band, but they are not rushing to turn it over to the government or to bring it to commercial use themselves. At present, it is difficult to define what the army can or cannot provide in terms of commercial satellite services. This is due to both the unavailability of information on exactly what they possess and to the need to use some of that capacity for still existing military purposes. The army also has a financial question to solve: as a nonprofit organization, it cannot decide how to deal with the profits collected from operating commercial satellite services.

Since the early 1990s when the military stopped funding and support for the space program, Russia has been in the process of trying to organize the space assets of the former Soviet Union into commercial ventures. The process is slow and chaotic, taking into account the magnitude of the task and the fact that it is the first time that government-paid employees have become entrepreneurs. It does seem possible, however, to combine Russian inexpensive launching facilities with Western finances to produce commercial communications satellites for about one-half of the present cost in Europe or the United States.[13] This may lead to an interesting competition in the world satellite industry and have a significant effect on present and planned international and domestic satellite systems. Many hope that with certain coordination efforts in place, chaos can be replaced with lower-cost spacecraft and launchers for international use.

Another big problem is that of connecting satellite stations to the existing public-switched telephone network (PSTN). It takes sophisticated and expensive technology to establish that connection. Most of the equipment for such connections was purchased abroad and not manufactured in Russia. Controls imposed by CoCom made that technology unavailable for export to Russia, which in turn impaired the functioning of Russian PSTNs. The Russian satellite industry suffered from CoCom bans on high-technology export and its control over the flow of information between East and West. Until 1993, the Western satellite companies were prohibited from using Russian facilities, implementing Russian technology, or employing Russian technicians.

Russia has several generations of Gorizont, Raduga, and Ekran geostationary satellites, as well as satellites Molnia and Loutch.[14] The Gorizont, Raduga, and Ekran satellites, which are called the Stasionar System in international registration books, cover the whole territory of Russia. The main traffic goes through Stasionar satellites while Molnia satellites are used for coverage of the far Siberian regions. Satellite systems Orbita, Ekran, Moskva, Loutch, Moskva Globalnaya, Rabita, Mars, Rossiia, and STV-12 deliver TV programs to the Russian population and provide minimal telephony services.

In 1971, Russia initiated the creation of a satellite organization, Intersputnik, which included 15 countries at that time. The main contributor to the organization was the former Soviet Union, and now Russia, whose ownership share is approximately 25 percent. Most of the earth stations in other countries were built with Russia's technical assistance and based on the equipment developed and manufactured in Russia. It is important to understand that Intersputnik has no satellites of its own. According to Intersputnik's charter, satellite time is leased from Russian satellites, and earth stations belong to the countries in which they are located.

The Soviet Union signed a treaty to join Intelsat in July 1991. It took a minimum share of 0.05 percent (costing $700,000) and became the international satellite consortium's 121st member.[15] Joining Intelsat will reduce the cost of using the organization's satellite circuits for Russia since it will share its profits, as do all of the other members. It is reasonable to expect Intelsat to extend its coverage to eastern Europe and Russia as those countries become members of the organization and have a voice in its decisions.

In 1992, the European satellite organization Eutelsat altered the beam of Eutelsat II F-IV to include more of European Russia, but as of today its capacity is fully booked. At the beginning of 1994, Eutelsat made another attempt to include Russia in its coverage and launched Eutelsat II F-5. The launch was unsuccessful.

Russia was a founding member of Inmarsat with its Morsviazsputnik in the late 1970s. Even during the Cold War, Russia continued its membership in the organization. Inmarsat has had earth stations in Odessa, Ukraine, and Nakhodka since the middle 1980s. Of all the satellite providers, Inmarsat has the smallest terminals, which are easy to move around. But they are also the most expensive. Inmarsat connections can cost $6 per minute and higher.[16] There are about 1,600 Inmarsat-A terminals in use in Russia, which were manufactured in the region.

To summarize, wireless communications have their own specific needs related to frequency spectrum allocations, control, and standardization. Russia needs to redefine the process of new spectrum allocation and establish federal standards in order to provide interconnection. The demand for telecommunications services is still higher than the supply. This presents a good opportunity for new joint ventures between Russia and other countries in order to find new solutions to the existing telecommunications challenges.

References

1 R. Famighetti, J. Foley, T. McGuire, C. Cheddar, M. Northrop, and D. Young, eds., *The World Almanac and Book of Facts* (New Jersey: Funk and Wagnalls, 1994).

2 H. Vilaggazdasag, "Eastern Europe: Governments Cautious in Relinquishing Telephone Monopolies," *Reuter Textline* (April 1994).

3 Y. Yudina, "State of Russia's Telecommunications," *Novecon* (June 1993).

4 H. Vilaggazdasag, op. cit.

5 "Eastern Europe: Waiting for a Telephone," *Euromoney Central European*, (January 1992).

6 H. Vilaggazdasag, op. cit.

7 N. Sokolov, "Telecommunications in Russia: The Need for Additional Capacity, Improved Quality, and Enhanced Technology Is a Serious Problem – and a Vast Opportunity," *I E Magazine*, no. 7, p 67.

8 "CTIA Notebook: Cellular Industry Concerned by Duplication of Standards," *Communications Daily*, no. 15:3 (1991).

9 R. Farish, "Mobile Networks: The Telephones of the Future or the Toys for the Mafia?" *Deloive Lyudi* (June 1944).

10 R. Farish, "Satellite Technology - The Choice When There Is No Choice" *Delovie Lyudi*, (June 1994) p 14.

11 *Satellite Communications Systems and Technology* (NASA/NSF Panel Report), vol. 1 (Baltimore: International Technology Research Institute, 1993), p.27.

12 R. Farish, "Satellite Technology," p 14.

13 *Satellite Communications Systems*, op. cit., p.27.

14 A. Curts, ed., *Space Satellite Handbook* (Houston: Gulf Publishing Company, 1994), p. 27

15 "Soviet Union to Join INTELSAT Next Month," *FinTech Telecom Markets*, (June 1991). 16 R. Farish, "Satellite Technology," p 10.

Mobile Communications Industry in Thailand

Donyaprueth Krairit
Professor, Department of Computer Science
Thammasat University, Thailand

Basic Regulatory Framework of Thai Telecommunications

This paper is intended to provide an overview of the wireless industry in Thailand through the presentation and analysis of its evolution and current marketing situation. In this first section, the study of basic legal entities and regulatory framework in communications is presented in order to provide background about telecommunications in Thailand. In addition, the current environmental status in telecommunications investment is also analyzed and presented. After that, in the second section, the evolution and current status of the wireless industry is shown. The last section is a conclusion that discusses business opportunities in the wireless communications market in Thailand.

Basic Regulatory Framework and Legal Entities

In Thailand, according to governing telecommunications laws discussed later, absolute rights in telecommunications are legally monopolized by the state. The state uses these rights through its four major legal entities involved in telecommunications. These agencies are: the Ministry of Transport and Communications (MOTC), the Post and Telegraph Department (PTD), the Telephone Organization of Thailand (TOT), and the Communications Authority of Thailand (CAT).

The MOTC is responsible for policy planning at a national level for both transportation and communications in Thailand. Currently, the MOTC has control over ten government departments and state enterprises, among which the major ones are the PTD, the TOT, and the CAT. In addition, the MOTC is also responsible for major telecommunications laws and enforces them through the departments and state enterprises subjugated to the MOTC.

The PTD is the main administrator of postal and telecommunications services in Thailand. The operations of the PTD are under the control of the MOTC and are mainly at the administrative level. Therefore, the department does not have much role in granting telecommunications concessions or in developing telecommunications projects. However, the PTD is the agency responsible for radio frequency administration and allocation.[1]

The TOT is a state enterprise separate from the PTD. The TOT is responsible for the provision of both wireline and wireless domestic telephone services, cellular phone, paging, and other services offered through telephone lines such as ISDN services.[2] The CAT is another state enterprise mainly involved in telecommunications. It is also separate from the PTD and is responsible for international telephone calls, cellular and private radio communications, and radio paging.[3]

These four organizations operate under four important laws involved in the provision of telecommunications services in Thailand. The first is the 1934 Telegraph and Telephone Act. This act is the most powerful act in telecommunications in Thailand, especially concerning wireline communications. Under this act, authority and exclusive rights within the kingdom to operate and own wireline telecommunications networks or provide wireline telecommunications services reside with the state. The 1955 radio communications act is quite similar to the 1934 Telegraph and Telephone Act but with concentration on wireless communications. This act authorizes the PTD to administrate and allocate the frequency resource for any requester as well as to control and monitor the usage of frequency that any agencies receive. Based on this act, the PTD has control over radio communication services and equipment, including pagers, mobile telephones, and satellite dishes.

The 1954 Telephone Organization of Thailand Act (TOT Act) and the 1976 Communications Authority of Thailand Act (CAT Act) are quite similar in content. They establish and determine the functions and scope of operations of the TOT and CAT. They transfer the authority in telephone and postal operations, formerly under the administration of the PTD, to the TOT and CAT, respectively. Due to the similarity in content of these two acts, there is an overlap in the responsibility of TOT and CAT. For instance, both of them can provide cellular and paging services and can grant the same kind of concessions to different private agencies.

Current Investment Situation

During the past five years, Thailand has been experiencing high economic growth, with a growth rate of 8.3 percent in 1994.[4] The economy is driven mainly by increased investment from both the government and private sectors in telecommunications infrastructure development projects.

In the area of telecommunications thus far monopolized by the government, there has been some relaxation of the government's control because of the urgent need to develop a telecommunications infrastructure to support economic development. Another reason is that telecommunications development projects by their nature require tremendous investment, which can be supported more easily by private investors.

Nevertheless, since the state holds the rights to build, own, and operate telecommunications networks and to provide telecommunications services, it is illegal to have any private agencies compete with the state unless there is an amendment to the current telecommunications laws. As a result, private investors participate in the telecommunications industry by bidding for concessions or joint-venturing with state enterprises. In either case, the private investors are not able to operate on their own and are bound legally by the agreements they make. The agreements and concessions established in the Thai telecommunications industry have some characteristics in common. These characteristics are summarized in *Table 1.*

Although the government controls the industry in Thailand, support for technological development from the private sector is seen as one solution to the telecommunications bottleneck that impedes economic development. Applying resources from the private sector as quickly as possible is needed. The Thai government has committed itself to liberalizing the industry by starting to grant many concessions in order to attract the country's telecommunications companies, which offer modern technology and quick service. Nevertheless, all agreements signed are more to the government's advantage than to that of the private partners.

Evolution of the Mobile Communications Industry

Mobile Telephone and Cellular Service

Mobile radio communications were first used in Thailand in the 1960s by the government, police, and traffic control authorities. Even though it had been

Subjects	Characteristics
Exclusivity	Government entities reserve the right to grant a similar concession to any other interested person or Government entities grant exclusive privileges to a private agency for a period of time, during which that agency enjoys the privilege of not having competitors for the services granted in the concession. However, after the expiration of the concession period, government entities reserve the right to grant approval to others.
Ownership of Assets	Government entities shall have absolute authority to operate and possess all priorities under the agreements at the beginning or expiration of the agreements or upon the completion of installation. Private entities have no ownership in these properties but are the only occupants using the properties during the term of the agreement. In addition, the owners shall be government entities regardless of the cause of termination of the agreement.
Renewal of Concessions	Renewal of the agreement depends on the decision of government agencies and shall be requested upon the expiration of the effective agreement.
Government Rights and Termination	Private partnerships are not entitled to terminate the agreements unless they fail to perform their obligations. The only entity empowered to do so is the governmental partner in that agreement. In some agreements, private partners are to be responsible for training officials from the government sector and to administer the training program.
Royalty or Revenue Sharing	Private entities have to pay royalties as a fixed percentage of annual gross revenue before deducting any expenses. The percentage might be fixed or vary throughout the concession period.
Tax Privileges[5]	There is no special tax treatment or privilege granted to non-government parties in all concessions. The non-government parties are responsible for all taxes, expenses, and fees arising out of the agreement.

Table 1 *Characteristics of telecommunications agreements and concessions in Thailand.*

present in Thailand for more than 30 years, Thai people did not have mobile radio communications services until 1983.[6] In 1982, there was a serious deficiency of telephone lines both in the metropolitan and rural areas due to the plans of the TOT to change the telephone infrastructure from an analog to a digital system in a short period of time. This plan hindered the installation of telephone lines at that time.[7]

As a result, the Communications Authority of Thailand started to provide mobile radio service to Thai people to relieve this problem. However, the increased use of this service created a problem of traffic congestion, which caused lower quality of service. A few years later, the TOT opened the first cellular service based on the NMT standard using a 450-MHz frequency (later changed to 470 MHz).[8] This service was more popular than the mobile radio service of CAT since its quality was better.

CAT decided to open its own cellular service using an 800-MHz (Band-A) frequency and the Advanced Mobile Phone System (AMPS) standard.[9] Although this system was better than the TOT service because the handset was smaller and more portable, it had the disadvantage of using 7-digit numbering, which made it difficult to expand service to provincial areas. TOT used the 9-digit numbering system and was able to provide service covering a wider area.

TOT subsequently decided to grant concessions in cellular service to Advanced Info Service Co., Ltd. (AIS), a subsidiary of Shinawatra Group – one of the most powerful players in Thai telecommunications. AIS uses the 900-MHz frequency band and the NMT standard.[10] This service, coupled with the numbering plan provided by TOT, made the NMT-900 cellular service better than other services at that time. CAT also granted concessions to provide cellular service in the 800-MHz (Band-B) frequency to the Total Access Communications Co., Ltd. (TAC), a subsidiary of the UCOM Group, using the AMPS standard.[11] However, TAC still had to use the same 7-digit numbering system in its service as did CAT. Because of these developments, mobile telephone services are currently provided by both public and private agencies, which gives more choices to Thai people. However, it is noteworthy that the rights in granting concessions are still held by TOT and CAT.

Paging Service

Paging service was first operated by CAT. However, because CAT did not have enough experience in marketing, this service was not very successful.

Subsequently, CAT granted concessions to the Pacific Telesis Co., Ltd. (PACTEL), a subsidiary of the Pacific Telesis Group. At that time, PACTEL was the only service provider in this area.[12] TOT granted two more concessions in paging service to Shinawatra Paging Co., Ltd. and Hutchison Co., Ltd., making the market more competitive. After that, both TOT and CAT granted more concessions to other companies, including UCOM Group, Matrix Com Co., Ltd., and Fonepoint.[13]

Trunk Mobile Radio Services

Trunk mobile radio was first used in Thailand in 1979 in the internal operations of the Post and Telegraph Department (PTD) and TOT.[14] However, CAT was the first organization to provide trunk mobile radio services to the Thai people. Nevertheless, the service was not very successful since TOT was providing NMT-470 cellular services at the same time, which were more popular than the trunk mobile service. In 1990, TOT granted the first trunk mobile radio concession to Jasmine International Group. This concession was implemented under a service named "Radio Phone." Subsequently, CAT granted another concession to the UCOM Group in order to provide services under the name "World Radio."[15]

Current Status of Wireless Communications Market

Mobile Telephone

Currently, there are four mobile telephone service providers in Thailand. They are in both the public and private sectors. Details about these services are shown in *Table 2*. The high degree of competition in this market is between the services provided by TAC's AMPS 800B system and AIS's NMT-900 system; both systems have their trade-offs. The NMT-900 system is better in terms of its area of coverage. It covers all provincial areas because it is based in the TOT central office. However, the quality of service is not good enough in the metropolitan area compared to TAC's system. As a result, users have to select the service that is most appropriate for their work.

One hot topic in the mobile communications market recently has been the movement of both AIS and TAC from an analog cellular system toward a digital one. After a long period of consideration, both companies decided to use the GSM standard but with different platforms. While AIS decided to use GSM in its NMT-900 digital system, TAC will use the DCS-1800 for its digital services.[17] However, implementation of both systems is only in the beginning stage.

Type	Concessionaire	Period	Signed	Owner	Service Started
Cellular NMT-400	–	–	–	TOT	1983
Cellular AMPS 800A	–	–	–	CAT	1984
Cellular NMT-900	Shinawatara Group (AIS)	20	1990	TOT	1990
Cellular AMPS 800B	UCOM Group (TAC)	15 (later changed to 22)	1990	CAT	1991

Table 2 *Mobile telephone projects in Thailand.*[16]

This system conversion came about after TAC changed its numbering plan from the 7-digit number to the 9-digit number in exchange for a seven-year extension of its concession. This is the main reason why AIS requested the same kind of extension from TOT, which is now under consideration. However, the change in TAC's numbering plan has made the cellular system in Thailand more compatible and eases the difficulty in the different numbering plans of TOT and CAT.

Another interesting topic is the granting of a new concession in mobile telephone services at the 1500-MHz frequency band. Due to high market demand, this new service is expected to reach a top value of US $2.4 billion within the next five years, with the total number of mobile telephones increasing to 2.6 million.[18] As a result, it is not surprising that every major telecommunications player is participating in the bidding for this concession. However, the new concession is still under consideration by the government.

This new concession, according to Mr. Pinit Jarusombat, Deputy Minister of Transport and Communications, will open more choices in technology for Thai users. The most probable backup technology for this concession is anticipated to be the new Personal Digital Cellular System (PDC) at the 1500-MHz frequency developed by the Nippon Telegraph and Telephone Company (NTT), a major developer of this technology in Japan.[19] Accordingly, there are many Thai companies trying to form a joint venture with NTT in order to increase their potential both in terms of marketing and concession bidding. NTT, nevertheless, has not decided to cooperate with any particular companies at this time.

Type	Concessionaire	Owner	Period	Year Started
Paclink	Pacific Telesis	CAT	10	1987
Phonelink	Shinawatra Paging	TOT	15	1990
Paclink	Percom Telesis	CAT	15	1991
Pagephone	Hutchison	TOT	15	1991
Page Call	Fonepoint	TOT	10	1991
Easy Call	Matrix Communications	CAT	15	1992
World Page	UCOM	TOT	15	1994

Table 3 *Paging services in Thailand*[20]

In conclusion, the potential of the mobile telephone industry in Thailand is very high, and the market is expected to continue to grow at a very high rate in the near future. However, the monopoly in telecommunications by the state is still the main obstacle to expansion in this sector.

Paging Services

Currently, seven concessions in paging have been granted. The details of these concessions are summarized in *Table 3*. The paging market has become very competitive since the UCOM Group entered this market by providing the "World Page" service. This is because UCOM is the biggest formal supplier of Motorola products. This company controls more than 80 percent of the current paging market share.[21]

The entrance of UCOM into the market has stimulated previous providers to develop new technology and offer more choices to users in order to retain their market share. Before World Page started its services, the market share was divided as shown in *Figure 1*.

At the same time that UCOM opened its new paging services, Fonepoint – the only provider of telepoint service in Thailand who has faced a deficit continuously due to its weakness in cellular services – decided to include the paging service "Easy

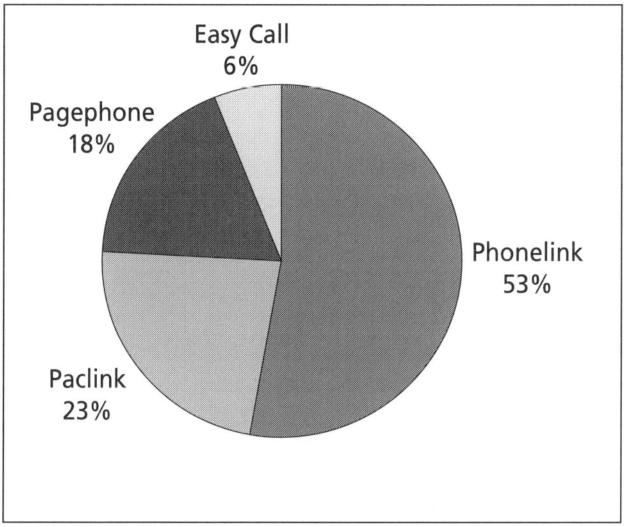

Figure 1 *Paging market share as of August, 1994*[22]

Call" into its telepoint service. This new type of service, called "Page Call," enables two-way communication for users instead of the call-out-only communication provided by telepoint service.

As a result of this competition, Phonelink, the biggest provider in this market, started a very attractive "buy-one-get-one-free" promotion and added supplementary services to its pager. Such services include the ability to keep information even when the pager is turned off and a photo sensor that enables users to read messages in the dark.

Paclink, another powerful player, has offered additional services that enable users to forward messages from telexes to pagers. This service is called "Telex to Paclink," and it has received quite satisfactory reactions from the business sector. Another interesting strategy of Paclink is an international paging service that provides roaming services in neighboring countries. Hutchison Pagephone entered this competition by reducing the price of its services by 30 percent and adding new services such as "LoxNews," which provides news reports on the stock market and economic information to its users.[23]

The competition in the paging service market is beneficial to Thai users since it not only introduces new technologies but also provides subscribers with more choices and better quality of service.

Type	Concessionaire	Owner	Period	Year Started
Radio Phone	Radio Phone	TOT	15	1993
World Radio	UCOM	CAT	10	1993

Table 4 *Trunk mobile radio services in Thailand.*[24]

Trunk Mobile Radio Services

Concessions in trunk mobile radio services were finally granted in 1993. There are only two service providers to date. The information about these services is presented in *Table 4.* Both Radio Phone and UCOM started their services last year with the first phase of implementation. These services are only in the metropolitan area and its surroundings. However, both companies expect to expand their services to the main provinces in each region of Thailand during 1995.

Since trunk mobile service is considered to be a supplementary service for remote areas, the main goal of these two companies is to create more market demand for their systems. One strategy that is used by both companies is direct sale to a targeted group of users, mostly small businesses.

Due to the fact that trunk mobile services are in the beginning stages of development, the competition is still not very strong. However, an expansion of this market can be expected in the future.

Business Opportunities in Thai Telecommunications

Recently, the Thai economy has expanded dramatically through increased investment in telecommunications development projects. These projects have attracted not only domestic companies but also foreign ones. Even though Thai law does not allow private agencies to operate or own telecommunications networks, the investment of these companies is needed by the government, which grants them exclusive rights through the concession system.

On the other hand, the profit that these companies can harvest during the period of their concessions is so great that many foreign companies are trying to form joint ventures with Thai-owned companies. Recently, more foreign companies

Foreign Companies	Thai Company	New Companies Established in Thailand	Types of Business
Siemens A.G.	–	SYSTEMS Co., Ltd.	Telephone manufacuring
ALCATEL	–	ALCATEL C.I.T. (Thailand) Co., Ltd.	Manufacture switching and signal communication components
AT&T	–	AT&T (Thailand) Co., Ltd.	Produce switching systems and telephone equipment
NOKIA	–	NOKIA (Thailand) Co., Ltd.	Sell telecommunications equipment
SINOBRIT	–	SINOBRIT Telecom	Telecommunications service provider
NEC	–	NEC Communication System (Thailand) Co., Ltd.	Switching installations, plan to establish its own manufacturing plant
L.M. Ericsson	–	Ericsson (Thailand) Co., Ltd.	Telecommunications equipment manufacturing plant
NYNEX	TelecomAsia	TelecomAsia	NYNEX is a 25% shareholder in the TelecomAsia group

Table 5 *Movement of foreign companies in the Thai telecommunications industry*[25]

have attempted to expand their operations into Thailand. Their movement is shown in *Table 5*.

The investment from these foreign companies is a main factor in the progress of Thai telecommunications. The expansion of their business coupled with strong support from the government has shown the high potential of the Thai telecommunications market. At the same time, the strong competition also demonstrates the profitability of this market in the near future.

Nevertheless, investment in Thai telecommunications is still hindered by the monopoly of the state. The monopolization creates uncertainty for both domestic and foreign investors since all the equipment is under a build-transfer-operate type of concession and therefore will be transferred to the state either at the termination of the concession or after installation is complete.

Another important factor that concerns investors is the instability of Thai politics. Since the telecommunications industry has long been monopolized by the state, there is always politics involved in the consideration of telecommunications concessions. This fact has lately become one of the most controversial topics in the industry. One possible solution that is being considered is the liberalization of the telecommunications market; another solution is to allow the state enterprises, TOT and CAT, to compete with private agencies.

Future Directions of Thai Telecommunications

As a result of the many problems created by the state's monopoly of the industry, there have been some changes in the government sector lately. The government has established a committee to conduct a feasibility study for a telecommunications laws amendment. The result of the study will be presented to the Cabinet and the House of Representatives. The amendment of these laws will open the telecommunications industry to private agencies and therefore increase investment and improve the economic situation.

Another change that is of interest is an attempt to privatize TOT and CAT by the government. The feasibility of this is currently being studied. It is expected that after the privatization of these two state enterprises, there will be more business opportunities for the private sector. In addition, strong competition and more variety of technology are also expected to result. The maximum benefit will be at the consumer level and not at the governmental level as at present since the concession system will be reduced or even eliminated. Therefore, even though there are still few opportunities for investment in Thai telecommunications, except in joint ventures with Thai companies or by investing in manufacturing plants, it is reasonable to expect changes and more opportunities in the near future.

References

1 Setthaporn Kusripitak, The Roles of the PTD in Development, paper presented as a part of the symposium Thai Telecommunications in the Next Decade, Bangkok, Thailand, November 15, 1989, p. 1.
2 39 Years of the TOT, *Telcom Journal,* February 15, 1993, p. 11.
3 16 Years of the CAT, *Telcom Journal,* February 15, 1993, p. 14.
4 G.F. Research Department, *Economic and Stock Situation in the First Half of 1994* (Bangkok, 1994), p. I-10.
5 Dej-Udom Krairit, *Tax Privileges* (Bangkok: Dej-Udom & Associates Co., 1992),p. 45, photocopied.
6 Mobile Radio Communications in Thailand (Part 1), *Telcom Journal,* October 1, 1993.
7 Mobile Radio Communications in Thailand (Part 2), *Telcom Journal,* October 15, 1993.
8 Ibid.
9 Mobile Radio Communications in Thailand (Part 3), *Telcom Journal,* November 1, 1993.
10 Ibid.
11 Mobile Radio Communications in Thailand (Part 4), *Telcom Journal,* November 15, 1993.
12 The (Under)Development of Cellular Phone in Thailand, *Telcom Journal,* March 1, 1994.
13 Paging Service in 1994, *Thai Financial,* January 4, 1994.
14 Trunk Mobile Radio in Thailand, *Telcom Journal,* January 16, 1994.
15 Two Companies Compete in Trunk Mobile Market, *Telcom Journal,* January 16, 1994.
16 The Nation, January 24, 1994.
17 Today's Digital Cellular Phone Market, *Telcom Journal,* May 1, 1994.
18 The Third Cellular Phone Concession, *Telcom Journal,* August 16, 1994.
19 Digital Cellular Phone Technology in Japan, *Telcom Journal,* March 1, 1994.
20 *The Nation,* January 24, 1994.
21 Pager Market is Preparing for the Newcomer, *Telcom Journal,* October 16, 1993.
22 Ibid.
23 Paging Service War at the End of 1993, *Telcom Journal,* November 1, 1993.
24 *The Nation,* January 24, 1994.
25 Thai Telecommunications Business, *Telcom Journal,* October 1, 1993.

Evolution of Wireless Applications in Mexico

José L. Kruyff
Executive Assistant to President
AT&T México

Abstract

The telecommunications industry in Mexico is undergoing a complex process of deregulation, modernization, and introduction of competition. Since the 1950s, when the national operator first became the dominant player, and eventually the only telephone service provider in the country, cellular telephony was the first major area of telecommunications competition.

As new service providers obtain licenses to offer basic telecommunications services, wireless technologies will play a major role in their business and implementation strategies. Therefore, the evolution of the cellular industry in Mexico can provide valuable insight into the future not only of telecommunications in Mexico, but in Latin America as well.

Introduction

Basic telephony was introduced in Mexico in 1878, and despite an 18-year intermission (1972 to 1990), most telecommunications services have traditionally been provided by the private sector. The current structure of the telecommunications industry in Mexico originated in 1947 when Teléfonos de México (Telmex) was incorporated. Telmex acquired the assets of Compañía Telefónica y Telegráfica Mexicana (Mextelco) and Compañía Telefónica Ericsson and eventually became the single provider of telephone service in Mexican territory by 1981 (service in Baja California Norte is provided by Teléfonos del Noroeste, a Telmex subsidiary).

The Communications and Transport Secretariat (SCT) has the main telecommunications regulatory responsibility in Mexico. The regulatory framework is supported by:

- the Mexican Constitution of 1917
- the General Means of Communications Act of 1940
- the Telecommunications Bylaws of 1989
- the Telecommunications Act, enacted June 7, 1995
- concession titles with authorized service providers

Through the SCT, the Mexican government operated a large part of the national microwave network and the Morelos Mexican Domestic Satellite System (Morelos) until 1989 when, as part of a process that included deregulation of customer premises equipment (CPE) and value-added services (VAS), the SCT merged telegraph, telex, and satellite operations and spun them off to Telecomunicaciones de México (Telecomm) – a newly created government-controlled organization. The federal microwave network was transferred to Telmex to enhance this company's value prior to privatization. The SCT only kept its telecommunications regulatory and spectrum management functions. At the same time, the Government of Mexico (GOM) introduced cellular telephony, privatized Telmex, and announced that basic telecommunications services would be opened to competition late in 1996.

The first wireless telephone services were offered in Mexico in the late 1950s when Servicios Organizados Secretariales (SOS) of Alejo Peralta – now Chairman of the Board of Grupo Iusa – and Radio Móvil Dipsa, a Telmex company, offered radio access to the telephone network. The service had small market penetration due to its limited coverage, high cost, and technology-imposed quality limitations.

Cellular telephony started operations in Mexico late in 1989. It is interesting that 100 years earlier, basic telephony started almost simultaneously in the United States and Mexico, while it took more than five years after its introduction in the United States for cellular service to be available in Mexican territory.

Through the SCT, in 1989 the GOM opened the process to grant concessions to provide cellular telephony services in Mexico. The following measures were taken for the introduction of the service:

1) The country was divided into nine cellular regions (*Table 1*).
2) It was determined that only two competing cellular licenses would be granted for each region.

Region	Area	Zone	Population	Bids
1	Baja California Norte Baja California Sur	Baja California	2.5 million	8
2	part of Sonora Sinaloa	Northwest	4.2 million	12
3	Chihuahua Durango part of Coahuila	North	5.1 million	13
4	part of Coahuila Nuevo Leon Tamaulipas	Northeast	6.2 million	15
5	part of Jalisco Nayarit Colima Michoacan	West	10.5 million	16
6	Aguascalientes San Luis Potosi Zacatecas Guanajuato Queretaro part of Jalisco	Central	8.8 million	16
7	Puebla Tlaxcala Veracruz Oaxaca Guerrero	Gulf/South	17 million	7
8	Chiapas Tabasco Yucatan Campeche Quintana Roo	Southeast	6 million	6
9	Mexico City Morelos Estado de Mexico	Mexico City	—	N/A

Table 1 *Cellular regions*

3) A-band licenses for Regions I to VIII were to be granted through open bidding (*Figure 1* and *Table 2*).

SOS (Iusacell) and Radio Móvil Dipsa (now Telmex's cellular operation under commercial brand Telcel) were not involved in the bidding process. When cellular technology became available, Alejo Peralta argued that its 1957 license for radio telephony allowed SOS to provide cellular service anywhere in the country (Peralta, of Grupo Iusa, had received a 50-year concession for radio mobile services with national coverage).

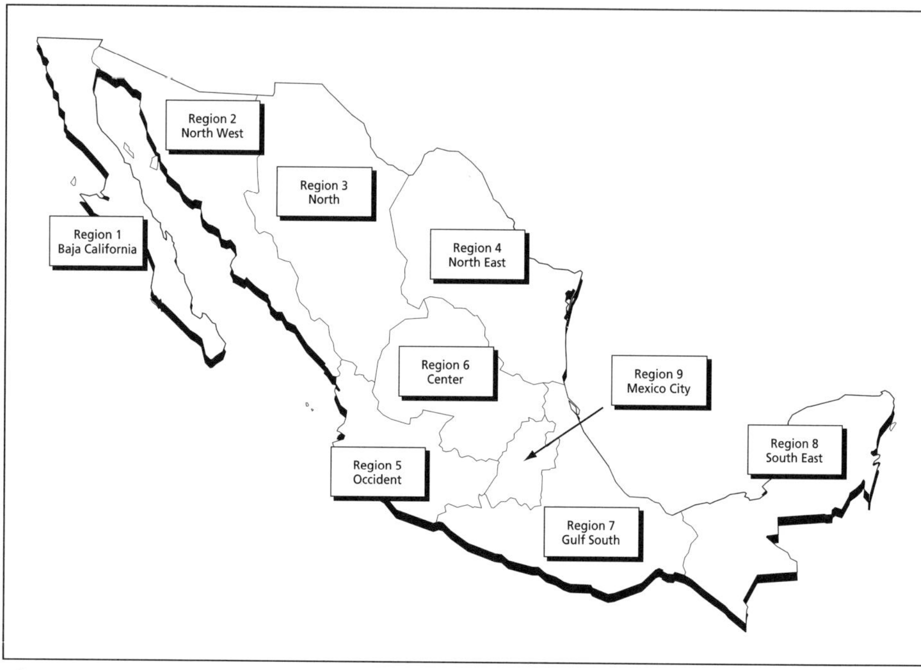

Figure 1 *The cellular industry in Mexico*

Because of the way things developed, the government was faced with a critical decision. Telmex had assumed that its monopoly in basic services would extend to cellular telephony. This approach conflicted with SOS's assumption of a *de facto* monopoly in more conventional radio-based mobile services of which, Iusa claimed, cellular was a part. Telmex and its main network supplier, Ericsson, had already conducted extensive planning activities to initiate service in a limited number of cities by the end of 1990. Original projections by Telmex called for 30,000 subscribers in Mexico City (50,000 overall) by 1992.[1]

Rather than challenge the government and confront Telmex, negotiations allowed SOS to obtain theA-bandcellular license in Mexico City, Morelos, and Edo. de Mexico (Region 9) and the opportunity to start offering services in September 1989. SOS agreed not to claim rights on additional regional licenses. Telmex eventually obtained licenses to provide service in the B-band in the nine cellular regions in the country.

Delays in defining policy in the cellular arena created new sources of conflict. One of these arose in the field of standards – the longer the process took, the higher

Region	Band	Operator	National Partners	International Partners
1	A	Baja Cellular	Tecelmex	General Cellular
2	A	Movitel	Tamsa Bachoco	McCaw Cellular Contel Cellular
3	A	Norcel	Grupo Domos	Motorola Centel
4	A	Cedetel	Protexa	Millicom
5	A	Comcel	Grupo Hermes Banamex Local investors	Racal South BellSouth
6	A	Portacel	Grupo Alarcon Local investors	Bell Canada
7	A	Telcom	Grupo Mexicano de Desarrollo Grupo Iusa	Bell Canada
8	A	Portatel	Nino de Rivera Local investors	Associated Communications
9	A	Iusacell	Grupo Iusa	N/A
1 to 9	B	Telcel	Telmex	N/A

Table 2 *Key cellular players in Mexico (1990)*

the external pressure on the implementation of digital cellular platform. Finally, the SCT determination was to adopt AMPS standards where possible, therefore allowing bidders to make provisions for migration to digital technology. This addressed concerns about the continuing delays in the introduction of cellular service, which was perceived as the only viable short-term alternative to Telmex, while allowing operators the flexibility to migrate to more advanced technology as international standards gained acceptance.

Demand in the large metropolitan areas immediately picked up because of the long lead time required to obtain traditional telephone service. Cellular operators also benefited from the lack of basic infrastructure and the need for mobile communications. Once the systems were in operation, sales surpassed all original estimates (*Table 3*), making Mexico one of the fastest growing cellular markets in the world. Iusacell reported that 1,000 subscribers contracted for the service during the first day of operation, despite monthly fees of over U.S. $50 and portable phone costs of U.S. $1,000 to $3,000.[3] Iusacell alone registered more than 12,000 subscribers during the first 12 months – a sharp contrast to most estimates

Year	Forecast[6]	Actual[7]
1989	8,000	1,150
1990	18,000	65,800
1991	52,000	175,350
1992	100,000	208,500
1993	175,000	471,192[8]
1994	275,000	565,000
1995	400,000	675,000

Table 3 *Cellular subscriber growth in Mexico*

which had reported a market potential of not more than 10,000 per year.[4] With a solid market in the Mexico City metropolitan area, the stage was set for other participants to expand the service nationwide.

After clearing the last of the major hurdles in November 1989, the SCT issued the official call for bids for the eight remaining cellular A-band licenses. Besides competition, the process was intended to provide more competitive tariffs, increase participation from private investment, and carrier financial self-sufficiency.

Mexican organizations, foreign equipment suppliers, U.S. cellular operators, European and even Japanese firms sent their offers for consideration by the Mexican authorities. Multiple multinational consortia were proposed and, in all, more than 100 proposals were received for the eight available licenses. (Note: in the cellular industry the National Commission for Foreign Investment can authorize foreign ownership in excess of 49 percent)

Through the bidding process, in March 1990 the SCT introduced new players to the Mexican market. According to the established mechanism, eight winning groups were awarded 20-year regional licenses as detailed in *Table 2*.

With the advantage of early access to the single biggest market, Iusacell was the most aggressive of the cellular operators. It was the first to introduce cellular service, and the first to offer value added services like voice mail, call waiting, call forwarding, and secretarial services in conjunction with its basic service offering. It became actively involved with foreign cellular operators in the United States, Canada, and Latin America to offer international roaming and participated in bids

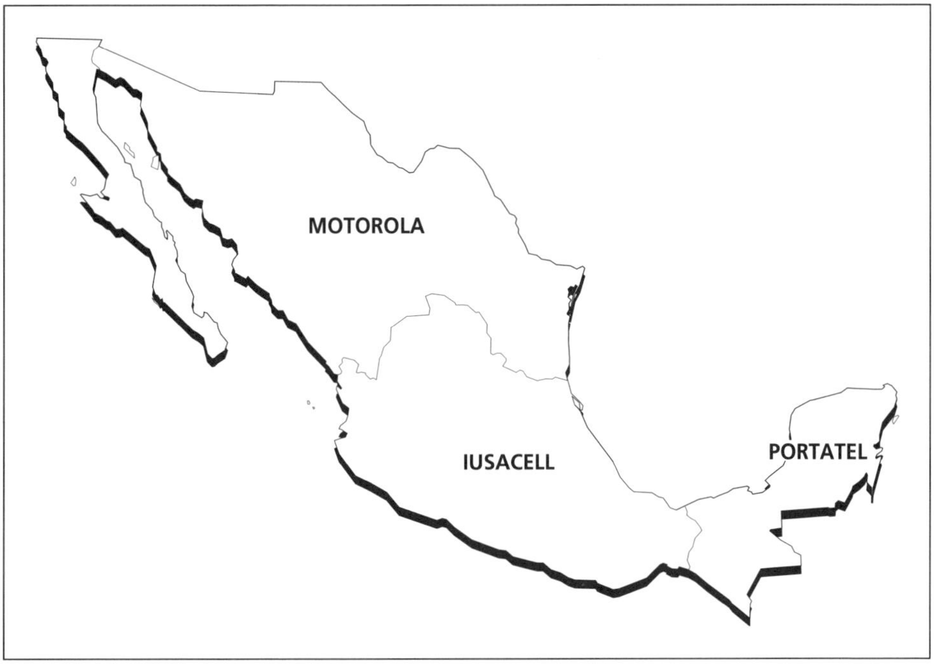

Figure 2 *The cellular industry after consolidation (1995)*

to obtain operating licenses in South American countries.

By December 1990, cellular service was in operation in each of the nine cellular regions. Although in the early stages cellular service was mostly a status symbol, the fad was quickly replaced by more practical uses as the cellular phone became an indispensable business tool and a device that provided security and peace of mind. By the end of 1993, the service had expanded to more than 200 cities and included features such as free emergency public telephones on major highways of the country.

Current Developments

The early stages of uncertainty have been replaced by a period of solid growth and consolidation in the industry. In 1992 Grupo Iusa, now a $1.7 billion "vertically integrated machine" of more than 100 companies (and parent company of Iusacell) incorporated Grupo Iusacell to encompass all telecommunications-related activities for the group.[5] In 1993, Iusacell acquired control of Cellular

Regions 6 and 7; in March of 1994, it acquired Region 5. With these three new regions and its original operation in Region 9, Iusacell has access now to nearly 70 percent of the Mexican population and a strong presence in the center of the country (*Figure 2*).

In 1993, Iusacell sold 46 percent of its equity to Bell Atlantic for an amount close to one billion dollars in the first of a number of alliances on preparation for competition in basic telecommunication services. With this venture, Iusacell acquired cash to reduce its debt and to support its expansion program. In July 1994, Iusacell signed a memorandum of understanding with Sprint to establish a partnership to enter the Mexican long distance market; however, Sprint decided to team up with Telmex, leaving the Iusacell/Bell Atlantic alliance at the end of 1994.

With a solid base in its cellular home turf, Iusacell represents a serious threat to Telmex. It has expanded into value-added services in private business data networks (Iusanet), satellite-delivered services (Satelitrón), medical services via cellular telephone (Cardiotel), and rural telephony and wireless domestic fixed service. Iusatel (as well as Telcel) conducted extensive field trials of wireless residential services, and they have been lobbying the government to allow them to launch the service in the areas of its cellular coverage. Following the strategy that allowed them to introduce cellular in 1989, Iusacell's position is that its original 1957 authorization and the lack of regulations for local service leave the door open for them to provide wireless local service. The regulatory authorities disagree with Iusa's interpretation and the dispute has extended over a period of one year. Other companies are closely watching these events, as it would be detrimental to potential competitors if Iusacell could get a head start.

Even with its aggressive growth and tremendous expansion, Iusacell is not the only big independent player in the cellular business in Mexico.

- Motorola gained control of cellular Regions 1, 2, 3 and 4, and has indicated a strong interest in getting involved in operational activities. The coverage under Motorola's control covers the northern part of the country and has key economic importance because of its tremendous growth and the proximity to the U. S. border. With trade becoming increasingly important, a significant part of the national economic activity is moving north of Mexico City to Monterrey, where the old industrial power base is combining with financial groups to create a new center of gravity at least as important as the Mexican capital. Motorola executives have indicated their belief that a public service, based on microcell technology, can reach 25 percent to 30 percent home penetration by the year 2000 and 80 percent to 90 percent penetration by 2010.

- Niño de Rivera's organization cannot be counted out. A former player in still-independent Cellular Region 8, it has expressed interest in obtaining a license for basic local telephone service. In addition, it is a strong participant in Cuba, where through Cubatel in partnership with the Cuban government, it is the authorized provider of cellular service in the island nation.

Mexican organizations are also becoming important international players in the industry:

- Grupo Domos, which controlled Cellular Region 3 before selling to Motorola, acquired 49 percent of the Cuban wireline telecommunications operator. In addition, this company is reported to be associated with Bell South and has expressed interest in applying for a license to provide basic telecommunications services Mexico. With the participation of Bell South, a strong wireless player in the United States and South America, wireless technologies would be an important element in the strategy of this alliance.

- Iusacell has direct involvement in cellular operations in South America (Conacel in Ecuador) and a long distance license in Chile (Iusatel Chile).

- Telmex has indicated interest in getting involved with telecommunications operators in Central America.

- Monterrey-based Pulsar invested in British wireless technology provider Ionica. Although this company has announced that they will no longer pursue a facilities-based strategy in Mexico, they are still analyzing the possibility of acting as service resellers.

- Radio Beep, also based in Monterrey, announced an alliance with Westel (U.S.) and Teleglobe (Canada). Although the original line of business was in the radiolocation industry, this group has also announced interest in obtaining licenses for basic telephone services in Mexico.

- Another alliance has been announced between Mexico's Mobilcom and U.S.-based Nextel.

Opportunities and Challenges

The future of the industry in Mexico is still unclear; in particular, there are many big challenges for wireless applications. Some of the major areas of concern are:

- pending regulation
- timing under which the SCT will issue the required regulations
- the consolidation process which will follow the opening movement
- increased competition and the consolidation of the "super-regions" under new ownership
- spectrum saturation in the main population centers and its impact on the quality of service
- reduction of tariffs to internationally competitive levels, especially with respect to access charges levied by Telmex for interconnection to the switched public telephone network
- migration from traditional analog cellular to next generation technologies
- spectrum allocation process, management, and enforcement of proper use
- limits of foreign ownership of telecommunications companies
- evolution of cellular carriers to full-fledged telecommunications services providers
- introduction of GSM and PCS
- impact on low and medium earth orbit satellite systems on more traditional channels of service distribution

With barely one year in office, the administration of President Zedillo has already been faced with several major challenges: short-term liquidity, problems to meet its national debt, devaluation of the Mexican Peso, a confidence crisis, and recurrent social problems in the State of Chiapas in southern Mexico.

With these problems demanding immediate attention, competition in telecommunications keeps being delayed. Originally scheduled to be available in mid-1994, the rules for competition were first delayed by the Presidential Elections (August 1994) then delayed again until the new administration took over (December 1, 1994), and then again by the crisis resulting from the devaluation of December 20.

Current delays have been attributed to the massive amount of work required to produce a coherent and comprehensive set of rules. The new administration has had to produce new laws on telecommunications, railroads, airports, and sea ports, while at the same time government officials go through the learning process associated with the development of new regulatory frameworks, the introduction of new technologies, and changing market dynamics.

Interested groups would like the government to make public all the information related to the process: the regulatory framework, when the licenses will be bid, how many licenses will be available, the criteria under which the decisions will be made, etc. With competition originally scheduled for August 1996 (and interconnection starting January 1, 1997) new entrants feel they are running out of time to plan, implement, and test a network that will give them the required competitive advantages to be successful in the Mexican market.

Yet, significant legislative and regulatory progress was achieved through the first part of 1995. The new telecommunications law enacted in June 1995 has been praised as progressive and world class and represents a major breakthrough to the current market structure. The law opens up to competition all segments of the public telecommunications service market.

The new law will allow investors to freely chose the segment(s) they want to address (including local service, domestic long distance, international long distance, or any combination thereof) with an open-ended selection of services to be provided using the technology and geographical coverage of their choice. It sets itself at the forefront of the changing regulatory paradigm in telephony by eliminating traditional boundaries between local and long-distance services, creating incentives for carriers to become facilities-based service providers, and setting the basic rules to guarantee a healthy competitive environment.

In the midst of these circumstances, 387 organizations registered with the SCT by August 4, indicating their interest to apply for 870 telecommunications licenses and permits. Of these, 36 were for long distance, 66 for local fixed wireless service, 11 for cellular, 75 for PCS, 41 for satellite services, and 134 for value-added services.

The application process in some areas has already started. Avantel (a Banamex-MCI company) received approval for its application for long-distance service and has started building its fiber backbone while the concession document is finalized. Iusatel (Iusa-Bell Atlantic), Marcatel (Radio Beep, IXC/Westel and Teleglobe) and Unicom (Bancomer-GTE) have all officially applied. The SCT is required to respond to applications in no more than 120 days. With the volume of applications expected, and assuming no problems or delays, most applicants will be receiving their licenses well into 1996, although by then the main players in long-distance telephony will probably have their concession titles.

In the wireless arena, recent information from the government of Mexico states that bidding for narrowband PCS allocations will start late in 1995, while wideband PCS will be introduced by the end of 1996. These time frames are determined by their workload as well as by the time and effort required to negotiate with existing users to vacate the required frequency bands.

The short implementation time frames, increased pressure on containing costs, the difficulty involved in obtaining rights-of-way, and the need to start providing service before the competition, define environmental conditions that set the stage for wireless technologies to become a key (if not the primary key) in the new telecommunications order in Mexico. This situation will no doubt put additional strain on the SCT, especially for those involved with the administration of the electromagnetic spectrum, as most frequency bands in major metropolitan centers are saturated and inefficiently used.

In areas such as these, the role of regulation as an incubator for healthy competition will be critical. PCS, for instance, will face a mature cellular market with entrenched players enjoying a dominant position resulting from the installed infrastructure, recognized brands, existing customer base, and so forth. Market entrants will have to invest significant amounts of time and money, first to obtain frequency allocations and then to build the network and market presence to compete. In the meantime, cellular operators will have time to adjust their marketing strategies to deal with the new players. Enabling regulation will be required as an equalizer to give PCS (restricted TV, some value-added services and applications, among others) a fair chance.

There are multiple opportunities, but not for the weak of heart. Although most financial experts believe the Mexican economy has a sound base, an economy in recovery and the lack of complete definition in the regulatory arena will challenge those interested in securing access to the potentially lucrative Mexican telecommunications market. International factors are likely to play a role, too. Some variables to consider include changing telecommunications regulation in the U.S., demand for investment capital in world markets, experiences obtained in competition in other telecommunications markets around the world (e.g., Chile and the United Kingdom), and alliances between international telecommunications carriers.

The doors are about to open for telecommunications equipment suppliers, entrepreneurs, and services providers to actively participate in the challenge to increase the teledensity in the Mexican territory from a current level of 9 per 100 inhabitants to an estimated 20 per 100 by the year 2000. Based on history, it is expected that most of the participants will be from the United States, with Canada and Europe following close behind. Let us just hope that these new doors are not of the revolving variety.

References

1 Pyramid Research, Inc., *Mexico's Telecom Strategy*, 1989, p.104.

2 General Cellular Corporation, 1989.

3 "Selling Cellular", *Business Mexico*. July 1992. P 26.

4 L. Luxner, "Mexico reaches for new telecom heights," *Telephony*, February 3, 1992, p 26.

5 Craig McCaw del Sur, *Forbes*, April 11, 1994, p 56.

6 Pyramid Research, Inc., 1989, p 106.

7 SCT

8 Alacel, 1994.

History of Wireless in Korea and Present Opportunities for Foreign Companies

Jong Wook Myung
University of Sussex

Wireless Industry in the 1980s

Due to the special political and military situation in the Korean Peninsula, namely the division of North and South Korea, the usage of radio waves was restricted for national security until the late 1970s. At that time, there was not even a radio-related manufacturing industry in Korea. Meanwhile, the radio industry in developed countries reached the full-development stage through continuous technology advancement, since it was considered an important industry for economic development.

In the early 1980s, the demand for diverse telecommunications services from all social and economic sectors began to increase, owing to Korea's success in economic development. As a result of the changes in internal and external environments, the Korean government recognized the necessity for changes in radio waves regulatory policy from its traditional control-oriented regulation to promotion-oriented administration.

Accordingly, the Ministry of Information and Communications (MIC) of Korea suggested the innovative open policy for radio waves promotion in the early 1980s. Since then, the MIC has oriented its energetic policies, such as introduction and promotion of new services, promotion of the radio industry, and technology development, to support this radio waves usage promotion policy.[1]

Improvement of Radio Waves Usage System

The cordless phone was authorized by the MICs in 1983 as one of promotion, and customers could freely use the cordless phone through simple notification by 1988.[2,3] There were many amendments made to the Radio Waves Regulation Law of 1989 to meet the increasing demand for radio waves usage due to Korea's huge success in economic development and the formation of an information society. According to its amendments, the purpose of the Radio Waves Regulation Law focused on the development of wireless telecommunications and the promotion of public welfare rather than by simply maintaining radio waves control systems.[4] This was accomplished through promotion of radio waves usage. In addition, a radio station emitting weak waves or one for reception only could be opened by simple notification.[5]

Introduction of New Services

According to the MIC's promotion policy, a variety of mobile telecommunications services including radio paging, cellular telephone, trunked radio communication, and airport radio telephone services were provided in the 1980s. The first radio paging service was provided in Seoul in December 1982. An additional 10,000 circuits were extended in 1984. In 1985, new signaling that also displayed the telephone number was introduced (see *Table 1*).

In the case of cellular telephone service, 3,000 circuits were first supplied to the Seoul metropolitan area, including Seoul, Anyang, Suwon, and Sungnam, in 1984. With continued expansion of the facilities during 1987 and 1988, the number of service areas reached 19 by the end of 1988. Service to another 36 municipalities was established in 1989; thus, car phone service became available to a total of 55 cities by the end of 1989 (see *Table 2*).

In the case of airport radio telephone services, the first service in Korea was established at the Kimpo International Airport on December 30, 1985, and the

Description	1988	1989	1990
Number of facilities	150,000	287,000 (91.3%)	535,000 (86.4%)
Number of subscribers	100,373	198,286 (97.5%)	410,480 (107.0%)
Number of service areas	32 cities	73 cities + 71 towns	73 cities +117 towns

Note: Figures in parentheses represent percent of increase over the previous year.

Table 1 *Radio paging facilities and number of subscribers*[6]

Description	Number of Facilities (Exchange)	Car Phone Subscribers	Portable Phone	Total	Operation Ratio (%)
Seoul metropolitan area	45,000	27,324	3,430	30,754	68.3
Other areas	22,500	6,729	2,235	8,964	39.8
Total	67,500	34,053	5,665	39,718	58.8

Table 2 *Cellular telephone facilities and number of subscribers in 1989*[7]

Korea Airports Authority took charge of the service operation.[8] By the end of 1988, Kimpo International Airport had been equipped with both earth-to-air and earth-to-earth facilities. Eight units of earth-to-air and 62 units of earth-to-earth communications facilities were being operated, but demand was on the increase because of the increasing number of foreign airlines being commissioned. This situation necessitated the extension of the earth-to-air and earth-to-earth systems by 1 unit respectively and an additional installation of 50 units of portable earth-to-earth terminals in 1989. This resulted in 9 units of earth-to-air terminals and 80 units of earth-to-earth terminals in operation as of the end of 1989. *Table 3* shows the status of facilities and subscribers of airport radio telephone services.

In the case of trunked radio communication services,[9] starting with the installation and operation of a repeater with 10 channels (service capacity: 800 units) of radio frequency as communication support for the 24th Seoul Olympics by the Korea Telecom (KT), this service began its second introduction of 5 channels (service capacity: 400 units) for frequency-sharing communication in Pusan on October 4, 1988. By using the support system provided for the Seoul Olympics, the service

	Earth-to-Air	Earth-to-Earth		Total
		Fixed	*Portable*	
Systems	2	2		4
Terminals	18	7	106	131
Subscribers (company)	13	7	14	34

Table 3 *Status of facilities and subscribers to airport radio telephone services*
Source: Kimpo Airport Telephone Station

Description	Area	End of 1988	Installation 1989	End of 1989	Service Capacity
Local Channel (Local phone call)	Pusan	5 Ch.	3 Ch.	8 Ch.	64 units
	Masan	–	3 Ch.	3 Ch.	240 units
Wide Area Channel (Toll call)	Pusan and Masan	–	2 Ch.	2 Ch.	160 units
	Ulsan	–	–	–	–
Total		5 Ch.	11 Ch.	16 Ch.	1,280 units

Table 4 *Frequency-sharing communication facilities*[10]

area has been extended to Masan, Ulsan, Changwon, Jinhae, and Kimhae since the end of 1989 as shown in *Table 4*. As a result, the capacity of frequency-sharing communication facilities totaled 1,280 with 18 companies having subscribed as of the end of 1989, and a total of 97 terminals in operation.

Korea Mobile Telecommunication Company (KMTC) was established on March 29, 1984, to meet the nation's need for mobile telecommunications services and to induce specialization in mobile telecommunications. But the Korea Telecom (KT) was still responsible for investment in mobile telecommunications facilities while the newly established KMTC had responsibility for maintenance, sales, and management. As a result, this division of responsibilities caused many difficulties. On the other hand, pressure from the developed countries to open the Korean mobile telecommunications market prompted the MIC to assign KMTC as an independent common carrier in accordance with the government's policy to foster specialization of public communication service that became effective in April 1988. KMTC has independently managed business functions related to mobile telecommunications services since July 1988 and it went public in October 1989.[11]

Wireless Industry in the 1990s

Due to the MIC's promotion policy of the 1980s, the market size of cellular and paging services and the sales volume of mobile terminals reached $323 million and $151 million, respectively, by the end of 1992. These figures indicate that the Korean mobile telecommunications market has been in the beginning stages of growth.[12] Since the late 1980s, pressure from the developed countries to open the

Korean market has only been increasing; furthermore, the high dependence of Korean mobile terminal manufacturers on foreign forerunners – the United States and Japan – is considered undesirable because this reliance implies that Korea, like most developing countries, has a weak mobile telecommunications manufacturing base.

In order to overcome these difficulties, the MIC decided to introduce competition into the market through restructuring of the telecommunications industry, and it chose a second cellular carrier in February 1994 to reinforce international competitiveness.[13] In addition, the MIC announced the Medium & Long-Term Plan of Radio Waves Promotion to facilitate continuous development in the mobile telecommunications industry. Based on this plan, the policies for digital cellular telecommunications system development, education, and training of mobile telecommunications engineers have been driven energetically.

Introduction of a Competitive Market System

In July 1990, the MIC announced an ambitious blueprint for restructuring the Korean telecommunications industry. The main focus of the plan was to introduce competition in some sub-markets traditionally operated by monopolists: Korean Telecom (KT) and the Korea Mobile Telecommunications Corp. (KMTC). According to this plan, the local call market is to remain a monopoly while competing entrants will be allowed into the markets for long-distance, international, and mobile telecommunications services. No particular entry restriction is to be imposed in the value-added service markets.

Through amendments to the relevant laws and regulations, such as the Restructuring Plan of the Telecommunications Industry and Telecommunications Business Law, a new entrant competing for a cellular service market is not restricted by service areas. For the paging service market, ten new paging service providers were permitted to compete against KMTC in each divided service area since 1993, which could establish a solid foundation for the competitive mobile services market in Korea. The main reason why nationwide service licenses were given to new cellular operators was to assure level competition circumstances with existing service providers such as KMTC. Furthermore, the inflow of foreign capital was prohibited for KMTC, while a second operator was permitted to form a consortium with foreign telecommunications companies.[14]

Two or three years after the introduction of competition, full-scale competition for market share among cellular operators is expected to be realized. This should lead to the provision of high-quality services and reductions in call rates for Korean cellular service users and encourage an increase in demand for mobile telecommunications services through a wide variety of marketing strategies.

Extensive Support for Radio Industry

Assuming that growing demand boosts related industries, the MIC has been trying to develop and promote new services for cellular telecommunications, satellite communications, and personal communications so as to expand the market size. To induce specialization in the manufacturing sector of radio equipment, the MIC urges the large-sized companies to concentrate on production of main systems and urges the small and medium-sized companies to concentrate on production of parts. In addition, the MIC has been restructuring sales and distribution for product promotion. In 1990, the Radio Industry Promotion Association was established to promote the industry's joint technology development projects.[15]

Mobile Telecommunications Technology Development

Currently, the government seems to have given up on localizing analog cellular systems. Instead, it is focusing on the development of digital cellular systems. In April 1992, the MIC announced the idea of developing core parts in digital cellular systems and terminals in order to secure international competitiveness. According to the plan, eight core parts of a digital cellular system, fourteen core parts of cellular phones, and four core parts of other mobile telecommunications equipment are targeted for development between 1992 and 1996. The investment will be 44 billion won (US $55 million).[16]

In particular, the joint development project for a CDMA-based digital cellular system is considered important for full localization of a digital cellular system technology. To meet the increasing demand for cellular services, a new cellular operator is scheduled to launch a digital cellular system in the Seoul metropolitan area in 1996. Under the guidance of the MIC, the Electronics and Telecommunications Research Institute (ETRI) has been working hard on the successful development of a digital cellular system jointly with KMTC and three Korean telecommunications equipment suppliers: SamSung Electronic, LG Information and Communications, and Hyundai Electronic.[17] The development of the portable phone has been led by ETRI jointly with those three manufacturers and Maxon Electronics.

Each participant in the development of the digital cellular system is obliged to

Phase		Items to Develop	Items to be Transferred
Basic phase	*1st phase ($1.9 M U.S.)*	• Define program and system • Initialize design of CDMA network	• Technology orientation • CDMA digital CAI (computer aided instruction) standard
Design phase	*2nd phase*	• Development of cellular system	• Transfer of RTS (research test system)
	($10 M U.S.)	• Test cells and mobile stations	• Install RTS in Korea • High-level design specifications of HW and SW
	3rd phase ($5 M U.S.)	• Support final product • Joint agreement with network facility manufacturer and prepare for manufacturing	• Low-level design specifications of HW and SW, such as cell HW and SW, cell RF and controller, and cell O&M system
Manufacturing phase	*4th phase (optional)*	• Analyze final testing data of cellular network	• Testing model of cellular network

Table 5 *CDMA development schedule of ETRI and Qualcomm, Inc.*
Source: M.J. Choi, "Dispute over conversion to digital cellular system," Monthly Cellular, November 1992, p. 789.

pay 12 billion won (US $15 million) as a share of expenses. Hyundai Electronic, which does not have any experience in TDX-10, has participated in limited development areas, such as hardware and software for value-added facilities of TDX- 10.[18,19]

In November 1994, ETRI announced that the so-called CMS-2 (CDMA Mobile System), a commercialized test product for digital cellular systems, had been developed. Its subscriber capacity is up to 1 million users, and for a base station system two to eight channels can be operated on each board. The development technologies of CMS-2 are to be transferred to participants such as SamSung, Hyundai, LG, and Maxon so that each manufacturer will be able to produce the CDMA cellular system competitively from 1995 on (see *Table 5*).

Personal Communication Service Technology Development

The MIC is promoting the PCS technology development plan as a first step toward realizing its vision of future public land mobile telephone services (FPLMTS). The objective is to look for ways to implement communications services anywhere at any time to any person.[20]

In the development of PCS technology, the main responsibility for control and coordination of all research participants, such as cellular carriers, university research centers, and industries, was given to ETRI. According to the plan, the fundamental research jobs were scheduled to be completed by 1993, and the PCS system models were to be placed in service by 1996.[21]

The government is planning to award the PCS licenses to Korea Telecom and two other domestic companies in 1996. In the case of the two cellular carriers, KMTC and Shinsegi Telecomm are automatically entitled to the PCS license without applying for it. An impassioned debate is raging over the PCS standard regarding whether to introduce TDMA or CDMA. To avoid overlapping investment, the domestically developed CDMA technology is likely to be adopted as the PCS standard technology by the MIC.

Opportunities for Foreign Telecommunications Companies in the Korean Wireless Market

The number of cellular service subscribers in Korea reached a total of 900,000 by the end of 1994. Cellular services were first introduced to Korea in May 1984, and there were 2,658 subscribers.[22] Only 10 years later, that number increased more than 340 times. As a result, in 1994 more than 40,000 cellular phone terminals on average were sold every month. In particular, in August 1995 the market share rates of Motorola and SamSung were 40.6 percent and 51.5 percent, respectively. Meanwhile, LG Electronic and Hyundai Electronic secured 4.1 percent and 3.8 percent shares, respectively. But Japanese and European cellular phones such as Nokia and Ericsson had their combined market share reduced by 3.8 percent. Recently, Motorola began to strengthen its direct sales activities by greatly reducing the price of Motorola products. Hence, the restructuring process in the cellular phone market has begun to reflect a new dual market system consisting of Motorola and SamSung.[23]

The year of 1993 brought an end to the government-owned KMTC's eight-year monopoly. Throughout the peninsula, ten new paging operators have serviced their own individual areas. According to a recent market survey of KMTC and new operators, in six out of nine paging service areas more than 55 percent of new subscribers tended to choose new operators.[24] The reasons for the success of these new operators include a customer-oriented business attitude and additional services such as voice mail systems. The cellular operator Shinsegi Telecomm, Inc. (STI) will begin operations in April 1996 and is aiming at a 42 percent market share by 1998, which would put the newcomer in fierce competition with KMTC.

Business Coalitions with Foreign Telecommunications Companies

KMTC intends to transform itself into an international telecommunications service company in the near future by focusing its efforts on sharpening international competitiveness. To cope with changing market situations and to prepare for the impending globalization of the industry, KMTC has begun to make strategic coalitions with domestic and foreign telecommunications companies. In 1995, KMTC will make strategic coalitions with Alcatel of France in the field of transmission network technology, Bell Atlantic in the field of frequency interference elimination, and Bell South and NYNEX in the field of network optimization technology.[25]

For mobile data communications services, licenses are likely to be granted to existing telecommunications operators such as KT, KMTC, and DACOM rather than to new operators because of the difficulty in forecasting demand. These operators are planning to start nationwide commercial mobile data communications services by installing additional equipment in the present cellular network beginning in 1997. The biggest reason for fierce competition in the mobile data communications services market among these operators is that the operator, who can prove its superiority in technology and service is expected to gain an advantage over other applicants for the licensing of PCS business in 1996.[26]

Major foreign telecommunications operators have been trying to draw attention of Korean telecommunications operators and equipment manufacturers for the purpose of forming partnerships in the LEO satellite communications business. As a result, Hyundai Electronic and DACOM decided to participate in the Global Star Project of LQSS Co., KMTC decided to participate in the Iridium Project of Motorola, and KT, Shinsegi Telecomm and SamSung Electronic have formed a consortium in a move to jointly participate in Project21 of Inmarsat.[27]

A nationwide license of trunked radio system (TRS) service will be granted to Korea Port Telephone Co., a subsidiary of KT. Competition in localizing TRS digital equipment of 380 MHz is heating up among Korean wireless equipment manufacturers. Hyundai Electronic, SamSung Electronic, and LG Communications have been working on developing TRS equipment with E. F. Johnson Co. of the United States, Ericsson of Sweden, and Motorola.[28]

Potential of the Korean Wireless Industry

According to the 1995 symposium on forecasting in the Korean information and telecommunications industry led by the Ministry of Information and

Communications, market size of the Korean wireless industry will reach 1.2 trillion won (US $1.5 billion).[29] Until the year 2000, the Korean wireless market is expected to grow by 16 percent annually on average due to investment in the information superhighway network, the spread of multimedia products, the expansion of contracts for information services, and the establishment of private communications networks for companies.

Starting in 1995, economic development cooperation between South Korea and North Korea will become active. Prior to manufacturing by South Korean firms, the establishment of telecommunications infrastructure in economic development areas designated by the North Korean government is considered very important for efficient support of manufacturing activities. KT is considering the establishment of a telecommunications network centering on wireless telephone systems because of undeveloped wireline telecommunications networks in North Korea. It is known that the North Korean government has a network establishment plan with a capacity of 40,000 wireline telephone lines, 1,200 mobile telephone lines, and 1,500 paging lines in the designated areas. The North Korean government hopes foreign telecommunications companies will participate in this network establishment project. AT&T and Alcatel of France have shown an especially strong interest in the project. Additionally, for ten years after unification, KT is planning to make a 16 trillion won (US $20 billion) investment in the modernization of telecommunications networks in North Korea.[30]

Notes

1. S. D. Yi, Policy Suggestions for Promotion of Korean Mobile Telecommunications Industry" (Seoul: KISDI (Korea Information Society Development Institute), 1992), p. 44.
2. Cordless phone users filled out the notification form at the telephone shop.
3. Ministry of Information and Communications, *Korean Telecommunications*, November 1988, p. 227.
4. Ministry of Information and Communications, *1990 Yearbook of the Ministry of Information and Communications*, November 1990, pp. 280-281.
5. In principle, foreigners were forbidden to open a radio station. However, in accordance with the principle of reciprocity, radio stations such as those installed at an embassy or other diplomatic establishments, amateur stations, some mobile stations on land, and radio stations tentatively operated during international events are allowed.
6. Ministry of Information and Communications, *1990 Annual Report on Telecommunications*, May 1991, p. 93.
7. Ibid., p. 92.

8 Airport radio telephone service provided for a foreign airline is radio communication used among the airline's office in the airport, aircraft, and land mobile stations (work vehicles and operators).

9 A conventional walkie-talkie is a system in which a specific user exclusively uses a specific frequency. The frequency-sharing communication service is for multiple subscriber groups to share a specific frequency channel. Accordingly, this is considered a sort of mobile communication service that can be utilized by individuals or groups within a company unit among multiple vehicles belonging to a caller, and with transceivers by persons on the move using radio relaying facilities.

10 Ministry of Information and Communications, *1990 Annual Report on Telecommunications*, May 1991, p. 95.

11 S. D. Yi, "Policy Suggestions for Promotion of Korean Mobile Telecommunications Industry (Seoul: KISDI, 1992), p. 48.

12 "The Trend & Forecasting of Korean Mobile Telecommunications" (Seoul: Ssang Yong Economics Research Institute, l991), p. 31.

13 POSCO (Pohang Iron & Steel Co.), the largest steelmaker in Korea, was selected as a second cellular carrier.

14 Jae-Cheol Kim and Tae-Seok Ro, "Current Policy Issues in the Korean Telecommunications Industry," *Telecommunications Policy*, September/October 1993, p. 481.

15 S. D. Yi, "Policy Suggestions for Promotion of Korean Mobile Telecommunications Industry" (Seoul: (Korea Information Society Development Institute, 1992), p.55.

16 M. H. Yi, "Too many parts in cellular equipment are imported from overseas," *Monthly Cellular*, March 1994, p. s7.

17 J. S. Park, "Current status of Korean mobile telecommunications equipment market and counter measures," *Monthly Cellular*, November 1992, p.45.

18 A Korean digital cellular switching system will be a variation of TDX-10.

19 Electronic Shinmun-Sa., *Information & Communications Yearbook 1993*, June 1993, p. 1020.

20 A new type of wireless telephone system that would use light, inexpensive handheld handsets and communicate via low-power antennas. PCN is primarily seen as a city communications system, with far less range than cellular.

21 Ministry of Information and Communications, *1991 Annual Report on Telecommunications*, May 1992, p. 96.

22 S. C. Choi, "The Number of Cellular Subscribers reaches 500,000, *Electronic Newspaper*, 7 November 1994, p.3.

23 S. W. Park, Restructuring process in portable phone market," *Information and Telecommunications Newspaper*. 9 November 1994, p.8.

24 S. C. Choi, Success of New Paging Operators, *Electronic Newspaper*, 16 November 1994, p. 5.

25 "KMTC's Strategic Coalitions," *Information and Telecommunications Newspaper*, 23 November 1994, p. 10.

26 S, C Choi, "Why so hurry in mobile data communications business," *Electronic Newspaper*. 2 November 1994, p. 7.

27 C. Kim, "Satellite War Over Korean Sky," *ChoSun Ilbo*, 4 November 1994, p. 10.

28 S. C. Choi, "Heat up in Localization of TRS Equipment," *Electronic Newspaper*, 11 November 1994, p. 12.

29 W. M. Ku, "Bright Forecasting of Information and Telecommunications Industry," *Electronic Newspaper*, 30 November 1994, p. 5.

30 M. H. Lee, "South-North Telecommunications Network Plan," *ChuneAne Ilbo*, November 11 1994, sec. Economics, p. 6.

Cellular and Personal Communications in Europe

Aldo Paoletti
Interdisciplinary Telecommunications Program
University of Colorado at Boulder

Introduction

In Europe as in the rest of the world, the cellular market is increasing dramatically. This market currently accounts for more than 15 million users of mobile communications. Furthermore, it is forecasted a market of 50 million by the turn of the century and 100 million by the year 2005.[1] Also, the portion of the population serviced by fixed telephony may not exceed an average of 50 percent. On the other hand, personal mobile communication is expected to reach nearly 80 percent of Europe's population. The scenario is quite different for eastern and western Europe as the former has minor fixed infrastructure and started the provision of cellular service at a later time. In 1994, cellular service had little more than 1 million subscribers.[10] On the other hand, in western Europe the market as a whole grew by 60 percent over the course of the year – the highest annual rate for five years – topping 14 million subscribers at the beginning of 1995.[6]

Analog systems accounted in the early 1990s for the majority of subscribers, especially in eastern Europe. However, a particular growth is registered in the adoption of digital systems thanks to the launch in the early 1990s of a pan-European digital system known as GSM (global system for mobile communications) along with its higher-frequency derivation, DCS1800 (digital cellular system). GSM is being adopted as the unique digital cellular standard all over Europe and is clearly taking over analog systems. At the end of 1994, 25 percent of cellular subscribers in western Europe were using digital networks, and the figure is increasing at a very high rate. Studies show that 84 percent of cellular units sold in 1999 will be digital.[4] All western Europe countries are already offering GSM service while most eastern countries are planning on it.

At the same time, the mobile scenario is going through rapid evolution. Several

European Union projects currently involve the development of so-called third-generation systems, such as universal mobile telecommunication systems (UMTS) and mobile broadband systems (MBS) for the provision of a variety of services including voice, data, video, and multimedia. These efforts are parallel to studies by the International Telecommunication Union (ITU) regarding the implementation of future public land mobile telecommunication systems (FPLMTS). For the provision of these services, an evolutionary path from GSM seems likely but is not certain. A European union study group is considering both the use of code division multiple access (CDMA) and time division multiple access (TDMA) technology to see which one would better fit in the scenario of third generation mobile services whose introduction is expected around the year 2000. The ITU's recommendations, service specifications, performance, network, and radio issues for FPLMTS are planned to be made available by 1998.[3]

If the deployment of cellular service in Europe seems quite mature, the situation looks different for personal communication services (PCS), which are referred to in Europe as the personal communication network (PCN). The U.K. and Germany are the only countries so far that have implemented it. In countries other than the U.K., France, and Germany, PCN may not be launched until GSM networks are well established. This is partly due to the lack of competition in most markets. In fact, the provision of mobile services in most European countries is granted to operators who generally are in a monopoly or duopoly situation. Typical operators are local PTTs and partnerships of international companies joining local interests. To stimulate the growth of the market, a "Green Paper on Mobile Telecommunications" (GPMC) was published by the European Union in 1994 recommending that all restrictions on mobile communications be lifted.

Furthermore, the GPMC aims at a full liberalization of telecommunications, which is expected by 1998. The promotion of new mobile communication services all over Europe is also envisioned. The UMTS satellite component is also taken into account as a means for accomplishing global coverage. However, GPMC's recommendations are not compulsory, and it will be the market that eventually determines operators' strategies.

Market

Western Europe Overlook

In the early 1980s, analog cellular systems became operational in Europe. These were

Subscribers as of Jan. 1 (thousands)	1992	1993	1994	1995	1996
Western Europe	4,700	5,820	8,700	14,220	21,463
Eastern Europe and CIS subscribers	N/A	111	324.5	486*	1,300
Rest of the world	11,400	17,000	24,900	40,100*	N/A

Table 1 *Europe's cellular markets*
* Estimates

based on national or imported standards such as total access communication system (TACS) operating at 900 MHz (U.K., Italy, Spain), Nordic mobile telephone (NMT) operating at 450 MHz and 900 MHz (Nordic and Benelux countries), and some proprietary standards found in countries such as France and Germany (C-Netz). Because of the incompatibility among analog systems, it turned out that there was a need for a standard that would enable roaming throughout Europe, the GSM. Currently the majority of subscribers are still using analog networks, but the proportion of digital networks is increasing rapidly. In the last few years, cellular growth has been spectacular – especially in the U.K., Germany, and Italy, which accounted for 57.6 percent of the total market share in 1994. The total west European growth figure for 1994 is nearly twice as high as the 2.8 million recorded in 1993 and more than four times as many as the 1.3 million added in 1992.[6] In absolute terms, western Europe accounted in 1994 for 35 percent of world subscribership, as shown in *Table 1*. In *Table 2*, Europe's ten largest cellular markets are reported along with the number of subscribers and market share.[7] The U.K. accounted for nearly one-third of all the net new subscribers added to western European cellular telephone networks from January 1994 to January 1995, adding more than 1.5 million new subscribers. The U.K. thus accounted for more than 28 percent of the 5.43 million subscribers added across western Europe as the total subscriber base grew to 14.22 million. The market as a whole grew by 60 percent over the year – the highest annual rate for five years.

The U.K. has overtaken Germany, which was the fastest growing market in 1993. This reflects the fact that German growth has slowed in both relative and absolute terms. The German market accounted for only 12.9 percent of west European growth, putting it in third place behind Italy, which added 1.03 million subscribers and accounted for 19 percent of total growth. In relative terms, the German market grew by only by 39 percent, compared to 78 percent in the U.K. and 86 percent in Italy.[6] The low German figures are partly due to the fact that it is getting harder to find new subscribers because of the relatively high tariffs charged by operators for their GSM services, which account for virtually all new customers.

Country	Subscriber Base 1/1/95	Subscriber Base 1/1/94	Subscriber Base 1/1/93	Market Share 1993 (%)	Market Share 1994 (%)
U.K	3,520,000	1,999,000	1,397,200	22.8	24.7
Germany	2,480,000	1,775,584	951,890	20.2	17.4
Italy	2,210,000	1,207,000	780,800	13.7	15.5
Sweden	1,300,000	844,455	676,960	9.6	9.1
France	870,000	561,000	437,000	6.4	6.1
Finland	652,800	480,233	350,913	5.5	4.6
Norway	528,000	371,512	282,918	4.2	3.7
Denmark	509,600	316,261	212,400	3.6	3.6
Switzerland	310,057	258,227	215,061	2.9	2.2
Spain	316,000	258,078	180,707	2.9	2.2
Total	**12,696,457**	**8,072,250**	**5,485,849**	**91.8**	**89.1**

Table 2 *Europe's 10 largest cellular markets*

Even if these three countries account for the majority of subscribers, cellular penetration is much higher in Scandinavia. Despite already high penetration, Scandinavian markets showed impressive subscriber growth over the year. For instance, the Norwegian market showed its highest percentage growth in nine years. Netcom, the country's privately owned second GSM operator, made a considerable impact on the market by providing stiff competition to incumbent cellular operator Tele-Mobil, the mobile arm of state-owned Telenor. Both operators introduced tariff packages targeted at consumers, and handset prices came down significantly. This was a trend across a number of countries in the region. Strong growth helped the Scandinavian countries maintain their position as the most highly penetrated cellular markets in Europe. Sweden, which already had the highest penetration of the five countries, further increased its lead over nearest rival Finland, taking overall penetration to nearly 16 percent of the population.[6]

Western Europe accounts for the countries with the highest penetration rate. A comparison among the top nine European countries and the United States and Hong Kong is shown in *Figure 1*. These two countries account for the highest penetration rate in America and in the Pacific rim, respectively. The market is growing quickly as operators benefit from economies of scale. As a matter of fact, European cellular prices fell in 1994. The cost of mobile telephone ownership is set to fall by an average of 10 percent across western Europe in the 1994-1997 time frame as operators attempt to open up the consumer market.[14] Handset

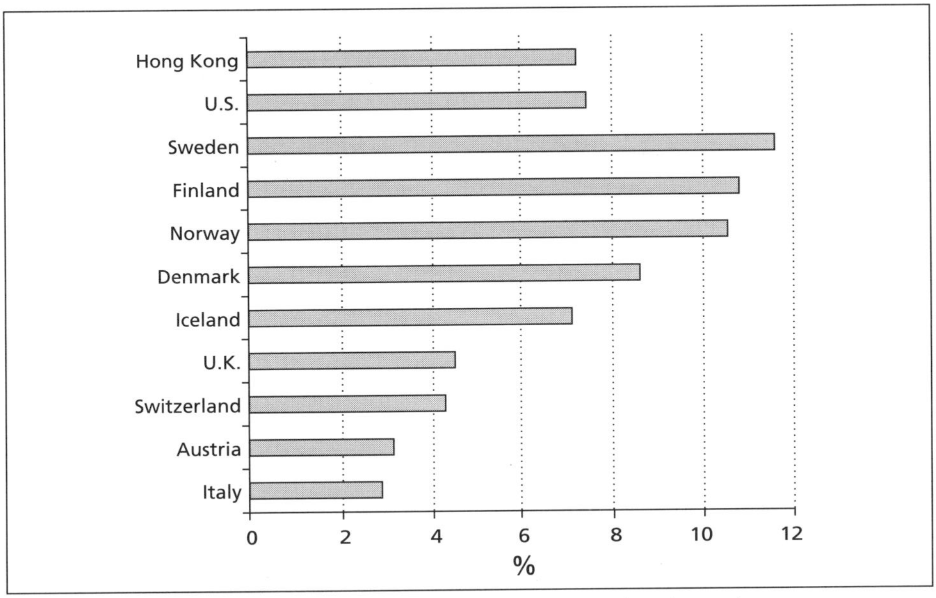

Figure 1 *Top cellular penetration rate in various countries (April, 1995)*

prices and usage prices are expected to fall as an increasing number of operators introduce equipment subsidies alongside a variety of low-user tariff packages and other pricing strategies. However, the fall in the cost of ownership will not be evenly spread. For example, Germany and Italy are expected to see above average price cuts. Handset subsidies will play a significant role because they represent a relatively easy way to attract customers. Subsidies may play a greater role than economies of scale in terms of reducing the retail price of handsets. Nonetheless, it seems that most operators are unlikely to take the subsidy approach as far as their U.K. equivalents have done.

A genuine mass market is still to be created in Europe. In order to achieve this, operators will have to make a number of changes to the way that they market their services. The ratio between the usage costs of cellular and fixed services, which currently stands about 3:1, will have to fall to something closer to 1.4:1 to 1.8:1. Operators will have to start promoting mobile telephony in the same way as other consumer goods and to create brand differentiation. They will have to put more emphasis on distributing through department stores and consumer electronics stores, as is already happening in the U.K. There is also a need for educating consumers about the limitations of cellular technology, and geographical coverage will need to be very good if those concerned about personal security are to be attracted. In such a case, residential users could account for as much as 55 percent of the 30 million cellular users predicted for western Europe by the year 2000. However, progress so far has been limited – only the U.K. and Scandinavia have made any significant moves in the direction of a mobile consumer market.

While an increasing number of European countries have introduced competition, this has not necessarily led to lower prices. In many duopoly markets, the costs of usage remain similar to or even higher than the prices that apply in monopoly markets. Analysts argue that there are several reasons for this.[8] Some monopoly operators, such as Telecom Italia, have implemented significant tariff reductions with the aim of establishing themselves in the consumer market before competitive services go into operation. At the same time, some of the former monopolists in newly liberalized markets have yet to reduce their prices because they have yet to feel any real impact from their emerging competitors. In Portugal, for instance, the state-owned operator TMN and the privately owned Telecel both offer relatively high tariffs. In yet other markets, the business demand for cellular services is so strong that both the new entrant and the incumbent can continue to grow their customer bases successfully without needing to target consumers with lower tariffs.

In order to be classed as competitive, a market needs to have at least three operators, as in the U.K., Sweden, and Germany. In fact, the average cost of usage in these markets is between 20 percent and 30 percent lower than in other markets, and prices are particularly low for emergency users with low call volumes. However, the average price in duopoly markets (Denmark, Finland, France, Greece, Norway, Portugal) is virtually the same as in monopoly markets (all others). Nonetheless, it is predicted that over the next three years the pattern of overall usage costs that include call costs, connection charges, rental costs, and handset prices will be significantly affected by the level of competition in the individual market.

There are wide fluctuations in the price of cellular telephones and services in western Europe. Scandinavia is the cheapest region in Europe, and southern Europe tends to be the most expensive. A cellular telephone subscriber in Denmark, Finland, Norway, or Sweden pays a total price for his or her service that is between a third and a half of that paid by a comparable subscriber in France or Germany. The Danish operator Sonofon has the cheapest cellular telephone services in absolute terms both for high users (defined as those making an average of 150 minutes worth of calls a month) and for low users (defined as those making 50 minutes of calls a month). In general, prices are lower in those countries where there is more than one operator. State-owned telephone operators face competition in Denmark, Finland, Norway, and Sweden. There is also competition in the U.K., which is the next cheapest country. Countries such as Cyprus, Belgium, and the Netherlands, where there is no competition, are among the most expensive.

The link between competition and price does not always hold true. Greece and Germany remain relatively expensive despite the fact that both countries have cellular competition. In both cases, the operators have similar quality of service levels on their networks, and there is a high level of pent-up demand. So there is no incentive for anyone to start a price war. The same was true of the U.K. when Cellnet and Vodafone operated in a duopoly. They have cut their prices only because of the emergence of competition from Mercury One-2-One and more recently from Orange.

Several operators in Europe now offer services geared to people who make only occasional use of their telephones. These have lower monthly charges but higher call rates than ordinary services. Cellular operators are beginning to offer innovative tariff packages to differentiate themselves from their competitors. Some include free calls in the monthly subscription charge. U.K. operator Orange has five different monthly subscription packages varying in price from about $25 to $160. The cheapest package includes 15 minutes of free calls and the most expensive one includes 540 minutes. Operators are also starting to offer regional and distance-related tariffs. The growing maturity of the cellular telephone business in Europe is evidenced by a more sophisticated approach to tariffing.

Most operators now have a range of packages aimed at different segments of the market. Nevertheless, cellular telephone services are still several times more expensive than fixed network services. *Table 3* shows some samples of average yearly costs for buying and using a cellular telephone. Figures are based on estimates of the average annual cost incurred by a user over the first three years of using a cellular service – the average length of a cellular subscription in the U.K. The total three-year cost includes the costs of acquiring a hand-portable telephone and of gaining connection to the network, plus total monthly subscription costs and call charges over three years. The call charges are calculated on the basis of two assumed patterns of usage: one for a high user and one for a low user. The high user is assumed to make an average of 150 minutes worth of calls each month, 140 at peak-time rates, and 10 minutes at off-peak rates. The low user is assumed to make 50 minutes of calls a month, 25 at peak-time rates, and 25 at off-peak rates.

A decrease in prices may be expected with the introduction of personal communication services. However, only the U.K. and Germany have implemented a PCN

Country	Average Yearly Cost for Low User* (U.S. $)	Average Yearly Cost for High User** (U.S. $)
Denmark	318	708
U.K.	452	824
Germany	808	1,728
France	1,020	1,874
Cyprus	1,205	2,597
Western Europe	728	1,302

Table 3 *Average yearly cost for cellular low and high users in Western Europe*

* average of 50 minutes of calls a month

** average of 150 minutes of calls a month

network so far. PCNs will achieve mass-market status in Europe by the year 2000 according to Frost and Sullivan.[15] Germany, the United Kingdom, and France will be the major countries spearheading the mass-market breakthrough by 2000. On the other hand, growth in countries other than Germany, the United Kingdom, and France will face a tougher time, especially if PCNs are not launched until GSM networks have become well-established.

It is likely that by the end of the decade, PCN will start to challenge the role of the fixed network.[12] This scenario has been encouraged by the attitude of the European Union Commission toward telecommunications industry liberalization. In 1994, the EU published a "Green Paper on Mobile Telecommunications" whose purpose was to establish the framework of future policy in the field of mobile and personal communications.[12] The Green Paper recommends that all restrictions on mobile communications be lifted. Moreover, it suggests that mobile network operators be allowed to provide their own fixed links and that each member country should establish a national PCN network, although no time scale is given for this, and it is not considered mandatory.

As far as PCN technologies are concerned, the digital European cordless telephone (DECT) market could achieve 30 million users. This figure is based on a 50 percent market penetration of the private branch exchange extension user base by the year 2000. In comparison to DECT, DCS 1800 could achieve a user base of between 40 million and 50 million users. PCN service will start to break through to a mass market once it reaches a threshold of about an 80 percent total cost premium compared with the fixed network.[12]

	1993	1995	1997	1999	2001	2003
Installed Base (Thousands)						
Wide area paging	3,158	3,739	4,420	4,663	4,874	5,002
PMR mobiles	4,574	4,834	5,013	5,129	5,214	5,245
PMR base stations	506	539	550	570	578	579
Trunked mobile radio	221	514	827	1,042	1,187	1,284
Analog cellular	7,193	8,815	8,520	6,690	4,170	2,000
Digital cellular	1,080	4,282	9,854	17,863	25,291	29,797
Annual Sales ($ Million)						
Wide area paging						
Unit Sales	776	766	818	825	771	758
Value	243	199	174	146	116	97
Private mobile radio						
Mobile Unit Sales	736	699	705	695	693	691
Value	1,091	987	958	916	879	832
PMR base stations	57	64	55	55	51	60
Value	443	497	425	428	399	468
Trunked mobile radio						
Unit Sales	95	213	236	216	204	203
Value	142	316	318	282	256	244
Analog cellular						
Unit Sales	2,381	2,065	1,496	1,273	815	270
Value	2,759	1,659	772	541	251	92
Digital cellular						
Unit Sales	955	2,350	4,498	6,914	7,759	7,496
Value	1,416	1,976	2,573	3,056	2,786	2,257
Total mobile equipment value	**6,094**	**5,634**	**5,220**	**5,369**	**4,687**	**3,990**
Service Revenues						
Wide area paging	796	900	930	918	888	872
Analog cellular	8,783	10,001	8,304	5,486	2,757	969
Digital cellular	1,160	4,947	9,555	15,206	20,156	22,376
Total mobile Service revenues	**10,739**	**15,848**	**18,789**	**21,610**	**23,801**	**24,217**
TOTAL MARKET	**16,833**	**21,482**	**24,009**	**26,979**	**28,488**	**28,207**

Table 4 *1993-2003 West European market for mobile communications*

The cellular market in western Europe is certainly booming, and the usage of digital technologies is already significant. *Table 4* shows in detail the results of this study performed in 1993 in terms of number of units, annual sales, and service revenues.[14]

Eastern Europe Overlook

In eastern Europe, cellular take-up has been rapid and penetration levels are rising much more steeply than in western Europe. Operators report high per-line call volumes and long average call duration.[10] This might be due to the fact that in many instances cellular is used as a fixed-link service providing public switched telephone network and international gateway access. A 1992 study carried out by Westel of Hungary showed that 50 percent of cellular subscribers did not have regular wireline telephone service. Trunked mobile radio networks and paging networks are now operational in several eastern European countries. All mobile communications networks in eastern Europe have been set up as partnerships between European and American operators joining local interests. This means that project financing is generally not a problem.

Table 5 shows the American companies currently involved in partnerships at writing time. Mobile networks are filling rapidly despite the high prices for terminals and calls compared to the price of using fixed communications. Therefore mobile networks are likely to produce an adequate return on investment and have a shorter payback period than fixed network projects. Cellular operators have kept tariffs high, particularly in Russia, Ukraine, and Belarus, by taking advantage of the poor availability of fixed infrastructure and long waiting lists. Nonetheless, tariff packages have proliferated and pricing strategies differ widely in order to take into account local market conditions. In Poland, for example, the operator has kept connection and monthly fees relatively low and has removed incoming call charges in order to encourage people to subscribe. In contrast, connection fees remain high in Hungary. Since network congestion causes service quality problems, connection fees stay at a high level. Estonia, Latvia, Bulgaria, Romania, and Hungary target a potentially large customer base via a relatively low tariff structure. Such structures included 1993 connection fees of more than $400, monthly fees of more than $35, and per-minute call charges of 35 cents domestically and $2.50 internationally. On the other hand, service providers in Belarus, Ukraine, and Russia aim their marketing efforts at the wealthiest. In 1993, connection fees were more than $600, monthly fees more than $60, and peak-hour per-minute calls cost more than 60 cents for domestic calls and more than $3 for international calls. Other countries such as Czech Republic, Slovakia, Slovenia, and Lithuania maintain mid-level pricing structures.[10]

NMT-450 has emerged as the main cellular technology used in eastern Europe. This is mainly due to the spectrum available in the 450-MHz bandwidth (the 900-MHz bandwidth is used for air traffic control) and because NMT 450 provides

Country	Operator	U.S. Companies
Belgium	Belgacom Mobile	AirTouch (25.0 %)
Czech-Slovak Republics	Eurotel	U S WEST, Bell Atlantic (24.5%)
Denmark	Dansk Mobil Telefon	BellSouth (49% with NordicTel)
Germany	Mannesmann Mobilefunk	AirTouch (29.17%)
Hungary	Westel	U S WEST (49%), Ameritech
Greece	Stet International	NYNEX (20%)
Ireland	PTT	Airtouch (bidding for license)
Italy	Omnitel-Pronto Italia	AirTouch (10.4%), Bell Atlantic (11.6%)
Netherlands	GSM Nederland Mobined	AirTouch (bidding for license) BellSouth (bidding for license)
Poland	Centel	Ameritech (minority stake)
Portugal	Telecel	AirTouch (23.0%)
Russia (Moscow)	Moscow's Cellular Comm.	U S WEST (22%)
Russia (St. Petersburg)	Delta Telecom	U S WEST (45%)
Spain	Sistelcom	AirTouch (16%)
Sweden	NordicTel	AirTouch (51.1%)

Table 5 *U.S. companies present in Europe for cellular service as of November, 1994*

cost-effective coverage. The use of NMT technology in the regions around the Baltic has the additional advantage of roaming among networks in Scandinavia, Estonia, Latvia, Lithuania, and St. Petersburg. Roaming among other NMT 450 networks in eastern Europe is planned. The older NMT 450 technology has been developed further to include features available with NMT 900. However, several countries are planning to build global system for mobile communications (GSM) networks. Hungary is the most advanced in its plans. Pogsag and Ermes paging licenses are also being awarded in several countries. The cost of installing a fixed mainline network is around $2,500 compared with $1,200 for a NMT 450 cellular network. However, terminal prices are very different. Fixed-wire telephones are available from $10, but NMT cellular phones cost $1,000 to $2,500.

Digital and Analog Cellular in Europe

In 1994, Germany remained the largest European market for digital services, although by a smaller margin than the year before. It accounted for nearly 42

System Type	Subscribers 1/1/93	Subscribers 1/1/94	Subscribers 1/1/95
Analog			
TACS 900 (U.K., Italy, Spain)	2,409,000	3,573,000 (+48.3%)	5,873,000 (+64.4%)
NMT-900 (Nordic, Benelux)	1,251,000	1,656,000 (+32.4%)	2,096,000* (+26.6%)
NMT-450 (Nordic, Benelux, Eastern Europe)	948,000	990,000 (+4.4%)	1,034,000* (+4.4%)
Others	1,192,000	1,215,000	Not Available
Digital			
GSM/DCS 1800	184,000	1,355,000 (+636%)	3,823,000 (+82.1%)
TOTAL	**5,984,000**	**8,789,000** (+46.9%)	**14,220,000** (+61.8%)

Table 6 *Western European cellular subscribers by system type*
* Estimate

percent of GSM and DCS-1800 subscribers, down from 74 percent at the end of 1993.[7] This reflects the fact that digital services are growing more quickly in other countries such as the U.K., France, and Sweden, each of which now accounts for more than 10 percent of the total market. Digital services make up almost 30 percent of the total number of European cellular customers. Just more than 2.5 million new customers joined the GSM and DCS-1800 networks during 1994, slightly more than the 2.3 million new additions to analog networks based on the TACS-900 technology. Digital growth was spread relatively evenly across Europe, while the vast majority of new TACS-900 customers were concentrated in two markets, the U.K. and Italy. These two countries accounted for more than 2 million of the new TACS-900 customers. The other two analog technologies, NMT-450 and NMT-900, added less than half a million new subscribers between them. Networks based on this technology are either shrinking or growing at a slow rate.

Table 6 shows a summary of the trend for various cellular systems as of January 1995. *Table 7* shows GSM viability and number of operators for each European country. Due to the GSM, the penetration of digital versus analog technology is growing in Europe; at the beginning of 1995, it made up a 27 percent share.[4] *Figure 2* shows the growth in the number of cellular subscribers and penetration rates of analog and digital systems in the European market during 1994.

The price of using digital GSM cellular services is becoming less than the price of

Country	Number of Operators	GSM
Albania	–	–
Austria	1	Yes
Belarus	1	No
Belgium	1	Yes
Bulgaria	2	Planned
Croatia	1	No
Cyprus	1	No
Czech	1	No
Denmark	2	Yes
Estonia	1	Yes
Finland	2	Yes
France	2	Yes
Germany	4	Yes
Greece	2	Yes
Hungary	2	Yes
Iceland	1	N/A
Ireland	1	Yes
Italy	2	Yes
Latvia	1	Yes
Lithuania	1	Yes
Luxembourg	1	Yes
Malta	1	N/A
Netherlands	1	Yes
Norway	2	Yes
Poland	2	Yes
Portugal	2	Yes
Romania	1	Planned
Russia (big cities only)	2	Yes
Slovak	1	No
Slovenia	1	No
Spain	2	Yes
Sweden	3	Yes
Switzerland	1	Yes
Ukraine	1	Planned
United Kingdom	4	Yes

Table 7 *Cellular operators in Europe as of Summer, 1995*

using analog services, particularly for heavy users. When GSM networks first opened in the summer of 1992, digital telephones were much more expensive than analog models. In a number of countries, the level of coverage on GSM networks was also inferior. However, GSM coverage has now improved and tele-

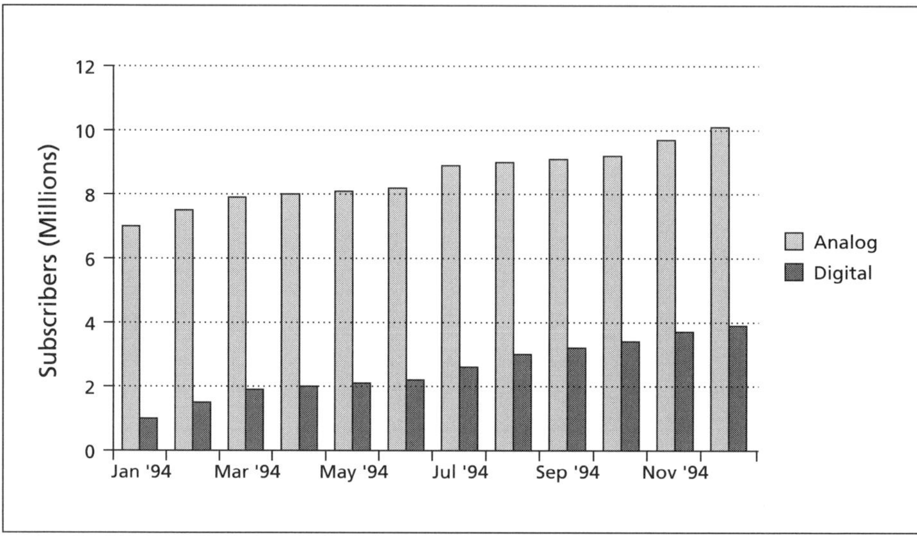

Figure 2 *European mobile cellular subscribers growth in 1994*

phone prices have fallen sharply: the average price of the cheapest GSM hand-portable telephone in each west European country is $557, which is about five percent less than the average price of the cheapest available analog hand-portable. However, GSM remains relatively uncompetitive in those countries that have networks based on the TACS-900 standard. TACS-900 cellular telephones are still much cheaper than GSM models. Furthermore, TACS-900 operators in Italy, Spain, and Austria have opened their networks only within the last five years. They want to maximize return on their investments in these networks before going full steam ahead with GSM.

A number of operators of analog networks have launched services aimed specifically at light users and, in particular, at consumers. For such services, monthly subscription charges are low but the per-minute cost of making calls is high. Where the analog networks are beginning to reach full capacity, operators are encouraging new high-spending customers to sign on to GSM networks and steering low-spending users toward the analog networks. Since most European countries have adopted the GSM as a standard, a key feature is wide roaming. As a result, roaming is today possible in a number of countries in and outside Europe. As of July 1995, roaming is possible in Australia, Austria, Belgium, Denmark, Estonia, Finland, France, Germany, Greece, Hong Kong, Hungary, Ireland, Luxembourg, Netherlands, Norway, Portugal, South Africa, Spain, Sweden, Switzerland, Turkey, and U.K.[16]

Technology: Third-Generation Mobile Systems

After implementation of the second-generation mobile system GSM, a massive effort has been undertaken in Europe to study possible enhancements to mobile systems that would bring about third-generation cellular systems around the year 2000. This effort comprises both research and standardization activities. Planned advanced systems under study include universal mobile telecommunication systems (UMTS) and a mobile broadband system (MBS). A significant part of the research activity on advanced mobile systems in Europe is undertaken within the framework of ongoing programs. Two technologies, TDMA and CDMA, are being evaluated. Two different RACE projects (research on advanced communication systems in Europe) are evaluating the effectiveness and the efficiency of the use of these techniques.[17]

Regardless of the access techniques, third-generation systems are intended to integrate all different services of second-generation systems and cover a much wider range of broadband services (including voice, data, video, and multimedia). *Figure 3* illustrates the evolution of technology since 1970 along with the range of services and bit rates supported by systems of different generations.[1] UMTS is a third-generation mobile system whose development is harmonized with the standardization activity at the global level on FPLMTS/AIMED undertaken in ITU. The FPLMTS work is being carried on by ITU members and regional standard organizations, such as RCR of Japan, the Telecommunication Industry Association (TIA) and Committee T1 of the United States, and the European Telecommunications Standards Institute (ETSI) of the European Union. In particular, ETSI's sub-technical committee SMG5 is specifying UMTS as a third-generation system based on FPLMTS. UMTS studies started in 1994, and draft standards are planned for 1998.[3]

The bit rate associated with the services supported by UMTS aligns with the forecast for FPLMTS/AIMED – that is, up to 2 Mbps operating in the 2-GHz bandwidth. *Table 8* illustrates the characteristics of some of the relevant UMTS bearer service types identified so far.[3] The studies of UMTS are related to an evolutionary path of the second-generation systems, without at the same time compromising the necessary innovation to meet the requirements of advanced systems. Higher bit rates are not generally considered feasible in the environment envisioned for UMTS. Support of higher rates are in the scope of MBS, a system designed for selected operating environments and intended to work in the 60-GHz bandwidth to support services up to 155 Mbps.

MBS is intended to extend to mobile users access to the services that will be available for fixed users. This will allow mobile communications to be fully integrated

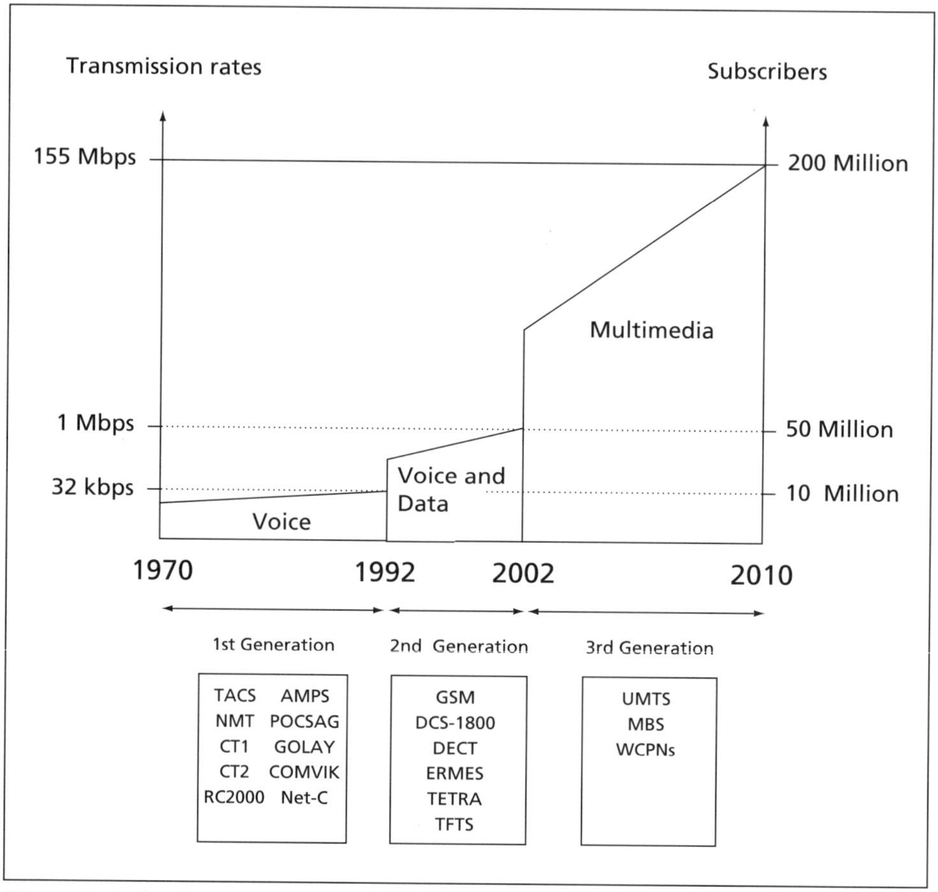

Figure 3 *Technology and subscriber evolution*

into B-ISDN. The bit rate required to support this service will be higher than the 2 Mbps maximum supported by UMTS. However, MBS will not replace UMTS but will be an enhancement of it where the traffic is high enough for it to be economical and where users will require higher bit rates than UMTS can support. *Figure 4* illustrates MBS in relation to other systems.[2]

Conclusion

Mobile communications are booming in Europe. The deployment of second-generation digital systems is already significant. More than one-fourth of cellular subscribers in western Europe were communicating through a digital network in

Teleservice	Throughput (kbps)	Residual Error Rate	Delay
Speech telephony/terrestrial	8-32	10E-4	40
Public voice announce	8-32	10E-4	90
High quality audio	940	10E-5	200
Voice band data	2.4-64	10E-6	200
Program sound	128	10E-6	200
Short messages/paging	1.2-9.6 (1.2-2.4 typical)	10E-6	100
Electronic mail	1.2-64	10E-6	100
Telefax (G4)	64	10E-6	100
Broadcasting/multicast	1.2-9.6 (1.2-2.4 typical)	10E-6	100
Public A/N announce	1.2-9.6 (1.2-2.4 typical)	10E-6	100
Unrestricted digital data	64-1920	10E-6	100
Database access	2.4-768	10E-6	200+
Electronic newspaper	2.4-2000	10E-6	200
Remote control services	1.2-9.6	10E-6	100
Location and navigation	64	10E-6	100
Telewriting	32-64	10E-6	90
Teleshopping	2.4-768	10E-6/10E-7	90
Video telephony	64-384	10E-7	40-90

Table 8 *UMTS teleservice specific quality of service values*

1994. Eastern Europe is catching up: most countries are either implementing a GSM network or planning to do so in the near term. On the other hand, the presence of only one or two operators in most European countries is slowing the deployment of PCS networks. Only the U.K. and Germany have implemented a PCN network so far.

However, it seems that, due to appropriate government agreements, Europe is ahead in the implementation of an integrated mobile network. This is allowing Europe to prepare for the advent and implementation of third-generation cellular systems to support services up to 155 Mbps around the year 2000.

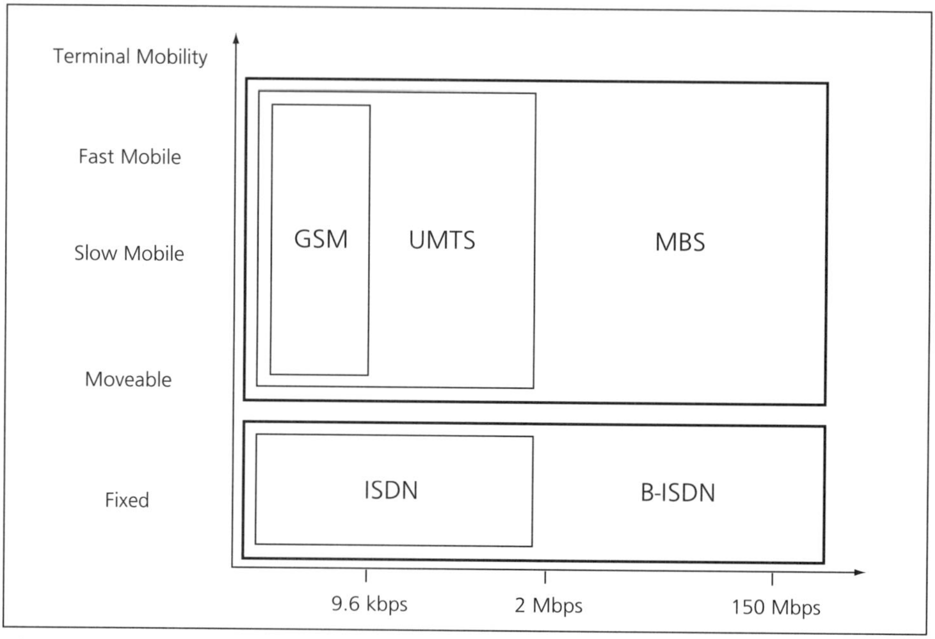

Figure 4 *MBS concept*

Note

1 A copy of the EU Green Paper may be obtained by the EU Commission Information Service, Washington, D.C.

References

1 J. S. DaSilva, and B. E. Fernandes, "The European Research Program for Advanced Mobile System," *IEEE Personal Communications Magazine*, February 1995, pp. 14-19.

2 L. Fernandes, Developing a System Concept and Technologies for Mobile Broadband Communications," *IEEE Personal Communications Magazine*, February 1995, pp. 54-59.

3 J. Rapeli, "UMTS: Targets, System Concept, and Standardization in a Global Framework," *IEEE Personal Communications Magazine*, February 1995, pp. 20-28.

4 "Digital Cellular Will Dominate European Mobile Communications," *FinTech Mobile Communication*, May 5, 1994.

5 "World Cellular Population Reaches 33.8 Million," *FinTech Mobile Communication*, May 5,1994.

6 "UK Holds Pole Position as German Growth Slows," *FinTech Mobile Communication*, January 26, 1995.

7 "German Mobile Market Finally Becomes the Biggest in Europe," *FinTech Mobile Communication*, March 9, 1995.

8 "European Cellular Tariffs," X25 Partnership, London, UK, 1994.

9 "Worldwide Cellular Growth Sustained in Mature Markets," *Communications Daily*, May 8, 1995.

10 S. Zehle, "The Growth of Mobile Networks in Eastern Europe; Telecommunications," *Telecommunications*, vol. 28., no. 3, pg. 47.

12 "Cellular Achieves 245 Percent Annual Growth in Central, Eastern Europe," *Mobile Phone News*, vol. 12. no. 18.

12 "Frost & Sullivan Report Highlights PCN Markets," *Mobile Phone News*, vol. 13, no. 6.

13 E. Buitenwerf, G. Colombo, H. Mitts, and P. Wright, "UMTS: Fixed Network Issues and Design Options," *IEEE Personal Communications Magazine*, February 1995, pp. 30-37.

Europe, A Crucial Part of Any Global Wireless Strategy

Malcolm H. Ross
Senior Industry Specialist
Arthur D. Little, International Inc.

Introduction

Over the past decade, the European mobile communications markets have opened up from being the realm of monopolistic state-owned network operators. This has led to an unprecedented opportunity for new entrants into the world's largest and one of the most dynamic telecommunications markets. While U.S. policy still restricts foreign ownership of U.S. telecommunications operators, some of these same U.S. operators have major participation in several of the new European firms. Many more line up to take part in the licensing of further new operators. Driven by both the liberalization policies of the European Union and by the adoption of a single standard (GSM) across Europe, this process of opening up is continuing. It has finally encompassed even the most reluctant of the European countries and is now spreading into Eastern Europe. A second wave of PCS licensing based on the DCS standard is now multiplying the opportunities for new licensees. Mobile communications is perceived to be the vanguard of a much broader liberalization of many sectors; by 1998, this will also open up opportunities in the fixed networks across Europe.

At the same time, several European markets lead the United States: more than one in four inhabitants of the Nordic countries (Sweden, Norway, Denmark, and Finland) has a mobile phone – a penetration rate twice as great as in the United States. In the U.K. and Germany, PCS operators have been in operation for several years, while the United States has just completed the first round of licensing. Although not all of the European experience will be applicable to the United States, an astute U.S. PCS operator should carefully examine the European experience for both opportunities to profitably participate and to learn from the mistakes and successes already evident in Europe.

While some U.S. operators are participating in successful mobile operators in some European markets, some have withdrawn and others have been mainly unsuccessful in gaining entry. The attractive opportunities in Europe hide some minefields, and participating in Europe is not right for every U.S. operator. Also, the pioneering PCS experiences in Europe have not been spectacular successes, which might perhaps dampen the recent enthusiasm for PCS licenses in the United States.

This essay examines the European opportunity from two angles:

- the opportunities in Europe are in most cases more attractive than in the United States, but there are some dangerous traps

- the PCS lessons sound some alarm signals for U.S. operators but pose more questions than answers

Nature of the Opportunities in Europe

Growth Rates

Although the technology of cellular telephony was developed first in the United States, regulatory hurdles held back commercial implementation so that Scandinavia launched services first. Growth rates in most countries including the United States have followed Sweden. The United States still lags some three years behind the Swedish curve, as shown in *Figure 1.* The penetration curves in this figure have been time-shifted to the point at which 0.3 percent penetration was achieved.

There is an extraordinary similarity between cellular growth rates across many countries. This pattern is repeated in countries as diverse as Japan, Thailand, Malaysia, Columbia, Mexico, Australia, and Hong Kong. This is remarkable in light of the differences in GNP per capita, the industrial structure, and the cultural variety of these countries. The reason for this is mainly that risk-averse operators and investors are not prepared to explore over-investing in the mobile network, which would allow for faster growth than shown in *Figure 1.* Our experience in working with operators across this spectrum leads us to a self-fulfilling prophesy explanation: given mobile's embryonic nature, all business planners know there is substantial risk involved in predicting growth.

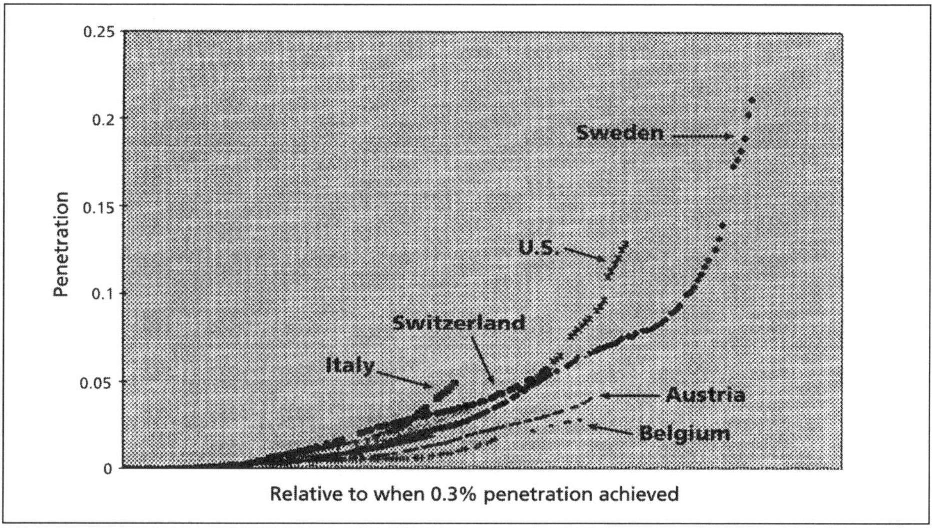

Figure 1 *Penetration relative to Sweden*

The Swedish experience is used as the basis for market projections and, more importantly, it is the basis for business plans concerning infrastructure investments. Prognoses above the Swedish curve are therefore seen as entering into unknown territory and are rejected as unnecessarily risky. Prognoses below the Swedish curve are seen as even more conservative than the Swedes and so again are rejected. As a result, investments in many new networks are made with reference to the past growth of Sweden. If growth exceeds expectations, the investment is insufficient to handle the excess demand, which results in congestion and slower growth rates.

The result of this is that the Swedish curve has become a self-fulfilling prophesy – whenever growth exceeds expectations, the threat of running out of capacity leads to reduced advertising and promotion. As a result, up until mid-1994 nearly all networks worldwide were growing close to the historical Swedish curve. The United States (widely seen as an innovative and techno-friendly society) began service in 1984 and has remained over a nine-year period consistently within 0.3 percent penetration points of Sweden and is growing with a lag of three years behind.

In the past eighteen months, a dramatic spurt in growth has taken place in Sweden, as well as in many of the other major cellular markets. Our analyses of some 20 possible causes for this growth spurt have identified a very strong correlation between this accelerated growth rate and the increase in spending on advertising for mobile. The cause of the increased spending has been different between countries; in many European markets, it coincides with both the hype surrounding the launch of GSM and the appearance of a second or third operator. The correlation between growth rate and the launch of lower-tariff packages is much poorer, which shows that the spurt in new customers is not primarily due to pricing.

The European Market Potential

Despite the acceleration in growth rates, our analyses of the European markets shows that as with the U.S. markets, the European market potential has hardly been tapped. Although a fast-track product by telecommunications benchmarks, mobile is a slow-track consumer product.

Market research in several European countries and in the United States shows a very high acceptance rate for mobile and a strong intent to buy. As many as eight out of ten consumers and business people are interested in purchasing mobile service within the first three to five years of its availability – even at today's prices and tariffs. This would indicate that market growth for mobile at least rivals the rate for medium-track consumer products such as color TVs or VCRs, which took seven to ten years in order to penetrate 50 percent of households. Perhaps mobile's penetration rate will even reach the growth rate of products such as hand calculators and video games, which reached 50 percent penetration within two years.

Although the mobile industry seems surprised and delighted with what it perceives as high growth rates, in practice mobile telephony is a slow-track product, little better than unexciting products such as electric carving knives or trash-compactors (see *Figure 2*). Cellular telephones first reached Swedish customers in 1981, and after fourteen years penetration has only reached 21 percent.

Accelerating European Growth

Although exciting for a telecommunications market, by consumer electronics standards, mobile telephony's growth has been slow and unexciting. However, this slow growth rate is more due to the industry approach to marketing, sales, and distribution rather than an intrinsic characteristic of the market. As we will discuss in greater detail, most service providers and network operators do little more than generic advertising and order taking, rather than the sophisticated approach to marketing characteristic of the consumer electronic business. In that industry,

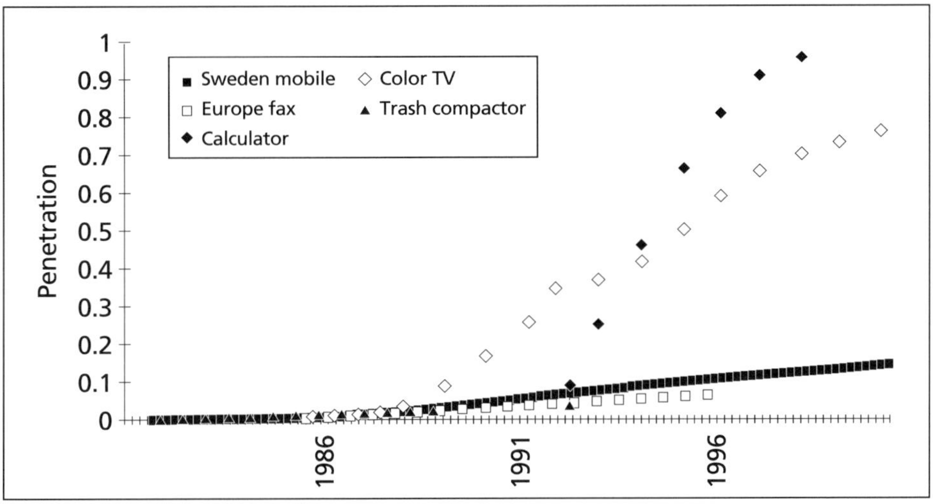

Figure 2 *Mobile is a slow-track product by consumer standards*

detailed analysis of segments and needs are conducted and target segments are prioritized. Specific campaigns to excite each segment are developed, and a portfolio of tariff/service packages to maximize product and service yield are offered.

Growth in at least Sweden, the U.K., the United States, Portugal, and Australia has finally broken out of the twelve-year pattern of slow steady growth established by the Nordic countries, as shown in *Figure 1.* It seems clear, however, that markets could grow even faster than these latest trends. In the last six months of 1994, penetration increased in Sweden by 2.9 percent of the population and by 4.2 percent in the first six months of 1995; at this rate, 50 percent of the Swedish population would be customers by the end of the year 2000. By consumer electronics standards, this is only a medium-track growth rate, similar to VCRs or color TV. However, our market research indicates that acceptance among consumers and business users points to a fast-track rate of growth, similar to hand calculators, video games, etc. Under a fast-track scenario, the 50 percent level would be reached in eighteen to twenty-four months.

Two pieces of evidence indicate that conditions are beginning to be right for stimulating market growth into at least a medium-track growth rate. First, customers who stay on the network for at least twelve months (i.e., those who are not seduced by free terminals and then churn off the network when high monthly bills start arriving) have developed a value story – that is, at least one application whose value to them far exceeds the cost of subscribing. For them, mobile

Customer Group	Application	Benefit	Value
Plumber	Can be called while at previous customer	Can send spouse out to work	Additional family income
Rank-based taxi	Can call, be called when at taxi-rank	Can run a second business from taxi	Additional income
Foreign-language speaking taxi driver	Hotel concierge can call for foreign guests	Differentiation, exploit a niche market	Bigger tips
Suburban taxi driver	Can be summoned direct by customer	Can send spouse out to work	Additional family income
Elderly relations living alone	Sons/duaghters want to be able to check on elderly parents	Peace of mind	
Washing machine service technician	Customer calls direct	Can better schedule service calls	Doubles income

Table 1 *Value stories for mobile communications users*

communications has become an indispensable necessity. It is encouraging that such value stories have been developed across a wide range of users – both business and private. This indicates that all sections of business and society could perceive value-for-money from using mobile communications. Mobile communications does not just benefit a few elite segments but is potentially of interest to the entire population. The variety of value stories is broad, as shown in *Table 1*. The differences even within one group, such as taxi-drivers, is interesting to note. Taxi driving is a multiple micro-niche business, but each of these different riches perceives enough value in mobile telephony to make it worthwhile becoming customers.

The second piece of evidence that mobile growth rates are set to increase is the fact that most factors cited during market research as being key prerequisites before particular segments would buy are rapidly being achieved. This means that barriers to wider adoption are being overcome. We have examined these barriers for eight types of business users and six types of private users, listed in *Table 2*. A close analysis of the prerequisites for each of these segments shows that price is no longer a barrier for most of the business segments, and the expected price reductions in terminals and in monthly charges will eliminate these as barriers to all business and most private segments by 1999. The rate at which barriers to the purchase of mobile communications will be reduced are shown in *Table 3*. Call charges will still be seen as expensive by low-income private users and by employers of blue-collar workers until 2002. Coverage will also meet the requirements of all segments by 1999. Convenience of purchase, availability of enhanced

Business Users	Private Users
Managers in medium and large corporations	Well-off singles
Managers and owners in small businesses	Well-off families
Transport workers and drivers	Privileged senior citizens
Field sales representatives	Students and lower income singles
Professionals	Working class families
White collar workers in large corporations	Pensioners
Mid-level government employees	
Blue collar workers in large corporations	

Table 2 *Business and private users studied*

	1996			1999			2002			2005		
Cost of hardware (DM)	1,000	500	100	1,000	500	100	1,000	500	100	1,000	500	100
	100%	10%	0%	100%	85%	20%	100%	95%	50%	100%	??	??
Cost of subscription (DM)	80	40	20	80	40	20	80	40	20	80	40	20
	100%	5%	0%	100%	50%	5%	100%	70%	25%	100%	??	??
Cost per minute of communication (DM)	1.4	0.8	0.3	1.4	0.8	0.3	1.4	0.8	0.3	1.4	??	??
	100%	5%	0%	100%	50%	5%	100%	75%	50%	100%	95%	??
Coverage	33%	66%	100%	33%	66%	100%	33%	66%	100%	33%	66%	??
	100%	95%	0%	100%	100%	90%	100%	100%	100%	100%	100%	??
Network quality	poor	avg.	high	peer	avg.	high	poor	avg.	high	poor	avg.	??
	100%	50%	0%	100%	90%	60%	100%	90%	70%	100%	??	??
Enhanced services	none	basic	plus	none	basic	plus	none	basic	plus	non	basic	??
	100%	20%	0%	100%	80%	30%	100%	100%	80%	100%	100%	??
Ease of use	complx	avg.	simple	complx	avg.	simple	complx	avg.	simple	complx	avg.	??
	100%	40%	10%	100%	100%	70%	100%	100%	90%	100%	100%	??
Convenience of purchase	difficlt	avg.	easy	difficlt	avg.	easy	difficlt	avg.	easy	difficlt	avg.	??
	100%	30%	0%	100%	70%	20%	100%	100%	40%	100%	100%	??

Table 3 *Rates at which barriers to the purchase of mobile communications will be reduced*

services, and network quality (especially in-building coverage) will be the main barriers to wider adoption.

It may be that the industry already has gone too far in reducing tariffs for the current stage of market development. We have discovered that at the present stage of market development in many European countries, spending on advertising is about twenty times more effective at increasing profitable sales than reducing tariffs. Unilaterally reducing tariffs, as some operators have been doing, seems to only slightly accelerate market growth over how it would have grown with the higher tariffs. The resulting lower revenues more than wipe out expected gains in revenue from the marginal increase in customers. However, spending the same money on targeted advertising seems to create an incremental additional market growth twenty times greater than that of the tariff reduction. The message seems to be at this stage of market development that awareness is a more important driver/barrier than price sensitivity.

Intrinsic Profitability of European Operators

With a few exceptions, the profitability of most European mobile operators will remain high. This is a result of several favorable factors:

- European mobile tariffs are higher than U.S. tariffs in most countries – exceptions are Italy and Sweden. Countries with above-average tariffs do not seem to suffer slower market growth or lower usage, which indicates that penetration of mobile has not seriously encompassed price-sensitive segments yet.

- The biggest single operations cost item is the cost of interconnection and links from base stations to switches. In most European countries, these costs are artificially high due to the lack of competition in fixed networks and because they subsidize the artificially low local exchange services. Liberalization forced under European Union law at the latest in 1998 will probably lead to these costs falling to match internationally competitive levels.

- Population densities are high and distances are short, leading to economical network designs.

- Cost of sales is currently over 25 percent of revenue. As mobile becomes a more mainstream product/service, it will begin to be distributed through more conventional distribution channels more cost-efficiently.

- The economies of scale in GSM and DCS equipment are only just beginning to filter through to procurement prices. As GSM further penetrates the international markets and volume builds, these economies of scale will decrease investment cost.

- Licensing fees are low relative to what U.S. PCS bidders have been offering.

Profitability is also restrained by the influence of many of the attitudes of traditional monopolist operators facing competition for the first time. Many of these have found the new operator can quickly enter and build in excess of 50 percent market share. Their response is usually aggressive pricing rather than looking at becoming competitive in their service offerings and in their approach to promotion. At the embryonic phase of development of these markets, the principal thrust of any operator should be to build the market rather than to compete for the existing customer base. Constructive competition is the name of the game, but traditional monopolist operators equate competition with a winner/loser battle. A consistent message from all operators that mobile is good for you, because ... rather than you should chose our service over theirs because ... is much more beneficial to all operators and will grow the total market faster.

PCS Lessons and Questions

The U.K, German, and French governments have licensed so-called PCS operators using the 1800-MHz band and DCS 1800, a standard very similar to GSM. Other countries are expected to follow within the next few years. These PCS networks were initially seen as offering PCS as something distinct from cellular services. In the few years that PCS networks have been in operation, it has become rapidly clear that there is little if any distinction.

The original thought was that the more limited range of DCS 1800 would allow for smaller cells and therefore be more suitable for hand-held terminals and local services. In practice, the leading existing GSM operators have used exactly the same engineering principals to arrive at the same result. They are using small cells in city centers for capacity reasons, and as a result they have good in-building coverage, matching that of the PCS operators. The features as well as the equipment costs of DCS and GSM are virtually identical. The entry of PCS operators has stimulated the cellular operators to provide equivalent network capabilities, services and tariff packages.

Going one step further, incumbent operators have seen they can also upgrade their analog networks to match the performance of the digital networks, at similar costs. After all, only the air-interface of a TACS or NMT network is analog – the switching, fixed transmission, and all of the intelligent network features are digital and based on the same platforms as GSM and DCS. The choice of investing in analog or digital for PCS is more of an economic, regulatory, and business question rather than a technical issue.

If a European operator is prepared to invest in small cells and the latest infrastructure, most network platforms (analog and digital) can provide a similar level of service to customers. Essentially this means that all mobile operators are PCS operators, and within a short period there will be little differentiation due to the platform. The only remaining technical advantages of GSM and DCS over analog platforms are wide international roaming and speech security through encryption. These are important to only a minority of European customers.

What also has become clear is that whatever platform is used today, it does not meet customer expectations in terms of geographic coverage or coverage quality. Customers expect to be able to initiate calls and to finish their conversation without a break and without either party noticing a speech quality degradation. They expect to receive this quality wherever they are, including tunnels, underground parking, deep wide buildings, and whether stationary or moving in a fast car. Neither analog or digital delivers this level of service yet.

The disappointing performance of half-rate codec brings an important realization to European PCS operators. Trials with customers are showing an unacceptable MOS (mean opinion score) for the quality of the half-rate technology, particularly in certain languages such as Dutch. This, and complaints by users of tinny speech quality, is leading operators to press instead for an enhanced full-rate codec. The abandonment of the half-rate codec will result in a massive increase in network investment requirements, and so a much lower return than originally planned.

This and other challenges of GSM are teaching European operators to be conservative in their choice of both technology and supplier. Confidence remains that most of the bugs have been ironed out of GSM, but it is also recognized that mobile is an embryonic technology and much remains to be done before it matches customer expectations. GSM is now seen as a proven, richly featured and reliable architecture, and continued development keeps it state-of-the-art. Given the time it has taken to shake down the end-to-end network architecture of GSM, European operators are wary of experimenting with the unproved CDMA air-interface.

The Financial Community and Operator Strategy

Once an operator has decided to invest in PCS network capabilities, the levels of investment required become so high that recourse is necessary to external financing. Several consortia of financial institutions have formed which are specifically geared to examining the funding of telecommunication investments. This is having a profound influence on European operators who are far enough along in their

PCS rollouts to require phase-two funding, and it should also benefit other operators not yet at this stage. One such influence is that the fundamental assumptions operators have made about their businesses are being examined in fine detail. This examination shows that in general the level of professionalism in business planning in the industry is low. In many cases, this has not led to banks insisting on more conservative plans, but rather that risk management (both identification of risk and contingency planning) become a much more important part of business planning. It has also become clear that most operators have grown fast and have not put in the corresponding basic business processes (many of which are not specific to mobile). This is leading to a tightening up of the control and auditing paths within such operators and the establishment of a more professionally run business.

These financial analyses are drawing attention to areas of the business that are sensitive to external factors and those that are not. For example, subscriber numbers and usage is far less important than cost of sales, churn, and the negotiation of interconnect costs. This is assisting top management to focus their attention on the parameters that most determine the future of the business. The profitability of such operators and those still working in the traditional way is dramatically different. The former are bringing in professional managers to set up sound internal processes; the latter focus on increasing subscriber number, profitable or not!

The dilemma is that most operators, service providers, financiers, and industry observers agree that eventually everyone will have at least one hand telephone, but as far as the necessary investment of cash, marketing resources, and commitment, few are willing to make it happen. Clearly, there is disagreement as to how quickly it will happen, but only a few operators and service providers are looking for ways to explore to what extent it can be driven by their efforts.

Often financial risk is cited as the barrier. However in Europe, once national coverage has been reached, investments in expanding cellular networks are incremental. The risks are therefore very controllable. For example, one possible approach is to select a trial city which is typical of the country as a whole and try to push penetration in that city as high and as fast as possible. This is the approach used in many other established markets: try it first in a test market, and if it works expand it to the total market. A first step would be to double advertising expenditures in the test city and double the sales staff. Ensure that terminal suppliers will guarantee that the test city never runs out of sets (they are of course also interested in exploring how fast the market can be made to grow). Given these measures, does the market growth at least double? Early indications from one such trial is yes. Once increased demand is detected, additional capacity can be added at minimal risk. The financial risk of the test city is limited mainly to the increased marketing budget and is miniscule compared to the total business plan investment.

Unfortunately, only a minority of operators have yet gone through such processes. The continued growth of the industry as a whole is hiding the inefficiencies and vulnerabilities of these operators. A chilling observation from similar industry booms, such as in the semiconductor, PC, or software industries, was that such inefficiencies were tolerated or unnoticed until the first bankruptcies occurred. Given the irrational business behavior of many mobile operators in Europe, there is clearly a risk of history repeating itself. A few indications that this might be the case include the following:

- Connection bonuses now subsidize new customers to the extent that it takes two to three years before the investment in a new customer breaks even. Half of the new customers churn before they discover a value story, and so are never profitable.

- Connection bonuses are being leveraged to invest in future growth. Any stumbling of market growth will therefore eliminate working capital from distribution chains and resellers.

- Planning capacity, usage, costs, and revenue are based on average customers, but new customers profiles are dramatically different from those of the installed base. For example, networks are still being expanded assuming 0.03 erhlangs per new subscriber, which has been the historical average. However, new users call less than half this amount and have a different peak/off-peak profile. Operators are therefore unknowingly over-investing in capacity and reducing returns.

- There is little understanding of the composition of existing users, what value they get out of mobile, what attracted them to mobile, and how many other similar potential high users are out there. Instead, there is a shotgun approach to use average channel targeting to bring in average customers. Although on average a customer is worth having, more than half of new customers are non-profitable.

Non-Selective Promotion and Profitability

European operators are promoting mobile on the basis of cheapness rather than on value. This non-targeted approach is leading to a mix of profitable and

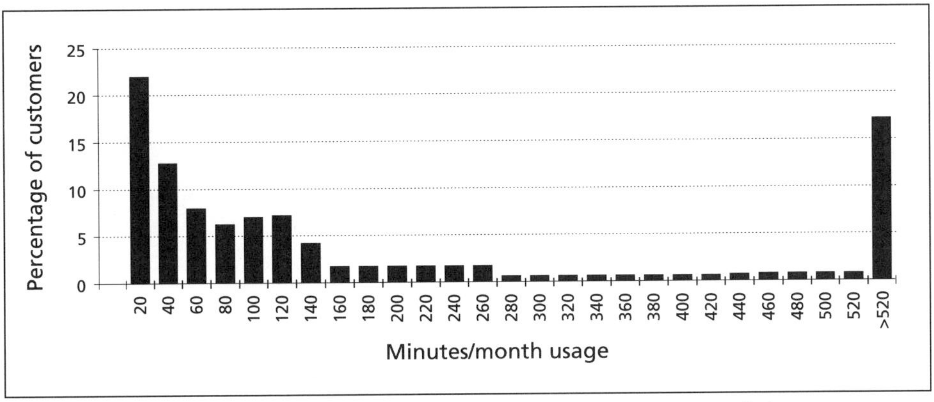

Figure 3 *There are no average customers; they are very heavy users or light users*

nonprofitable customers that is seriously eroding previously high profit margins. This is analogous to offering a one-tariff cheap airfare across the Atlantic, even though there are certain segments willing to pay for first class and business class seats. Rather than targeting those customers likely to be loyal and heavy users because the value of mobile communications to them is so high, European operators promotion and tariff policies are attracting many low users who have not discovered a value story and who churn off when they get one or two unexpectedly high monthly bills.

Operators in the future will not just need to dramatically increase penetration; they will also need to persuade customers to use their mobile telephone more. At present, there are some customers who are heavy users, calling for over 500 minutes a month. However, an increasing percentage of users are light users, as shown in *Figure 3*. If we can get customers to see their mobile phone as their principal telephone, we should be able to raise private usage of residential phones closer to the 300-minutes-per-month level and business usage closer to 600 minutes per month, rates that are typical in fixed telephony.

Most observers have been assuming some sort of smooth distribution of usage, such as a Poisson curve. In practice, we find an inverse curve, with a substantial number of heavy users with over 500 minutes per month. The other 60 percent of users have less than 120 minutes per month with a mass of customers using 60 or less minutes per month.

Examining this distribution over time, we noticed that at market penetrations of under 2 to 3 percent, the heavy user peak was much more dominant. Once market penetration reaches 6 to 8 percent, few additional heavy users are added to the customer base, and the new growth is dominated by low-usage customers. The net effect is a slow, steady decline in average usage. However, this hides the big difference between new and existing customers. We have noticed, for example, that for many (but not all) operators with both analog and digital networks, the average usage on the GSM network is much lower than on their analog networks. For such operators, the new customers are much lower users, and so less profitable. A PCS operator that enters a market already 8 percent penetrated may therefore only attract marginally profitable customers!

Another contributing factor to decreasing average usage is that there is a high percentage of phantom customers on many of the networks. Operators have been so intoxicated by measuring their performance in terms of subscriber numbers that they have not aggressively culled nonpaying customers from their claimed customer base. Such phantom customers include all the promotional free SIM cards given out at service launch, which are now inactive. There are also many SIMs that have been deactivated when a customer does not pay (an increasing trend) but the operator has not yet removed the customer from the database. The situation is in fact far worse than officially admitted, with both operators and service providers having many phantom customers and not daring to eliminate them for fear of losing the publicity battle or net new connection commissions.

Such phantom customers can make up almost 20 percent of an operator's claimed customer base. Recently, some operators have bitten the bullet and started to aggressively cull their apparent customer base. They have been rewarded by negative press reports. Others who continue to hide these phantoms are praised for their growth but also suffer from a faster decline in average usage.

A detailed audit of an operator's customer base can also provide other interesting insights. For example, we determined that 10 percent of one operator's new customers never made a call. We arranged for some sample interviews with such customers and found out that they had not been told to press the send button after entering a number. They had not complained about their terminal not working because they assumed the network was not yet available in their area and were patiently waiting for the service to arrivve. Some of our customers are better than we deserve!

Growth Rates and the Approach of Operators and Channels

Given higher than expected growth rates, the approach of operators so far has been dominated by short-term technical and operational priorities: a scramble to install network coverage and capacity as fast as possible and to install systems to accept and handle the new customers. Long-term planning has meant sales tactics for the next six weeks and network investments for the next eighteen months. Awash in a flood of new customers stimulated by untargeted generic advertising, the industry has been driven by an order-taking approach to sales with little time or resources available for longer-term marketing strategy or planning.

As pointed out in *Figures 1* and *2*, growth in mobile markets has been unexciting. We believe that a major element in this has been the inadvertent creation in most markets of a well-organized and coherent SPO – sales prevention organization. As seen through the eyes of a consumer, it seems that most operators and service providers have erected barriers to make it as difficult as possible to purchase. Characteristics of the SPO include:

- locating dealers inconveniently far away from where most people work or shop, presumably to keep costs low

- allowing dealers to be open only from 10am to 5pm so business people have to take time off from work to shop

- providing a limited variety of heavy, unattractively shaped terminals, all in black

- requiring the purchaser to wait for a long, complicated credit check for a purchase that to the customer is like buying a VCR

- preventing instant gratification for the customer by providing uncharged NiCad batteries that require 16 hours of charging before first use and which do not provide a first-day lithium battery equivalent

- providing a network that has congestion, variable speech quality, dropped calls, and unreliable service in 10 percent of the claimed coverage area, including all the places a user expects to be able to use a phone. Such places include underground stations, road tunnels, steel structures with elegant silvered windows such as modern office buildings, shopping malls, airports, and the customer's own car

What comes as a surprise is the extent to which the existing customer base accepts these deficiencies. Fortunately, during the early stages of development of any market, the pioneer customers and early adopters are very keen to buy and will accept all of these barriers. In even the most penetrated countries, the majority of existing customers are innovators or early adopters. Individuals with these psychological profiles tend to be the first to try any new product or service and are the opinion leaders. They make up 10 percent to 15 percent of the total potential market, are more tolerant of deficiencies, and actively seek out the product or services despite barriers, inconveniences, cost, and poor performance. As the market moves beyond these, the next wave of customers will neither understand nor accept these barriers. The message is clear – in markets reaching the 10 percent penetration level the SPO must be dismantled as quickly as possible.

Another analysis can reveal some unexpected insights regarding which type of customer is most profitable. For one operator with relatively low tariffs, we determined that nearly half the customer base was unprofitable when the cost of sales and billing as well as costs of providing the service were taken into account. For most operators, the non-user (who pays a monthly fee but never makes calls) is the most profitable because capacity or customer service does not have to be provided for this customer, assuming this fact is recognized and the infrastructure has not been overbuilt.

Providing we have negotiated our interconnect agreements aggressively, the next most profitable customer is someone who rarely makes calls but often receives calls from the fixed network. This includes segments such as plumbers, craftsmen and many small businesses primarily getting business through Yellow Pages. However, think about the incentive structure we have set up for service providers – they receive a percentage of out-going call revenue. They have little incentive to sell to what could potentially be our most profitable customers. Fortunately for operators, the marketing approaches of most service providers are not yet sophisticated and selective enough to prevent them from selling to customers who are unprofitable to them.

Recently a couple of operators have recognized that the fixed-network operators are encouraging them to provide an end-to-end bypass. I am sure that the fixed operators have not yet understood this. In countries where the fixed operator has managed to keep interconnect and link charges high, mobile operators can save this substantial fee if a mobile customer calls another mobile subscriber because the call is kept inside the mobile operator's network. Again, detailed analysis of calling patterns showed that for several operators, 80 percent of outgoing calls were made to one of three numbers. Such a mobile operator might realize that it would remain profitable to offer lower tariffs on mobile-to-mobile calls or give each subscriber three free or subsidised terminals to give to his or her three most often-called communications partners. The mobile operator can afford this because he is saving on interconnect charges.

Sources of Performance Benchmarks and Know-how

As an industry, we can be proud internally of our technical achievements. But for the next stage of market development, we have to move beyond the Model-T "any color as long as it's black" mentality in order to building a professional, customer-driven business.

In assignments with mobile operators over the past two years, we have been cataloging the business processes, and through discussions with customers, service providers, and operators, we have been gauging how important each process is to success in the business. Two key points emerge:

- there are several hundred processes, many of which are basic and not specific to the mobile business and most of which are rated as important or very important

- both traditional operators and newcomers either do not have most of these processes in place, or the processes are far from optimal

Our hypothesis is therefore that although some operators may be better than others, none is nearly good enough to satisfy the majority of potential users (see *Table 4*). This has not held back growth or usage to date, but it will become a major problem as the market matures – that is, as penetration exceeds 10 percent to 15 percent. Installing and optimizing this large number of business processes will become an urgent issue for the leading, most-penetrated markets during 1995, and in most other markets in the 1996 to 1998 time frame.

Corporations in other industries have solved many of the problems that face the mobile industry and have many of the key factors required for future success in this business. Every new industry has many attributes and characteristics that are unique. One characteristic that seems universal for all new sectors is the belief that the sector is so unique nothing can possibly be learned from any other segment. Twenty years ago the semiconductor industry had this mentality, ten years ago it was the PC industry, and now it is the mobile industry.

However, despite the differences and uniqueness of the young and fast-growing

Key Processes	Weight Embryonic Business	Weight Mature Business	Weight Mature Consumer	Traditional Operator	New-comer	"Best in class"
Grow market						
Understand user needs	**	***	**	–	–	System Houses
Value-oriented PR	**	***	**	*	*	Unilver; P&G
Recognized consumer brand	*	**	***	*	–	Sony; Swatch
Accessible sales points	***	***	***	*	*	High street; mail order
Instant sales	***	***	***	–	–	Lottery
Variety of devices	*	***	***	–	–	Sony; Swatch
Easy to use	*	***	***	–	–	Sony; Swatch
Coverage (99% in building)	***	***	***	–	–	Radio; TV; paging
Links to sales partners	*	**	***	–	–	U.S. airlines
Market Share						
Excellent customer care	***	***	***	–	–	Mail order; U.S. airlines
Coverage quality	***	***	***	–	0	Radio; paging
Voice mail; directory	***	**	*	0	*	Medidian (NT)
Marketing Intelligence System	***	*	*	–	–	Claritas
Flexible billing	*	***	**	–	–	Hertz; AMEX
Profitability						
Optimized network design	***	**	**	0	*	FedEx
Churn control	***	***	***	0	*	VISA; AMEX
Credit/fraud control	***	*	*	–	*	VISA; AMEX
Own IN broadband network	***	**	**	*	–	MCI
Low cost of sales	***	***	***	–	–	Sony; Swatch

Table 4 *Some operators may be better than others, but none are nearly good enough*

mobile industry, there are also many lessons, procedures, processes, and tricks that can be learned from other industry sectors. Rather than wasting time making old mistakes and learning old lessons, these lessons should be learned from whatever source is appropriate so that those in the mobile industry can use their time, energy, and resources to make new mistakes and learn new lessons in the areas that truly are unique to this particular industry.

For example, if the goal is to open up the mass markets for mobile communications, shouldn't some lessons be learned from the marketing power and experience of Swatch, Sony, Matsushita, Nintendo, Nike, Virgin, and De Beers? These companies have experience that seems, at least at first-sight, to be what those in the industry are looking for; they have the reputation, image, distribution channels, "share of voice," and perhaps most importantly, shelf-space. They have a track record of creating mass market demand.

Given that good customer care is crucial to churn-control, can the mobile industry learn faster from the existing excellence in customer care that characterizes mail-order, courier, airline, and car rental companies? What can be learned from the credit card industry to deal more effectively with fraud, bad debt, and credit-control problems? Think about where a dealer keeps unissued SIM cards – would a bank store its unissued credit cards like that? Would a credit card company have mechanisms for ringing alarms if unusual buying behavior was noticed, or would they wait (as the mobile industry does) for the customer to complain about a higher than expected bill before looking for possible fraud?

For many of the basic processes that are crucial to the mobile business – but which are not necessarily specific to the mobile industry – there exist excellent models for us to adopt, copy, and adapt. In these areas, the mobile industry should be bench-marking itself not against the best within industry, but rather against the best in that class of operation, from whichever industry is appropriate.

Marketing will become a complex task with multiple objectives: to grow the overall market and to preferentially gain market share prioritized according to the most profitable mix of customers. The goal is to retain such customers and increase their usage. Credit card companies carefully target specific groups of potential customers based on extensive demographic data regarding buying habits, credit risks, and a good understanding of what makes specific types of customers profitable. Slowly, leading mobile operators are beginning to understand that they also have to take a much more multimicro-niche approach to the mobile business. This requires a sophisticated marketing intelligence system (MIS) to collect, collate, and analyze data regarding potential and actual customers.

Once such a system is in place, it can be used for both short-term tactical as well as strategic marketing. For example, suppose marketing knows that network capacity is under pressure in a particular town and that technical planning cannot solve the problem within the next eight weeks. On one hand, it would be bad for the operator's image to continue selling capacity in that town, since congestion would result. However, the service providers and dealer networks cannot remain idle. The MIS can be used to identify which groups can still be targeted in that city without adding to the capacity problems – for example, long-distance transport, field sales forces, international business people, low usage groups, and so on.

Such an MIS that focuses dynamically on current priorities, linked to a broad portfolio of service/tariff packages, becomes a core element to yield management – a business process that has become crucial to the profitability of other service industries that the sell time of using a "perishable" service: airlines, hotels, car rentals, etc. An airline aims to fill first-class and business-class seats with full-fare passengers first and then progressively offers reduced tickets to fill remaining seats. One airline might have hundreds of thousands of different tariff/destination combinations in order to fill its daily capacity of tens of thousands of seats. The mobile industry is, by comparison, at a primate stage of tariff development, worrying about possibly confusing customers by offering a handful of airtime tariff options.

As an industry, mobile telecommunications providers must pro-actively seek opportunities to kick-start medium or fast-track growth, and this must be done before early unfavorable symptoms grow serious.

The European experience has therefore many lessons and mistakes that us players can learn from. It also demonstrates that there are many attractive opportunities, but also some dangers. Mobile in Europe, as in the United States, is still at a pioneering stage, with more questions, uncertainties, and unknowns than answers. By combining the opportunities, skills and experiences, we can all grow this industry faster and more cost-effectively.

Wireless Systems in Japan and Associated Issues

Bryce W. Hunt
Graduate Student, Telecommunications
University of Colorado at Boulder

Abstract

This essay reviews the history of analog cellular and digital cellular development in Japan. As technology development unfolds, the cellular service providers that are competing for market share in Japan are introduced. Cellular subscriber growth history is reviewed, and growth projections are presented. With its service debut in 1995, a new micro-cellular technology that promises to attract a vast segment of the cellular market is discussed. Finally, the future of NTT is discussed.

Introduction

The origins of cellular telephone development can be traced back to Japan as well as to America. Though Americans may contest, the Japanese are pioneers in cellular systems development and deployment. Japan's dominant telecommunication's provider, Nippon Telephone and Telegraph (NTT), researched cellular systems for nearly 20 years before deploying the world's first commercial system in 1979.

Analog and Digital Cellular Development

At first, NTT developed much of their cellular technology and systems from their own proprietary designs since at the time few cellular system specifications existed in the international community. As cellular technology developed across the globe, NTT and others upgraded their systems to better comply with cellular standards adopted by the international community.

By 1994, NTT (DoCoMo) along with its old and new competitors successfully deployed digital cellular systems throughout the more urban areas of Japan. Unlike previous analog developments, the Japanese government was more involved this time in the initial development of system specifications so that certain compatibility issues associated with the existing analog systems were resolved. Not only did the digital systems enhance voice quality and data services, but they relieved many analog systems of the excessive traffic constantly prevalent in the more urban areas of Japan.

Once there were only one or two service providers in Japan soliciting subscribers – now there are three or four all competing for market share in the same service areas. This development along with additional Japanese government deregulatory legislation has decisively fueled subscriber growth even in the midst of a stagnant or recessive Japanese economy. As stated by Seppo Laukkanen, Nokia Mobile Phones Vice-President, "The rapid growth has been beyond everyone's expectations."

Surprisingly enough, there is a new micro-cellular technology being implemented in the larger markets of Japan that seems to have an even brighter future than conventional cellular systems. It is called the personal handy phone system (PHS). With service available in the latter half of 1995, PHS promises to attract millions of customers before the year 2000 with service prices significantly lower than conventional cellular service prices. PHS stands to carve its own market niche for those who do not require mobile communications while in vehicular transit.

First-Generation Cellular: Analog Systems – NTT Service Only

NTT Laboratories began research on analog cellular systems in the early 1960s. Efforts toward an 800-MHz band analog system were accelerated by 1967. By the end of 1975, all technical trials of an analog system in Tokyo were completed. By 1979, the world's first commercial analog cellular system was introduced in Tokyo. Before this time, there were no public land-mobile voice communication systems in Japan.

NTT called their first cellular system "Mobile Communication System - L1" (MCS-L1). The frequency band in which NTT's MCS-L1 system operates is shown in *Figure 1*. Additional technical characteristics of MCS-L1 are compared with those of other analog systems in *Table 1*.

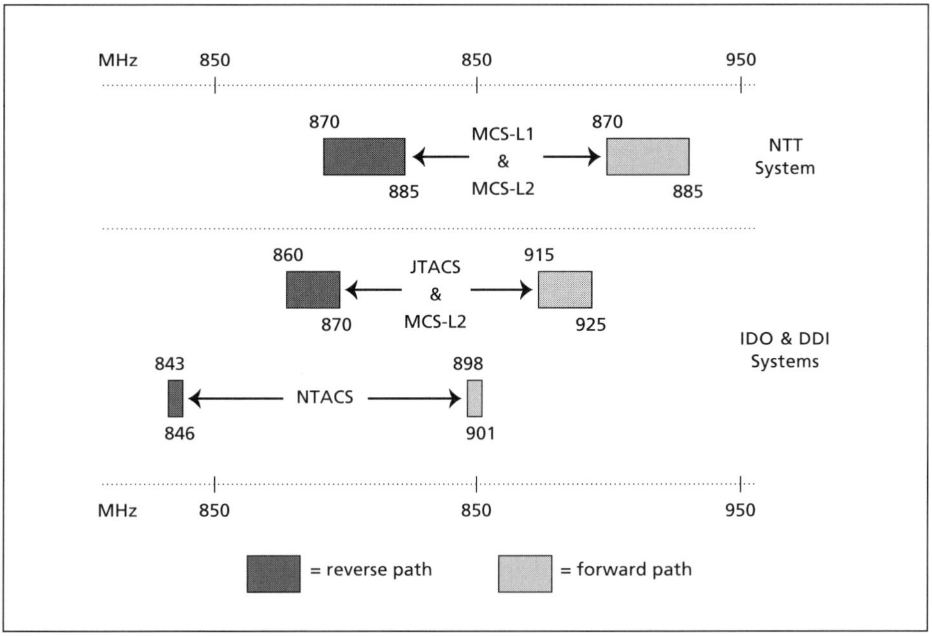

Figure 1 *Frequency allocation of all analog cellular systems in Japan*

Within several years, the MCS-L1 system became congested in specific areas around the main Tokyo business districts. NTT soon developed another system that provided an increase in radio channel density, rather than an increase in frequency reuse efficiency. Retrofitting the old base station equipment with new equipment was more economical than installing a whole new base station infrastructure. The second-generation system, called MCS-L2, was introduced in May 1988. It doubled the channel density provided by MCS-L1.

When a MCS-L1 cell area became consistently overloaded, the MCS-L1 base station servicing that cell area was retrofitted with MCS-L2 equipment. This allowed the more frequency-efficient MCS-L2 base stations to gradually displace the MCS-L1 base stations when demand warranted the need. The MCS-L2 system shares the same frequency band as the MCS-L1 system as shown in *Figure 1*. The technical characteristics of MCS-L2 are compared with those of MCS-L1 and other analog systems in *Table 1*. Today MCS-L1 and MCS-L2 systems are referred to in the same breath as the NTT system since they were originally developed by NTT. Any reference to an NTT system or NTT compatibility implies MCS-L1 and MCS-L2 functionality and compatibility.

Item	MCS-L1	MCS-L2	JTACS	NTACS
Bandwidth allocated (in MHz) for each path:				
NTT	15	15	–	–
IDO[1]	–	10	–	2
Transmit and receive Path separation (MHz)	55	55	55	55
Channel separation (kHz)				
without interleaving	25.00	12.50	50.00	25.00
with interleaving	(none)	6.26	25.00	12.50
Number of channels (without interleaving)				
15 MHz bandwidth	600	1200	–	–
10 MHz bandwidth	–	800	200	–
02 MHz bandwidth	–	–	–	80
Number of channels (without interleaving)				
15 MHz bandwidth	(none)	2400	–	–
10 MHz bandwidth	–	1600	400	–
02 MHz bandwidth	–	–	–	160
Modulation	PM	PM	PM	PM
Max. freq. deviation (kHz)	+/- 5.0	+/- 2.5	+/- 9.5	+/- 5.0
Compandor technique	2:1	2:1	2:1	2:1
Signaling[2] (in bps):				
Common control	300	2400	8000	8000
Channel-associated	(none)	100	8000	100

Table 1 *Comparison of all analog cellular systems in Japan*

1 The service areas of IDO and DDI do not overlap, which permits them to share the same frequency band

2 All signaling is Manchester bi-phase encoded

NTT's Corporate Restructuring

With pressure from the government, NTT separated their cellular service division in 1992 from the rest of NTT operations. NTT called the subsidiary NTT Mobile Communications Network Inc., but more commonly named DoCoMo by NTT and the public (Do Communications over the Mobile network). By 1993, DoCoMo split their cellular operations into nine separate regional service subsidiaries. *Figure 2* shows where these nine DoCoMo service providers are located in Japan.

Second-Generation Cellular: Analog Systems – Some Service Competitors

Following in the footsteps of the U.S. deregulatory trend, the Japanese govern-

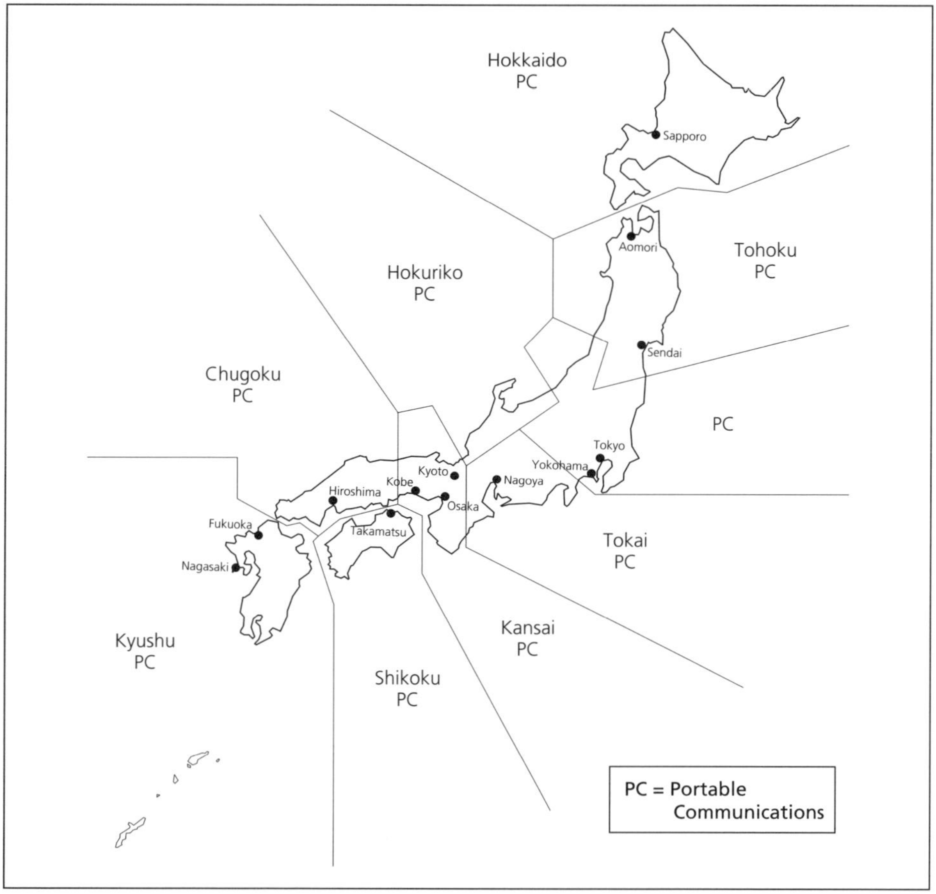

Figure 2 *Regional boundaries of DoCoMo subsidiaries in Japan*

ment deregulated their telecommunications industry with similar consequences in April 1985. NTT no longer had exclusive rights to provide the cellular service that they provided across the most populated areas of Japan. Two companies soon emerged to take advantage of the opportunity to compete with DoCoMo in the cellular service market: Nippon Idou Tsushin Corp. (IDO) and Cellular Telephone.

With approval from the Ministry of Post & Telecommunications, IDO focused on competing with DoCoMo in the Tokyo-Nagoya service area while Cellular Telephone focused on competing with DoCoMo in all other strategic service areas of Japan. Cellular Telephone is broken up into eight separate operations, each serving a different geographic region of Japan. *Figure 3* shows where IDO and Cellular Telephone have operations.

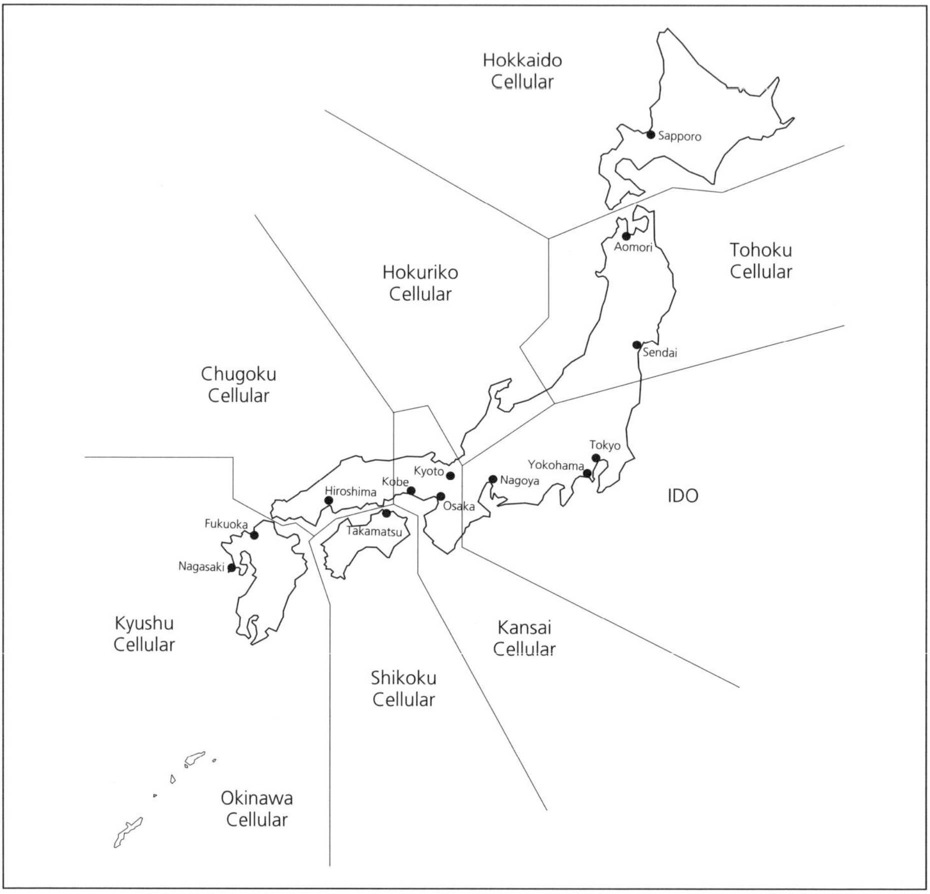

Figure 3 *Regional boundaries of IDO and cellular telephone (DDI operations)*

IDO was started by a consortium of investors, none of which had a majority stake in the business. The largest investors are a long-distance carrier, automakers, and power utilities. *Table 2* lists IDO's investors. Like IDO, each operation of Cellular Telephone was started by a consortium of investors. Unlike IDO, one particular investor has a majority stake (at least 60 percent) in each Cellular Telephone operation: Daini-Denden, Inc. (DDI).

DDI is a Type I domestic long-distance carrier – the first to compete with NTT in 1985. The second largest investor (20 percent stake) in each of the regional Cellular Telephone operations is the regional power utility respective to the region. *Table 2* lists the investors of each regional Cellular Telephone operation. Since each Cellular Telephone operation is controlled by DDI, this essay refers to

Service Provider	Incept Date	Service Date	Investors	N[1]	Business	%
IDO	Mar '87	Dec '88	Nihon Kousoku Tsushin	J	Carrier (Type ?)	14.40
			Toyota Jidoosha	J	Automaker	13.10
			Tokyo Denryoku	J	Power utility	10.50
			Nissan Jidoosha	J	Automaker	5.20
Kansai Cellular Telephone	Jun '87	Jul '89	DDI (Daini-Denden)	J	Carrier Type 1 - Domestic	64.25
			Kansai Denryoku	J	Power utility	20.00
			Osaka Gasu	J	Gas (utility?)	2.50
			31 others[2]			<2.00
Chugoku Cellular Telephone	Nov '87	Dec '89	DDI (Daini-Denden)	J	Carrier Type 1 - Domestic	61.65
			Chugoku Denryoku	J	Power utility	20.00
			Mazda Motors	?	Automaker	5.00
			30 others[2]			<2.00
Kyushu Cellular Telephone	Oct 87	Dec 89	DDI (Daini-Denden)	J	Carrier Type 1 - Domestic	63.45
			Kyushu Denryoku	J	Power utility	20.00
			36 others[2]			<2.00
Tohoku Cellular Telephone	Apr '88	Apr '90	DDI (Daini-Denden)	J	Carrier Type 1 - Domestic	64.10
			Tohoku Denryoku	J	Power utility	20.00
			39 others[2]			<2.00
Hokkkaido Cellular Telephone	Jul '88	Aug '90	DDI (Daini-Denden)	J	Carrier Type 1 - Domestic	?
			Hokkaido Denryoku	J	Power utility	?
			29 others[2]			
Hokuriku Cellular Telephone	May '88	Sep '90	DDI (Daini-Denden)	J	Carrier Type 1 - Domestic	63.00
			Hokuriku Denryoku	J	Power utility	20.00
			33 others[2]			<2.00
Shikoku Cellular Telephone	Apr '88	Dec '90	DDI (Daini-Denden)	J	Carrier Type 1 - Domestic	62.00
			Shikoku Denryoku	J	Power utility	20.00
			36 others[2]			<2.00
Okinawa Cellular Telephone	Jun '91	Oct '92	DDI (Daini-Denden)	J	Carrier Type 1 - Domestic	60.00
			51 others[2]			<2.00

Table 2 *IDO and cellular telephone investors*

1 Nationality of company: J=Japanese; A=American; B=British

2 Other companies have invested 2% or less into the cellular service provider

their operations simply as DDI operations.

The Original IDO & DDI Systems

In December 1988, seven months after DoCoMo deployed their first MCS-L2 system, IDO deployed a system similar to the MCS-L2 in the Tokyo-Nagoya area. The key difference between the two systems is in frequency spectrum use and number of channels provided. *Table 1* provides a more detailed comparison between the DoCoMo MCS-L2 system and the IDO MCS-L2 system.

DDI (actually Kansai Cellular Telephone) deployed its first system called JTACS in

July 1989. While IDO's MCS-L2 system is similar to DoCoMo's MCS-L2 system, DDI's JTACS system is similar to the British TACS system. The only difference between DDI's JTACS system and the British TACS system is in frequency spectrum use and number of channels provided. The JTACS system is completely incompatible with either MCS-L1 or MCS-L2 systems.

The frequency band in which IDO's MCS-L2 and DDI's JTACS systems operate is shown in *Figure 1.* Since originally IDO and DDI operated in geographically separate areas of Japan, their original systems share the same frequency spectrum. The technical characteristics of JTACS are compared with those of other analog systems in *Table 1.*

NTACS Deployment

In August 1991, a pair of 2-MHz blocks in the 800-MHz band were newly allocated to IDO and DDI to begin providing cellular service with Motorola's NTACS system. The NTACS system, which doubles the channel capacity provided by JTACS, is similar to the NAMPS system developed by Motorola for the American cellular infrastructure but also backward compatible with JTACS. NTACS doubles channel capacity by narrowing channel separation from 25.0 KHz to 12.5 KHz. The frequency band in which NTACS operates is shown in *Figure 1.* The technical characteristics of NTACS are compared with those of the other previously discussed analog systems in *Table 1.*

Like those handsets leased by DoCoMo that can operate with both MCS-L1 and MCS-L2 systems, new cellular handsets leased by DDI are compatible with both JTACS and NTACS systems. Today, JTACS and NTACS systems are referred to in the same breath as the TACS system. Any reference to a TACS system or TACS compatibility implies JTACS and NTACS functionality and compatibility.

Roaming Issues for Analog Cellular Systems

The only cellular service provider that offers seamless service throughout Japan is DoCoMo. If customers desire the ability to roam freely throughout the entire country without disrupting their service, they must subscribe to DoCoMo's analog cellular service and posses a mobile unit compatible with MCS-L1 and MCS-L2. DoCoMo has no analog cellular roaming agreements with IDO or DDI to allow IDO or DDI subscribers the ability to roam beyond their service provider's coverage areas. This impasse is due more to the significant technical incompatibilities between the NTT and TACS systems than to any business leveraging strategy devised by DoCoMo – though DoCoMo has yet to be motivated to address the roaming dilemma.

Many potential subscribers refrain from subscribing with DoCoMo's competitors because of the inability for non-DoCoMo subscribers to roam freely throughout Japan and retain seamless service. Even with the enticing lower service fees and rates offered by DoCoMo's competitors, many potential subscribers and DoCoMo subscribers are unwilling to subscribe with anyone else but DoCoMo. Thus, by March 1994 when digital service was becoming activated by many service providers, DoCoMo accounted for 62 percent of the total cellular market in Japan, and DDI and IDO accounted for 22 percent and 16 percent, respectively.

Third-Generation Cellular: Digital Systems and Providers

Inadequacies in the analog cellular systems in Japan along with a shortage of available frequency spectrum have led to the development of a digital cellular specification in the country. Unlike the analog cellular system development, the digital cellular specification was carefully planned by the Ministry of Posts and Telecommunications to ensure roaming compatibility between multiple service providers. They wanted to avoid the roaming dilemma that exists between the NTT and TACS systems. By doing so, cellular subscribers are no longer obligated to subscribe only to a nationwide provider like DoCoMo in order to roam freely.

Digital Specifications

The Ministry of Posts and Telecommunications appointed the Digital Mobile Telephone System Committee of the Telecommunications Technology Council to devise a digital cellular plan for Japan. The ministry specifically instructed the committee to develop a plan that addressed and resolved the cellular roaming issue. In June 1990, the committee reported on the technical conditions for a digital cellular system that provided an air-interface specification.

By April 1991, the Research and Development Center for Radio Systems published a detailed standard for Japanese digital cellular (JDC). The standard provides a general system architecture and the air-interface specification. The major specifications for the unified air interface are shown in *Table 3* and compared with the accepted American and European specifications.

	JDC (Japan)	ADC (American)	GSM (European)
Forward frequency (MHz)	810-826 1477 1501-1513	869-894 – –	935-960 – –
Reverse frequency (MHz)	940-956 1429-1441 1453-1465	824-849 – –	890-915 – –
Carrier spacing (kHz) with interleaving	25	30	200
Modulation	π/4 shifted QPSK	π/4 shifted QPSK	GMSK
Multiple access (TDMA) Full-rate speech Codec Half-rate speech Codec	 3 channel 6 channel	 3 channel 6 channel	 8 channel 16 channel
Carrier bit rate (kbps) Full-rate Half-rate	 11.2 (VSLEP) 5.6	 13 (VSLEP) 6.4	 22.8 (RPE-LTP) 11.4
Equalizer	Optional	–	Mandatory
Diversity	Optional	Optional	–
Frequency hopping	–	–	Optional

Table 3 *Comparison of international digital cellular system specifications*

Modulation Technique

The development of narrowband modulation techniques and low bit-rate codecs enables more effective use of available frequency spectrum compared to analog cellular systems. In addition, digital signal processing, such as FEC and ARQ, promise to improve speech and data communication quality. Selection of π/4-shifted QPSK modulation is attributable to its efficient spectrum utilization and ease of linear power amplification and detection.

Multiplexing Technique

TDMA, adopted for digital mobile communications in Europe and North America, was also recommended for use in the JDC system. The major reasons for this are:

- time division multiplexing offers greater savings in base station equipment, cost, and space
- monitoring the receiving level from surrounding base stations in unused TDMA

slots by the mobile station enhances call control, thus enabling the construction of smaller cells

Initially, the number of TDMA slots was three. In the future, this will increase to six once a half-rate speech codec has been developed. An 11.2-Kbps traffic channel bit rate was selected based upon voice quality and adjacent carrier interferences.

Speech Encoding Technique

The speech codec implementation that was chosen uses the vector sum excited linear prediction (VSELP) coding algorithm because of the favorable voice quality opinion test (mean opinion score evaluation) results and the low hardware/software implementation costs.

Data Transport Technique

Error-free communication is achieved in facsimile transmissions by adopting the high-throughput ARQ transmission scheme. If facsimile capability is desired, then a signal processing unit is added to mobile stations and switching equipment in the cellular infrastructure in order to provide error correction. Applying error correction to the 11.2-Kbps channel results in an effective throughput rate of 4.8 Kbps. For data communications, an additional signal processing unit is added to the mobile stations and switching equipment in order to implement the error correction algorithm associated with the widely accepted modem protocol, micro-networking protocol (MNP).

Frequency Spectrum Allocation

Since the frequency bands for the existing analog cellular systems have already been allocated, new bands had to be assigned to digital systems. They are the 800-MHz and 900-MHz bands with 130 MHz of duplex separation, and the 1.5-GHz band with 48 MHz of duplex separation as shown in *Figure 4*. Because of the plan to swap out the analog channels carrier by carrier with those of the digital system, the carrier frequency separation was chosen to be 25 KHz with interleaving and 50 KHz without interleaving.

The digital cellular systems in Japan that utilize the 810-MHz to 826-MHz and the 940-MHz to 956-MHz bands are referred to by the cellular industry as the 800-MHz systems. The digital cellular systems that utilize the 1429-

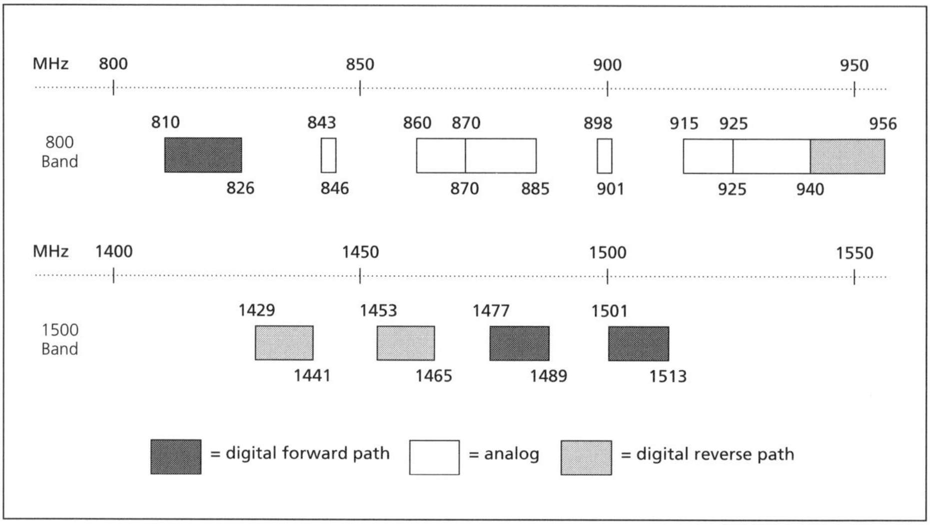

Figure 4 *Frequency allocation of all digital cellular systems in Japan*

MHz to 1441-MHz, 1453-MHz to 1465-MHz, and the 1477-MHz to 1489-MHz bands are referred to by the cellular industry in Japan as the 1.5-GHz systems.

Service Providers

Both new and existing service providers entered the digital cellular service market after the government allocated spectrum for digital cellular. All the existing analog providers (DoCoMo, IDO, and DDI) entered the digital market in service areas where analog systems became loaded with traffic. *Figure 5* shows all the analog and digital systems DoCoMo currently has deployed in all of their service operations. *Figure 6* shows all of the analog and digital systems IDO and DDI have currently deployed in all of their service areas of Japan.

Also, new service providers entered the digital cellular service market in areas where analog systems became loaded with traffic: Digital Phone, Tu-Ka Cellular, Tu-Ka Phone, and Digital Tu-Ka. Like IDO and Cellular Telephone, these service providers were started by a consortium of investors. And like Cellular Telephone, Digital Phone and Tu-Ka Cellular were further divided into regional operations. *Table 4* lists their investors. *Figure 7* shows where these new digital service providers operate in Japan and the type of systems they implement.

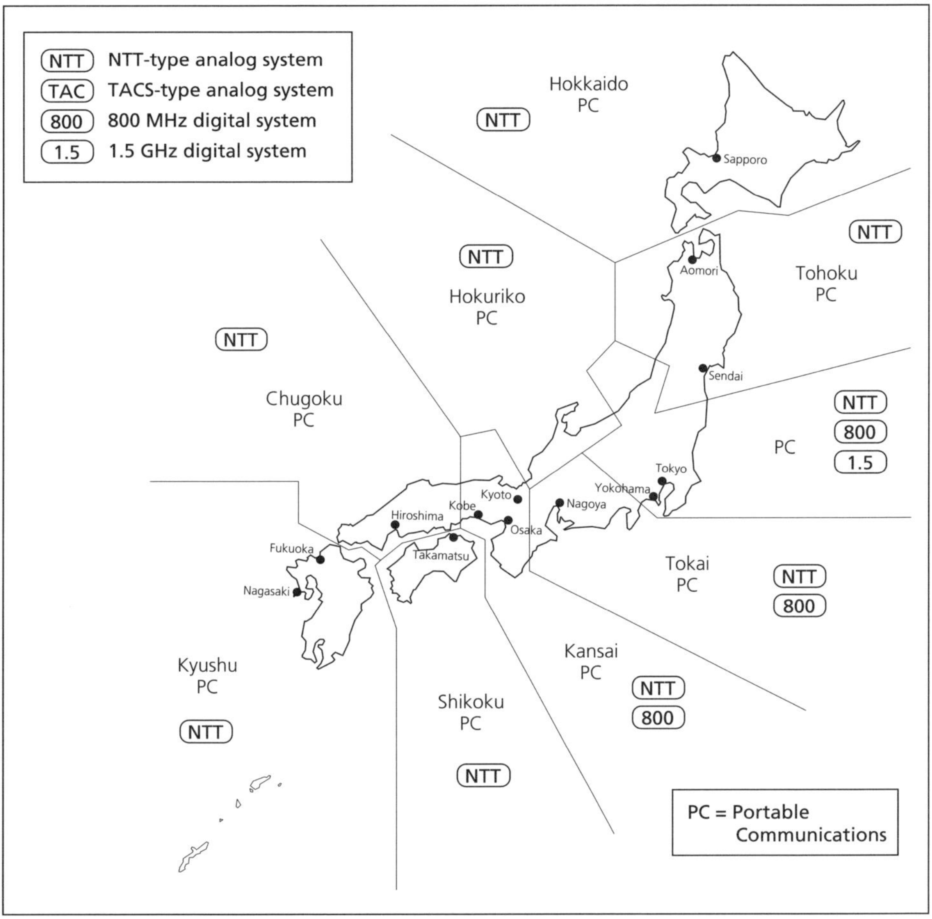

Figure 5 *All DoCoMo analog and digital cellular systems in Japan*

Subscriber Growth in Japan

The growth of cellular subscribers in Japan has swelled and ebbed in response to the economic forces at play in the country. Many of these forces have been significantly influenced by legislative movements enacted by the Japanese government. Two such movements have most significantly affected this growth: the reformation of the loan process within the banking industry has negatively affected growth, and the deregulation of telecommunication services and equipment has positively affected growth.

Aside from government legislation, other forces both macro and micro with

Service Provider	Incept Date	Service Date	Investors	N[1]	Business	%
IDO	Jul. '91	Apr. '94	Nihon Telecom	J	CarrierType 1 – Domestic	22.75
			Pacific Telesis	A	Comm. Service Provider	15.00
			Higashi Nihon R. Tetsudo	J	Railway operator	12.00
			Metro Phone Service	J	Carrier (Type ?)	12.00
			Cable & Wireless	B	Comm. Service Provider	8.00
			Toyota Jidoosha	J	Automaker	4.00
			Shin Nihon Seitetsu	J	Steelmaker	4.00
			CSK	J	Information Service Provider	4.00
			KDD	J	Carrier Type 1 – Int'l	2.50
			71 others[2]			
Toakai Digital Phone	Mar. '92	Aug. '94	Nihon Telecom	J	CarrierType 1 – Domestic	24.27
			Pacific Telesis	A	Comm. Service Provider	13.00
			Japan Railways	J	Railway operator	12.00
			Cable & Wireless	B	Comm. Service Provider	7.20
			Toyota Jidoosha	J	Automaker	4.00
			CSK	J	Information Service Provier	4.00
			Toenec	J	EE contractor (Chubu area)	3.00
			Chubu Nihon Building	J	Construction	2.50
			KDD	J	Carrier Type 1 – Int'l	2.50
			108 others[2]			
Kansai Digital Phone	Nov. '91	May '94	Nihon Telecom	J	CarrierType 1 – Domestic	?
			Pacific Telesis	A	Comm. Service Provider	?
			Japan Railways	J	Railway operator	?
			Cable & Wireless	B	Comm. Service Provider	?
			Toyota Jidoosha	J	Automaker	?
			CSK	J	Information Service Provider	?
			86 others[2]			
Tokyo Tu-Ka Cellular	Jul. '91	Jun. '94	Nissan Jidoosha	J	Automaker	26.42
			DDI (Daini-Denden)	J	CarrierType 1 – Domestic	26.42
			Nihon Motorola	A	Cellular + Manufacturer	8.00
			Sony	J	Electronics Manufacturer	5.50
			Hitachi	J	Electronics Manufacturer	5.50
			British Telecom	J	Comm. Service Provider	5.00
			29 others[2]			<2.00
Tokai Tu-Ka Cellular	Feb. '92	Jul. '94	DDI (Daini-Denden)	J	CarrierType 1 – Domestic	26.25
			Nissan Jidoosha	J	Automaker	26.25
			Nagoya Tetsudo	J	Railway operator	5.00
			Chunichi Shimbun	J	Newspaper publisher	5.00
			Nihon Motorola	A	Cellular + Manufacturer	4.50
			Sony	J	Electronics manufacuter	3.50
			Hitachi	J	Electronics manufacturer	3.50
			British Telecom	B	Comm. Service Provider	3.00
			KDD	J	Carrier Type 1 – Int'l	3.00
			41 others[2]			<2.00

Service Provider	Incept Date	Service Date	Investors	N[1]	Business	%
Kansai Tu-Ka Phone	Oct. '91	Apr. '94	Nissan Jidoosha	J	Automaker	36.95
			Kobe Steel	J	Steelmaker	9.00
			Hitachi	J	Electronics manufacturer	9.00
			Matsushita Denki Sangyo	J	Electronics manufacturer	5.25
			British Telecom	B	Comm. Service Provider	5.00
			Marubeni	J	General trader	5.00
			GTE	A	Comm. Service Provider	4.00
			Nichimen	J	General trader	3.00
			KDD	J	Carrier Type 1 – Int'l	2.50
			48 others[2]			
Kyushu	Mar. '94	??? '96	Nissan Jidoosha	J	Automaker	22.86
			Nihon Telecom	J	Carrier Type 1 – Domestic	22.86
			Hitachi	J	Electronics manufacturer	5.50
			Japan Railways	J	Railway operator	5.50
			Nishi Nihon Tetsudo	J	Railway operator	5.25
			Shin Nihon Seitetsu	J	Steelmaker	5.00
			Toyota Jidoosha	J	Automaker	5.00
			Pacific Telesis	A	Comm. Service Provider	4.50
			GTE	A	Comm. Service Provider	4.25
			Cable & Wireless	B	Comm. Service Provider	2.50
			British Telecom	B	Comm. Service Provider	2.50
			45 others[2]			

Table 4 *Investors of the new digital service providers*

1 Nationality of company: J=Japanese; A=American; B=British

2 Other companies have invested 2% or less into the cellular service provider

respect to the cellular service market have also affected subscriber growth in a fortuitous manner. One is the emergence of four new cellular service providers. They have intensified service competition and brought service fees and rates down.

Growth Trends

Figure 8 shows the total number of cellular subscribers in Japan since the introduction of service in 1979. Notice that from the beginning of 1984 until the end of 1990, the growth rate always exceeded 50 percent, peaking at 102 percent at the end of 1988. At the end of 1990, the average growth rate finished at 63 percent. Once economic trauma hit Japan in early 1991, the anticipated growth rate has dropped.

Figure 8 provides two maximum and minimum growth trends. The line with triangle milestones projects a more pessimistic growth trend for cellular subscriptions. This trend predicts that there will be approximately 10 million subscribers by the year 2000. The line with horizontal-bar milestones projects a more optimistic growth

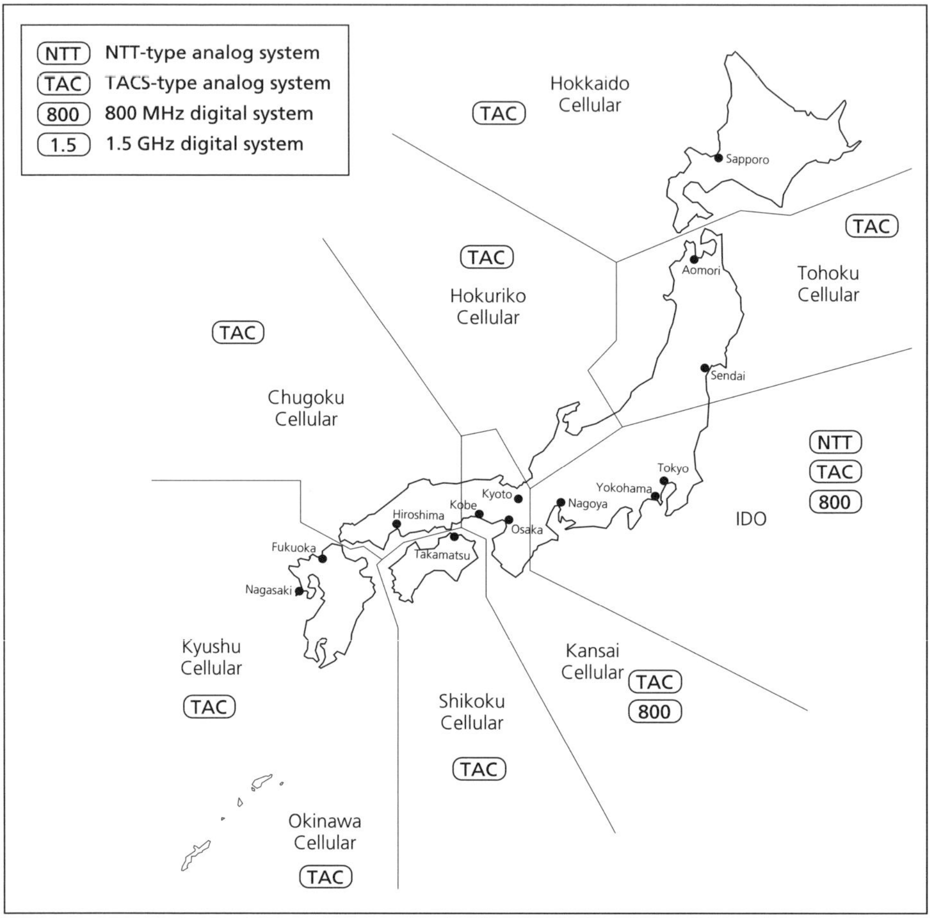

Figure 6 *All IDO and DDI analog and digital cellular systems in Japan*

trend for cellular subscriptions. This trend predicts that there will be approximately 20 million subscribers (i.e., one in every six inhabitants) by the year 2000.

A recent finding indicates that the actual growth trend is closely following the optimistic model. A 1995 finding indicates that there were 1,710,000 total cellular subscribers by the beginning of 1992. This is much closer to the more optimistic projection of 1,750,000 for that time period than to the more pessimistic projection of only 1,400,000. This finding is plotted in *Figure 8* with diamond milestones.

In the first four months of 1994, the total number of new cellular subscribers

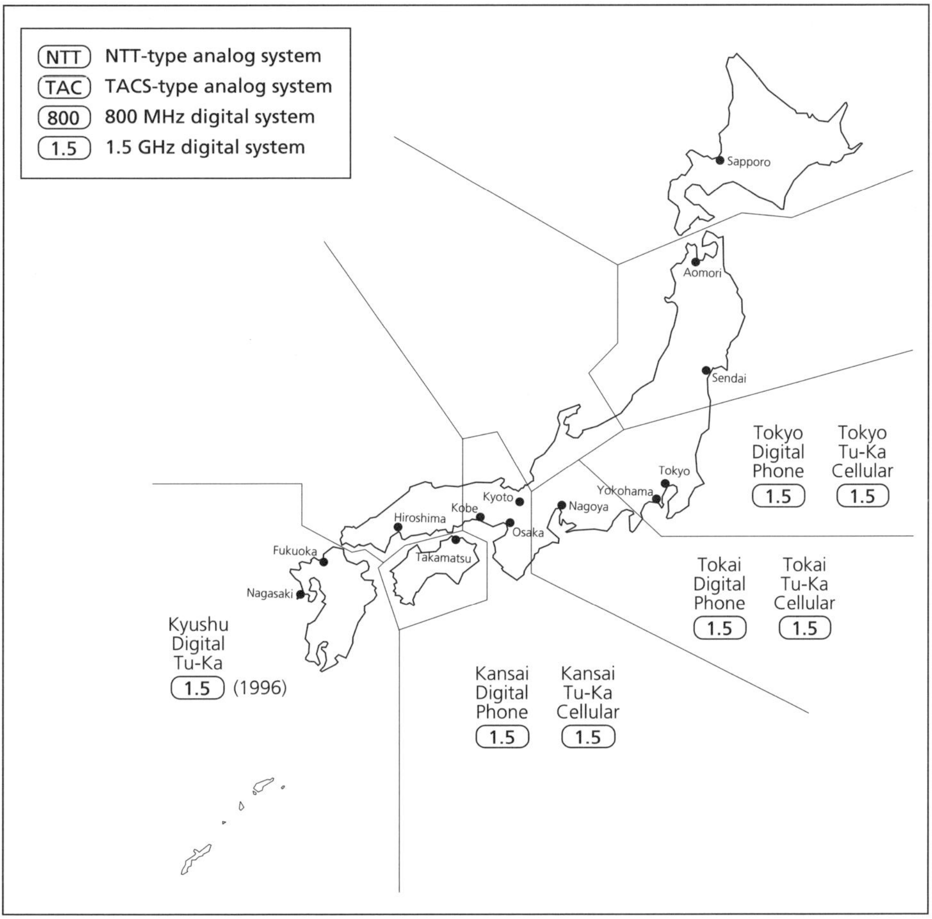

Figure 7 *Digital cellular systems of all new service providers*

(488,000) surpassed the total number (444,700) of new cellular subscribers for all of 1993. Three months after the Japanese government permitted cellular phone ownership (April 1, 1994), the number of new cellular phone subscribers surpassed the number of new POTS subscribers – the first time this has ever occurred. In the following month (July 1994), the same occurred but with an even larger disparity. These figures are shown in *Table 5*.

By the end of July 1994, the total number of new cellular subscribers was at 883,600. The total number of new subscribers was expected to top 1.4 million by the end of 1994. This unanticipated surge in new cellular subscriptions has left some industry strategists second-guessing, at least Isamu Kuru, President of

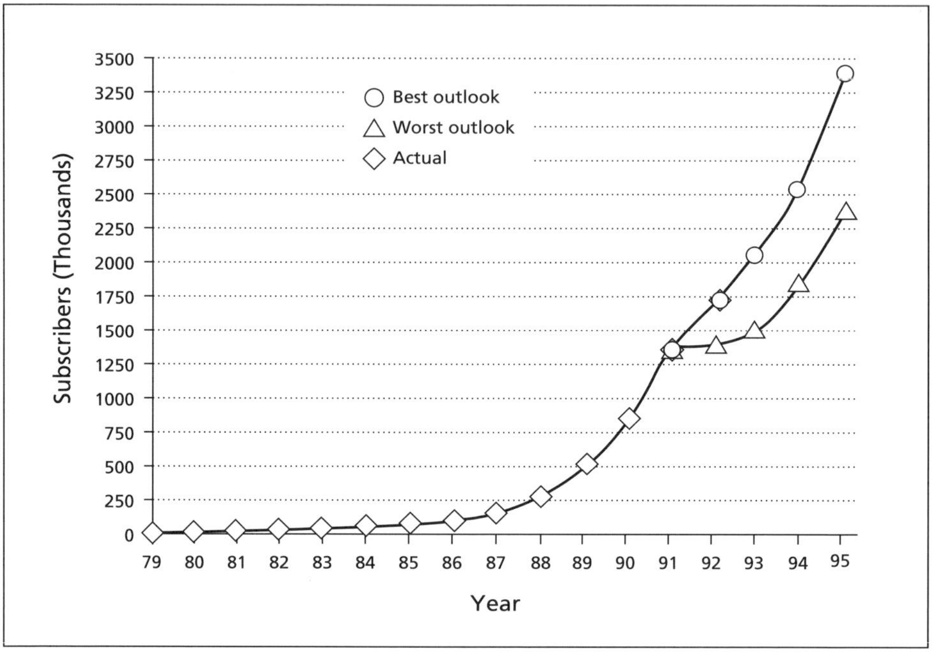

Figure 8 *Subscriber growth in Japan*

Nippon Motorola Ltd., who said,

> *Last year, I had estimated that the net increase in cellular phone subscribers would be about 1 million in 1994. In summer, I revised that prediction to 1.2 million. Now it's likely to top 1.4 million...I believe that the growth in cellular phone sign-ups in 1995 will equal that of this year.*

Stimulants to Growth in 1994

Isamu Kuru and other cellular industry strategists attribute this surge in cellular subscriptions in 1994 to the following reasons:

- the government allowing the ownership of cellular phones, which in turn introduced more competition in handset manufacturing, and driving handset prices down
- the recent introduction of four new service providers that have increased competition and caused service providers to lower service fees and rates
- the market introducing summer service discounts for subscribing

	POTS	Cellular
June '94	100,000	130,000
July '94	90,000	156,000

Table 5 *POTS versus cellular subscribers in Japan, June - July 1994*

- the government providing income tax rebates for cellular subscriptions

Emphasizing the first two reasons, Isamu Kuru said that "the boost [of new subscriptions in 1994] is due to competition. The price of cellular handsets dropped considerably when newcomers entered the market (both cellular service providers and handset manufacturers)." Motorola has been supplying IDO with cellular infrastructure equipment via its Nippon Motorola subsidiary since 1988. IDO has also retailed cellular handsets for Motorola since deregulation.

Overlooked by many industry analysts, the surge in cellular subscriptions can also be attributed to the introduction in 1994 of new cellular service packages that cater more to the personal subscriber than the business subscriber. Nokia says that massive growth in cellular subscriptions can be attributed to the cellular service market's transformation from a "business segment to a consumer market."

Fourth-Generation Cellular: Personal Handy Phone System and Providers

A new wireless technology made its debut in Japan around mid-1995 called the personal handy phone system (PHS). PHS is considered the next generation in public mobile communication for Japan, and it is expected to directly compete with current analog and digital cellular systems. According to a survey in 1994 conducted by the Ministry of Posts and Telecommunications (MPT), about 40 percent of conventional cellular users were interested in switching to PHS. The MPT predicts that within 10 years of service there will be anywhere from 10 million to 20 million PHS subscribers in Japan.

PHS Service Pricing

The reason that PHS is expected to become so popular so quickly is due to price. For the subscriber, PHS service will be far cheaper than conventional cellular service. Typical PHS service rates will be about 75 percent less than those for conventional cellular, and subscription fees and basic monthly charges will be 40 percent to 70 percent less than those for conventional cellular. PHS service prices can deeply undercut those for conventional cellular service for several reasons.

PHS Advantages

One advantage of PHS is that its infrastructure costs are cheaper. Compared to a conventional base station, a PHS base station is much smaller, cheaper, and easier to install in far more locations. Base stations can be installed indoors (i.e., inside buildings) and even in underground subway terminals. More importantly, a PHS base station can support a greater load of traffic at any given time than a conventional cellular system because of its denser distribution of micro-cells. Thus, PHS service can offer cheaper fees and rates to the customer compared to conventional cellular service.

PHS Disadvantages

When compared to conventional cellular, the only clear disadvantage to using PHS is that the subscriber must not move faster than 10 Km/hr while using the service; thus a subscriber cannot use the service while riding in or on any vehicle traveling faster than 10 Km/hr. Apparently the call hand-off process between PHS cells does not occur quickly enough for a subscriber to not notice any communication glitches while traveling faster than 10 Km/hr from cell to cell. This limitation plus pricing are the two factors that will sway subscribers either to PHS or conventional cellular.

PHS Development

PHS is a digital, micro-cellular type of technology. It is Japan's equivalent to PCS. To what degree PHS is based upon or compatible with the JDC specification is unclear. The development of PHS has lagged behind conventional digital cellular by about one to two years. The MPT has been heavily involved in its development, perhaps even more so than with conventional digital.

In 1990, the MPT issued a directive to various research groups to propose a micro-cellular type of system. By 1992, the Communications Research Laboratory issued a report recommending that PHS (labeled such by this group) "...build upon several existing technologies, while making various improvements to bring about a new standard for cordless digital telecommunications." The MPT allocated 12 MHz of bandwidth for PHS in the 1.9-GHz band. Under the authority of the

MPT, field trials were conducted by NTT and others in Tokyo and Sapporo from the fall of 1993 until 1994.

PHS Service Providers

The government initially permitted only three consortiums to provide PHS: NTT Personal, the Astel Group, and DDI Pocket. The government wanted to initially create just enough competition to ensure quality service yet not so much that a price war would result, in which only the strongest (i.e., NTT) would survive. Though only speculation, perhaps this strategy was in response to a seeming glut of conventional digital providers currently in the Tokyo and Kansai markets – the most lucrative markets. Perhaps the competition in conventional digital is so fierce that the newer providers are finding survival difficult.

NTT Personal

Seventy-six percent of NTT Personal is owned by NTT and DoCoMo. The remaining share is predominantly owned by Marubeni and Itochu (both in general trade), and Cable & Wireless. NTT Personal will have nine regional operations. So far, the operation servicing the Tokyo area has the most infrastructure installed. By mid-1996, NTT Personal plans to have PHS coverage in 70 cities and then in an additional 50 cities by 2000.

The Astel Group

Predominant shareholders in the Astel Group are Japan Railways, KDD, Daiei Supermarkets, Recruit Cosmos, Mitsubishi, Mistui, Nihon Telecom, and the power utility provider in each region that they service. Currently, the Astel Group plans to operate only in the Kansai and Tokyo areas. By March 1996, Astel Kansai anticipates 15,000 base stations will be installed in the Osaka, Kyoto, and Kobe areas. By March 1997, Astel Tokyo anticipates it will have 60,000 base stations installed in the entire Tokyo area.

DDI Pocket

DDI is the predominant shareholder in DDI Pocket. The identity of other investors is unknown, though DDI is soliciting Nichimen and Tomen (both in general trade) to invest. Unlike the PHS offered by the other two consortiums, the PHS offered by DDI Pocket will have a cell range of 500 meters per base station as opposed to only 200 meters associated with PHS offered by the other consortiums. DDI Pocket is implementing a different PHS technology to permit this.

NTT's Future

Prior to 1985, NTT was a totally government-financed utility that had complete control over the public telecommunications infrastructure. After witnessing the success of deregulation in America, the Japanese government chose to introduce competition into the long-distance and cellular market. At the same time, the Japanese government wanted to divest itself from NTT by offering NTT stock to the public. Though NTT is now a public company, the government (more specifically the Ministry of Finance) still owns 65.6 percent of NTT. The government had planned to sell more stock earlier, but when a strong recession hit the Japanese economy in the early 1990s, the government refrained from selling more stock.

At present, there are strong forces within the Japanese government that not only want to completely divest the government from NTT, but also wish to break up NTT. In April 1995, the Minister of Post & Telecommunications asked an advisory council to review NTT's status in order "to ensure fair and effective competition" in the industry. So far the council has recommended splitting NTT up into a stand-alone local service provider, a domestic long-distance carrier, and a mobile communications provider.

According to a senior official at the MPT, Japan's telecommunication industry requires substantial realignment to encourage new business growth. The president of a major Japanese public corporation said in off- the-record comments that the breakup of NTT was imminent, and that those who are planning to do this are being very careful to ensure that it comes about.

In an attempt to dissuade government officials from dismantling the status quo, NTT has demonstrated its willingness to streamline operations by eliminating 30,000 jobs over the past two years from its now current payroll of 200,000. NTT plans to eliminate an additional 30,000 jobs by the year 2000. Recently (September 1995), NTT decided to divest two software divisions from the corporation. The divestment plan that is still under development would make NTT's Information Systems Division (with about 5,000 employees) and NTT's Telecommunications Software Division (with about 4,000 employees) independent entities.

NTT seems to be playing a wait and see game with the MPT. In increments at a time, NTT has demonstrated (or at least announced) streamlining measures or sub-divestment measures, subsequently anticipating the MPT's reaction to such measures. NTT hopes that the MPT will eventually be satisfied by such self-imposed measures, short of dismantling the strategic components of corporate NTT. An NTT executive said, The software divisions could have been spun off a long time ago, but have been kept as a trump card to counter a breakup. Anticipate a strategic announcement from the MPT regarding the fate of NTT sometime in early 1996.

Conclusion

When it comes to strategic telecommunication legislation, Japan usually follows the American example. Still, Japanese bureaucracy can be even more inert than its American counterpart when it comes to big-change legislation. Thus, slaying the NTT monopoly-beast will be more difficult than was the breakup of AT&T, though the end is near. NTT will most likely be broken up as mentioned earlier with the local access business remaining heavily regulated as the RBOCs have been for the past ten years.

In the next fifteen years, PHS service could replace most POTS in all populated areas of Japan. Ultimately, the NTT local access provider will simply become the local access segment of the entire backbone infrastructure for micro-cellular services, interfacing more with competing micro-cellular service providers than directly with subscribers.

The local access wire connection into homes and businesses probably will not be abandoned because multimedia services (with voice service bundled within) will most likely be provided via ISDN technology across these wires. The NTT local access provider probably will not be permitted to directly offer multimedia services to subscribers; only competing multimedia service providers (like America On-line) will be able to offer this. Rather, NTT local access would simply provide a windowing kernel (like X-windows/Motif) via which competing multimedia service providers will offer their services.

PHS will most likely steal many of the personal, cost-conscious subscribers away from conventional cellular. The conventional cellular market will regress to an environment primarily consisting of business subscribers who require communication capability while even in vehicular transit. Most likely, the limitations in the hand-off process for PHS will be overcome within ten years. PHSs and conventional digital systems will then become integrated within fifteen years, and these wireless service providers will reunite their subscriber bases.

Bibliography

Dodwell Marketing Consultants. *The Structure of the Japanese Electronics Industry,* 2nd Edition. (Tokyo: Dodwell Marketing Consultants, December 1988).

Ikeda, Nobuo, ed. "NTT Plans to Cut 30,000 Jobs by 2000." The Nikkei Weekly, (Tokyo) 12 June 1995, 8.

International Publishing Corporation. "Japan's 'Personal Handy Phone' Moves Ahead." Electronic News 25 July 1994, 24.

Kageki, Norri. "Handy-phone firms jockey for market edge." The Nikkei Weekly, (Tokyo) 7 November 1994, 8.

Kageki, Norri. "Cellular-phone competition to intensify." The Nikkei Weekly, (Tokyo) 26 December 1994 and 2 January 1995, 8.

Kageki, Norri. "Global firms tie into handy-phone future." The Nikkei Weekly, (Tokyo) 27 February 1995, 1.

Kageki, Norri. "Handy-phone prices too deeply undercut cellular." The Nikkei Weekly, (Tokyo) 24 April 1995, 8.

Kageki, Norri. "'Fight fair,' rings rival challenge to NTT." The Nikkei Weekly, (Tokyo) 1 May 1995, 1.

Maezawa, Hidetada, ed. "New Cellular Subscriptions Top Those for Fixed Phones." *The Nikkei Weekly,* (Tokyo) 19 September 1994, 8.

Mansfield, Simon. "Japan's Personal Handyphone Systems." ATIP Tokyo, 31 May 1995, ATIP, ATIP95.26.

Masuko, Takashi. "Strong demand for cellular phones, PDA's brightens industry outlook." In Japan Economic Almanac 1995: *The Nikkei Weekly,* ed. Junichi Suzuki. 86-87. Tokyo: Nihon Keizai Shimbun, Inc., 1995.

Moller, Roland. "Finland: Japan Cellular Phone Market Growth Surprises Nokia." Reuter News Service - Western Europe, 15 February 1995, Reuter Textline.

Nakajima, N. "Japanese Digital Cellular Radio." In Cellular Radio Systems, ed. D.M. Balston and R.C.V. Macario, The Artech House Mobile Communications Series, ed. John Walker. 135-50. Norwood, Mass.: Artech House, Inc. 1993.

RIC Telecom. "4 Sha Kyoogoo Jidai no Makuake [The Beginning of an Era of 4 Competing Companies]." Keitai Denwa Hanbai Gaido Bukku [Portable (Cellular) Phone Sales Guidebook] (Tokyo), (Spring 1994): 14-15.

RIC Telecom. "Jidoosha Keitai Denwa Saabisu Jigyoosha Sooran [Car Portable (Cellular) Phone Service Providers Guide]." Keitai Denwa Hanbai Gaido Bukku [Portable (Cellular) Phone Sales Guidebook] (Tokyo), (Spring 1994): 43-53.

Sakamoto, M. "Analog Cellular Radio in Japan." In Cellular Radio Systems, ed. D.M. Balston and R.C.V. Macario, *The Artech House Mobile Communications Series,* ed. John Walker. 135-50. Norwood, Mass.: Artech House, Inc. 1993.

Toyo Keizai Inc. Japan Company Handbook : First Section : Spring 1995. Tokyo: Tolyo

Keizai Inc., 5 February 1995.

Toyo Keizai Inc. Japan Company Handbook: Second Section : Spring 1995. Tokyo: Tolyo Keizai Inc., 6 February 1995.

Ueda, Jinichiro, ed. *Asahi Shimbun Japan Almanac 1995*. Tokyo: Asahi Shimbun.

Dodwell Marketing Consultants. *The Structure of the Japanese Electronics Industry,* 2nd Edition. (Tokyo: Dodwell Marketing Consultants, December 1988).

Ikeda, Nobuo, ed. "NTT Plans to Cut 30,000 Jobs by 2000." The Nikkei Weekly, (Tokyo) 12 June 1995, 8.

International Publishing Corporation. "Japan's 'Personal Handy Phone' Moves Ahead." Electronic News 25 July 1994, 24.

Kageki, Norri. "Handy-phone firms jockey for market edge." The Nikkei Weekly, (Tokyo) 7 November 1994, 8.

Kageki, Norri. "Cellular-phone competition to intensify." The Nikkei Weekly, (Tokyo) 26 December 1994 and 2 January 1995, 8.

Kageki, Norri. "Global firms tie into handy-phone future." The Nikkei Weekly, (Tokyo) 27 February 1995, 1.

Kageki, Norri. "Handy-phone prices too deeply undercut cellular." The Nikkei Weekly, (Tokyo) 24 April 1995, 8.

Kageki, Norri. "'Fight fair,' rings rival challenge to NTT." The Nikkei Weekly, (Tokyo) 1 May 1995, 1.

Maezawa, Hidetada, ed. "New Cellular Subscriptions Top Those for Fixed Phones." *The Nikkei Weekly*, (Tokyo) 19 September 1994, 8.

Mansfield, Simon. "Japan's Personal Handyphone Systems." ATIP Tokyo, 31 May 1995, ATIP, ATIP95.26.

Masuko, Takashi. "Strong demand for cellular phones, PDA's brightens industry outlook." In Japan Economic Almanac 1995: *The Nikkei Weekly*, ed. Junichi Suzuki. 86-87. Tokyo: Nihon Keizai Shimbun, Inc., 1995.

Moller, Roland. "Finland: Japan Cellular Phone Market Growth Surprises Nokia." Reuter News Service - Western Europe, 15 February 1995, Reuter Textline.

Nakajima, N. "Japanese Digital Cellular Radio." In Cellular Radio Systems, ed. D.M. Balston and R.C.V. Macario, The Artech House Mobile Communications Series, ed. John Walker. 135-50. Norwood, Mass.: Artech House, Inc. 1993.

RIC Telecom. "4 Sha Kyoogoo Jidai no Makuake [The Beginning of an Era of 4 Competing Companies]." Keitai Denwa Hanbai Gaido Bukku [Portable (Cellular) Phone Sales Guidebook] (Tokyo), (Spring 1994): 14-15.

RIC Telecom. "Jidoosha Keitai Denwa Saabisu Jigyoosha Sooran [Car Portable (Cellular) Phone Service Providers Guide]." Keitai Denwa Hanbai Gaido Bukku [Portable (Cellular) Phone Sales Guidebook] (Tokyo), (Spring 1994): 43-53.

Sakamoto, M. "Analog Cellular Radio in Japan." In Cellular Radio Systems, ed. D.M.

Balston and R.C.V. Macario, *The Artech House Mobile Communications Series*, ed. John Walker. 135-50. Norwood, Mass.: Artech House, Inc. 1993.

Toyo Keizai Inc. Japan Company Handbook : First Section : Spring 1995. Tokyo: Tolyo Keizai Inc., 5 February 1995.

Toyo Keizai Inc. Japan Company Handbook: Second Section : Spring 1995. Tokyo: Tolyo Keizai Inc., 6 February 1995.

Ueda, Jinichiro, ed. *Asahi Shimbun Japan Almanac 1995*. Tokyo: Asahi Shimbun.

Section IV
The Future

An Ice Age is Coming to the Wireless World

Carl Robert Aron
National Director, Wireless Consulting Practice
EDS Management Consulting Services

The wireless world is about to undergo a very substantial change. That change will undermine common beliefs, upset widely held expectations, and cause substantial disappointment. The nature of that change is, perhaps, best communicated through the use of metaphor, and, thus the notion of a coming Ice Age with all of its implications in terms of climactic hostility and decreased food supply seems an appropriate device to assist in explaining the new conditions coming to the wireless world.

At any time, four fundamental forces determine whether inhabitants of the world of mobile telephony enjoy prosperity or starve into extinction: the nature and number of the inhabitants (competitors); the adequacy of the food supply (revenue); the skill level (operational effectiveness) of the population; and the availability of stores of food (the supply of capital).

The period from 1983 to 1995 was a period of plenty for the inhabitants of the wireless world because the population (of competitors) was small, the food supply (revenue) was more than adequate for the population, the skill level of the population (their operational effectiveness) was not critical to their survival, and the stores of food (supply of capital) were more than adequate to cover the consumption habits of the population. That period was the Golden Age of wireless.

The conditions for prosperity in the Golden Age were not created by the inhabitants of the wireless world because the smallness of the population (the cellular telephone duopoly) was a gift from the FCC, and the growth of food developed without an intense level of competition between the inhabitants in any locale. The inhabitants of the Golden Age became successful and often wealthy because they were carried by market forces, not because of their skills (and, sometimes, despite the lack thereof). Lack of skill was not severely punished in the Golden Age, because capital was available to the skilled and the unskilled alike.

The adequacy of food and the few inhabitants of the Golden Age together assured a low level of competitive intensity because duopolists have a significant disincentive to compete on price, and valuation methodologies minimized the importance of market share. Based upon their experience, the inhabitants of the Golden Age established a set of lessons that will be dangerous to follow or rely upon in the Ice Age to come. Those lessons were the following:

- we know how to compete because we've been doing it for years
- if you build it, they will come (always in adequate numbers)
- operational effectiveness affects profitability, not survival
- ownership of radio spectrum is an assurance of wealth

Following those lessons in the Ice Age will be a formula for extinction.

The coming wireless Ice Age will be the product of radical changes in the fundamental forces which shape the wireless world. The population (number of competitors) will increase significantly, and crop yields (revenue) will, in time, stop growing. The importance of skills (operational effectiveness) will increase a hundred fold, and the availability of stores of food (capital) will decrease.

In the wireless Ice Age, growth in the population will lead to a very significant increase in the intensity of competition. That competitive intensity will assure significant decreases in service pricing. The following factors will assure an extraordinary increase in the intensity of competition:

- the many new players entering the market
- the absence of a competitor with dominant market share
- the eventual slowing of market growth
- high churn rates
- the diversity of strategic interests and agendas
- competitors targeting the same market segments

In the wireless Ice Age, crop yield (revenue) growth will slow and will become inadequate to feed the whole population (all the competitors). Decreasing prices will contribute to revenue inadequacy, and this revenue inadequacy will contribute to the increase of competitive intensity. A downward price spiral may well develop.

Scarcity of food (revenue) will cause the starvation of some, perhaps many, of the inhabitants of the wireless Ice Age. In many wireless markets, the survival of a competitor will depend upon its ability to secure a disproportionate market share. Under conditions of revenue inadequacy, the survival of one competitor requires the extinction of another.

In the face of revenue inadequacy and intense competition, operational effectiveness in the Ice Age will be a critical success factor to a degree unprecedented in the Golden Age. Competitive skills will not only increase the likelihood of survival on the part of one competitor, but they also increase likelihood of extinction on the part of another.

In the wireless Ice Age, those who have provided capital to network operators will demonstrate much greater sensitivity than they did in the Golden Age to the results of the operations of the providers of mobile telephone service in which they invest. As capital providers become weary of prolonged price wars and the consequent negative cash flows, operators will find that capital will rapidly become much less available.

The following are Laws of Survival applicable to the Ice Age:

1. The food supply is not large enough to support all inhabitants, and the survivors will be those who act upon that dismal conclusion.
2. To survive, inhabitants must fight successfully for more than their share of the food supply and, by so doing, assure the starvation of those who fight less successfully.
3. No two inhabitants can both survive if they fight for their share of the food supply in the same way because one of them will have some advantage in that form of fighting and assure the extinction of the other inhabitant by depriving it of food.
4. An inhabitant will survive only if it is able to fight for its share of the food supply using a strategy others cannot use or cannot use as well.
5. A strategy to fight for food must not only be better executable by an inhabitant in principle, but also better executed by that inhabitant in fact in order to offer the prospect of survival.
6. A strategy to fight for food must be intended to capture enough food to assure survival, or the most effective execution of that strategy would be pointless.
7. Never lose access to stores of food (capital) in times of inadequate crop yields (annual revenue).

To survive the Ice Age, a competitor will need to create a strategy by integrating the sources of competitive advantage to which it has superior access. The sources of competitive advantage in wireless telephony include management talent and vision, size, functionality, time, physical resources, information resources, distribution resources, brand recognition, incumbency, and scope of offering. Those sources of competitive advantage can be classified as follows.

The process of integrating sources of competitive advantage to create a strategy appears to indicate four centers of gravity around which those sources can be focused:

Lowest cost service – The provider engineers its enterprise to offer a wireless telephony service of an acceptable quality at a cost lower than any of its competitors.

Focused service superiority – The provider concentrates on continuing development of its wireless service offerings, which offer noticeably superior value by reason of leadership in wireless service and wireless product innovation.

Horizontal service integration – The provider focuses on its ability to bundle other services with the wireless products, thereby offering a broad range of communications services.

Superior customer relationship – The provider focuses on managing every aspect of the customer interface, thereby reducing churn and attracting customers who value such a relationship.

While the selection of an effective competitive strategy will help to prepare a competitor to meet the challenges of the Ice Age, that selection will not alone assure success. Ice Age inhabitants, like their distant ancestors, will have to develop and use tools to eke out adequate sustenance under those hostile conditions. Tools exist to maximize fitness (operational effectiveness) and to provide advantage in the competition for sustenance (revenue). In strict business terms, the following approach is utilized to organize thinking about the strategy development and execution by wireless carriers in the Ice Age. A competitor must:

- develop an understanding of the sources of competitive advantage in wireless telephony
- understand the operator's relative access to the sources of competitive advantage in the market concerned
- examine the extent to which the relative access of the competitors to the sources of competitive advantage can be altered
- determine whether it has, at least, the minimal level of access to the sources of competitive advantage necessary to enter the competition

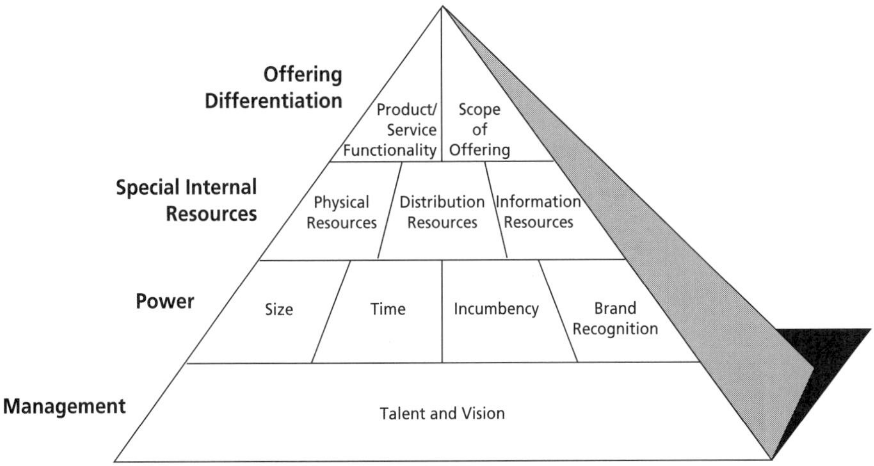

Figure 1 *Sources of competitive advantage*

- integrate the sources of competitive advantage into a chosen preliminary strategy;
- establish that it can execute its chosen preliminary strategy more effectively than that strategy could be executed by any other competitor
- test whether the segment of the market (to which its chosen preliminary strategy is directed) is large enough to support its revenue requirements
- acquire or develop the operational tools, processes, and methods which assure the best execution of the chosen strategy
- recognize that sustaining competitive advantage requires constant improvement and change
- act upon the realization that long-term survival requires not only a disproportional market share, but also access to capital to assure the continued viability to execute the chosen strategy

The Sources Of Competitive Advantage

Four sources of competitive advantage in wireless telephony exist: (I) the talent and vision of management, (ii) power, (iii) special internal resources, and (iv) product or service offering differentiation. This categorization is developed in greater detail in *Figure 1.*

Sources of Competitive Advantage Associated With Management Talent and Vision

The importance of management talent and vision cannot be overestimated. Knowledge of the design, implementation, operation, and marketing of mobile telephone systems is not innate. The increase in the number of competitors in the mobile telephone industry will require more skilled and experienced personnel than are presently available. Large organizations with substantial numbers of employees who are skilled in relation to wireless telephony or telecommunications in general have access to an important source of competitive advantage. Those organizations with truly visionary and talented leadership will surely be critically advantaged. The vision and talent of management is, particularly in changing times and under hostile conditions, the foundation of any organization's effort to integrate access to the sources of strategic advantage into an effective strategy.

Sources of Competitive Advantage Associated With Power

Four major sources of competitive advantage are associated with power: size, time, brand strength, and incumbency.

Size

Size is one of the most important sources of competitive advantage. In the wireless industry, there are several methods for attaining this advantage. One of the most prevalent means for increasing one's size in the wireless industry has been acquisitions (e.g., the RBOCs acquisitions of out-of-region, non-wireline carriers). Other providers, such as PageNet, have attained a size advantage almost entirely through organic/internal growth. Other means of achieving a size advantage include: (I) partnering with firms both internal and external to the industry (e.g., intercarrier agreements, alignments with software providers), (ii) integrating horizontally and vertically by acquisition or joint venture, and (iii) outsourcing functions which need not be carried out internally to take advantage of scale economies of the third-party processors.

The advantages deriving from size include:
- economies of scale
- economies of scope
- financial resources

Size can provide purchasing power, and the scale of that power allows economies to be achieved, for example, in advertising, billing, product development, and overhead. Economies of scale translate into higher margins and a greater ability to price aggressively. In the intensely competitive Ice Age, economies of scale will be a critical source of strength necessary to respond effectively to aggressive behavior on the part of competitors.

Size can make possible economies of scope which can reduce costs by sharing them among multiple product lines or increase revenue by leveraging geographic scope to gain an advantage. Utilizing a sales force to sell multiple products is an example of a cost-reducing economy of scope, and utilizing national or regional scope to effect alliances with major distribution channels which prefer to deal with fewer suppliers is an example of a revenue-increasing economy of scope. In times of revenue inadequacy, economies of scope can be particularly valuable.

Size often indicates access to financial resources, and financial resources will be critical to survival in the Ice Age. Financial resources not only provide the staying power to weather the harsh Ice Age conditions, but also supply the power to make those conditions still harsher for their competitors. Financial resources not only make withstanding aggressive price competition possible, but also permit a competitor to respond with even greater aggressiveness. Financial resources make possible the massive investments in infrastructure necessary to compete in wireless telephony, and also support even larger, preemptive commitments designed to drive competitive costs beyond the ability of some competitors to bear. In terms of the Ice Age metaphor, stores of food are particularly critical when crop yield is inadequate.

Time

Time or speed of delivering can also be a source of competitive advantage. A number of possibilities exist for translating time-based competition into an advantage. One possibility is to redesign processes to reduce the time (and the cost) it takes to accomplish critical tasks such as service order entry and provisioning. A second possibility is to decrease new product development cycle times to introduce more services faster and, thereby, increase marketing effectiveness. A third approach would be to engineer the company (e.g., its systems, structures, and culture) to respond rapidly to and take advantage of market and technology changes. Finally, the outsourcing of non-critical functions, such as collections and billing, will reduce the time to market and allow an operator to achieve economies of scale faster. The time-based advantage can be an important source of differentiation in a crowded competitive environment.

Incumbency

Incumbency is itself a critical source of strategic advantage, but one which combines advantages which derive from multiple sources. The cellular telephone incumbents certainly have advantages deriving from preferred access to information resources, distribution resources and physical resources. They may have all of the advantages that can derive from size. Incumbency may not be a source of product or service superiority, but the fact of incumbency does provide a significant advantage in terms of time. Of course, the most important sources of the incumbent's advantage are in terms of their customer base.

Brand Recognition

The reputation of a competitor can be a source of considerable strategic advantage, provided, however, that the reputation is positive. Few brand names are recognized in wireless telephony, and the RBOCs, which have strong brand recognition in some respects (particularly in-territory), have very little brand recognition as wireless service providers outside of their own territories.

Competitive Advantages Associated With Special Internal Resources

Special internal resources can include three distinct competitive advantages: physical resources; distribution resources; and information resources.

Physical Resources

Preferred access to physical resources will be another important source of competitive advantage. Such preferred access can take many forms. For example, in the early stages of PCS network construction, network infrastructure providers may not be able to meet the demand for equipment. Those competitors which can obtain assurance of delivery of infrastructure components will have an advantage. (The advantage may, of course, be seen as deriving from time or from size if the preferred access arose from large prior purchases.)

Preferred access to antenna sites, interconnection facilities, and other physical resources can provide competitive advantage as well. The nature of the advantage will depend upon whether the preference is measured in terms of price or in terms of exclusive or earlier access.

Distribution Resources

Access to customers, either directly or by reason of access to distribution channels, is an additional vital source of competitive advantage. Competitors which already have billing relationships with the customers to whom they intend to offer wireless telephony services may be able to effect significant reductions in marketing and sales cost. Competitors which have relationships with relevant distribution channels may similarly be able to effect such savings. However, the advantage inherent in such access is not expressed only in terms of cost savings. Subscriber growth may be the more important direct advantage of preferential access to distribution resources.

Information Resources

Access to information, particularly information concerning customers, can provide significant competitive advantage. Customer lists, information concerning credit, demographic information, and information concerning buying practices can all be used to increase marketing effectiveness. Additionally, quick access to customer information by customer service representatives, knowledge of a customers wants and needs, and complete and accurate billing information can prove to be effective in reducing a company's churn. If preferential access to such information is available to a particular competitor, competitive advantage can be created.

Sources of Competitive Advantage Associated with Offering Differentiation

Two sources of competitive advantage are a type of offering differentiation: functionality and scope of offering.

Functionality

Product and service superiority expressed in terms of greater functionality is another important source of competitive advantage. Cellular telephone service has not been exemplary in quality, and much room exists for the creation of competitive advantage simply in providing a service with fewer blocked calls, fewer dropped calls, and less cross-talk. Functionality will also be measured in terms of coverage, features, and the integration of voice, data, and information services.

The primary drivers of product and service superiority as a source of advantage will be quality and technical superiority which can serve to protect price (maintain price levels at a premium) or to attack the competition with more value at the same price level. Delivering demonstrably superior service on attributes of concern to customers, such as coverage and network quality, will attract customers willing to pay a premium. Similarly, providing greater functionality (e.g., integrated services) should yield an ability to charge a premium.

Scope of Offering

The scope of the service offering can be an important source of competitive advantage. Consumer market research has repeatedly indicated that a sizable segment of customers will prefer to buy a broader package of services from a single preferred provider. While some of the appeal lies in the consumer expectation that the provider will offer a discount price on the package, there is clear intrinsic appeal in the breadth of the offering. The customer can elect to receive a single bill and have a single vendor to contact, and all of the purchases can be made from the vendor best meeting the consumer's service preferences.

Competitors' Relative Strengths and Weaknesses

Based upon its understanding of the sources of competitive advantage, a competitor must reach an objective assessment of the relative access to the sources of competitive advantage of its competitors. Since competitors in each market will be different, the competitive dynamics could shift dramatically from market to market.

Table 1 illustrates a method of organizing the competitive assessment process. It records certain very tentative conclusions about each group's potential to establish strategic advantage as a result of their access to the sources of such advantage. The table, by necessity, is not market-specific and must be made so to be useful.

	AT&T	Wireless Co.	PCS Prime Co.	RBOC/ GTE Cellular	RBOC/ GTE PCS	Desig-nated Entities	Indepen-dents
Management Talent and Vision	•	•	•	•	•	?	?
Size							
Economies of Sales	•	•	•	•	•		
Economies of Scope	•	•		•/?	•/?		
Financial Resources	•	•	•	•	•		
Functionality	?	?	?	?	?		?
Time	?	?		?	?		?
Physical Resources	•	•	•	•	•		•
Information Resources	•	•		•	?		?
Distribution Resources	•	•	•	•	•		?
Brand Strength	•	?	?	•			
Incumbency	•/?	?		•	•		
Scope of Offering	•	•		•/?	•/?		

Table 1 *Preliminary competitive assessment*

Alteration of Access to Sources of Competitive Advantage

A competitor's access to the sources of competitive advantage is not a fixed characteristic. Competitors can and do change their access to the sources of competitive advantage. A competitor can change its access to these sources of competitive advantage rather quickly through acquisitions and partnerships.

AT&T secured the advantages of incumbency by its acquisition of McCaw Cellular Communications. Sprint and its cable television partners have sought to maximize their respective strategic advantages by a pooling of their resources. AirTouch, US West, Bell Atlantic, and NYNEX have sought to establish national presence by a pooling of their resources to support an aggressive PCS auction strategy and by an effort to maximize their incumbency advantage indirectly, aggregating it from among the partners.

Minimal Level of Access to Sources of Strategic Advantage

No competitor should make an attempt to enter the wireless telephone market without determining that it has, at least, the minimal level of access to the sources of competitive advantage necessary to become an effective competitor (approximately three or more).

A competitor which has no superior access to a source of competitive advantage must necessarily fail and will probably do so quite quickly. No strategy at all can be formulated for a competitor in that position. A competitor which has superior access to inadequate sources of competitive advantage will fail in due course because no effective strategy can be developed for such a competitor.

Implicit in the decisions of a number of prominent wireless providers not to bid for many PCS licenses must be the recognition that, under the expected market conditions, a winning strategy could not be created (i.e., a strategy offering the reasonably certain prospect of an adequate return on capital).

In the face of the Ice Age, the hard-nosed realism implicit in those decisions is precisely what is called for. The enthusiasm of certain organizations for their PCS prospects suggests, in light of their lack of obvious access to the sources of competitive advantage, that their business plans have not been subjected to the necessary level of hard-nosed realism.

Integration of the Sources of Competitive Advantage

Strategy development involves integrating the sources of advantage to which the competitor has preferential access into the optimal approach to the marketplace. The integration process is non-trivial. There are several sources of competitive advantages and a large number of ways in which a company can integrate them.

The process of integration in the wireless telephony industry tends to gravitate toward four themes which reflect the conditions of the Ice Age. The related prospects of intense price competition and inadequate revenue suggest that success requires:

(i) being able to meet intense price competition better than any other competitor; or
(ii) being able to avoid intense price competition by offering focused service superiority; or
(iii) being able to avoid intense price competition by a broad service offering which cannot be compared easily with other wireless telephony offerings on the basis of price or product superiority; or
(iv) being able to avoid intense price competition and reduce churn by maintaining customer relationships superior to those of the competitors.

These four themes reflect what would be the primary focus of a firm that concentrates its resources. As such, these themes provide the basis for four strategies which are discussed below.

- Lowest Cost Service – The provider engineers its enterprise to offer a wireless telephony service of an acceptable quality at a cost lower than any of its competitors, allowing it to engage in price competition as it chooses. A provider supporting this strategy is preparing to compete in a commodity market.
- Focused Service Superiority – The provider concentrates on continuing development of its wireless service offerings, which offer noticeably superior value by reason of leadership in wireless service and product innovation. A provider supporting this strategy is attempting to avoid competing in a commodity market by staying one step ahead of the innovation-commoditization cycle (i.e., creating the next cycle).

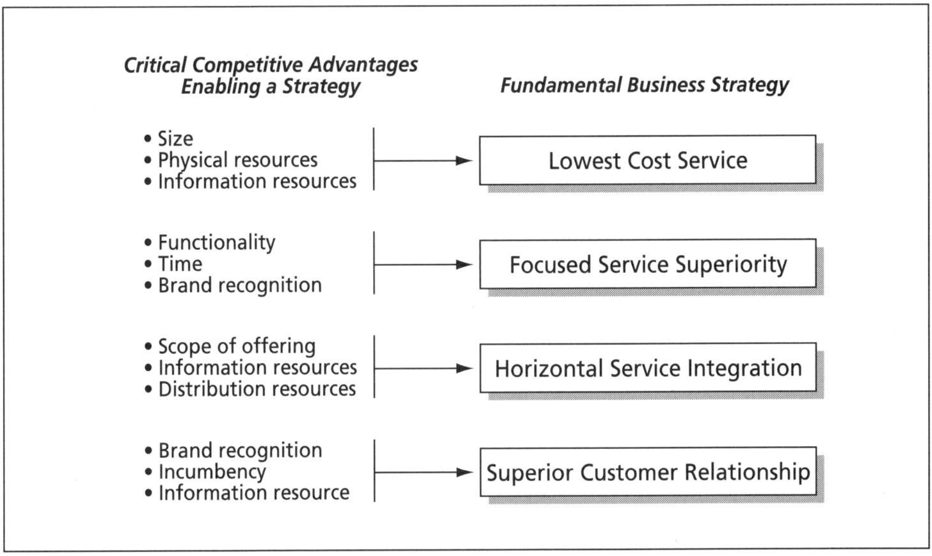

Figure 2 *Integration of competitive advantage into a strategy*

- Horizontal Service Integration – The provider focuses on its ability to bundle other services with the wireless products, thereby offering a broad range of communications services. A provider supporting this strategy is attempting to redefine the industry in which it is competing by broadening the services it provides and hopefully narrowing the competition.
- Superior Customer Relationship – The provider focuses on managing every aspect of the customer interface, thereby reducing churn and attracting customers who value such a relationship. A provider supporting this strategy attempts to avoid competing in a commodity market by providing value to the customer in the form of a superior customer relationship.

While these strategies may not be the only ones available, they are among the alternatives that will likely emerge during the wireless Ice Age. Each strategy tends to be enabled or enhanced by a particular set of competitive advantages, as shown in *Figure 2*. The vision and talent of management are, of course, critical to all strategies.

The majority of companies operating in multiple geographic markets will find that their effective competitive advantages are distributed across several of the strategies. In those cases, the effective execution of a chosen business strategy will obviously depend on the company's ability to develop and integrate the complete set of requisite competitive advantages.

Although the four fundamental business strategies appear, upon initial examination, to be separate and distinct, in fact they are not. In principle, there are elements common to more than one strategy. The strategies are, in effect, named after their theme which is given the greatest weight in the strategy (i.e., its center of gravity), but other themes are not absent, only assigned lesser weight.

Thus, for example, delivering service at the lowest cost is obviously the central theme of the lowest cost strategy, but controlling costs is still quite important to the focused service superiority, horizontal service integration, and superior customer relationship strategies. Similarly, while wireless service and product functionality is critical to the focused service superiority strategy, functionality is hardly irrelevant to the low cost service, horizontal service integration, and superior customer relationship strategies. In practice, the suggested strategies may represent points on a continuum as strategy development can integrate the organizing themes with infinite variety.

Lowest Cost Service

The one true lowest cost provider of service – and there can only be one in any given market – will have several distinct advantages. The lowest cost provider of service will be able to set the low price in its markets subject to the willingness and ability of its competitors to operate at a loss. The low cost provider of service will have the greatest ability to react to price competition in its markets and will have higher margins (or smaller losses) than any other provider offering comparable functionality and service.

The lowest cost service strategy recognizes that there is a large segment of the market that is not overly concerned with product and service innovations. This market segment consists of customers who care primarily about a core set of features in the product offering and want those features delivered at a bargain price. Low price is obviously valuable in targeting that customer segment.

Business and residential consumers attracted by low price will have minimum standards or expectations regarding product innovation and customer service, but their primary concern will be price. Being the low cost service provider establishes the ability to make large cuts in price when the situation warrants, as well as small cuts over extended periods of time. Such larger and more dramatic cuts in price can have a considerable effect on the competition.

Driving out all costs that do not contribute extra value to the target customer is essential to successful execution of this strategy. A company which pursues this strategy must be committed to both radical and continuous improvement across all of its business processes. Driving out unnecessary costs requires the development of operations that are streamlined, standardized, and uniform. Operations such as order entry, order processing, receivables management, and customer care must be designed with a focus on maintaining maximum productivity and effectiveness with a minimum number of people. A company which pursues this strategy must be committed to engineering its internal processes from the beginning with this focus in mind and maintaining that commitment as the company grows.

All areas of operating costs must be evaluated and minimized, but particular attention must be given to the cost of acquiring customers. Financial performance is very sensitive to customer acquisition costs, and that sensitivity increases with falling service pricing, such as will be in effect in the Ice Age. The present level of customer acquisition costs in the cellular telephone industry is simply too high for the service price levels which will exist in the Ice Age.

A competitor which is pursuing a lowest service cost strategy will have its main focus not on delivering the superior product or setting the lead in service innovation, but, rather, on more mainstream products and services that do not require high levels of specialized service and maintenance. A balance will need to be struck in this area. Bargain-basement priced products, combined with an inaccurate billing system, will drive up the cost of operations significantly as the provider is forced to expand customer care capabilities. Accordingly, the lowest service cost strategy must include an effort to minimize the number of customer service calls.

The ability of a competitor to utilize the lowest cost service strategy depends upon the superiority of its access to those sources of competitive advantage which translate into lower costs. Access to physical resources, information resources, and size are critical in this respect.

Focused Service Superiority

A company which commits itself to a focused service superiority strategy not only believes that wireless services are a rapidly evolving market, but also embraces and encourages that trend. Many changes will take place in the wireless market. User needs and wants will be evolving. Technological advances will occur. New user devices will be developed. Improved network infrastructure technologies will become available. Rather than viewing these developments as problems, a wireless provider selecting this strategy must see in them the opportunity to create competitive advantage. That opportunity consists not only in introducing new products but also in offering superior and valuable services and products substantially faster than the competition.

The focused service superiority strategy is based on there being a sufficiently robust base of potential customers for whom superior products and services are more important than price. The desired superiority may take many forms, but the essence of the focused service superiority strategy is focused in the wireless telephony dimension. Focused service superiority is based upon the belief that the superiority of services and products in the wireless telephony dimension can be for many customers more important than price and more important than the convenience of packaged offerings which extend beyond wireless.

Successful execution of this strategy requires investing a significant amount of resources into the product development stream. Many operators today do not explicitly recognize product development and introduction as key business processes in their enterprise. Operators exercising this strategy must do so. They must engineer their corporation to achieve two primary objectives: (i) make new product roll-outs an everyday event and (ii) achieve a rapid time-to-market for new products. In order to meet these objectives the enterprise – the network, business support systems, people, and business processes – must be viewed as a transaction engine and organized in such a way that recognizes and structures recurring business processes to facilitate their use in a variety of different circumstances. A new product is, therefore, simply another transaction to be run on the enterprise. A corporation can offer wireless POTS today and the Universal Number Service tomorrow.

Focused service superiority must not only secure but also sustain its competitive advantage over time. The strategy is most effective if it is executed relentlessly to maintain product and service leadership, rather than incrementally to make new offerings as competitors become able to match old ones. Continuing investment and the willingness to shorten the life cycle of older products are critical to this strategy.

The ability of a competitor to carry out a focused service superiority strategy depends upon the superiority of its access to those sources of competitive advantage which translate into or support the development of innovative services and products. Access to functionality, brand recognition, and time are critical sources of competitive advantage in this process.

Horizontal Service Integration

Those providers which have other service delivery capabilities such as cable, long distance, local telephony, cellular telephony, or paging will be able to package services together and derive competitive advantages in the market. The nature of the differential advantage created by service packaging will depend upon those core capabilities inherent in the company, but at the heart of the competitive advantage created by packaging is the value which the customer places upon the convenience of dealing with one source for multiple products and services. The horizontal service integration strategy depends upon the existence of a significant market segment which places greater value on the convenience of a single point of purchase, service, and billing than upon wireless product superiority. The horizontal service integration strategy can take the form of a price strategy, but one designed to avoid a service-by-service pricing comparison.

The ability of a competitor to carry out a horizontal service integration strategy depends upon its access to other relevant services and its ability to integrate the service offerings seamlessly in order to capture the element of convenience. The availability of packageable services has, of course, its source of superiority in the scope of the offering, although such scope can be secured through reselling (e.g., as proposed by MCI) and partnership arrangements (e.g., Sprint and its cable partners), as well as internal capabilities (e.g., AT&T).

The ability of a competitor to carry out a horizontal integration strategy depends upon the superiority of its access to those sources of competitive advantage which allows a provider to differentiate itself by developing customized packages of previously disparate services aimed at specific industry segments that may have unique needs. Access to scope of offering, distribution resources and information resources are critical in this process.

Superior Customer Relationship

The superior customer relationship strategy, if executed correctly, will have the distinct advantage of having a lower customer churn rate than other competitors in the market. By managing every aspect of the customer relationship, this strategy enables an operator to own the customer, thus raising the switching costs to the customer.

The superior customer relationship strategy recognizes that there are customers who value a relationship with their service providers and prefer to do business with these operators. As wireless services become more complicated, and customers become more dependent on their wireless service, the value of relationship between the provider and customer increases. Customers are likely to seek out providers with an excellent reputation and/or with people who can help them when needed. Executing this strategy for many customers could take as little as a good reputation, friendly and helpful service over the phone, and an absence of problems with the product. As products and services become more complex, however, an operator following this strategy will need to differentiate itself by the amount of time spent with the customer helping him or her to get the most value from the product and customizing or integrating the service into the customer's business or lifestyle.

Developing customer-for-life programs is essential to this strategy. Every contact with the customer must be treated as a moment of truth. All information known about the customer must be made readily available to the customer contact person and organized in a way that is most likely to be useful in supporting the joint objectives of the company and customer in their interaction. For example, if the customer has a problem, calls in, and resolution is still pending, this information should be of paramount interest and visibility to the contact person. If the customer has been solicited for a new product, this fact should be known by the contact person. If the customer falls within a segment that has benefited from a new offering or configuration, an appropriate invitation to try the service should be offered.

Operators executing a superior customer relationship strategy also need to be able to serve small market segments efficiently. In order to accomplish this task, the providers need to view their service offerings not as fixed blocks of services, but rather as many small components which can be mixed and matched to meet the needs of the specific customer.

One of the most critical risks of pursuing this strategy is that too many operators will claim to be customer-focused. No operator will find it easy to stand out in the early stages of competition. It takes time to demonstrate consistent customer responsiveness. Accordingly, those who choose this strategy must be prepared to make the commitment and have the patience to make the strategy work. The commitment can be defined in terms of the resources and investment to build the kind of flexible response capability necessary to attract and keep customers. A primary example of such investments include sophisticated customer care systems that deliver detailed information on customers to front-line company representatives.

The ability of a competitor to successfully execute a superior customer service strategy depends upon the superiority of its access to those sources of competitive advantage which translate into superior customer relationships. Information resources, brand recognition, and incumbency are critical in this respect.

Superiority in Execution

A strategy will be effective for a particular competitor only if that competitor can execute that strategy more effectively than the competition. Thus, for example, a lowest-cost service strategy will not be effective for a competitor which does not have the lowest-cost structure. For a competitor to chose a strategy the execution of which it cannot attain and sustain is nothing but a formula for failure starvation or extinction in terms of the Ice Age metaphor.

Before a competitor finally chooses a strategy, it must be subjected to a competitive audit to establish that the competitor can execute the strategy better than its competition. The audit should consider, among other matters, whether the competitor has superior access to those sources of competitive advantage which are critical to the chosen strategy.

Strategic Sufficiency

A strategy, however well executed, will not be effective unless it is addressed to a sufficiently large segment of the market to provide the prospect of adequate revenue for a particular competitor. Best execution is a necessary, but insufficient, condition for the effectiveness of a strategy. The sufficient conditions for the effectiveness of strategy are best execution, robustness, and value proposition (e.g., low price, product superiority) directed to a sufficiently large segment of the market and powerful enough to secure market share against alternative value propositions.

Before a competitor finally closes a strategy, it must be subjected to a sufficiency audit to establish the robustness of the strategy. The audit should include advanced modeling of market demand based upon primary market research and other market research methodologies designed to test the comparative appeal of the strategy's underlying value proposition and durability.

Tool Acquisition

Since the emergence of Homo Sapiens, the making and use of tools have been that species' method of contending with challenges. The power of tools endures to this day, and tools (methods, techniques, processes) can be of considerable assistance in effective strategy execution.

The utility of particular tools depends, quite naturally, upon the strategy being pursued, as some tools are more relevant to one strategy than another. *Table 2* shows the relative importance of methods, tools, and techniques for each of the strategies outlined above.

Tools for Improving Revenue Development

Given the fact that price levels will be determined by competition and will not likely be controllable by a particular competitor, the revenue development tools and techniques focus on the revenue factor which is most influenced by management's actions – net subscriber growth. Accordingly, the revenue development tools and techniques have been categorized in accordance with the purpose thereof.

- Customer understanding: The providers must have a solid understanding of the customer segments which they will be targeting. The tools in this category will assist the provider in obtaining this understanding.
- Customer attraction: Once the target segments have been identified, the potential customers must be attracted to subscribe to and use the service. The tools in this category will assist the provider in obtaining the maximum benefit from its marketing activities.
- Customer loyalty: With the customer churn rate expected to be very high in the Ice Age, churn reduction will be critical for all providers.
- The tools and techniques in this category will assist providers to develop loyal customers and manage churn effectively.

Tools for Customer Understanding

While most operators would say that they have segmented their markets, few have done the primary research needed to develop the rich understanding of market segments and price/feature trade-offs necessary to support their chosen strategy in the Ice Age. There are new approaches to understanding the customer which are non-linear, iterative, and oriented toward finding ways to change the rules of competition.

Two tools which can assist a competitor with these approaches are marketing decision support systems and market analysis modeling. Marketing decision

	Lowest Cost Service	Focused Service Superiority	Horizontal Service Integration	Superior Customer Service
Revenue Development Tools				
Customer Understanding				
Market Decision Support Systems	Low	**High**	Medium	Medium
Market Analysis Modeling	Medium	**High**	**High**	**High**
Customer Attraction				
Distribution Management	**High**	**High**	**High**	**High**
Brand Enhancement	Medium	Medium	Medium	**High**
Database Marketing	**High**	**High**	**High**	**High**
Customer Loyalty				
Customer Loyalty Programs	Low	Medium	Medium	**High**
Customer Management Systems	Low	Medium	Medium	**High**
Operational Effectiveness Tools				
Firm Operations				
Product Development Stream				
Integrated Product Development	Low	**High**	Medium	**High**
Revenue Stream				
Revenue Assurance	**High**	Low	Medium	Low
Operations Resources				
Information Technology Resources				
Information Systems Planning	**High**	**High**	**High**	**High**
Automation Evaluation	High	Medium	Low	Medium
Enterprise Data Sharing	Low	Medium	**High**	**High**
Organizational Resources				
Organizational Alignment	**High**	**High**	**High**	**High**
Vendor/Partner Integration	Low	**High**	**High**	**High**

Table 2 *Tools/strategy matrix*

support systems assist operators in obtaining a superior knowledge of customer needs and in recognizing emerging needs early. The relevant information can come from a wide variety of sources: focus groups and research studies, customer comments, customer usage patterns, success and failure of recent product innovations by customer segment, and related customer purchase behavior. Market analysis modeling assists an organization in understanding the value that different segments of market place on both current and future product features by ranking the demand for features across a range of prices.

Tools for Customer Attraction

In the wireless Ice Age, the relationship between the purchaser and the end user of mobile telephone service may be quite different from what it was in the past when these two roles were often performed by the same person. In the Ice Age, these roles will become more distinct, particularly for the business markets. Additionally, during the Ice Age, the consumer market will become a primary source of growth. These changing market dynamics will require providers to develop effective means to attract the targeted market segments. This requirement will be complicated by the fact that carriers can and will have overlapping target markets.

Three tools which can assist a wireless provider in managing these changing customer dynamics effectively are brand enhancement, distribution management, and database marketing. Brand enhancement is an iterative technique that assists in aligning the brand image to the competitive strategy. An audit is initially performed to determine whether the brand is properly aligned to the competitive advantage sought for leveraging. If it is not, new objectives are set to achieve this alignment, and communications are developed to build the desired image in the minds of the prospects. Distribution management techniques assist the wireless providers in determining the most effective methods of distribution and in managing those distribution methods to achieve a maximum return. Database marketing tools utilize a systematic approach to matching service offerings to prospects likely to be particularly receptive. Database marketing tools thus facilitate the direction of marketing efforts towards those potential customers which are most likely to respond to the competitor's value proposition.

Tools for Customer Loyalty

As both revenue and subscriber growth begin to slow down in the wireless Ice Age, reducing customer churn will become a critical issue. That criticality is heightened by the fact that more competitors will be vying for the same customers and will be increasing each other's churn. Thus, in order to survive and maintain a disproportionate share of the revenue, providers will need to develop effective means to retain the most valuable customers.

Two tools which can help develop this new approach are customer loyalty programs and sophisticated customer management systems. A customer loyalty program is a tool to provide incentives for customers to remain on the network and thus develop high switching costs. Whether the program rewards high-usage or continued subscriber status, its purpose is to reduce churn, particularly churn amongst the most valuable customers. Sophisticated customer management systems empower customer service personnel by providing them with the information and the authority to meet the customer's needs. Such systems also enable customer service personnel to pursue the agenda that the company wants to be associated with this customer. These systems will facilitate the most effective use of all interactions with the customer.

Tools for Improving Operational Effectiveness

Whatever strategy is chosen, the proponent of that strategy must be extremely effective in maximizing the value of the sources of competitive advantage underlying that strategy. Certain tools and methods can be quite useful in aligning business operations and technology/organizational resources with strategy.

Tools for Enhancing Business Operations

In the wireless industry, improvements in business operations are most easily identified by viewing the processes across functional lines. For the wireless industry, two cross-functional views are useful:

- The product development stream: As new products and innovations pour into the market place, providers must be able to manage the roll-out of new offerings effectively.
- The revenue stream: As the market becomes saturated and price wars ensue, a stabilized revenue stream which captures every potential dollar of revenue as soon as possible will become essential.

In relation to the product development stream, integrated product development methodologies support all aspects of the product creation and deployment process, including process facilitation and improvement, product delivery and support, and technology development. The methodologies can offer the wireless carrier significant competitive advantage by reducing product development and major program initiative cycle times, thereby aligning the product development process with the overall business strategy and reducing the costs that are necessary to introduce and support products and services in a competitive environment.

In relation to the revenue stream, revenue assurance techniques ensure that all calls placed on the network are appropriately accounted for and billed. The objective is achieved through four interrelated processes: establishment of operational control procedures at different points in the revenue stream, the definition of automated system strategies to assist in controlling the revenue stream, the establishment of new fraud control techniques, and the assignment of organizational responsibilities for each control.

Tools for Improving Operational Resources

During the Ice Age, wireless providers need to ensure that their operations are aligned to support the business strategy. They also need to ensure that the resources available to the provider (both technical and organizational) are properly allocated and aligned to support the strategy.

Three Information Technology (IT) tools, in particular, are useful in achieving this goal: information systems planning, automation evaluation, and enterprise data sharing. Information systems planning ensures that a company's IT plan is tightly integrated with the business strategy and therefore able to support the present and future needs of the company. Automation is a tried and true approach for using IT to drive down costs. Several opportunities exist with wireless carriers to use this approach for reducing costs: automation of payments processing, automation of order processing, automation of network engineering and control, automation of customer service, and automation of sales/distribution or point of sale. Enterprise data sharing is a sophisticated IT approach capable of accessing and delivering detailed information on customers, distribution channels, calling patterns, costs, and other parameters to multiple users within the organization. Each user within the organization uses the information for different purposes.

In relation to organization resources – two tools, organizational alignment techniques and vendor/partner integration – are particularly useful in maximizing the organizational resources of a wireless provider. Organizational alignment techniques can be utilized to assist in identifying and executing the required alignment of organizational structure to strategy. Vendor/partner integration can assist a service provider and its vendors align their strategies and business activities to accomplish their goals and objectives through mutual cooperation.

Need to Sustain Competitive Advantage

Figure 3 summarizes the sources of competitive advantage a competitor needs to obtain in order to effectively pursue a particular strategy and the tools which will help that provider execute that strategy effectively.

However, success is rarely achieved in one bold stroke. The creation of competitive advantage at any stage does not assure that such advantage will be sustained. In some industries, competitive advantage cannot be expected to last for more than one season and must, therefore, be recreated four times a year.

The accelerating rate of change in the wireless industry of which the Ice Age is a reflection presents a challenge to any competitor in terms of its ability to sustain strategic advantage, and thus the strategy it is pursuing. For a competitor to sustain a strate-

Lowest Cost Service

Critical Sources of Competitive Advantage
- Size
- Physical resources
- Information resources

Critical Tools
- Distribution management
- Database marketing
- Revenue assurance
- Information systems planning
- Automation evaluation
- Organizational alignment

Focused Service Superiority

Critical Sources of Competitive Advantage
- Functionality
- Time
- Brand recognition

Critical Tools
- Marketing decision support systems
- Market analysis modeling
- Distribution management
- Database marketing
- Integrated product development
- Information systems planning
- Organizational alignment
- Vendor/partner integration

Horizontal Service Integration

Critical Sources of Competitive Advantage
- Scope of offering
- Information resources
- Distribution resources

Critical Tools
- Market analysis modeling
- Distribution management
- Database marketing
- Information systems planning
- Enterprise data sharing
- Organizational alignment
- Vendor/partner integration

Superior Customer Relationship

Critical Sources of Competitive Advantage
- Brand recognition
- Incumbency
- Information resources

Critical Tools
- Market analysis modeling
- Distribution management
- Database marketing
- Customer loyalty programs
- Customer management systems
- Information systems planning
- Enterprise data sharing
- Organizational alignment
- Vendor/partner integration

Figure 3 *Critical resources of competitive advantage and tools for each strategy*

gic advantage over time, it must create new strategic advantages at a rate faster than the rate of change in the wireless industry because change dissolves past advantage.

Time has already been recognized as a source of competitive advantage, and time-based competition can take a variety of forms: faster time to market, faster response time to customers, and faster elimination of unnecessary costs. In the Ice Age with its condition of revenue inadequacy, the inability to sustain strategic advantage does not mean merely a lesser market share, but rather an inability to secure adequate revenue to survive.

No relief may be sought from a commitment to sustain competitive advantage, for that commitment and its fulfillment are conditions to survival. A competitor must adopt a program of continually applied pragmatism – an uninterrupted and uninterruptible effort to plumb all sources of competitive advantage at all times as a matter of policy and not in reaction to the moves of other competitors. At its best, continuously applied pragmatism is a program to sustain a leadership position in a chosen strategic dimension by becoming a driver of change rather than a reactor to change.

Assurance of Capital Availability

The need to assure continued capital availability is self-evident. The inability to provide that assurance is a good reason not to enter competition in mobile telephony in the Ice Age. No strategy, however well formulated or executed, will generate financial self-sufficiency under Ice Age conditions for a very long time.

Last Word on Strategy

Lessons from evolutionary studies serve as a reminder of the need to develop effective strategies and alter them with changed conditions.

Species well adjusted to one era may find themselves helpless in another if they cannot sustain the competitive advantage they previously enjoyed. Homo Sapiens has always relied upon their adaptive skills to meet new challenges. The evolutionary process is often simply too slow to provide a basis for responding to new conditions.

Copyright © 1995 EDS Management Consulting Services. Reprinted with permission.

Cellular Communications: What's Ahead

Jagdish N. Sheth
Professor of Marketing
Emory University

Rajendra S. Sisodia
Associate Professor of Business
George Mason University

Introduction

In this paper, we look at the future of the industry. We see several dramatic shifts that are becoming apparent. In particular, we have identified six key drivers of change *(Figure 1)*.

Technology

As the move towards digital technology gathers momentum, some fundamental characteristics of the business are going to change *(Figure 2)*. Consider the fact that GSM handsets today have up to 450,000 lines of computer code programmed into their memories. This illustrates that both the technological thrust, as well the economics of the wireless communication business, will increasingly derive from the computer industry rather than the radio industry.

Moving to digital transmission has a two-fold impact on carrier operations. First, carriers can station receivers at greater distances, which means they can buy less network equipment to cover an area. This makes it cheaper for carriers to deploy new services. Second, digital transmissions pack information in bits and bytes; these are simpler to manage than the electronic waves of analog techniques. With

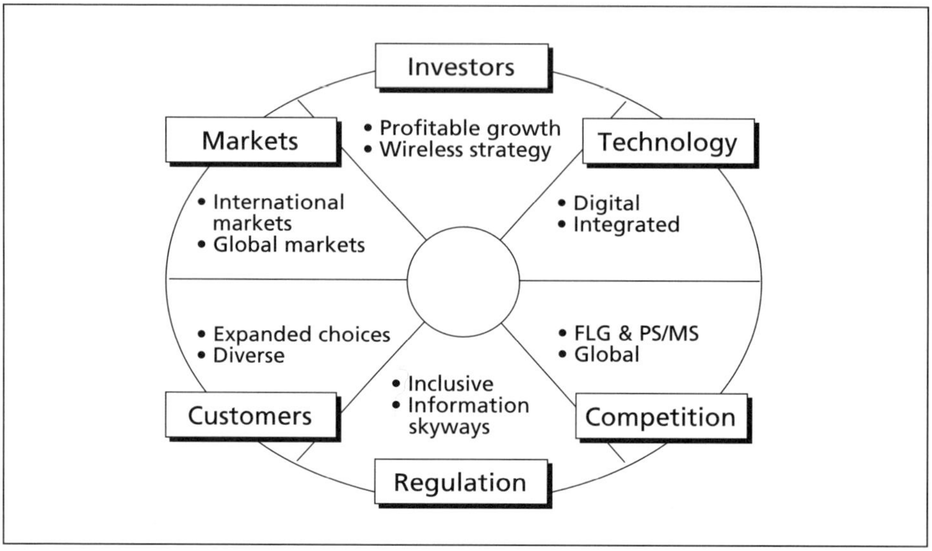

Figure 1 *Key drivers of change*

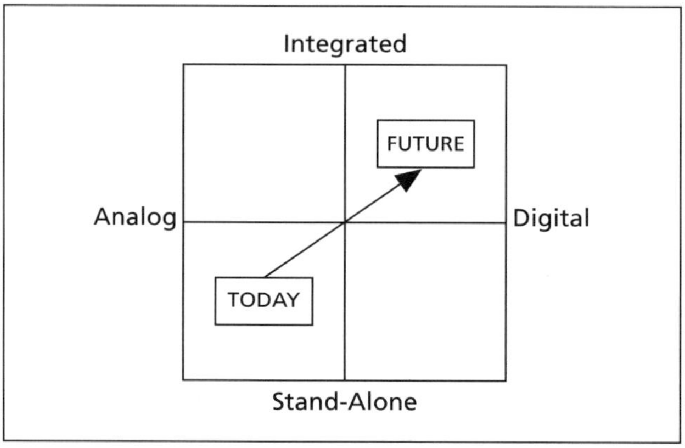

Figure 2 *Change driver: Technology*

digital transmission, carriers can more easily troubleshoot network problems remotely.[1]

Technological advances are permitting wireless technology to be integrated with wireline services to a much greater extent. Various networks are separated in a commercial sense rather than on technological grounds. Service providers are increasingly being pushed to offer a "one stop shop," or a variety of competitive services via a single billing mechanism; this trend is bound to continue and accelerate. We expect to see the emergence of communications that will integrate the PSTN and mobile services in a single, transparent, and seamless whole.[2]

For this to happen, a number of hitherto separate networks will have to come together in an integrated way: wireline, cellular, satellite, PCS, LAN, and broadcast. Currently, each of these networks has its own technology and is incompatible with many of the other networks.

Few of us could have predicted the all-pervasive influence that personal computers would come to have in business and personal life over the last decade. Yet we are seeing the integration of voice, data, image, and communications in a natural evolution towards the multimedia workstation. Similarly, emerging technologies in mobile communications have made wireless, mobile, and personal communications one of the fastest growing markets in the world. The range of technologies becoming available is immense. Future wireless and mobile communications systems will blur the demarcation lines between wireless and fixed telecommunications.

The speed of developments in the industry is accelerating. For example, it took cellular phone manufacturers eight hours to make a cellular phone in 1988; 30 minutes in 1992. Today it takes a mere 12 minutes and this is a telephone containing computer capacity of between 30 million and 40 million instructions per second, equivalent to the power of mainframes of only a few years ago.[3]

The Impact of Digital Electronics

In an electromechanical industry, experience curves move down slowly, and their lower asymptotes (marginal costs) are quite high *(Figure 3)*. Given these slow cost dynamics, such industries see no real need to speed things up; they focus on economies of scale rather than economy of time. They tend to emphasize "ability and willingness to pay" in determining the order of market penetration and exploitation. Because such industries tend to be capital-intensive, they allocate capacity to certain segments sequentially. Thus, most such industries first target the government market, followed by the industrial market, commercial sector (small businesses), and finally, the personal market.

This trickle-down approach worked quite well in an era defined by electromechanical technology dynamics, limited communications, and slow-changing technologies. Under the trickle-down approach, companies do not innovate for

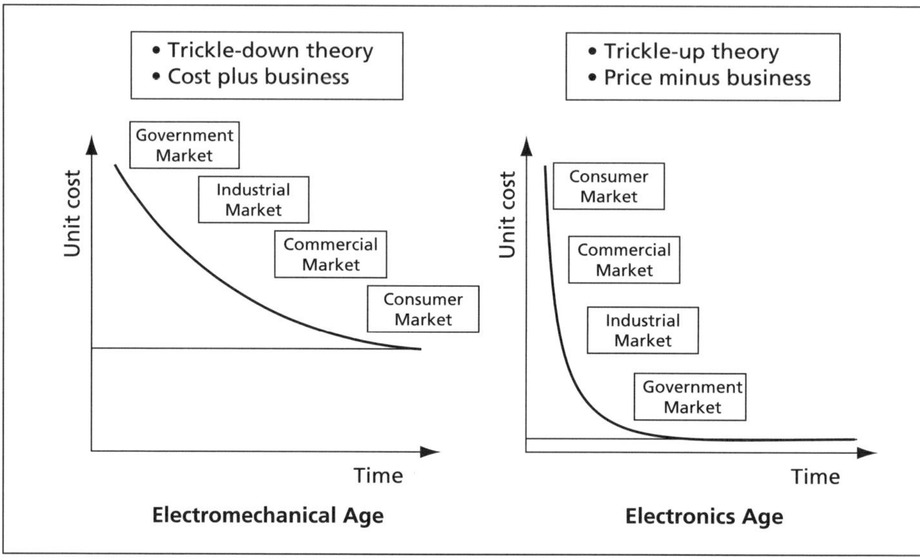

Figure 3 *Egalitarianism of electronics technology*

the personal market; these are treated as "hand-me-down" markets. The presumption has been that these companies can take care of the personal market (e.g., personal copiers) "down the road." This would be all right if the industry were moving slowly in terms of technology evolution or costs; however, when the industry becomes electronic, its dynamics change dramatically.

In electronics, a "trickle-up" approach is far more likely to succeed. Electronics-based industries, almost without exception, follow a different experience curve than electromechanical industries (*Figure 3*). Speed is the driver, much more than scale. The slope is much steeper, and the lower asymptote is close to zero. This effect is most dramatic for software (where the marginal cost is close to zero), but is also very pronounced for electronic components (especially chips) and a little less so for finished information technology products (which have significant electromechanical components; for example, PCs have electromechanical hard drives, floppy drives, power supplies, casings, keyboards, etc., VCRs have tape transport mechanisms etc.). The greater the extent of "solid-state" electronics and software-based operation and control, the greater this type of experience curve effect.

As a consequence of these cost dynamics, success in electronics-based industries derives in large measure from an ability to force the experience curve down rapidly by generating a large volume of demand in a short time. In essence, this calls for a "trickle-up" approach to market penetration and development: creating demand on a large scale. The largest part of the market is usually the segment that is least willing or able to pay a premium price.

Companies in an electronic industry that continue to operate as though they were in an electromechanical industry are doomed to failure. By the same token, companies that recognize the economics of electronics and alter their mindset accordingly will likely succeed.

When the technological base of an industry is both electronic (leading to the trickle up property described above) as well as digital (equating with versatility, i.e., the ability to handle any form of information interchangeably), the implications are dramatic. The former leads to an emphasis on personal markets, which is where the volumes necessary for cost reduction are, and the latter leads to integration of multiple services on the same platform.

Implications for Cellular Service Providers

As a result of the trickle-up theory described above, personal markets will become the lead markets for technology evolution, rather than business markets. As a result of the versatility and greater capacity resulting from digitalization, personal markets will become multimedia. Personal markets will become universal through global standards. Voice-video-data evolution is more attractive in personal markets than voice-data-video evolution.

The Economics of Electronics

Since technology performance-price improvements will likely continue at a rapid rate, investment cycles will also be rapid. As we evolve towards a "network of networks," the wirefree and wireline aspects of the industry will increasingly converge. The long-run evolution will be towards global multimedia networks.

As network capacities expand dramatically, service providers will have to revise their normal pricing formulas. Rather than use cost-based pricing, as has been the practice, service providers will use price-based costing, i.e., they will set prices with a view to maximizing revenues (since variable costs will be low) and then ensure that their cost structures enable them to make a profit. Market demand will have to be "exploded" through the aggressive use of penetration pricing at the low end of the market and value-based pricing for other customers.

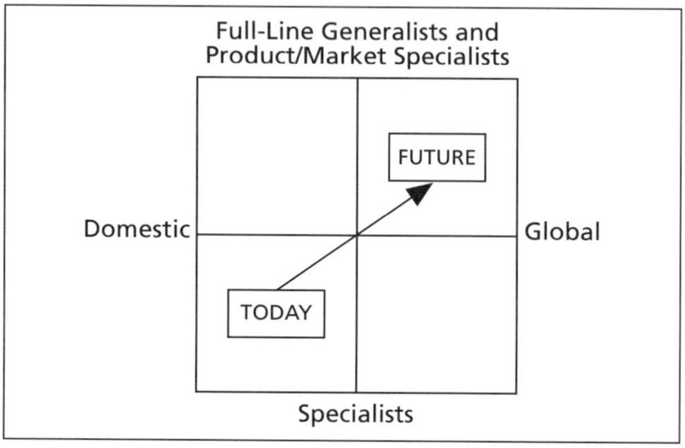

Figure 4 *Change driver: Competition*

Besides surpassing institutional markets in revenues in each information business, personal markets will also become multimedia in each business. Further, they will become universal through the emergence of global standards. Convergence will take place in networks as the various parallel systems overlap, interconnect, and ultimately join in a seamless structure.

Finally, compatible technology standards will result in a more level playing field, making it harder for any one company to achieve a relative competitive advantage. This shift to generally comparable systems will have a number of implications. Service providers will find it increasingly easy to substitute vendors. We are already seeing this phenomenon as carriers shift from one wireless switch provider to another and then back again. Off-the-shelf components will change the relative importance of hardware. "Moving iron" will give way to systems and solutions selling with commensurate changes in engineering, sales, and distribution. Software that customizes the basic platform will be more important than the platform itself.

Competition

Thus far, competition in the cellular business has come primarily from domestic players who are either pure cellular specialists or have a limited range of communications offerings (such as cable television or local telephony). As we go forward, however, this is likely to change *(Figure 4)*.

Increasingly, we are likely to see the presence of global competitors in the U.S. market. British Telecom is already involved, via its ownership in MCI. NTT has a stake in Nextel. France Telecom and Deutsche Telecom are seeking to acquire a

Generalists	Specialists
Volume-driven, Asset turns advantage	Margin-driven, exclusivity advantage
One-stop shop, full line products and services	Separate stops shop, limited line of products
Mainstream market positioning, single corporate identity	Target market positioning, multiple brand indentity
Value competition, shared operations	Image/service competition, dedicated operations
Multiple channels, integrated organization	Focused channels, multibusiness organization

Table 1 *Generalists versus specialists*

stake in Sprint. As other countries liberalize their telecommunications regimes and allow U.S. carriers entry into their domestic markets, carriers from these countries will also be given greater access to the U.S. market. The protection afforded wireless providers via the 1934 Communication Act provisions on foreign ownership will likely diminish over time, and no such regulatory advantages are likely to persist.

That this is likely to occur can be seen from the strong position already enjoyed by U.S. carriers (mostly the RBOCs) in providing wireless services abroad. Global carriers will demand, and eventually receive, similar access to the U.S. market.

Over the next few years, after Congress completes its overhaul of the 1934 Act, some of the larger local and long-distance carriers will be increasingly likely to become "full-line generalists," i.e., providers of a range of communications services to a variety of different markets, including local and long-distance calling, wireless voice, data, and video, and interactive cable television. Other players, lacking the scale and scope to compete as full-line generalists, will position themselves as either product specialists or market specialists.

Strategies For Generalists vs Specialists

Successful generalists and specialists follow different strategies and have very different operating characteristics, as depicted in *Table 1.*

Generalists are large, volume-driven players, dependent on economies of scale and scope for much on their competitive advantage. They have large fixed assets in place, and their success depends heavily on their ability to maximally utilize those assets. In other words, the ability to generate substantial revenues on a given asset base is crucial.

Specialists, on the other hand, eschew volume and velocity in favor of high margins (revenue minus cost). It is important to point out that high margins do not necessarily imply high prices. Southwest Airlines, for example, is a specialist that achieves its attractive margins (far above market averages) through aggressive cost cutting. Other successful specialists have achieved high margins by combining average cost structures with an exclusivity advantage – a form of differentiation others cannot emulate.

Regulation

About the state of competition in the cellular industry, a former senior official of the FCC, Albert Halprin, the nation's top telephone regulator in the mid-1980s, said: "The amount of competition is not what I would have hoped for. 'Mistake' is far from the right word to describe our action, but it was not perfect. With hindsight, we should have authorized a third carrier."[4]

A report in 1993 by the General Accounting Office, the investigative arm of Congress, concluded that the current duopoly arrangement in place in most areas was "unlikely to provide a product at a competitively set price."

Likewise, in the United Kingdom, the government concluded after a lengthy "duopoly review" (which covered wired as well as wireless communications), that it would be necessary to open the market to many more players in order to foster greater competition. As a result, the UK already has five cellular/PCS networks, and dozens of companies competing in various parts of the wireline market.

As a result of these developments, the regulatory regime in the U.S. has become strongly driven by the concept of inclusion rather than exclusion *(Figure 5)*. In other words, regulators will take every possible action to ensure that no telecommunications service providers of any kind are given any kind of exclusivity over the provision of any services. Rather, regulators will try to facilitate the entry of as many players into as many markets as possible.

For incumbents, this represents an opportunity as well as a threat. It is virtually certain that they will face heightened competition from additional players in their core business. At the same time, they will be presented with numerous new

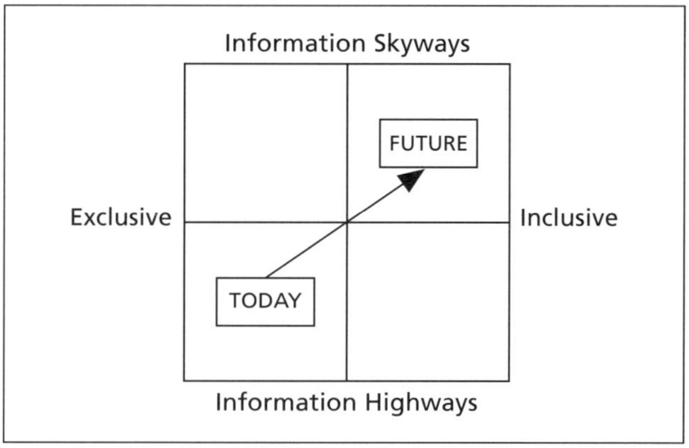

Figure 5 *Change driver: Regulation*

opportunities to extend their competencies into new products and new markets. Those unable to compete on these terms will no longer be protected from decline.

European telecommunications regulators, who have traditionally lagged the U.S. in opening markets and fostering competition, are moving rapidly to deregulate cellular services and increase competition. The European cellular industry, deregulated in the late 1980s, is now a role model for Europe's regulated businesses. The European Union recently recommended that any remaining exclusive mobile licenses be eliminated by 1998. It also recommended allowing mobile operators to build their own lines to connect cellular stations with land-based telephone networks, avoiding hefty leasing fees from state-owned carriers. Such changes may not be legislated anytime soon, but they signal a sea change in attitudes among regulators.[5]

A shift in regulatory mindset is contingent on some technological vectors developing as expected. The widespread assumption that wireless communication will always be inherently suited to narrowband communication is starting to be challenged. The assumption owes much to an assertion by Nicholas Negroponte of MIT's Media Lab that there will be a switch in how voice and video information is promulgated. Negroponte suggested that what goes over the air will go under the ground, and what goes underground will go over the air.

This zero-sum view of spectrum is being challenged by technological changes. Thomas Wheeler of the CTIA has expressed this view as follows:

> Wireless digital devices have a basic limitation: they operate on the current

> narrowband wireless and wireline networks. They cannot handle the high speed broadband digital data of many modern computer applications. A fiber optic network is only part of the solution to the information infrastructure needs of the 21st century. While there is a great deal of policy focus on building a broadband, wired fiber network, there is no such equivalent focus on a wireless broadband network. In fact, there is the possibility that pending government action could inhibit the construction of the wireless equivalent of optical fiber. Broadband, 21st century wireless networks can be built quickly and in the public interest.[6]

There are huge amounts of frequency spectrum available above 20 GHz. Spectrum scarcity will thus become less of a factor, and we will eventually see broadband on wireless. For example, the Teledesic plan propounded by Bill Gates and Craig McCaw will provide "gobs of spectrum" – 400 MHz per carrier.

We believe that this vision of "information skyways" will be accepted by regulators, who will take actions to facilitate the creation of broadband interactive wireless networks. For example, wireless cable provider Cellular Vision's system has around 300 times the bandwidth of conventional broadcast television, and even more than the average cable system.

Customers

The present state of the mobile phone market has been likened to "a thin sandwich in search of a thick filling. Two thin slices of the business market are currently being serviced – namely senior executives on the one hand, and traveling sales people plus the self-employed 'roamers' (plumbers, electricians, and so on), on the other."[7]

In between lies the bulk of the population, the vast majority of which is still largely untouched by this technology. In the future, wireless telephony are expected to penetrate both the business and the consumer sectors in parallel, one reinforcing the other as prices fall. Ultimately, and the time is not too distant, mobile phones will come to be regarded as necessities.[8] We are thus moving from a situation in which the market is relatively homogeneous to one in which the market will consist of many different segments *(Figure 6)*.

Customers in coming years will find that they have many more choices for wireless communication than they do today. We will also find that customer needs will

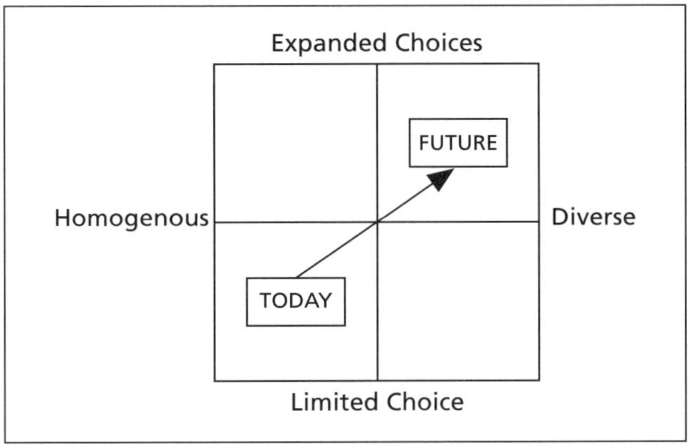

Figure 6 *Change driver: Customers*

vary greatly. This will be true for residential as well as business customers. In this section, we will examine the divergent needs of consumers and businesses separately.

Consumer Markets

The primary source of new demand for full-line telecommunications companies (such as an RHC) will shift from business customers to consumer markets. Supported by the underlying technological change and investor growth expectations, cellular carriers will have to look increasingly to the consumer market for revenue growth.

Future consumers will be dramatically different from past or even present consumers. They will be more demanding, more time driven, more information intensive, and highly individualistic. Their communication needs will thus become highly divergent. It is no longer possible to stereotype consumers into a single image or category such as "middle class America." Some of them are happy with basic service. Many more of them seek information products and services that free them from being hostages of time and place. Their interests will drive product and service development; the cordless phones and answering machines of today will become the personal communicators and voice mail services of tomorrow.

The following discussion covers shifts in behavior and economics that will shape consumers and the implications for telecommunications service providers and ultimately equipment suppliers. Our examples are drawn largely from the United States since that is where the process is most advanced. However, the same forces are at work throughout the industrialized world.

Behavioral and Economic Shifts

As a consequence of lifestyle and demographic changes, the following will be some of the fundamental behavioral and economic shifts among consumers:

Growth in service sector – The older affluent population's concerns for health, security, and recreation combined with single-person household requirements will maintain the growth in services.

Scarcity of time – Time will become the most precious commodity. As activities compete for time, we will redesign tasks that consume too much time, and embrace time-shifting technologies. Customers will demand hassle-free ("get it right the first time") service .

Decline of homogeneous markets – Population diversity in age, geography, income, household characteristics, and ethnic mix point to the emergence of segmented, niche markets. The days of the mass market are over.

Development of individualistic lifestyles – As single-person or dual-career households proliferate, the need to define a separate existence or space will result in highly individualistic lifestyles and behaviors even within family units. This will increase the need for personalized attention to each household; telcos will thus have to learn the skill of mass customization.

Scale of buying power – Growing buying power will put consumers in the driver's seat, dictating the development of new markets. While business markets will continue to expand, consumer spending will become an ever larger part of GDP and economic growth.

Increase in regional differences – Population move to the Sunbelt and small towns with their respective differences in climate, value structure, and even occupation will widen the cultural differences by parts of the country. The United States is apt to more closely resemble Europe, where regions vary significantly in terms of growth, employment, language, and consumption values.

Growing dichotomy between rich and poor – Although we will still have a sizable middle class, the nation will manifest a sharp dichotomy between the rich and poor. A large percentage of the population will be affluent and a sizable group will be below the official poverty level.

Increased stress – The blurring of traditional family roles, increase in autonomy, older age, and need to manage time, all point toward a society that will have higher levels of stress both at home and work. Stress, in turn, will generate productivity issues and behavioral problems, such as drug and alcohol abuse.

Greater concern for privacy – People will become more aware of their lack of privacy and potential loss of individual rights. As the social norms of a previously homogeneous society give way to pluralistic and diverse values, we will emphasize legal rights of individuals.

Emphasis on safety and security – There will be a sharp rise in concern for personal and public safety, partly due to the aging population, and partly due to income redistribution. Additionally, as more people live alone, they will feel more vulnerable. Law enforcement will remain a major social issue.

Entrepreneurial spirit – Opportunities created by the rise in niche markets will encourage personal entrepreneurship. As a result, small businesses will continue as the dominant component of societal change, in terms of new business formation, employment growth, political power, regulatory policy, and personal wealth.

Expanding Choices

Consumer choices will expand rapidly due to the impact of greater competition and new technologies. For some consumers, cellular already represents a viable alternative to the local and long distance phone companies. The digitalization of cellular services and the widespread deployment of PCS will greatly accelerate this trend. A variety of new wireless services will also become available to consumers, increasing the ubiquity of computing and communications. Some of the wireless services will rapidly migrate to broadband, presenting an alternative to both the cable TV as well as telco wired delivery of information. Consumer choice will also increase through the proliferation of video-on-demand and video-based shopping services.

Implications for Service Providers

A major implication for service providers of the above changes is that they will have to develop an entirely new definition of need. *Table 2* gives several examples of the new definition of consumer needs.

What Customers will Demand

Service providers will have to play an active role in creating demand for their new

Country	Subscribers (January 1993)	Subscribers (January 1992)	Growth (%)
Australia	559,000	336,000	54
China	160,000	38,000	321
Hong Kong	233,000	180,000	29
Indonesia	31,500	16,000	97
Japan	1,614,000	1,251,000	29
Malaysia	207,000	135,000	53
New Zealand	93,900	69,200	36
Philippines	55,000	40,000	38
Singapore	111,000	70,000	59
South Korea	271,000	173,000	56
Taiwan	428,000	198,900	116
Thailand	256,800	154,000	67
TOTAL	**4,021,800**	**2,688,400**	**50**

Table 2 *SE Asia and Australasia: Cellular subscriber growth in the 12 largest markets*

services initially, rather than simply responding to demand as its materializes. Utilization levels for even wired telecommunications networks are extremely low: only an average of 30 minutes a day in the U.S., and only about five minutes a day in the U.K. Cellular usage levels are far lower.

As capacities become largely unconstrained in coming years, wireless communication providers will have to create demand for their services.

Telecom companies must develop ways to greatly increase the usage level, first in terms of time spent, then in terms of bits exchanged. In fact, there is almost no limit to the use of communications technologies; certainly, the limits are far beyond anything that has been approached anywhere thus far. Telecom companies have the great advantage that information technologies in general tend to be highly addictive; once consumers become accustomed to using them and design their work and/or leisure activities around them, they are extremely hard to give up. To capitalize on the opportunities available to them to dramatically grow their business, service providers will have to understand not only purchase behavior, but also consumption behavior. By understanding (through the use of focus groups and other qualitative techniques) how a small number of their customers may be using their service highly innovative ways, they can create much larger markets.

The shift from a middle class, homogeneous society to a diverse, rich-poor, and highly individualistic society with time poverty, high stress, and tremendous buying power has other significant implications for the telecommunications industry.

The need for better segmentation – First, it will be essential to segment the consumer market and develop specific products and services and offer them on a targeted basis. In other words, the industry must abandon the shotgun approach in favor of the rifle approach to serving consumer markets. Not only service providers will be committed in this process; equipment suppliers also must understand their customers' customers if they are to succeed. Consumers in the future will demand and receive personalized attention: segment-of-one marketing, when needed.

Trickle-up technology – Second, a demand for instant access to the world from anywhere and at all times will require faster deployment of the next generation of technologies to consumer markets. Technologies must be not only seamless to the consumer, but transparent as well, since they do not have the same degree of technical expertise as business customers.

Focus on customer satisfaction – Finally, future consumers will not only demand reliable products and services that work but will insist on quality support services that are hassle free, convenient, and easy to do business with. The twin processes of customer satisfaction (product quality and customer service) will become key in the future. Their implementation will require an unprecedented partnership effort between service provider and equipment supplier.

Business Markets

Though much of the incremental growth in demand for wireless services will come from consumers, business customers will also demand more services and generate considerable growth.

Large Business Customers

Large business customers will be increasingly globalized, and will thus require uniform global provisioning of advanced services. More and more companies rely on mobile communications as a key competitive weapon. In many industries, these capabilities will become a strategic necessity. The movement to re-engineer business processes in light of available untethered voice and data communications capabilities will accelerate and become much more widespread.

Wireless Data

For traditional phone companies, data transmission revenues are growing at six times the rate of voice traffic, and already represent 20 percent of industry revenues. According to International Resource Development, a consultancy based in Canaan, Connecticut, the U.S. market for wireless data and services is already worth over $2 billion a year, and will reach $15 billion by 2000. Consulting firm Arthur D. Little estimates that, given the opportunity, 40 million Americans in 130 occupations might use wireless communications for eight hours or more per week.

The two dominant providers of mobile data services, RAM Mobile Data (partly owned by BellSouth) and Motorola's Ardis have both done poorly so far. Part of the reason is that they both use proprietary hardware, with resulting high prices (for example, a modem costs $800). Now they are moving to an open standard to stimulate market growth. The newest entrant, cellular digital packet data (CDPD) is expected to be much more successful.

Small Business Customers

Small business customers will probably represent the fastest growing segment of the business market. In fact, total employment in firms employing less than 100 employees each now exceeds total employment in Fortune 500 companies. The former number is growing further, while the latter continues to decline. Small firms lack in-house telecommunications expertise, and are thus far more dependent on service providers. Their unique entrepreneurial culture and lack of bureaucracy make them attractive customers for wireless service providers. Selling wireless services to small business customers will require a different approach. In this case, marketers must lead rather than follow the customer. In other words, it is important to educate customers as to the role this technology can play in their business. Doing so requires a depth of expertise and a commitment to consultative relationship marketing.

Markets

> We are going global in a wireless world.
>
> – The Financial Times

In 1993, the North American cellular market grew 43 percent to 16 million and western Europe grew 47 percent to 8.8 million. By contrast, eastern Europe grew

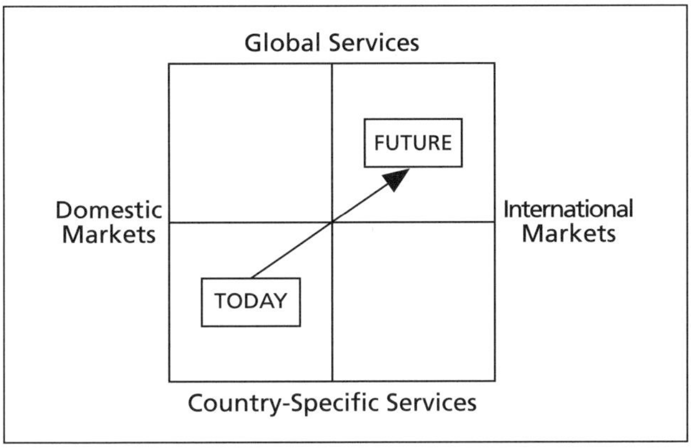

Figure 7 *Change driver: Markets*

129 percent (but still only to 110,000), while Malaysia grew by 64 percent, Argentina by 220 percent, and China by 300 percent. China, however, still has barely 700,000 subscribers for one billion people – fewer than Sweden, and only three times more than Switzerland *(Figure 7)*.[10]

International Markets

Cellular technology is playing a pivotal role in how people communicate around the world. In 1983, less than 10 countries had cellular systems. By 1988, this number had jumped to about 50 countries and by 1993, there were approximately 140 countries with cellular service, representing the vast majority of the world's countries. According to EMCI's report on world cellular markets, there will be 53.6 million subscribers worldwide by 1996. The U.S. will account for approximately 45 percent, Western Europe 25 percent, Asia will almost catch up with Western Europe, and Latin America, Africa, Eastern Europe, the Middle East and the Caribbean will account for 5 percent."

In the global market, cellular is clearly not just "a rich man's toy." It is available in countries with average GNP per capita incomes ranging from $100 to $25,000. In large industrial economies, cellular is moving from business use to a greater intensity of consumer use. In lower income countries, cellular is moving from the wealthy elite to burgeoning middle classes and entrepreneurs. As prices for cellular decline, and lower income countries economically develop, cellular will continue to penetrate new market segments. Digital implementation will be particularly important in regions and countries with high demand densities.

In developed economies, cellular is a valuable productivity-enhancing tool and is rapidly evolving into a personal communications service. In developing economies, cellular provides access to telecommunications that is not possible, in many cases, with existing landline networks. EMCI believes that cellular growth will continue to be strong over the next few years, reaching nearly 90 million subscribers by 1998.

The Wireless Local Loop

> The technology of the local loop is fundamentally unchanged from the Victorian era.
>
> – George Calhoun

As cellular technology matures, copper is being increasingly recognized as an inefficient way to make the local connection. The 130 million lines installed in the U.S. cost between $900 to $1200 each to install. Local phone companies are profitable only because of their ability to write off this spending over extended periods of time. A wireless link to the local exchange costs less than $1000, and is falling rapidly, unlike the costs of digging up roads. These falling costs, and the rapidity with which wireless technology can be deployed, are increasingly making it viable for phone companies to replace copper with radio (as in Quitaque, Texas), or to forego copper altogether in favor of wireless local access (as in several developing countries). In China, for example, the cost of adding a new wired subscriber is estimated at over $2000, and numerous cellular projects are underway to provide primary telecommunications services. Likewise, most East European countries have some form of wireless networks, most of which are being used in place of conventional telephones.

Investors

Cellular companies have shown poor profitability during this decade of low competition. Most of this is deliberate as they wrote off their equipment very rapidly. Cellular started out using 10-year depreciation; many companies later switched to 3-year depreciation schedules. They only had to satisfy their auditors, most of whom agreed due to the impending conversion to digital *(Figure 8)*.

Now when PCS comes in, cellular will be able to lower prices dramatically in order to compete.

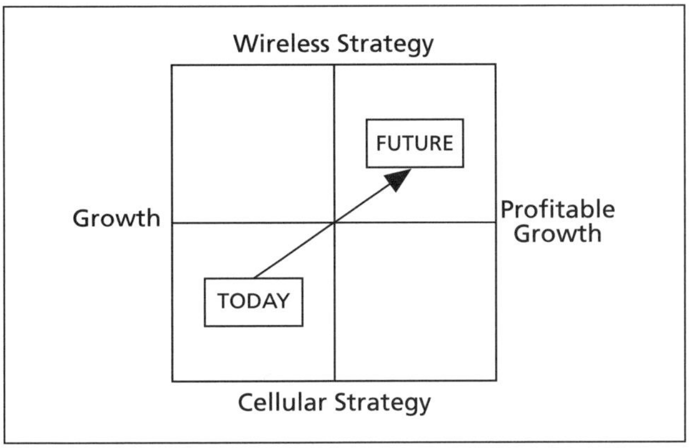

Figure 8 *Change driver: Investors*

In the 1983-92 time frame, the market essentially gave the cellular industry a "free ride"; now investors will demand profitability as well. This may make the cellular industry vulnerable to PCS pricing strategies, which may undermine cellular profitability just when investors start expecting dividends.

In the future, investors will expect the industry to generate healthy earnings as well pay dividends. They will also expect today's cellular providers (especially the larger one) to become more broad-based providers of wireless services, relatively independent of technology or architecture.

Conclusion

The wireless communications industry clearly has an exciting and explosive future. There is little doubt that the worldwide subscriber base and revenues will grow at double-digit rates or better for the foreseeable future. The challenge will be two-fold: how to become the customer's choice in an increasingly pluralistic market, and how to operate profitably. A clear understanding of the forces driving competition in the industry, and those that promise to transform it in the future, is a prerequisite to succeeding in the marketplace.

Endnotes

1 Paul Korzeniowski, "CDMA Vs. TDMA: Which Way Should Carriers Turn? – Cellular Service Providers Need To Prepare Their Networks For Growth, But They Are Faced

With A Standard Dichotomy," *Communications Week,* February 7, 1994, Pg. 4A.

2 Martyn Warwick, "Cellular Mobile Radio Changes Colors; Service Providers Adapt To Changing Market," *Communications International,* January, 1994, Vol. 21; No. 1; Pg. 4.

3 Paul Quigley, "Watchword For The Future – Global Mobile Communications," *Financial Times,* March 16, 1994, Survey of Information and Communications Technology (14); Pg. VIII.

4 Anthony Ramirez, "Mapping Out The Wireless-Phone Future," *The New York Times,* November 12, 1992, Section A; Page 1; Column 2.

5 Gail Edmondson, "Wireless Terriers," *Business Week,* May 23, 1994, No. 3373; Pg. 117.

6 Cellular Marketing, "CTIA: Remarks Excerpts, Thomas E. Wheeler, Federal Communications Bar Association, June 24, 1993," September 1993, Vol. 8; No. 9; Pg. 16.

7 Andrew Adonis, "Survey of Telecommunications in Business," *Financial Times,* June 15, 1994, Pg. III.

8 Adonis, op. cit.

9 Quigley, op. cit.

10 Adonis, op. cit.

11 Jonathan R. Tarlinj "The Global Phenomenon; Global Cellular Communications Markets," Cellular Marketing, February 1994, vol. 9; no. 2; p.18

12 Source: Financial Times Mobile Communications

Acronym Guide

AAL	ATM Adaption Layer
ADC	Analog Digital Converter
AGC	Automatic Gain Control
ALE	ATM Link Enhancement
AM	Amplitude Modulation
AMPS	Advanced Mobile Phone Service
ANSI	American National Standards Institute
ARQ	Automatic Repeat reQuest
ATM	Asynchronous Transfer Mode
AWGN	Additive White Gaussian Noise
BFSK	Binary Frequency Shift Keying
BISDN	Broadband Integrated Services Digital Network
BPSK	Binary Phase Shift Keying
CAI	Common Air Interface
CAP	Competitive Access Provider
CAT	Communications Authority of Thailand
CCM	Continuous Conduction Mode
CCSO	Cellular Centralized Service Organization
CDMA	Code Division Multiple Access
CDPD	Cellular Digital Packet Data
CPE	Customer Premises Equipment
CPU	Central Processing Unit
CTIA	Cellular Telecommunications Industry Association
DAC	Digital Analog Converter
DAMA	Demand Assignment Multiple Access
DAT	Digital AudioTape
DBIR	Directed Beam Infrared
DCM	Discontinuous Conduction Mode
DCME	Digital Circuit Multiplication Equipment
DCS	Digital Cellular System
DECT	Digital European Cordless Telephone
DES	Data Exchange System
DFIR	Diffused IR
DMH	Data Message Handlers
DSA	Decision Support Architecture
DSMA	Data Sense Multiple Access
DSSS	Direct Sequence Spread-Spectrum
EAS	Extended Area Service
ECBS	Enhanced Cordless Base Station
ECL	Effective Code Length
EIA	Electronic Industries Association
EIRP	Equivalent Isotropically Radiated Power
ESMR	Enhanced Specialized Mobile Radio
ETSI	European Telecommunications Standards Institute
FCC	Federal Communications Commission
FDDI	Fiber Distribution Data Interface
FDMA	Frequency Division Multiplex Access
FDTDM	Finite-Difference Time Domain Models
FEC	Forward Error Correction
FED	Free Euclidean Distance
FHSS	Frequency Hopping Spread-Spectrum
FM	Frequency Modulation
FPLMTS	Future Public Land Mobile Telecommunications Service
FSK	Frequency Shift Keying
GEO	Geosynchronous Earth Orbit
GII	Global Information Infrastructure
GMSK	Gaussian Minimum Shift Keying
GPMC	Green Paper on Mobile Communications
GSM	Global System for Mobile communications
HALE	High-Altitude Long-Endurance
HDLC	High-level Data Link Control
HEC	Header Error Correction
HLR	Home Location Register
IAM	Initial Address Message
IEEE	Institute of Electrical and Electronic Engineers
IMTS	Improved Mobile Telephone Service
IN	Intelligent Network
INAP	IN Application Protocol
ISC	International Switching Center
ISI	Inter Symbol Interference
ISL	Inter-Satellite Link
ISO	International Standards Organization
ISDN	Integrated Services Digital Network
ISUP	Integrated Services digital network User Part
IT	Information Technology
ITU	International Telecommunications Union
IXC	Inter-eXchange Carrier
LAN	Local Area Network
LBT	Listen-Before-Talk
LCD	Liquid Crystal Display
LCR	Level Crossing Rate
LEC	Local Exchange Carrier
LEO	Low Earth Orbital
LMR	Land Mobile Radio
LOS	Line-of-Sight
MA	Multiple Access
MAC	Media Access Control
MAP	Mobile Application Part
MBS	Mobile Broadband Systems
MDBS	Mobile Database Station
MEO	Medium Earth Orbital
MIC	Ministry of Information and Communications
MIN	Mobile Identification Number
MIRS	Motorola Integrated Radio System
MFJ	Modification of Final Judgment
MNP	Micro-Networking Protocol
MOTC	Ministry of Transport and Communications
MPSK	M-ary Phase Shift Keying

MPT Ministry of Posts and Telecommunications
MSA Metropolitan Statistical Area
MSC Mobile Switching Center
MTCM Multiple Trellis-Coded Modulation
MTSO Mobile Telephone Switching Office
NACN North American Cellular Network
NAMPS Narrowband Advanced Mobile Phone Service
NANP North American Numbering Plan
NCC Network Control Center
NIST National Institute of Standards and Technology
NMT Nordic Mobile Telephone
NPA Numbering Plan Area
NTIA National Telecommunications and Information Administration
NTT Nippon Telephone and Telegraph
OAM&P Operations, Administration, Maintenance, and Provisioning
PACS Hybrid PHP and WACS
PRBS Pseudo-Random Binary Sequence
PBX Private Branch eXchange
PCME Packetized Circuit Multiplication Equipment
PCN Personal Communications Network
PCS Personal Communications Services (or System)
PD Product Distance
PDA Personal Digital Assistant
PDH Plesiochronous Digital Hierarchy
PHP Personal Handy Phone
PHS Personal Handy Phone System
PL Path-Loss
PMM Personal Mobility Management
POP Per Unit of Population
POTS Plain Old Telephone Service
PSTN Public Switched Telephone Network
PTD Post and Telegraph Department
PTN Personal Telecommunications Number
QAM Quadrature Amplitude Modulation
QL Quasi Longitudinal
QOS Quality of Service
QPSK Quadrature Phase Shift Keying
RAM Random Access Memory
RF Radio Frequency
RLP Radio Link Protocol
ROM Read Only Memory
RSA Rural Statistical Area
SAR Specific Absorption Rate
SCM Satellite Connection Manager
SCP Service Control Point
SCT Communications and Transport Secretariat
SDH Synchronous Digital Hierarchy
SDMA Spatial Division Multiple Access
SI Scintillation Index
SIM Subscriber Identification Module
SMR Specialized Mobile Radio
SOS Servicios Organizados Secretariales
SSP Service Switching Point
SS7 Signaling System 7
STP Signal Transfer Point
TACS Total Access Communication System
TCAP Transaction Capabilities Application Part
TCM Trellis-Coded Modulation
TDMA Time Division Multiple Access
TEC Total Electron Content
TETRA Trans-European Trunked Radio
TIA Telecommunications Industry Association
TLDN Temporary Local Directory Number
TMM Terminal Mobility Management
TMN Telecommunications Management Network
TOT Telephone Organization of Thailand
TR Transmitter-Receiver
TRS Trunked Radio System
UF Urban Factor
UHF Ultra-High Frequency
UMTS Universal Mobile Telecommunication Systems
VAS Value-Added Services
VLR Visited Location Register
VSAT Very Small Aperture Terminal
VSELP Vector Sum Excited Linear Prediction
WACS Wireless Access Communications System
WAN Wide Area Network
WLAN Wireless Local Area Network
WLL Wireless Local Loop

Index